Edited by Baldev Raj,
Marcel Van de Voorde,
and Yashwant Mahajan

Nanotechnology for Energy
Sustainability

Further Volumes of the Series "Nanotechnology Innovation & Applications"

Axelos, M. A. V. and Van de Voorde, M. (eds.)

Nanotechnology in Agriculture and Food Science

2017
Print ISBN: 9783527339891

Cornier, J., Kwade, A., Owen, A., Van de Voorde, M. (eds.)

Pharmaceutical Nanotechnology

Innovation and Production

2017
Print ISBN: 9783527340545

Fermon, C. and Van de Voorde, M. (eds.)

Nanomagnetism

Applications and Perspectives

2017
Print ISBN: 9783527339853

Mansfield, E., Kaiser, D. L., Fujita, D., Van de Voorde, M. (eds.)

Metrology and Standardization for Nanotechnology

Protocols and Industrial Innovations

2017
Print ISBN: 9783527340392

Meyrueis, P., Sakoda, K., Van de Voorde, M. (eds.)

Micro- and Nanophotonic Technologies

2017
Print ISBN: 9783527340378

Müller, B. and Van de Voorde, M. (eds.)

Nanoscience and Nanotechnology for Human Health

2017
Print ISBN: 9783527338603

Puers, R., Baldi, L., Van de Voorde, M., van Nooten, S. E. (eds.)

Nanoelectronics

Materials, Devices, Applications

2017
Print ISBN: 9783527340538

Sels, B. and Van de Voorde, M. (eds.)

Nanotechnology in Catalysis

Applications in the Chemical Industry, Energy Development, and Environment Protection

2017
Print ISBN: 9783527339143

Edited by Baldev Raj, Marcel Van de Voorde,
and Yashwant Mahajan

Nanotechnology for Energy Sustainability

WILEY-VCH
Verlag GmbH & Co. KGaA

Volume Editors

Prof. Baldev Raj
National Institute of Advanced Studies
(NIAS)
Indian Institute of Science Campus
Bangalore 560012
Karnataka
India

Prof. Dr. Dr. h.c. Marcel H. Van de Voorde
Member of the Science Council
of the French Senate and National
Assembly, Paris
Rue du Rhodania, 5
BRISTOL A, Appartement 31
3963 Crans-Montana
Switzerland

Dr. Yashwant Mahajan
International Advanced Research for
Powder Metallurygy and New Materials
(ARCI)
Centre for Knowledge Management of
Nanoscience and Technology (CKMNT)
Balapur P.O.
Hyderabad 500005, Telangana
India

Series Editor

Prof. Dr. Dr. h.c. Marcel H. Van de Voorde
Member of the Science Council
of the French Senate and National
Assembly, Paris
Rue du Rhodania, 5
BRISTOL A, Appartement 31
3963 Crans-Montana
Switzerland

Cover credits: Planet earth: fotolia_Romolo
Tavani Forest,
background: fotolia_© doris oberfrank-list

Library of Congress Card No.: applied for

British Library Cataloguing-in-Publication Data
A catalogue record for this book is available from the British Library.

Bibliographic information published by the Deutsche Nationalbibliothek
The Deutsche Nationalbibliothek lists this publication in the Deutsche Nationalbibliografie; detailed bibliographic data are available on the Internet at http://dnb.d-nb.de.

© 2017 Wiley-VCH Verlag GmbH & Co. KGaA, Boschstr. 12, 69469 Weinheim, Germany

Print ISBN: 978-3-527-34014-9
ePDF ISBN: 978-3-527-69614-7
ePub ISBN: 978-3-527-69611-6
Mobi ISBN: 978-3-527-69612-3
oBook ISBN: 978-3-527-69610-9

Cover Design Adam Design
Typesetting Thomson Digital, Noida, India
Printing and Binding Markono Print Media Pte Ltd, Singapore

Printed on acid-free paper

Thanks to my wife for her patience with me spending many hours working on the book series through the nights and over weekends.
The assistance of my son Marc Philip related to the complex and large computer files with many sophisticated scientific figures is also greatly appreciated.

Marcel Van de Voorde

Series Editor Preface

Since years, nanoscience and nanotechnology have become particularly an important technology areas worldwide. As a result, there are many universities that offer courses as well as degrees in nanotechnology. Many governments including European institutions and research agencies have vast nanotechnology programmes and many companies file nanotechnology-related patents to protect their innovations. In short, nanoscience is a hot topic!

Nanoscience started in the physics field with electronics as a forerunner, quickly followed by the chemical and pharmacy industries. Today, nanotechnology finds interests in all branches of research and industry worldwide. In addition, governments and consumers are also keen to follow the developments, particularly from a safety and security point of view.

This books series fills the gap between books that are available on various specific topics and the encyclopedias on nanoscience. This well-selected series of books consists of volumes that are all edited by experts in the field from all over the world and assemble top-class contributions. The topical scope of the book is broad, ranging from nanoelectronics and nanocatalysis to nanometrology. Common to all the books in the series is that they represent top-notch research and are highly application-oriented, innovative, and relevant for industry. Finally they collect a valuable source of information on safety aspects for governments, consumer agencies and the society.

The titles of the volumes in the series are as follows:

Human-related nanoscience and nanotechnology

- *Nanoscience and Nanotechnology for Human Health*
- *Pharmaceutical Nanotechnology*
- *Nanotechnology in Agriculture and Food Science*

Nanoscience and nanotechnology in information and communication

- *Nanoelectronics*
- *Micro- and Nanophotonic Technologies*
- *Nanomagnetism: Perspectives and Applications*

Nanoscience and nanotechnology in industry

- Nanotechnology for Energy Sustainability
- Metrology and Standardization of Nanomaterials
- Nanotechnology in Catalysis: Applications in the Chemical Industry, Energy Development, and Environmental Protection

The book series appeals to a wide range of readers with backgrounds in physics, chemistry, biology, and medicine, from students at universities to scientists at institutes, in industrial companies and government agencies and ministries.

Ever since nanoscience was introduced many years ago, it has greatly changed our lives – and will continue to do so!

March 2016

Marcel Van de Voorde

About the Series Editor

Marcel Van de Voorde, Prof. Dr. ir. Ing. Dr. h.c., has 40 years' experience in European Research Organisations, including CERN-Geneva and the European Commission, with 10 years at the Max Planck Institute for Metals Research, Stuttgart. For many years, he was involved in research and research strategies, policy, and management, especially in European research institutions.

He has been a member of many Research Councils and Governing Boards of research institutions across Europe, the United States, and Japan. In addition to his Professorship at the University of Technology in Delft, the Netherlands, he holds multiple visiting professorships in Europe and worldwide. He holds a doctor honoris causa and various honorary professorships.

He is a senator of the European Academy for Sciences and Arts, Salzburg, and Fellow of the World Academy for Sciences. He is a member of the Science Council of the French Senate/National Assembly in Paris. He has also provided executive advisory services to presidents, ministers of science policy, rectors of Universities, and CEOs of technology institutions, for example, to the president and CEO of IMEC, Technology Centre in Leuven, Belgium. He is also a Fellow of various scientific societies. He has been honored by the Belgian King and European authorities, for example, he received an award for European merits in Luxemburg given by the former President of the European Commission. He is author of multiple scientific and technical publications and has coedited multiple books, especially in the field of nanoscience and nanotechnology.

Contents

Volume 2

Foreword

The variety of fields on which nanotechnology had and/or has a significant impact is remarkable. It ranges from health, environmental, and social issues to nanomedicine, nanoelectronics, and energy applications of nanotechnology. In the past decades, energy applications of nanotechnology have been recognized as one of the most promising and important facets of nanotechnology. In fact, even today, nanotechnology plays a key role in producing, storing, and distributing energy. The fast growth of nanotechnology in energy applications keeps rapidly expanding and diversifying our knowledge in that area. As a result, it becomes increasingly difficult to keep up with all of these developments. Similar situations in other fields indicate that an efficient way to overcome this difficulty is by means of reviews written by a team of internationally known experts discussing critically the present state of knowledge as well as conceivable visions and perspectives. This is the motivation and the approach used in this book on energy applications of nanotechnology.

If one looks at nanotechnology as a whole, it becomes evident that many of the key developments in nanotechnology – in basic science as well as in all kinds of technological applications – were initiated by discoveries or developments of new nanometer-sized structures. In turn, these new structures resulted in new properties that opened the way to new applications of nanotechnology.

Historically, the earliest examples – dating back to the fourth century – are Roman glasses containing nanometer-sized gold precipitates. In modern times, new properties were discovered when nanometer-structured materials were developed, for example, in the form of suspensions of colloidal crystals, in the form of buckyballs, or in the form of nanotubes. Basically, the same applies to solid materials with macroscopy external dimensions. About 40 years ago, solid materials consisting of nanometer-sized crystallites that are connected by grain or interface boundaries pioneered the field of nanocrystalline materials. Today, this field has expanded to more than 100 000 publications. All of the nanostructured materials that are available today are characterized by specific atomic arrangements such as the specific atomic arrangements in buckyballs, in nanometer-sized crystals of catalysts, or in nanocrystalline materials.

A world of nanometer-structured solids that have attracted little attention so far for applications in nanotechnology are nanostructured glassy solids. One reason may be that the methods available today to control the microstructure of glasses on a nanometer scale are very limited. However, the few studies of nanometer-structured glassy solids – called nanoglasses – have evidenced promising new features of nanoglasses such as unforeseen new magnetic, mechanical, or biological properties as well as the option to generate alloys in the form of nanoglasses that consist of components that are immiscible in crystalline materials. As already pointed out, the discovery of new kinds of nanostructured materials with new properties stood frequently at the beginning of new developments in nanotechnology. Hence, nanoglasses may represent the beginning of a family of new developments in the area of nanotechnology.

In conclusion, this book gives an overview of research results on structural and functional materials in the wide field of energy technology and is likely to become an important source of information for students, researchers, and industrialists. Moreover, it highlights the needs for nanomaterials research in the field, provides roadmaps for innovation, and pinpoints toward new nanofabrication techniques. Hence, the book is expected to be a stimulus for the further developments of the energy technology for the benefit of industry and economy and the world society.

Distinguished Fellow, Honorary Advisor   *Prof. Dr. Dr. hc. Mult. Herbert Gleiter*
Karlsruhe Institute of Technology
(Germany)
Director, Herbert Gleiter Institute of Nanoscience
Nanjing University of Science and Technology (China)
Fellow, Institute for Advanced Studies,
City University of Hong Kong

Karlsruhe, March 21, 2016

Foreword

The exploration of the nanoworld in recent decades has truly led to a phase transformation in both scientific understanding and technology. The driver has been the much augmented impact of interfacial zones owing to their greater volume fraction, as well as the fact that many nanostructures show mesoscopic behavior, that is, a qualitatively different behavior from the macroscopic bulk. Furthermore, not only local properties can vary, even mechanistic changes can appear.

There is almost no field in science and technology that remains untouched by such phenomena. Referring to energy technologies is highly relevant in this context, not only because this is a field of major relevance, but also because it is particularly rich in nanoscale effects. The whole of Volume I is devoted to energy "production," Volume II deals with energy storage, distribution, and conversion, and Volume III includes special materials issues and related issues referring to environment and society.

Apart from its relevance, this three-volume book also comes at the right time: The saying (ascribed to Dyson) that a book that intends to cover the time up to $t_0 - \Delta t$ (t_0: present time) will be outdated at $t_0 + \Delta t$ stresses the point that a field must have reached a certain degree of maturity, should it be worth to be treated comprehensively and sustainably. Even though still a hot topic, nanoscience has reached a degree of maturity that allows one to identify major directions in the scientific treatment. Equally, nanotechnology has provided advanced tools in terms of preparation and analysis and has opened the pathway for various applications.

A representative example is offered by the influence on electrical properties, be it the field of nanoelectronics that is affected or be it – the author's favorite – the field of nanoionics. What the first has achieved in terms of information technology is expected for the latter in terms of energy technology.

In both fields, size reduction can lead to variations by orders of magnitude and, in a variety of cases to qualitative changes as far as transport or storage properties are concerned.

All in all, this book provides a collection of pertinent contributions by well-known experts on nanomaterials with strong emphasis on energy and environmental aspects.

The author wishes the book not only many buyers but also many readers.

Director, Max Planck Institute for Solid State Research   *Professor Joachim Maier*
Stuttgart, Germany

Stuttgart, July 2016

Foreword

Nanoscience is one of the main thrust areas of science today, and the way the subject has blossomed in the last two decades is truly impressive. One of the reasons for the fast development of this field is the variety of applications of nanomaterials. Some of the important applications are in nanomedicine followed by electronics. While the developments in nanomedicine are truly impressive, applications in the area of energy devices are equally noteworthy. A large number of papers have been published in the last few years on the use of nanomaterials in various energy devices.

Among the nanomaterials, special mention must be made of carbon nanotubes that have found many uses. Two-dimensional materials have been widely used in the last few years because of their unique properties. One of the important two-dimensional materials is graphene that has found many applications, although not in pure form but in a state where it is doped suitably or functionalized in an appropriate manner. Inorganic analogues of graphene such as molybdenum disulfide and related chalcogenides have found a variety of possible applications in energy devices and other areas. These materials, unlike graphene, have a band gap.

In battery R&D, nanomaterials have been employed to improve or modify performance. This is specially true of lithium and sodium batteries. Two-dimensional nanomaterials have yielded excellent results in the area of supercapacitors. Thus, graphene and nitrogen-doped graphene have shown good performance as supercapacitor electrodes. Borocarbonitrides, $B_xC_yN_z$, are also very good supercapacitor materials. In the case of fuel cells, several 2D materials have been found to be good catalysts for the oxygen reduction reaction. Specially noteworthy is the performance of borocarbonitrides.

The hydrogen evolution reaction has assumed great importance because of the wide interest in hydrogen economy. Photochemically induced generation of hydrogen has been achieved by using heterostructures of semiconducting nanomaterials or by dye sensitization employing nanosheets of two-dimensional MoS_2 and other materials. Photoelectrochemical generation of hydrogen using nanomaterials has also been accomplished. Electrochemical hydrogen evolution generally employs a platinum catalyst and there have been recent efforts to substitute platinum by nonmetallic materials. Borocarbonitrides and a few other materials are found to be effective for this purpose. Production of hydrogen by employing a solar thermochemical cycle based on nanoparticles of metal oxides (e.g., Mn_3O_4) is an attractive possibility. In photovoltaics, there has been much progress in recent years, particularly in organic PVs, and there are aspects where nanoscience has made a difference.

The story of nanoenergy goes on, and the subject has become a vital component of nanoscience and technology. Clearly, some of the important solutions to the energy problem will emerge by the application of nanomaterials.

The book by Dr. Baldev Raj, Prof. Marcel Van de Voorde, and Dr. Yashwant Mahajan is a comprehensive knowledge base for students, academicians, industries, and policymakers to look at the totality and to understand current status and search for opportunities in a broad spectrum of energy domains. Nanotechnologies can play an important evolutionary or paradigm change contributions to clean energy realization. These volumes cover all energy sources and allied subjects such as storage and environment.

Professor C. N. R. Rao, F. R. S.
National Research Professor
Linus Pauling Research Professor & Honorary President
Jawaharlal Nehru Centre for Advanced Scientific Research
Jakkur Campus, Jakkur
Bangalore, India

Jakkur, July 2016

"Perspective" on the Book on Nanotechnology for Sustainable Energy

Since the dawn of the twenty-first century, the human society has been experiencing the global transformation and facing the grand challenges of the population increase and unprecedented growth in energy demand. Substantial reduction in fuel consumption and environmental pollution, and harnessing the renewable energy are vital to people's well-being worldwide. Nanotechnology is one of the enabling approaches to tackle these issues and can be widely employed to improve energy sustainability.

This three-volume book covers a variety of topics on nanotechnology and sustainable energy, and contains many recent research findings, achievements, and industrial applications contributed by the scientists and researchers from nearly 20 different countries. Nanotechnology fundamentally represents a convergence of many sciences and technologies at the nanometer scale. Its multidisciplinary nature draws from physics, chemistry, biology, and engineering, and has the potential to address the problems of reliance on fossil energy, pollution damage to the environment, and reduction of the cost of renewable energy.

The introductory section of the book provides an overview of nanotechnology for conventional and renewable energies by the world-renowned scientists, namely, Prof. Herbert Gleiter, a pioneer in the field of nanoscience and nanotechnology, and Prof. Dr. Joachim Maier, who is credited with developing a new scientific field, nowadays termed as nanoionics. Additionally, Prof. C.N.R. Rao, a mentor of chemistry and materials science and engineering, has been working on nanotechnology for many years and has already delivered a well-rounded presentation of various aspects of nanomaterials in his monograph entitled *Nanocrystals* published in 2007. As we all know, hydrogen energy is a perspective clean energy with a bright future and beneficial to the human society, this time Rao shares his vision on foreseeable future directions of nanotechnology in hydrogen energy applications.

The first part of Volume 1 is on energy production. Global warming and air pollution are the direct impact of combustion of fossil fuels resulting in degradation of plant growth and threatening human health. This part is devoted toward the development of cost-effective nanotechnologies for efficient utilization and storage of the existing traditional energy along with the novel approaches for usage of nonfossil and renewable resources. The effect of the nanoparticle catalysts on enhancing the efficiency of petroleum recovery, refining, and natural gas conversion backs up the theme of nanotechnology for sustainable energy. It is apparent that the understanding and application of nanomaterials have made considerable progress in upgrading of traditional enterprises and economy escalation.

Photovoltaics is one of the most important aspects of renewable energy. There are seven chapters in this part related to photovoltaics and solar thermal energy conversions, which indicate nanomaterials and nanotechnologies will improve efficient utilization and conversion of the solar energy.

Bioenergetics, nuclear fission and fusion, and electric and wind powers are treated separately in four chapters, including biotechnology, nanotechnologies in advanced nuclear energy systems, nanotechnology applications for electric power, and lightweight nanostructured materials for wind energy technology where one may recognize nanotechnology can reduce cost, reduce waste, and reduce risk in various kinds of energy applications.

Part Two of Volume 2 on energy storage includes six chapters related to batteries, supercapacitors, hydrogen storage, and thermal energy storage. Nanomaterials for Li-ion battery are discussed, and nanotechnology development of potential cathodes and anodes in the next-generation battery applications is also thoroughly articulated.

The nanostructured supercapacitors are reviewed in two chapters, which can be exploited and utilized in many sustainable energy fields such as wind power, solar energy, and electric vehicles. There is at least one bus fleet, as far as I know, equipped with supercapacitors running in the south China city Ningbo.

Power transmission by CNT wires and cables and nanostructured thermoelectric materials for near term and future applications are described in Part Two.

Part Three of Volume 2 focuses on energy conversion and harvesting. Nanotechnology combined with superconducting materials offers to make superconductor technology cost-competitive. Fuel cells and nano-photocatalysts converting solar energy to hydrogen and energy harvesting are also presented in Part Three. One would expect that nanomaterials and nanotechnologies enable more efficient and cheaper fuel cells, and therefore utilization of hydrogen and fuel cells is certainly an option for powering the next-generation vehicles.

Part Four of Volume 3 deals with nano-enabled materials and coatings for energy applications where nanostructured bainitic steel, nanocomposites for transportation, graphene, inorganic nanotubes, and fullerene-like particles are addressed. The use of graphene is wide and rife with potential to disrupt the *status quo* whose energy applications seem well suited to accept its physical and chemical advantages. It is interesting to know that photovoltaic and thermoelectric energies can be generated by semiconducting nanowires.

Part Five of Volume 3 on energy conservation and management includes nanotechnology in architecture, aerogels for energy conservation, and nanofluids for efficient heat transfer that has potential of energy saving and microchip cooling systems. Nanoliquid metals and nanofluids as the promising materials are also discussed in this part. The readers may find various nanofabrication techniques in this part that can lower the manufacture cost and reduce carbon emission.

In Part Six of Volume 3, several chapters are devoted to the intellectual property, market, and commercial opportunities, which are the highlights of the book and inform us that not only the technical innovation but also the business modal will play a significant role in developing the sustainability and shaping our destiny.

Carbon dioxide is a major contributor to the greenhouse effect and global warming. Part Seven of Volume 3 addresses environmental remediation issues, especially on conversion of carbon dioxide into renewable fuels and value-added products and on cleaning contaminated water in soil spill sites.

In summary, this book can be definitely considered as one of the most influential books on solving the difficult problems that mankind faces in the present century. The developed countries in the past century have boosted the economy and industries by injudiciously utilizing the fossil energy at the expense of the environment. However, China and India, with nearly three billion people, and other developing countries cannot follow and copy the same economic development model. The editors intend to ascertain the appropriate way – integrating the state-of-the-art technologies into many industry sectors – to tackle the aforementioned scientific, economic, and social issues. This novel approach may embolden the researchers and engineers worldwide to further strive for the better quality of life in the future.

Moreover, we must point out that the book has been systematically designed and Prof. Baldev Raj as the ex-President of Indian National Academy of Engineering along with Prof. Van de Voorde Marcel and Dr. Yashwant Mahajan have put in enormous efforts in organizing this *magnum opus*. We can imagine that there has been tremendous work for the three editors to arrange such a wide range of topics and to select the authors with right credentials. In the sense of sustainable society, this book offers a clear strategy and helps the readers embrace the scientific innovation to overcome the challenges in the energy sector.

One thing that I think is also crucial to the sustainable society relates to education that is particularly imperative in BRIC countries. We need a huge number of experts, young scientists, and engineers to implement and promote the nanotechnology applications. In addition, an awareness on control of energy consumption and protection of environment should be developed generation-by-generation so that it becomes a custom, habit, and obligation.

Finally, I am certain this book will interest a broad audience, from professors, scientists, researchers, engineers, students, policymakers to people concerned about future generations. Perhaps the real meaning of the book is to introduce nanotechnologies and nanomaterials as the possible innovation path to exploring new ways of living for many billions of people in the developing as well as the developed countries.

Member, Chinese Academy of Engineering

Honorary President, General Research
Institute for Nonferrous Metals

Beijing, July 2016

Dr Tu Hailing

A Way Forward

The development of cleaner, cheaper, and more efficient energy production and its utilization requires an interdisciplinary effort involving various scientific domains such as physics, chemistry, life sciences, and technological fields (energy, chemical engineering, biotechnologies, sensors, electrical and electronics, information technologies, and materials science and engineering). All these domains belong to science, technology, engineering, and mathematics (STEM). Indeed, measurable progress is being accomplished in clean energy production and access at affordable cost to most of the citizens of the world. Among these, nanomaterials and nanotechnologies are identified to impact the energy systems to deliver sustainable goals expeditiously with technoeconomic considerations. Figure 1 schematically illustrates the current scenario of the nanotechnology-based product development.

During the last few decades, nanomaterials and nanotechnology have demonstrated evolutionary and constructive disruptive innovations and research possibilities. Nanotechnology has been at the basis of many technological advances with many new devices or applications being realized or in developmental stages. Nanotechnology is being pursued to improve the functionality of existing devices, to reduce the cost of devices, and to bring new capabilities and functions that did not previously exist. As described in the book, nanotechnology is destined to play a key role in the coming decades to change the energy scenario of the world. The projected applications will result in efficient energy usages, by not only being environment-friendly but also with huge impact on the world economy and the social well-being. Figure 2 illustrates important domains where nanotechnology is destined to play a key role in the energy sector. This commentary is an overview of the technology developments in frontier areas of hot pursuits, research reviews along with patents landscaping analysis.

On a scientific note, the most striking effect of nanomaterials for energy applications can be ascribed to a larger number of atoms on the surface compared to that present in bulk material. The large surface area leads to a high reactivity, which is desirable for improved catalysts leading to higher reaction rates, lower processing temperatures, and reduced emission and for improving effectiveness. As an example, the nanoparticle-based catalytic and cathodic systems in conjunction with room-temperature ionic liquids are making progress in CO_2

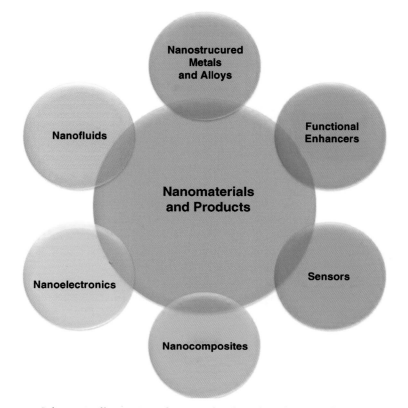

Figure 1 Schematic illustration of nanotechnology-based materials products.

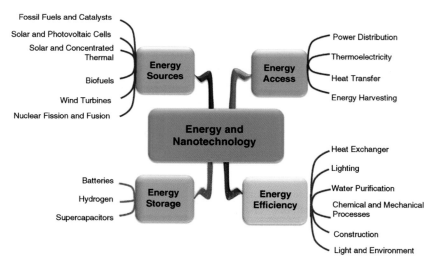

Figure 2 Schematic of the application of nanotechnology in energy sector with existing and futuristic applications.

reduction process for production of fuel chemicals. Computer stimulation and modeling efforts are being directed to hasten successes in this important domain. For example, recent studies in electroreduction of CO_2 by nanosized cobalt oxide structures below 2 nm have shown enhancement of catalytic reactions by a factor of more than 25. Metal organic frameworks (MOFs) that are self-assembly of metal ions and organic linkers have specific advantages of large surface areas, adjustable pore sizes, and controllable surface properties. These materials also exhibit the highest CO_2 capture capacities of up to 2.5 mol CO_2/kg. The inorganic nanotubes (INTs) and fullerene-like nanostructures (IFs) based on layered materials, including WS_2, MoS_2 BN, BNC, and so on, are being pursued as possible candidates for energy-related applications such as energy conversion and storage, catalysis, and so on. Coke-free methane-forming catalysts in gas-to-liquid fuels through the design of multifunctional alloys and doping are defining the frontier. This is important for achieving lower temperatures and reduced operating costs. This requires further work on stability and kinetics of optimally designed catalysts with new characterization and analytical tools. Compositions, nanostructures, and morphology are to be tailored during synthesis and postsynthetic treatments, which are salient features of design and performance optimization of zeolites for methanol-to-hydrogen conversion. It is an active area of research.

Higher efficiency or higher absorption rates for photons in electromagnetic spectrum and so on are at the center of quantum improvements and realization of disruptive technologies. For example, the surface effect is also relevant for energy storage applications, for example, hydrogen storage or electrodes for lithium batteries or energy transformation through electrodes for fuel cells. To discuss this point further, the analysis of nanostructured materials applications in lithium-ion batteries reveals that anode, cathode, electrolyte, separator, and collector (in the order of high percentage to low percentage) are being pursued to improve the performance and life. Looking at materials, the use of carbon nanotubes, graphene, and graphene oxide is emerging as frontiers in materials science and technology. The synthesis and doping of WS_2 and fullerenes with enhanced tribological performance can be accomplished through reduced wear and tear of solid lubricants to enhance life and safety of efficient machines.

Fractals and energy are divisible and analyzed in domains of space and time, considering the fact that the electrochemical reactions involved in microelectricity generation in batteries involve fractals. The fractal-based approach has validated and established its value in macro- and microscales. However, its value and applicability in nanoscale is an area of significant interest and research. For example, the perovskite ceramic microstructures with fractal-structured configuration of grains and pores constitute a favorable system for getting insights into the phenomenology and mechanistic aspects.

In the application domain, computer electronics and electrical vehicles remain as the main drivers for developing Li-ion batteries. The safety of lithium-ion batteries remains a worthy pursuit considering varying sizes, variants, and quantities. The safety vents, positive temperature coefficient elements, shutdown separators, less flammable electrolytes, and redox shuttles have been employed to improve safety. The coating of active materials with nanoscopic particles has been found to improve the thermal stability. The processing and packaging, environment fallout, safety, and so on are other frontiers toward large-scale commercialization of Li-ion batteries.

The surface area-related property enhancement in nanomaterials is also utilized in other functional applications such as sensors, thermoelectric devices, and so on. Nanosensors, because of high surface area of materials, selectivity, sensitivity and tailoring to functionalization, less power use, and wireless possibility, offer a number of possibilities for monitoring performance over life cycle. The nanosensors for physical, chemical, and mechanical measurements are becoming available for accurate and reliable measurements such as chemistry of sodium, composition of gases, wastes, and progression of combustion, degradation of structures and machines, and so on in energy systems. Nanopores and their morphology control in double-layer supercapacitors, superhydrophobic surfaces for corrosion mitigation and improving steam condensers performance, specificity of nanoparticles along with understanding taxonomic and physiological types of microbial species for enhancing efficiency of conversion of biowaste and biomasses to gaseous energy recovery, and so on are a few examples. Hybrid organic/inorganic thermoelectric materials offer significant degrees of freedom and promise and thus better trajectory of progress for waste heat recovery at low temperatures.

Bulk thermoelectric materials-based devices with power output of 20 MW have efficiencies (ZT) as low as 0.1% for temperature difference of 1.5 K in low temperature domain. These could be used in portable storage of vaccines and food. Miniature Peltier coolers can enhance well-being of many societal communities at the bottom of the economy. High-temperature thermoelectricity with promise can be achieved through confinement in nanowire superlattices for glass, aluminum, steel, and power industry for waste heat recovery.

Due to conflicting requirements of Seebeck coefficient and electrical and thermal conductivities, ZT hardly surpasses a value of 1 in case of bulk engineered materials. In order to compete with current dynamic technologies, ZT is required to be above 3. This is the main reason that thermoelectric devices are yet to be found in commercial applications. Chalcogenides, such as $CrSb_3$, β-Zn_4Sb_3, skutterudites, clathrates, and Zintl phases are being pursued for higher ZT to realize tangible applications. Nanotubes with additional phonon scattering on inner and outer surfaces also show good promise. However, the challenges remain to ensure electrical contact to all the nanowires in a high-density arrangement with minimizing the heat leakages.

Another material, SbSn, is reported to achieve ZT of 2.6. Cross-disciplinary efforts can offset slow pace through synergy of synthetic chemistry, solid-state physics, material science, fabrication, device design, validation, and new measurement technologies to enhance the pace. The application temperatures for thermoelectric materials up to 1000 °C are being explored. Materials to manufacturing engineering, scaling up to commercial production, are a big challenge.

Thus, achieving high ZT of 3 can be considered as only the beginning of the challenge.

As alternative materials, the topological insulators and polymer thermoelectric are also in the race for commercial thermoelectric materials. For example, the superlattice of $SrTiO_3/SrO$ has shown ZT of 2.4 at room temperature. The independent control of electrons and phonon in oxides as well as tailoring of nanostructure are pathways to enhance ZT to a higher value. The influence of processes and consistent results with acceptable variations for large-scale production during manufacturing are widely regarded as the challenges need to be addressed. For high-temperature thermoelectrics, nanocomposites offer a viable approach because thermal conductivity is suppressed by the phonon scattering without hindering the electrical properties. For example, phonon glass–electron crystal (PGEC) materials are estimated to have ZT of 4 at 77–300 K.

Another emerging application domain, wherein enhanced physical properties (thermal conductivity) of nanoscale particles dispersed in fluid, is the nanofluids. Current petroleum nanotechnologies involve modifications of various technological fluids by addition of solid nanosized particles. The industrial applications of technological nanofluids have matured for enhanced efficiency and effectiveness in oil/gas well drilling, water influx prevention, reservoir rock fracturing, and enhanced oil recovery. We know that oils are themselves natural nanofluids, and thus there is a promise of using native crude oils as association nanofluids and move toward nanoecology of petroleum. To substantiate this point, the challenges for purification of oil-contaminated water demand significant research as toxicity and cost are considerations along with effectiveness. The inorganic or organic ferro-oxides, nanocomposite aerogels, magnetic nanoparticles, and natural or waste porous hydrophobic materials are promising candidates for many of the futuristic applications. Magnetic nanoparticles have the advantage that the spills with particles can be magnetically eliminated from water–oil mixture. However, all these developments are at laboratory scale and need to move to pilot and higher technology readiness levels.

Concerning other functional applications of nanomaterials, the spectral absorption of light can be tuned by tailoring the size/morphology dimension of the absorbing nanoparticles. Future approaches involving the composition, morphology, hybrid structures, and so on could be used to improve light absorption in solar cells to reach higher and higher efficiencies. The strong optical scattering improves absorption of light in thin films of solar cells. These conditions also help current extraction. Well-designed studies and modeling efforts are pathways to enhance efficiency. The enhancement of light scattering for different configurations of metal and dielectric nanoparticles in silicon, GaAs, CdTe, and CIGS is being pursued as alternative approaches.

Theoretical framework approaches for a broadband light absorber made with simple binary lattice of metallic nanoparticles embedded in a transparent host material are useful concepts for understanding and gaining insights. This kind of approach is considered as rationale approach for optimal metamaterials. These approaches are useful for photo and thermophotovoltaic energy conversion (plasmonic nanoparticle networks). In all these systems, nanostructures are introduced in solar cells as these enhance improvements in one of the main electronic or optical tasks of solar cells stacks.

Solar cells with clear nanocrystalline character even today are not considered as nanomaterials. The examples are photonic crystals as intermediate reflectors for tandem solar cells, composite materials consisting of nanocrystalline silicon embedded in amorphous SiO_2, and/or organic heterojunctions. Thus, nanostructures stay relevant and rewarding pursuits in search of economical and rugged solar electricity. The same is true about many other materials, where nanostructures were responsible for functionality enhancement, but these had matured before a visible field of nanoscience and technology appeared on the horizon. Some examples include plasmonics providing bright colors to church windows and carbon nanotubes/cementite nanowires providing exceptional mechanical properties of damascene steel until the late eighteenth century, when the secret of its processing conditions got lost.

Energy consumption in information processing, be it computing, memory, or communication, is of strategic and vital importance for improvements in functions combined with efficiency. It is known that of about 22 TWh of energy produced, 8–10% is consumed in ICT. The fundamental limits for energy consumption in different components of computing can be calculated and one finds the time for exceedingly challenging aspects for further improvements in energy consumption. The paradigm changes are being pursued taking inspiration from living cells to achieve high functional and efficient performance. Flexible on-demand 3D connections, routing, and topology optimization for energy reduction are a few of the possibilities. As solar cells progress, factors such as geometric component effects effectively reduce the thickness of light-absorbing layers. Also, concept for harvesting underutilized ultraviolet and unutilized infrared photons remains as continued areas of research.

Coatings for solar thermal power generation need to have self-cleaning attributes and such multilayered coating is desired to be stable, durable, as well as less expensive. The pretreatment of the substrates along with carefully designed oxides and/or nitrides combination is the chosen pursuit path. Optical simulation and optimization of performance of multiplayed coatings can cut down the cost and time of development of solar thermal power system.

Cost reductions of solar photovoltaic through stable perovskites dye-sensitized polymers with metallic nanoparticle, novel carbon nanostructures such as graphenes: carbon nanotubes, I–D metal oxide semiconductor nanostructures, low-cost stable functional dyes, quantum dots, and so on are promising pathways to enhance efficiency and cost reduction for next-generation solar cells. Summarizing, nanoelectronics, which have been flourishing in the last few decades, will have a greater impact in the development of renewable energy sources such as photoelectrics, fuel cells, supercapacitors, hydrogen production and storage, and so on.

Apart from functional properties, the mechanical properties could change by a large measure in the nanoscale region compared with the same material with coarser microstructural characteristics. For example, carbon nanotubes are stronger than steel of the same dimension. In the context of wind energy conversion, an understanding of the aeroelastic, aerodynamics, and aeroacoustic characterization with damage initiation and propagation validation is important. While pursuing such aspect, the key will be to introduce polymer–carbon nanotube-based advanced composites and nano-enhanced coatings. The material fabrication-related challenges, such as avoidance of agglomeration, densification, and so on, remain to be addressed in a satisfactory manner.

Engineering the structures at nanoscale is a way to enhance properties in a significant manner, particularly in steels that are used in transport, power, defense, nuclear, space, and many other industries. It has been demonstrated that carefully designed composition without costly and less abundant elements can be converted to nanostructured bainitic steels without severe deformation or complex heat treatments. Nanostructures of bainitic ferrite with a thickness of 20–60 nm interwoven by austenite are a promising structural material, which is on the path of translational research from laboratory-scale coupon samples to functional components.

Magnesium (density of 1.74 g/cm^3, slightly higher than plastics)-based composites provide excellent weight advantage in transport. Mg-based nanocomposites with metals and ceramic reinforcements are a worthy direction of pursuit. Regulatory concerns with respect to inflammability and corrosion attributes are mitigated with Al_2O_3 reinforcements in Mg matrix and these materials have shown early promise. The overall improvement in absolute and specific mechanical properties, energy absorption capability, and corrosion mitigation are characteristics and attributes to be enhanced and achieved.

After discussing several nanotechnology-enabled research and development in multiple facets of niche area of technological relevance, it is instructive to analyze its impact on modern society. The buildings in urban society currently consume about 40% of global energy. Nanotechnologies can contribute toward enhancing energy efficiency by making improvements in thermal insulation, glazing, lighting, and controls through Internet of Things. Nanotechnologies can also improve the structural durability and the safety of structures in the building. For example, nanoepoxy bubbles containing self-healing polymer adhesive in concrete can be designed to rupture, and as the cracks develop, adhesives are released, thus providing self-healing and repairing capability. Beyond LEDs, OLED and quantum dots can further improve the efficiency of lighting. Aerogels have begun to contribute to making the buildings sustainable by using less material, but more value to the well-being of the residents or visitors in shopping malls or entertainment places. In particular, silicon aerogels with a cross-linked internal pattern of silicon dioxide chains create a network of air-filled nanoscale pores and these materials are estimated to be five times more effective compared to currently used polystyrene panels. To prevent moisture degradation of structure, aerogels are subjected to surface functionalization to impart hydrophobicity and are also sandwiched between wall panels. Nano-enhanced silver (approximately 8 nm nanoparticles) as well as functional nanoparticles can remove odor, noxious vapors, and others pollutants.

In future, the houses and public places are expected to be designed with special walls and structures to contain wireless, batteryless sensors and radio frequency identify tags to collect data on stresses, vibrations, temperature, humidity, gas levels, unwarranted persons, robots, terrorists, and so on. Such provision will provide prior and calibrated warning for taking precautionary measures to escape and thus saving the lives. Aerogels are finding applications for moisture (sweat) management of body, clothing for fire fighters, and tourists in subzero temperature terrains.

Finally, the perceived risk, real risk, and benefits of technologies are important aspects of communication and policy frameworks for research-based and validated facts. It is an important domain but often neglected in research, and efforts are to be directed to connect the research base to society. These aspects are considered vital to realize large-scale use of nanomaterials and technologies for energy systems.

Baldev Raj
Marcel Van de Voorde
Yashwant Mahajan

Introduction

Baldev Raj,[1] Marcel Van de Voorde,[2] and Yashwant Mahajan[3]

[1]*National Institute of Advanced Studies (NIAS), Indian Institute of Science Campus, Bangalore 560012, India*
[2]*TU Delft, Faculty of Natural Science, Eeuwige Laan 33, 1861 CL Bergen, The Netherlands*
[3]*International Advanced Research for Powder Metallurygy and New Materials (ARCI), Centre for Knowledge Management of Nanoscience and Technology (CKMNT), Balapur P.O., Hyderabad 500005, Telangana, India*

The world is facing unprecedented challenges in meeting ever-increasing demands of fundamental necessities, that is, water, food, energy, clean air, clothing, shelter, health care, employment, and so on. In this context, the products, such as mobile phones, cars, computer, television, and so on, to improve quality and comforts of life are embedded in the society of twenty-first century. At the same time, the efforts need to be focused on minimizing the negative effects of human activities on climate and environment and enhancing the well-being of society. The principles of sustainable development are feasible only when we enhance the three key pillars: social progress, economic growth, and environmental protection. The challenge remains as how to achieve a right balance among these three pillars.

We are in the midst of a global transformation relating to energy resources, production, storage, and usage. The increasing world population accompanied by the rapid economic growth and aspirations for better quality of life – especially in emerging economies like China, India, Brazil, Africa, and Indonesia –is a major driving force for an unprecedented growth in global energy demand. According to Royal Dutch Shell plc sustainability report 2011, the global energy demand is projected to rise by 60% (i.e., >400 million barrels of oil equivalent a day) by 2050. It has been emphasized in that report that fossil fuels and nuclear energy together can meet at least 70% of global energy demand in 2050. We face major concerns over dwindling fossil fuel reserves and their adverse influence on climate change, security of uninterrupted energy supply, local environmental pollution, the emission of greenhouse gases linked to climate change, and water shortages. The nexus among energy–food–water–land–forests is becoming more visible with progress in economies of many of the developing countries. The growing energy need of the world in a sustainable manner is among the grand challenges of our times and this challenge is to be pursued neither harming the environment nor contributing to global warming.

Climate change is one of the high-priority challenges facing humanity. The increasing emissions of greenhouse gases such as CO_2, nitrous oxide, and sulfur dioxide from the combustion of fossil fuels are the key drivers of the global climate change. These realities have led to worldwide efforts to develop cost-effective path-breaking technologies for efficient use of existing fossil fuels along with novel approaches for energy production and the usage of nonfossil and renewable resources such as solar, wind, biomass, wastes, and so on. Also of equal importance are hydrogen production and storage, transmission and distribution of energy, and so on.

Replacing fossil energy fuels with renewable energy sources poses great challenge due to the large variations in production during day, seasons, and the year, energy cost relative to fossil fuels, reduced reliability over the life cycle, and so on. Thus, it is a widely accepted fact that no single source of energy has the possibility to provide a comprehensive solution to achieve the goal of technoeconomically viable sustainable energy. A robust mix of various energy sources such as nuclear, fossil fuels, solar, wind, and so on, along with inventions and innovations in storage, transmission, and distributions will be required to meet the growing demands of energy and availability to the end users. Moreover, Germany, for example, says no to nuclear energy, whereas France and the United Kingdom embraces nuclear energy in the energy mix. The judicious energy mix of each country varies widely depending on aspirations, competences in technologies, and public acceptance.

Nanotechnology is an opportunity to unleash new solutions for energy inclusiveness and growth, while meeting the challenges of sustainability with a market potential of over 1 trillion US dollars. The breakthroughs in nanoscience and technology, especially for applications in the energy sector, have opened up new vistas of moving beyond conventional energy generation approaches and introducing technologies that are more efficient, sustainable, environmentally benign, and cost-effective. Nanotechnology has a significant possibility to reduce reliance on fossil fuels, together with the ability for enhanced energy efficiency, storage, and conservation.

This three-volume comprehensive book takes the reader through multiple and diverse, but fascinating, facets of nanotechnology-enabled energy sustainability by state-of-the-art reviews and visions of the future. In several chapters of these three volumes, the whole energy spectrum is covered, including energy generation, storage, transmission, and distribution. The conventional energy sources as well as relatively new renewable energy technologies attracted attention in these volumes. The role of nanomaterials and nanotechnologies has been highlighted with the current status revealing the advantages and limitation of nanotechnologies, and so on. The pointers to the future are discussed to give a perspective on research directions for researchers and policymakers.

The energy production covers nanotechnology-based solutions for the efficient utilization of fossil fuels. The special attention is given to nanocatalysts that result in improving the efficiency of petroleum recovery, refining, natural gas conversion, and coal liquefaction. The nanocatalysts also help gasoline and diesel fuels burn more efficiently, thereby increasing vehicle mileage and reducing air pollution. New discoveries of nanomaterials have been highlighted, such as resorcinol-formaldehyde-activated carbon xerogel, which could help in removing contaminants such as N_2 or CO_2 from natural gas to produce high-purity methane.

A number of chapters in this book describe opportunities for nanotechnologies in renewable energies, such as photovoltaics, concentrated solar power (CSP), and wind power. The clear applications of nanotechnologies in nuclear fission and fusion, with a focus on reduced cost, enhanced safety, and substantially reduced high-level waste making nuclear as an attractive choice, are described. Other novel energy sources, such as turbo electric, are particularly relevant for energy harvesting in nano- and micropower devices. Another important category of fuels is biofuels derived from biowaste. Nanotechnology is effective in enhancing their efficiency, leading to higher yields and thus reducing their processing costs.

The book devotes much attention to energy storage systems, which are essential for increased utilization of renewable energy, especially solar and wind energy. This is particularly relevant as these are interrupted inconsistent power resources depending on the time of the day and the year. Energy storage systems can also reduce the amount of electrical power that needs to be transmitted with power lines, which in turn reduces power loss. Nanoengineered materials are destined to play an ever-increasing and important role in the development of high-energy density systems, renewable and low-cost Li-ion, Li-S, and other types of batteries (e.g., supercapacitors and hydrogen storage materials). Thermal storage materials are being engineered to incorporate nano-enhanced phase change materials (NEPCMs). At present, the storage of large quantities of electrical energy is expensive and cumbersome. Therefore, novel approaches are being developed worldwide, including hydrogen, liquid air energy storage, and so on. Nanomaterials and technologies are likely to accelerate these innovations and harnessing these for society.

Nanotechnology is also improving the performance of power transmission cables. Currently, there is a requirement for installation of long-distance transmission lines for electricity, but that is associated with a significant cost due to high energy loss. Nanoengineered cables along with smart grids have the possibilities to reduce such power loss and to improve the performance. Carbon nanotubes possess extraordinary electrical conductivity, low weight, and ultrahigh strength, thus making these materials ideally suited for applications in power lines and robust cables enabling low transmission and distribution loss. The materials concept to technologies is the fascinating journey.

Superconducting cables have a high current carrying capacity, almost five times higher compared to copper wires of the same cross section. However, further improvements are necessary to make this technology commercially viable. Nanotechnology combined with superconductors offers to make superconductor technology a reality and especially cost-competitive. Other major developments involve hybrid nanostructures to simultaneously conduct and to store energy.

Thermoelectrics, fuel cells, hydrogen energy, and energy harvesting are important domains of high growth and significance in energy systems. Nanostructured thermoelectric materials enable energy to be recovered from waste heat in the form of electrical energy. The products based on thermoelectric materials offer next revolution by realizing a promise to develop a cost-effective and pollution-free energy conversion system. Hydrogen is an ideal energy carrier. The utilization of hydrogen and fuel cells is an attractive option for powering future vehicles. Hydrogen production has been acknowledged as the key element in a future hydrogen economy. In particular, nanotechnology has significant potential to enhance the efficiency of various hydrogen production methods. Nanomaterials find applications as nanostructured electrodes for photoelectrochemical water splitting and also as nano-photocatalysts for hydrogen production. Fuel cell-based conversion devices can provide power for diverse applications, including vehicles, stationary power generation systems, and portable power devices. The noteworthy benefits of fuel cell technology include high conversion efficiency, reduced CO_2 emission and air pollution, and capability of utilizing diverse fuels, such as hydrogen, natural gas, propane, and diesel. However, the development for achieving a rapid pace is focused at cost-effectiveness, stability, and life cycle costs. Nanomaterials enable more efficient, cheaper, stable, and long-life fuel cells that are commercially viable.

Energy harvesting refers to the processes by which unused or waste energy in the form of heat, light, sound, vibration, or movement is extracted from the local environment and transformed into more useful form of electrical energy. A wide variety of energy harvesting technologies enables powering mobile electronics and power sensors without the use of batteries. This technology also addresses the issue of climate change because it mitigates environmental pollution. The above discussion clearly exemplifies the transformative role played by nanotechnology in energy harvesting devices.

Nanostructured materials for energy components and systems play a pivotal role in realizing mandates of efficiency, sustainability, and cost-effectiveness, which are key ingredients for the sustainable goals of Earth planet. As an enabling technology, nanotechnology-enhanced bulk materials and coatings offer new and unique properties leading to the development of state-of-the-art cost-effective components and systems with superior energy efficiency and improved performance. The experienced authors have provided sufficient coverage of nano-enabled materials and coatings for energy applications, including nanostructured and nanofunctional materials, such as nanoengineered steels; nano-enhanced lightweight materials especially for transport applications; nanostructured surfaces and nanocoatings; nanolubricants; nanosensors; and modeling and simulation. These approaches and accomplishments make us confident to set our goals high such as reduction in weight of an automobile to achieve fuel saving and decreased CO_2 emissions; conservation of energy by low-emissivity nanocoatings for smart windows of a building; nanocoatings, thin films, and nanolubricants with reduced friction for improved efficiency, longevity, and maintenance for industrial machines and plants; nanocrystalline and nanolayered coatings and highly functional coatings for improved efficiency of solar cells and concentrated solar power (CSP). Novel nanofluids such as nanoliquid metal and nanoionic fluids are promising materials for energy applications, especially in the domain of enhanced heat transfer.

Several issues pertaining to energy conservation and management by using nanofluids and nanoporous superinsulation materials are described in the book. The improvement of heat transfer by means of nanofluids has huge possibility in energy saving, reduction of process time, and improving the service life of equipment. Nanofluids find a variety of applications in electronic chip cooling systems, cooling of automobile engines, welding equipment, direct absorption solar thermal collectors, and as a futuristic candidate for primary coolant in pressurized water reactors (PWRs).

Aerogels are an emerging ultralow-density superinsulation material having applications in thermal protection systems for space vehicles, nuclear reactors, and pipe insulation in steam generation facilities. Nanotechnology offers significant opportunities for enhancing cooling and heating efficiency and producing less waste and less greenhouse gases, thus mitigating global warming. Nanotechnology can realize these benefits by the use of high-strength lightweight materials, improved building insulation, nanocoatings, improved air filtration, low-energy lighting devices, thin film solar cells, energy-efficient nanosensors for environment computing machines and architectures, sensing, and so on.

Nanotechnology offers a number of advantages in energy technology and new findings coupled with innovations will lead to worldwide competence and competitions. Much attention is given to technological and intellectual property-related issues and potential markets. Patents and their landscaping provide, apart from up-to-date information on scientific and technical areas, an insight into technological and competitive trends and chronological development of patented inventions. A few chapters in the book present patent landscaping and product scenarios, as examples, in the domains of Li-ion batteries, wind energy, and fuel cells. The chapters cover commercial aspects, products, and markets for wind energy sector. Bibliometric analysis of nanotechnology research trends in the energy sector is another important area, which is addressed in the book.

Energy is strongly linked to environmental aspects and these issues have merited authoritative coverage in the book. Nanotechnology is promising to lessen the impact of energy production and distributions on environment throughout the value chain of the energy sector. For environmental remediation, nanotechnology has a variety of applications, including the capture and sequestration of carbon dioxide emitted from thermal power stations and its conversion into renewable fuels, nano-based systems for control and clean-up of oil spills, nanomaterials for water filtration, and so on. Although nanotechnology offers possibilities to prevent, remediate, and treat environmental pollutants, there is always a risk involved as nanoremediation could release manufactured nanoparticles into the environment and ecosystems and thereby pose unintended consequences. In the above backdrop, the safety aspects in the ecosystem and for workers in the industry are presented in the book. The editors have ensured coverage of marketing aspects to highlight the importance of nanomaterials and nanotechnology in energy applications and economic growth.

The book concludes with authoritative commentary on issues pertaining to acceptance for nanotechnology related to various facets of energy by the society together with future directions and a policy framework in a comprehensive way and in a manner that can be appreciated and utilized to maximize harnessing of nanotechnologies for cleaner and sustainable energy for the world.

In conclusion, the book gives a state of the art of nanomaterials and nanotechnology in the whole sector of energy technology. The three volumes of encyclopedic book capture current status and provide important directions for research, development, and innovations for environmental and eco-friendly energy technologies. Nanotechnology success shall be measured in terms of bringing solutions in a variety of energy technologies, which are presently in use and also those that are in the process of maturity through breakthroughs. The technologies are always uncertain, but the pursuits are fascinating and rewarding. It is clear that nanotechnologies are new, fascinating, promising, and thus attracting attention of bright young students, researchers, industry, and policy makers and thus demands careful understanding and comprehension. The combination of imagination, financial implications, risk taking capacity, and meaningful collaborations between researchers, manufacturing industry, and the government has a vital role in achieving aspiration of the world community in the years up to 2025.

The book is designed to provide a robust basis for science and engineering students and researchers in energy-related pursuits and a comprehensive resource for the whole spectrum of professionals in the energy domain. The book is also a resource of validated expertise for governmental bodies, which are responsible for the well-being of the society and funding agencies.

The editors would like to thank everyone without whose help this book would not have become a reality. We express our sincere gratitude, especially to all the eminent authors for their outstanding contributions. Last, but not the least, we would like to take this opportunity to express our deepest gratitude and appreciation to all the experts who have painstakingly reviewed the manuscripts on highly specialized topics pertaining to their domain expertise.

Part Four
Nanoenabled Materials and Coatings for Energy Applications

29
Nanocrystalline Bainitic Steels for Industrial Applications

C. Garcia-Mateo and F.G. Caballero

Materalia Research Group, Department of Physical Metallurgy, National Centre for Metallurgical Research (CENIM-CSIC), Avda. Gregorio del Amo 8, Madrid, 28040, Spain

29.1
Introduction

One of the most recent developments in the physical metallurgy of steels is the so-called nanocrystalline bainite. The unique and exceptional microstructural features that define such bainite result in improved mechanical properties. Such an improvement results from the fine scale of the microstructure consisting of bainitic ferrite slender plates, tens of nanometer thick, closely interweaved with retained austenite. One of the earliest objectives of this developmental work was to obtain such a microstructure and properties without applying complex and expensive manufacturing processes, but only by alloying with control aiming at low bainite transformation temperatures in the range of 125–350 °C [1,2]. Although impressive improvements in mechanical properties [3,4] as well as transformation kinetics were achieved in different generations of nanocrystalline bainitic steel grades [5], but still there remained some issues to be addressed before the industrialization of such microstructures could be undertaken. This chapter describes the design and characterization of properties followed during the development of what could possibly be the last generation of nanocrystalline bainitic steels ready for industrial production.

29.2
Design of Nanocrystalline Steel Grades: Scientific Concepts

For the industrialization of these novel microstructures, the design process of the steels grades must invariably take into account the essential scientific principles of the bainitic transformation as well as certain industrial demands.

The joint efforts of the scientific, industrial, and final user communities have reported a finite set of relevant parameters to be considered during the design

Nanotechnology for Energy Sustainability, First Edition. Edited by Baldev Raj, Marcel Van de Voorde, and Yashwant Mahajan.
© 2017 Wiley-VCH Verlag GmbH & Co. KGaA. Published 2017 by Wiley-VCH Verlag GmbH & Co. KGaA.

process. Based on this, the blueprint for a new generation of nanocrystalline bainitic steels could be summarized as follows:

i) *A simple alloy system*: In order to avoid being limited to niche applications, the alloys that have to be produced must be cheap and elements that have been used in the past for different applications of these alloys, such as Ni for hardenability or Al and Co to accelerate the transformation [1,5–9], were not considered during designing, always aiming at a cheap and lean system, for example, Fe–C–Si–Mn–Cr.

ii) *Carbide-free microstructure*: Any effort in designing a strong, ductile, and tough microstructure would be wasted if it contains coarse cementite, an undesirable hard and brittle phase. Further, since the aim is to produce a microstructure consisting of bainitic ferrite and to retain austenite, any cementite, if present, would act as a C sink, making the austenite more prone to martensitic transformation during cooling after bainitic transformation [10]. It is therefore imperative to retard and, if possible, to prevent the precipitation of cementite from austenite during bainite reaction. For this purpose, Si has to be added to the chemical composition in quantities of at least 1.5 wt% [11].

iii) *Low austenite to bainite transformation temperature*: Reducing this temperature T ($M_S < T < B_S$, where M_S and B_S represent the martensite and bainite start temperatures, respectively) is a way to attain higher fractions of finer bainitic ferrite plates [12,13]. It can also enhance the thin-film morphology of retained austenite as opposed to blocky [14]. The benefits of this are increased strength and toughness.

iv) *Faster rate of reaction*: Unquestionably, a major concern in the development and production of these novel bainitic steels is the time taken for the transformation. The original time scale of "days" is unthinkable from the point of view of production efficiency. Most common approaches rely on thermodynamics, for example, increase of the critical free energy for transformation by chemical composition control, and also on the reduction of the prior austenite grain size (PAGS), by increasing the density of potential nucleation sites for bainite [5,15].

v) *Increased hardenability*: There are two reasons underlying this concept. First is that the formation of other phases prior to bainite transformation might impair the targeted properties. Second, the design process relies on having austenite with the bulk chemical composition transforming into bainite, but if other transformations take place before bainite, the austenite from which it will form would not have the bulk chemical composition.

vi) *Additional refinement of the microstructure*: The scale of the microstructure, that is, bainite plate thickness, is mainly controlled by the strength of the austenite from which it grows [12,13,16,17]. The easiest way to do so is by controlled additions of austenite solid solution strengtheners such as Si and Mo. It is estimated that an increase of 1 wt% of Si implies an increase of 4% of the YS of austenite, while a 0.1 wt% increase of Mo represents a 0.3% increase of the YS [16].

Table 29.1 Chemical composition of designed alloys and their corresponding experimental M_S and A_{C3} temperatures and critical cooling rate (CCR).

Family	Designation	C	Si	Mn	Cr	Mo	Nb	M_S (°C)	A_{C3} (°C)	CCR (°C/s)
MC	0.6C	0.60	1.50	1.25	1.50	—	—	217	838	5
	0.6CCr	0.60	1.50	0.75	2.50	—	—	240	872	5
	0.6CNb	0.60	1.50	1.25	1.50	—	0.03	207	840	5
	0.6CMo	0.60	1.50	1.25	1.50	0.1	—	220	853	3
HC	1C	1.00	1.50	0.75	0.50	—	—	130	880	20
	1CNb	1.00	1.50	0.75	0.50	—	0.03	123	881	20
	1CMo	1.00	1.50	0.75	0.50	0.1	—	130	880	15
	1CSi	1.00	2.9	0.75	0.50	—	—	165	893	20
	0.8C	0.80	1.50	0.75	0.50	—	—	165	808	13

The tools used for the design process comprise bainite transformation theory together with some physical metallurgy concepts [18–20], aided by commercial software for thermodynamics calculations as well as other devoted free access software libraries and in-house tools [21,22]. The final results of the design process based on the above-listed guidelines are shown in Table 29.1. This shows two sets of alloys, one with a medium C content (MC), the 0.6 wt% C family, and the other with higher C content (HC), the 0.8–1.0 wt% C family. As required, all these alloys contain sufficient amount of Si (at least 1.5 wt%) to suppress the precipitation of cementite from austenite. The levels of C content are justified keeping in mind that it is the most effective way of reducing the B_S and M_S temperatures. Also added, but to a lesser extent, are solutes such as Mn and Cr that in the present context increase the stability of austenite relative to ferrite. The disadvantage of a high C concentration is that it slows down bainite transformation kinetics, and therefore it was decreased to 0.6% in some of the alloy grades (MC). Since reduction of C content leads to an increase of the B_S and M_S temperatures, it has to be compensated with higher Mn and Cr contents as compared to the other alloy grades (HC), see Table 29.1.

In earlier experiments on nanocrystalline steels, bainite transformation kinetics was boosted by controlled addition of Co and/or Al [5]. The present alloys lack such "exotic" elements, and therefore, by keeping Cr and Mn as low as possible, the transformation times could be maintained within the range of the steels containing Co and Co + Al [5], while at the same time ensuring sufficient hardenability to avoid transformation during cooling from the austenitization temperature to the bainite transformation temperature. Additional acceleration of the transformation kinetics was sought to be achieved by controlling the PAGS: (i) by keeping the austenitization temperature as low as possible [5,20] and (ii) by the introduction of microalloying additions such as Nb and V that could slow down austenite grain growth by pinning the grain boundaries through their stable carbonitrides [23]. With this purpose in mind, small quantities of Nb, 0.03 wt%, were added in selected alloys of both families, see Table 29.1.

Extra additions of Si (>1.5 wt%) and Mo have also been attempted in the quest for finer bainitic ferrite plates through the strengthening of the parent austenite [12,13,16,17].

29.3
Microstructure and Properties

29.3.1
Bainitic Transformation and Microstructure

Steel grades designed as described above were produced in the laboratory using vacuum induction melting to obtain ingots of 20 or 35 kg. The cast ingots were reheated to 1200 °C and forged into bars of approximately 40 mm diameter. The forged bars were slowly cooled in a furnace to avoid cracking.

The heat treatment to obtain nanocrystalline bainite – bainitizing – consists of complete austenitization followed by cooling to the required isothermal transformation temperature, $M_S < T < B_S$. The critical cooling rate depends on the hardenability of the steel grade since the only requirement is to avoid transformation during cooling. Therefore, as a first step, dilatometry was used to determine the temperature of complete austenitization, the critical cooling rate (CCR), and the M_S temperature, see Table 29.1. Then, the same technique was used to track the bainite transformation at different temperatures. Together with a detailed microstructural characterization, this procedure allowed for the definition of the isothermal transformation temperature and time, which were the most appropriate for the subsequent study of the mechanical behavior [19,24,25].

Thus, for the HC and MC families the austenitizing temperature was adjusted to keep it as low as possible, that is, 950 and 890 °C, respectively, in order to avoid exaggerated austenite grain growth and enhance bainitic transformation. In terms of hardenability, it was found that 30 and 5 °C/s were the optimal cooling rates for the HC and MC families, respectively. The M_S was typically well below 200 °C for the HC grades, thus allowing bainitizing at relatively low temperatures of 200–350 °C. In the MC grades, however, bainitizing temperature range was slightly higher at 240–350 °C.

Regardless of the alloy chemical composition and the temperature of bainitization, the microstructure consisted of a mixture of two phases, bainitic ferrite (α) and carbon-enriched regions of austenite (γ). Figure 29.1 shows a representative example of the microstructure.

The obtained microstructure and its evolution with the transformation temperature did not differ from that already reported elsewhere [1,2,5,6,26,27]. Bainitic ferrite (α) is the dominant phase, with fractions, V_α, in excess of 0.60 and plate thickness, t_α, ranging from 21 to 65 nm. Carbon-enriched austenite is the

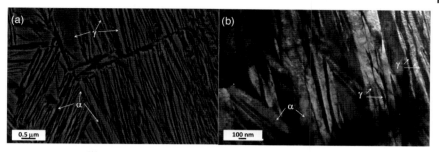

Figure 29.1 Example of the typical nanocrystalline bainite obtained in the HC 1CSi alloy after isothermal transformation at 250 °C: (a) SEM and (b) TEM micrographs.

dispersed second phase, with two distinct morphologies: (i) thin films between the plates of ferrite, also in the nanorange, see the identified features in Figure 29.1b; and (ii) blocks separating sheaves of bainite (groups of bainitic ferrite plates sharing a common crystallographic orientation), see Figure 29.1a. High-magnification observation of the microstructure failed to reveal the presence of any cementite, which is the consequence of the presence of 1.5 wt% Si in the alloys [11,20].

As expected from the mechanism of bainite transformation and the incomplete reaction phenomena ruling the transformation [20], the general tendency is that when the transformation temperature is increased, the amount of austenite (V_γ) and the coarseness of the microstructural features also increase (Figure 29.2). Further as is to be expected, the HC grades show a more refined microstructure (Figure 29.2b) and slower reaction kinetics (Figure 29.3) compared to the MC grades [19,25]. For the HC grades, the plate thickness ranges between 21 and 38 nm while for the MC grades the range is higher at 43–65 nm. Finally, as one of the most important industrial requirements is that the transformation must be achieved in a realistic time, it is noteworthy that both the

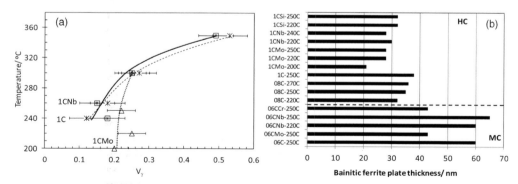

Figure 29.2 (a) For selected grades evolution of the austenite fraction as a function of transformation temperature and (b) for both families, MC and HC, variations of the bainitic ferrite plate thickness at different transformation temperatures.

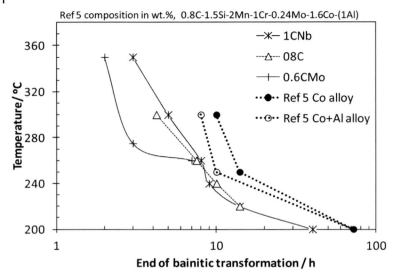

Figure 29.3 For selected grades and as a function of transformation temperature the time needed to complete the transformation. For comparison purposes, alloys from Ref. [5] are also presented.

approaches used in the present work to accelerate the transformation, namely, chemical composition control and reduction of the PAGS [5,18,19], lead to a significant decrease in the transformation time when compared with the best in the class till date [5]. For example, for HC grades the bainite transformation takes about 32–55 h at 200 °C and only a few hours, 3–6 h, at 350 °C. These times are lower than those reported by Garcia-Mateo *et al.* [5], namely, ~72 h at 200 °C and between 7 and 10 h at 250 and 300 °C, respectively, see Figure 29.3. For the MC grades and in general terms, the alloys exhibit a faster transformation kinetics than their higher C counterpart, the difference being bigger as the transformation T increases, for example, at 350 °C, no more than 2 h is needed to finish the transformation, while at 260 °C the time needed is similar to that of the HC grades.

29.3.2
Tensile Properties: Strength and Ductility

When explaining the strength of the low-temperature bainite, one must look at the hardest phase in the microstructure, namely, bainitic ferrite. Without doubt, the reported high strength levels are related to both volume fraction and scale of the bainitic ferrite and also the dislocation density introduced during the transformation [14,28]. Other factors such as the C trapped in dislocations and twins or the amount of C that remains in solid solution also play an important role in the strength of nanocrystalline bainite. However, the influence of high carbon retained austenite on strength properties cannot be ignored. It is well known that the transformation-induced plasticity (TRIP) effect is an important factor enhancing the ultimate tensile strength (UTS) and uniform elongation simultaneously in steels containing high carbon retained austenite within the nanocrystalline bainite microstructure at room temperature. During the plastic deformation of such a microstructure, retained austenite changes to martensite from nucleation sites created by the strain and the TRIP effect, besides the potential sites for nucleation of the athermal martensite [29]. Thus, replacing the ductile austenite with hard martensite can further increase the strength level. In addition, the plastic deformation accompanying the displacive martensitic transformation generates a high dislocation density that in turn increases the work hardening rate [30]. Thus, the TRIP effect can influence the mechanical properties by affecting the work hardening and delaying the necking during straining the sample [31–34]. This is the point where the mechanical stability of retained austenite must be considered.

In the present case, the tensile tests resulted in UTS values always in excess of 2 GPa, with nonnegligible ductility (Table 29.2). A detailed analysis of the results clearly brings out three features: (i) MC grades do not exhibit significant difference in tensile properties as a function of composition. (ii) For the same transformation temperature higher carbon grades (HC) exhibit higher tensile strengths (typically >2200 MPa) and lower elongations. (iii) Finally, the 1CSi alloy exhibits surprisingly high ductility, with >10% uniform elongation for a UTS of 2 GPa [24]. As a measure of comparison, the standard grade steel 100Cr6 does not exhibit measurable ductility when heat-treated to 2 GPa and above.

For the morphologies at hand, the dependence of yield strength on lath thickness (t_α) is expected to follow an inverse relationship, that is, $YS = f(1/t_\alpha)$. Thus, the yield strength correlated well with the ratio V_α/t_α, see Table 29.2 and Figure 29.4a, but the effect of the same parameter on UTS is less clear. The Longford-type bainitic ferrite plate contribution to strength, $(t_\alpha)^{-1}$ [35,36], instead of the typical Hall–Petch $(t_\alpha)^{-1/2}$, is used in this analysis because of the grain size being well below the submicrometer level where the mechanism of yielding involves the initiation of dislocation sources in grain boundaries.

Table 29.2 Tensile test and microstructural characterization results on selected grades and conditions.

Alloy	Transformation temperature (°C)	YS (MPa)	UTS (MPa)	UE (%)	TE (%)	V_γ (%)	t_α (nm)
0.8C	220	1931	2329	3.19	4.1	22	32
	270	1701	2036	4.44	12.64	24	36
1CMo	200	2019	2091	0.37	0.38	20	21
	250	1852	2164	2.86	8.29	22	32
1CSi	220	1704	2287	7.37	7.37	36	32
	250	1698	2068	11.62	21.32	34	32

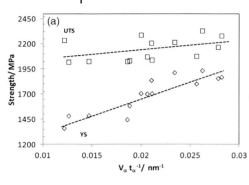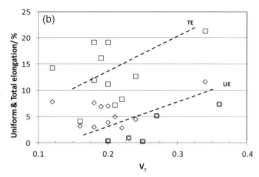

Figure 29.4 (a) Strength of different samples as a function of the ratio V_α/t_α and (b) uniform and total elongation, UE and TE, respectively, as a function of the austenite fraction.

In a similar scheme, and for the described reasons, attempts to understand the origin of ductility in most specimens have, however, been met with only partial success. Indeed, the selected microstructural parameters (retained austenite or bainite content and bainite plate thickness) fail to provide a clear picture of the ductility, see the data scatter in Figure 29.4b [37]. This is particularly visible with the results shown in Table 29.2, where similar parameters lead to significantly different ductility, very evident when transforming at 200/220 or 250/270 °C. As an example, the 1CSi steel exhibits identical retained austenite and bainite lath thickness values after heat treatment at 220 and 250 °C, but its tensile ductility values are quite different. This difference was not seen in microstructural parameters, and it is thus likely that retained austenite stability, and not its quantity, is the key to ductility in those microstructures [37].

29.3.3
Toughness

In general terms for carbide-free bainitic microstructures, it is safe to assume that toughness is mainly controlled by the carbon-enriched retained austenite [38,39] through several mechanisms, namely, crack blunting effect, stress relief, and martensite transformation ahead of the cracks [40]. It has been reported that there is an improvement in the toughness if the TRIP effect of "moderately" stable austenite takes place at the tip of the crack. Although in this type of microstructures, the tendency of the newly formed martensite to crack might depend on its absolute size but, if the TRIP effect takes place in more stable austenite (richer in C), the newly formed high carbon martensite could affect toughness detrimentally [41]. Additionally, the individual strength and hardness of each microstructural constituent influence the toughness properties, where the large mismatch might facilitate stress concentration. Bainitic ferrite also plays an important role during the propagation and growth of a crack, as it is capable of arresting/deflecting the cracks in the packet and plate boundaries.

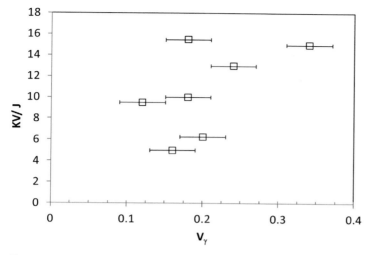

Figure 29.5 Charpy test results for selected compositions and their corresponding austenite fractions.

Moreover, the ability to resist the propagation of cracks is related to the size of the microstructural units.

A clear example of the complexity in explaining toughness is the behavior of the 0.8C alloy shown in Figure 29.5. It seems from this figure that a higher volume fraction of retained austenite solely cannot guarantee better toughness. Thus, for a similar microstructure it has been suggested [6] that superior impact energy can be achieved by optimizing several factors: (i) a matrix composed of smaller plates and therefore smaller (crystallographic) packets, (ii) a milder matrix in terms of strength (smaller mismatch of properties between the matrix (α) and the second phase (γ)), and (iii) retained austenite size and its mechanical stability.

29.3.4
Wear

In terms of sliding/rolling wear resistance, several studies have confirmed the excellent potential of carbide-free bainitic steels. Specifically, Clayton and Jin [42,43] have reported exceptional wear resistance in high silicon cast bainitic steels and confirmed that carbide-free structures gave the best wear resistance, which increased with the amount of carbide-free bainite. From their results, the authors clearly anticipated that increasing the hardness while retaining the carbide-free microstructure should enable outstanding wear resistance with the added bonus of better toughness and ductility due to the lower carbon concentration.

Since then, there have been numerous additional reports of the good rolling/sliding wear performance of carbide-free bainitic steels, Vuorinen *et al.* [44],

Table 29.3 Hardness and specific wear rate (SWR) for investigated alloys and conditions.

	HV	SWR (mm³/N m)
1C–250C	660	1.13E−04
1CSi–250C	630	8.67E−05
08C–220C	693	4.62E−05
08C–250C	639	8.94E−05
08C–270C	621	9.63E−05
06C–250C	589	9.91E−05
06CMo–250C	608	1.08E−04
06CNb–250C	600	8.38E−05

Chang [45], and Leiro *et al.* [46] have reported excellent results on medium carbon high silicon steels isothermally transformed to bainite, attributed to the absence of carbides, the ability of the microstructures to tolerate a large degree of plastic strain, and the mechanically induced martensitic transformation of retained austenite. Vuorinen *et al.* [44] found that it was possible to match the rolling/sliding wear performance of surface-hardened steels by isothermal treatment of high-Si commercial steel with a carbide-free bainitic microstructure. They also studied the effect of retained austenite content on the rolling/sliding wear rate of high-Si steel with a carbide-free microstructure [46]. It was found that at higher transformation temperatures and retained austenite content the wear rate increased. Results presented in Table 29.3 for the present alloys are in line with these findings. The wear resistance of nanocrystalline bainite is indeed significantly superior to that of standard material of similar hardness. For example, the combination of HV (hardness) and SWR (specific wear rate) for commercial steel 100Cr6 is 714/1.7×10^{-4} mm³/(N m). The same material with bainitic microstructure obtained by isothermal transformation at 250 and 300 °C reported HV/SWR combinations of 712/1.5×10^{-4} and 600/2.3×10^{-4} mm³/(N m), respectively. Table 29.3 shows that in the present alloys having nanocrystalline bainite, with hardness values of approximately 600 HV, the wear rates are much lower even though varying over an order of magnitude [47].

Thus, the wear performance of nanocrystalline bainite is found to be largely improved compared to bainitic 100Cr6. Performance in rolling/sliding tests appeared to be improved by the presence of significant quantities of retained austenite in the nanocrystalline bainitic alloys as opposed to conventional bainite microstructures. The highly refined microstructure in addition to the transformation of retained austenite to martensite during wear testing contributes to increasing surface hardness (well above that measured on conventional bainitic 100Cr6), thus reducing the wear rate as explained in Refs [24,25,47].

29.3.5
Fatigue

Most applications of structural materials involve cyclic loading and life of a component is generally limited by fatigue and not by static strength. In order to fully exploit the potential of nanocrystalline bainitic steels in industrial applications, a better understanding must be achieved of the relationship between microstructure and in-service properties, as fatigue behavior is one of the least investigated aspects of the bainitic steels. Peet *et al.* [48] were the first to report on the results of fatigue tests on this type of microstructure. Using a stress ratio of $R = 0.1$, they conducted the test up to a maximum of 10^5 cycles and then extrapolated the results to 10^7 cycles to estimate the fatigue strength. The validity of this method is undermined when considering the two-step or duplex $S-N$ curves observed in high-strength steels [49–54]. Subsequent investigations on high-cycle bending fatigue of nanocrystalline bainitic steels [55] revealed that the fatigue limit for no failure in 10^7 cycles is influenced not only by the hardness or strength but also by the microstructure so that the retained austenite and secondary cracks in fatigue fractures have a positive effect on the fatigue resistance. Most recent work of Leiro *et al.* [56], comparing the rotating bending fatigue behavior of quenched and tempered (QT) steels with that of nanobainitic steels, showed an improvement in fatigue properties of the latter, mainly as a result of the refinement of the microstructure, suppression of carbide precipitation, and the TRIP effect. The beneficial effect of retained austenite in the fatigue performance of bainitic steels has been thoroughly investigated [48,55,57]. Results showed that it increases the threshold intensity [57], and the stress- or strain-induced transformation of bainite to martensite can blunt the crack by absorbing more energy than necessary for fatigue crack propagation [55].

Many engineering components contain a variety of stress concentrators such as grooves, fillets, holes, or nonmetallic inclusions. It has been observed that fatigue failure usually occurs as a result of crack initiation and growth from these stress raisers. Therefore, the correct estimation of stress/strain concentration and crack development in the critical region is essential for practical machine design in service loading. Hence, notched tension–tension fatigue specimens were used to study the behavior of nanocrystalline bainitic microstructures for actual engineering applications.

The fatigue limit of most metallic materials, for both notched and unnotched specimens, is determined by the propagation condition of nucleated small cracks. Crack closure and microstructural barriers are the two mechanisms for blocking small crack propagation at the fatigue limit [58]. Microstructural barriers in bainitic steels come from the nature of the displacive transformation itself, where it is generally assumed that the close-packed {111} planes of the austenite (face-centered cubic (FCC)) are exactly parallel to the {110} planes of the bainitic ferrite (body-centered cubic (BCC)). In nanobainitic steels, most of the bainitic plates and the parent austenite interfaces are found to have an orientation relationships (OR) close to the Nishiyama–Wassermann (N–W), $\{111\}\gamma\|\{110\}\alpha$ with

$\langle 110 \rangle \gamma || \langle 001 \rangle \alpha$ [27,59]. The transformed products, however, possess the lattice-invariant line rather than the plane and direction parallelism between bainitic ferrite and parent austenite. Therefore, the relative orientation can never be precisely the N–W, but may lie between the Kurdjumov–Sachs (K–S) and N–W values depending on the ratio of the lattice parameters between the parent austenite and bainitic ferrite [60]. From a single austenite crystal, 12 crystallographic variants can be formed with an N–W orientation relationship due to the symmetry of cubic systems. A crystallographic packet is defined as a group of crystallographic variants with a common {111} austenite plane (i.e., the same habit plane). Each bainite packet can be divided into bainite blocks of the three variants of the N–W relationship satisfying the same parallel plane relationship [61]. As a consequence, the microstructure obtained is highly disoriented featuring a wide variety of potential microstructural barriers.

For the alloys and microstructures at hand, the evaluation of fatigue performance was done by means of rotating–bending type of fatigue tests on notched specimens. Reasons for choosing such tests are as follows: (i) end-user applications generally exhibit stress concentrators, therefore notched condition is of great relevance; (ii) notched bar fatigue may be more sensitive to the presence of retained austenite as the stress concentration under the notch may enhance its decomposition; and (iii) it is a well-established fact that high-cycle fatigue performance of high-strength material (UTS > ~1200 MPa) is strongly influenced by defects present in the material bulk or surface, and because cleanliness of laboratory casts is not controlled, the use of notched rotation bending specimens is again justified. The combination of the rotating–bending method and notching would significantly limit the sampling volume since, first, the rotating–bending specimens are only at the maximum stress in their outer thickness and, second, the notch will further localize the volume under load along the sample length. Since no literature was available for the standardized testing conditions for rotating–bending fatigue, results could only be compared with one another.

A comparison of the fatigue strength achieved in MC and HC materials suggests a significant difference between these two grades, see Table 29.4. While satisfactory for the former, the fatigue strength of the latter was below expectation. Reasons for this are still under investigation but a preliminary microstructural examination at Stage I indicates that the crack deflects at the interphase boundaries between blocks, packets, and twins of the ferritic phase, but not to a significant extent in the interphase boundaries within a single

Table 29.4 Rotating–bending fatigue strength results for 10^7 cycles (K_t is the stress factor).

Material	Transformation temperature (°C)	UTS (MPa)	Sa, 50% at 10^7 cycles (MPa) for $R = -1$, $K_t = 1.6$
0.6C	250	2023	665
0.8C	270	2036	440
1CNb	220	2073	430

block [62]. It suggests that the bainite block size is the crystallographic parameter controlling the crack propagation at this stage. However, its effect on the fatigue strength remains unclear, partially because it cannot be isolated from other microstructural features that are also likely to be affecting the mechanical behavior, for example, the TRIP behavior of austenite during fatigue crack propagation as a function of its morphology (composition) or the favored places for fatigue damage (crack initiation) [62].

29.4
Summary

The bijective relationship between materials and technology is leading to the development of metals with a strength–ductility tradeoff out of what was feasible a short while ago. One of the key strategies seeking advancements in the steel industry relies on engineering the structures at the nanoscale. In this sense, nanocrystalline bainite was conceived to be obtained by conventional manufacturing methods merely by heat treatment of carefully designed high-carbon high-silicon steels. Nowadays, bulk material can be produced with a nanometer-scale structure throughout the three dimensions without the use of severe deformation or complex heat treatments, being executed in an affordable manner with no cost penalty for manufacturers.

The structures achieved consist of nanoscaled crystals of bainitic ferrite, with a thickness of 20–60 nm, interwoven by austenite. The amount of these crystals per unit volume is so high that nanocrystalline bainite contains one of the highest density of ferrite/austenite interfaces known.

The concept of nanocrystalline bainitic steel has moved from the laboratory scale up to full-scale industrial production and component trials. The first achievement of such process has been to demonstrate that reasonable transformation kinetics could be achieved through tailoring of the alloy composition without the use of sophisticated alloying elements such as Co and Ni, thus leading to a reasonably inexpensive material. The second achievement was to demonstrate the full potential of nanocrystalline bainite in terms of its properties, the extraordinary strength–ductility combinations accompanied by nonnegligible toughness and service properties, such as wear and fatigue, showing considerable improvements.

Acknowledgments

The authors gratefully acknowledge the support of both the European Research Fund for Coal and Steel, the Spanish Ministry of Economy and Competitiveness, and the Fondo Europeo de Desarrollo Regional (FEDER) for partially funding this research under contracts RFSR-CT-2008-00022 and MAT2013-47460-C5-1-P. The authors would like to thank all the partners in the RFCS consortium, that is, T. Sourmail and V. Smanio from Asco Industries, M. Kuntz from Robert Bosch, R. Elvira from Gerdau I + D, A. Leiro and E. Vuorinen from Lulea University of Technology, T. Teeri from Metso Minerals Oy and C. Ziegler, and V. Heuer from ALD Vacuum Technology GmbH. Finally, the authors would also like to acknowledge Prof. H.K.D.H. Bhadeshia and Prof. C. Garcia de Andres for their support and countless scientific discussions.

References

1 Garcia-Mateo, C., Caballero, F., and Bhadeshia, H. (2003) Development of hard bainite. *ISIJ Int.*, **43**, 1238–1243.

2 Caballero, F.G., Bhadeshia, H.K.D.H., Mawella, K.J.A., Jones, D.G., and Brown, P. (2002) Very strong low temperature bainite. *Mater. Sci. Technol.*, **18**, 279–284.

3 Garcia-Mateo, C. and Caballero, F. (2005) Ultra-high-strength bainitic steels. *ISIJ Int.*, **45**, 1736–1740.

4 Caballero, F.G., Bhadeshia, H.K.D.H., Mawella, K.J.A., Jones, D.G., and Brown, P. (2001) Design of novel high strength bainitic steels: Part 2. *Mater. Sci. Technol.*, **17**, 517–522.

5 Garcia-Mateo, C., Caballero, F., and Bhadeshia, H. (2003) Acceleration of low-temperature bainite. *ISIJ Int.*, **43**, 1821–1825.

6 Avishan, B., Yazdani, S., and Nedjad, S.H. (2012) Toughness variations in nanostructured bainitic steels. *Mater. Sci. Eng. A*, **548**, 106–111.

7 Hu, F. and Wu, K.M. (2011) Nanostructured high-carbon dual-phase steels. *Scr. Mater.*, **65**, 351–354.

8 Yoozbashi, M.N., Yazdani, S., and Wang, T.S. (2011) Design of a new nanostructured, high-Si bainitic steel with lower cost production. *Mater. Des.*, **32**, 3248–3253.

9 Amel-Farzad, H., Faridi, H.R., Rajabpour, F., Abolhasani, A., Kazemi, S., and Khaledzadeh, Y. (2013) Developing very hard nanostructured bainitic steel. *Mater. Sci. Eng. A*, **559**, 68–73.

10 Bhadeshia, H.K.D.H., and Edmonds, D.V. (1980) The mechanism of bainite formation in steels. *Acta Metall.*, **28**, 1265–1273.

11 Kozeschnik, E. and Bhadeshia, H.K.D.H. (2008) Influence of silicon on cementite precipitation in steels. *Mater. Sci. Technol.*, **24**, 343–347.

12 Singh, S.B. and Bhadeshia, H.K.D.H. (1998) Estimation of bainite plate-thickness in low-alloy steels. *Mater. Sci. Eng. A*, **245**, 72–79.

13 Cornide, J., Garcia-Mateo, C., Capdevila, C., and Caballero, F.G. (2013) An assessment of the contributing factors to the nanoscale structural refinement of advanced bainitic steels. *J. Alloys Compd.*, **577**, S43–S47.

14 Bhadeshia, H.K.D.H., and Edmonds, D.V. (1983) Bainite in silicon steels: new composition-property approach. Part 1. *Met. Sci.*, **17**, 411–419.

15 Caballero, F.Garcia., Santofimia, M.Jesus., Capidevila, C., Garcia-Mateo, C., and Garcia De Andres, C. (2006) Design of advanced bainitic steels by optimisation of TTT diagrams and T-0 curves. *ISIJ Int.*, **46**, 1479–1488.

16 Young, C.H. and Bhadeshia, H.K.D.H. (1994) Strength of mixtures of bainite and martensite. *Mater. Sci. Technol.*, **10**, 209–214.

17 Peet, M.J. (2001) *Neural network modelling of hot deformation of austenite. Materials science and metallurgy.* M. Phil. University of Cambridge.

18 Garcia-Mateo, C., Caballero, F.G., Sourmail, T., Smanio, V., and Garcia de Andres, C. (2014) Industrialised nanocrystalline bainitic steels. Design approach. *Int. J. Mater. Res.*, **105**, 725–734.

19 Garcia-Mateo, C., Caballero, F.G., Sourmail, T., Cornide, J., Smanio, V., and Elvira, R. (2014) Composition design of

nanocrystalline bainitic steels by diffusionless solid reaction. *Met. Mater. Int.*, **20**, 405–415.

20 Bhadeshia, H.K.D.H. (2015) *Bainite in Steels: Theory and Practice*, 3rd edn, Maney Publishing.

21 Laboratory, N.P. (2003) MTDATA. Teddington, Middlesex, UK.

22 Bhadeshia, H.K.D.H. (2015) Materials Algorithms Project (MAP). University of Cambridge, Cambridge.

23 Gladman, T. (1997) *The Physical Metallurgy of Microalloyed Steels*, Institute of Materials, London.

24 Sourmail, T., Caballero, F.G., Garcia-Mateo, C., Smanio, V., Ziegler, C., Kuntz, M., Elvira, R., Leiro, A., Vuorinen, E., and Teeri, T. (2013) Evaluation of potential of high Si high C steel nanostructured bainite for wear and fatigue applications. *Mater. Sci. Technol.*, **29**, 1166–1173.

25 Sourmail, T., Smanio, V., Ziegler, C., Heuer, V., Kuntz, M., Caballero, F.G., Garcia-Mateo, C., Cornide, J., Elvira, R., Leiro, A., Vuorinen, E., and Teeri, T. (2013) Novel nanostructured bainitic steel grades to answer the need for high-performance steel components (Nanobain). RFSR-CT-2008-00022, European Commission, Luxembourg.

26 Timokhina, I.B., Beladi, H., Xiong, X.Y., Adachi, Y., and Hodgson, P.D. (2011) Nanoscale microstructural characterization of a nanobainitic steel. *Acta Mater.*, **59**, 5511–5522.

27 Beladi, H., Adachi, Y., Timokhina, I., and Hodgson, P.D. (2009) Crystallographic analysis of nanobainitic steels. *Scr. Mater.*, **60**, 455–458.

28 Bhadeshia, H.K.D.H. and Edmonds, D.V. (1983) Bainite in silicon steels: new composition-property approach. Part 2. *Met. Sci.*, **17**, 420–425.

29 Olson, G.B. and Cohen, M. (1975) Kinetics of strain-induced martensitic nucleation. *Metall. Mater. Trans. A*, **6**, 791–795.

30 Jacques, P.J. (2004) Transformation-induced plasticity for high strength formable steels. *Curr. Opin. Solid State Mater. Sci.*, **8**, 259–265.

31 Jiménez, J.A., Carsí, M., Ruano, O.A., and Frommeyer, G. (2009) Effect of testing temperature and strain rate on the transformation behaviour of retained austenite in low-alloyed multiphase steel. *Mater. Sci. Eng. A*, **508**, 195–199.

32 Matsumura, O., Sakuma, Y., Ishii, Y., and Zhao, J. (1992) Effect of retained austenite on formability of high strength sheet steels. *ISIJ Int.*, **32**, 1110–1116.

33 Olson, G.B. and Cohen, M. (1972) A mechanism for the strain-induced nucleation of martensitic transformations. *J. Less Common Met.*, **28**, 107–118.

34 Sugimoto, K.I., Kobayashi, M., and Hashimoto, S.I. (1992) Ductility and strain-induced transformation in a high-strength transformation-induced plasticity-aided dual-phase steel. *Metall. Mater. Trans. A*, **23**, 3085–3091.

35 Langford, G. and Cohen, M. (1970) Calculation of cell-size strengthening of wire-drawn iron. *Metall. Mater. Trans. B*, **1**, 1478–1480.

36 Langford, G. and Cohen, M. (1969) Strain hardening of iron by severe plastic deformation. *ASM Trans. Quart.*, **62**, 623–638.

37 Garcia-Mateo, C., Caballero, F.G., Sourmail, T., Kuntz, M., Cornide, J., Smanio, V., and Elvira, R. (2012) Tensile behaviour of a nanocrystalline bainitic steel containing 3 wt% silicon. *Mater. Sci. Eng. A*, **549**, 185–192.

38 Sandvik, B.P.J. and Navalainen, H.P. (1981) Structure-property relationships in commercial low-alloy bainitic-austenitic steel with high strength, ductility, and toughness. *Met. Technol.*, **8**, 213–220.

39 Miihkinen, V.T.T. and Edmonds, D.V. (1987) Fracture toughness of two experimental high-strength bainitic low-alloy steels containing silicon. *Mater. Sci. Technol.*, **3**, 441–449.

40 Wu, R., Li, W., Zhou, S., Zhong, Y., Wang, L., and Jin, X. (2014) Effect of retained austenite on the fracture toughness of quenching and partitioning (Q&P)-treated sheet steels. *Metall. Mater. Trans. A*, **45**, 1892–1902.

41 Chatterjee, S. and Bhadeshia, H.K.D.H. (2006) TRIP-assisted steels: cracking of high-carbon martensite. *Mater. Sci. Technol.*, **22**, 645–649.

42 Clayton, P. and Jin, N. (1996) Unlubricated sliding and rolling/sliding wear behavior of

continuously cooled, low/medium carbon bainitic steels. *Wear*, **200**, 74–82.

43 Jin, N. and Clayton, P. (1997) Effect of microstructure on rolling/sliding wear of low carbon bainitic steels. *Wear*, **202**, 202–207.

44 Vuorinen, E., Pino, D., Lundmark, J., and Prakash, B. (2007) Wear characteristic of surface hardened ausferritic Si-steel. *J. Iron. Steel Res. Int.*, **14**, 245–248.

45 Chang, L.C. (2005) The rolling/sliding wear performance of high silicon carbide-free bainitic steels. *Wear*, **258**, 730–743.

46 Leiro, A., Kankanala, A., Vuorinen, E., and Prakash, B. (2011) Tribological behaviour of carbide-free bainitic steel under dry rolling/sliding conditions. *Wear*, **273**, 2–8.

47 Leiro, A., Vuorinen, E., Sundin, K.G., Prakash, B., Sourmail, T., Smanio, V., Caballero, F.G., Garcia-Mateo, C., and Elvira, R. (2013) Wear of nano-structured carbide-free bainitic steels under dry rolling-sliding conditions. *Wear*, **298**, 42–47.

48 Peet, M.J., Hill, P., Rawson, M., Wood, S., and Bhadeshia, H.K.D.H. (2011) Fatigue of extremely fine bainite. *Mater. Sci. Technol.*, **27**, 119–123.

49 Sakai, T., Takeda, M., Shiozawa, K., Ochi, Y., Nakajima, M., Nakamura, T., and Oguma, N. (2000) Experimental reconfirmation of characteristic S–N property for high carbon chromium bearing steel in wide life region in rotating bending. *Zairyo/J. Soc. Mater. Sci. (Japan)*, **49**, 779–785.

50 Murakami, Y., Nomoto, T., Ueda, T., and Murakami, Y. (2000) On the mechanism of fatigue failure in the superlong life regime (N > 107 cycles). Part I: Influence of hydrogen trapped by inclusions. *Fatigue Fract. Eng. Mater. Struct.*, **23**, 893–902.

51 Murakami, Y., Nomoto, T., Ueda, T., and Murakami, Y. (2000) On the mechanism of fatigue failure in the superlong life regime (N > 107 cycles). Part II: A fractographic investigation. *Fatigue Fract. Eng. Mater. Struct.*, **23**, 903–910.

52 Murakami, Y., Yokoyama, N.N., and Nagata, J. (2002) Mechanism of fatigue failure in ultralong life regime. *Fatigue Fract. Eng. Mater. Struct.*, **25**, 735–746.

53 Shiozawa, K. and Lu, L. (2002) Very high-cycle fatigue behaviour of shot-peened high-carbon-chromium bearing steel. *Fatigue Fract. Eng. Mater. Struct.*, **25**, 813–822.

54 Sakai, T., Sato, Y., and Oguma, N. (2002) Characteristics S-N properties of high-carbon-chromium-bearing steel under axial loading in long-life fatigue. *Fatigue Fract. Eng. Mater. Struct.*, **25**, 765–773.

55 Yang, J., Wang, T.S., Zhang, B., and Zhang, F.C. (2012) High-cycle bending fatigue behaviour of nanostructured bainitic steel. *Scr. Mater.*, **66**, 363–366.

56 Leiro, A., Roshan, A., Sundin, K.G., Vuorinen, E., and Prakash, B. (2014) Fatigue of 0.55C–1.72Si steel with tempered nano martensitic and carbide-free bainitic microstructures. *Acta Metall. Sinica*, **27**, 55–62.

57 Wenyan, L.I.U., Jingxin, Q.U., and Hesheng, S. (1997) Fatigue crack growth behaviour of a Si–Mn steel with carbide-free lathy bainite. *J. Mater. Sci.*, **32**, 427–430.

58 Akiniwa, Y., Tanaka, K., and Kimura, H. (2001) Microstructural effects on crack closure and propagation thresholds of small fatigue cracks. *Fatigue Fract. Eng. Mater. Struct.*, **24**, 817–829.

59 Sandvik, B.P.J. (1982) The bainite reaction in Fe-Si-C alloys: the primary stage. *Metall. Trans. A*, **13**, 777–787.

60 Christian, J.W. (1990) Simple geometry and crystallography applied to ferrous bainites. *Metall. Trans. A*, **21**, 799–803.

61 Marder, A.R. and Krauss, G. (1967) The morphology of martensite in iron-carbon alloys. *Trans. ASM*, **60**, 651–660.

62 Rementeria, R., Morales-Rivas, L., Kuntz, M., Garcia-Mateo, C., Kerscher, E., Sourmail, T., and Caballero, F.G. (2015) On the role of microstructure in governing the fatigue behaviour of nanostructured bainitic steels. *Mater. Sci. Eng. A*, **630**, 71–77.

30

Graphene and Graphene Oxide for Energy Storage

Edward P. Randviir and Craig E. Banks

Faculty of Science and Engineering, Manchester Metropolitan University, Chester Street, Manchester, M1 5GD, UK

30.1
Graphene Hits the Headlines

The field of nanotechnology has no doubt taken over modern technology, and was perhaps kick-started in 1999 when Richard Smalley told Congress that "the impact of nanotechnology on health, wealth, and lives of people will be at least the equivalent of the combined influences of microelectronics, medical imaging, computer-aided engineering, and manmade polymers" [1]. We are sure even he was doubting the considerable perpetuation of nanotechnology since then, not least since 5 years on in 2004 when the famous report emerged of the endearingly simple "Scotch tape" method for thin film isolation [2]. The paper carefully reported the electric field effect in thin carbon films – an effect that is not normally observed in a bulk conductive material. However, thin materials exhibit the field effect when privy to an external electric field, allowing some properties of the material to be manipulated. Novoselov and coworkers were the first to report this effect in thin carbon materials, and later they were able to fabricate nanographite, or graphene, for the same purpose, which triggered a nanomaterial "gold rush" shortly afterward, and has continued until the present day. In 2014 alone, there were 20 935 accepted peer-reviewed articles (Web of Knowledge) containing the word graphene, which is an indication of the relative level of research activity within the many applications of this material.

The surge in interest surrounding graphene was kick-started by two famous research papers coauthored by Novoselov and Geim in 2004 and 2005, respectively. The 2004 article is often incorrectly cited as being the paper that reported "graphene", because the focus of that study was few-layered graphene [3]. Despite this, it is considered pivotal toward opening the door for today's graphene research. Novoselov and coworkers subsequently reported the field effect in single-layer graphene in 2005, reporting a charge carrier mobility of $2000-5000\,cm^2/V/s$, far

Nanotechnology for Energy Sustainability, First Edition. Edited by Baldev Raj, Marcel Van de Voorde, and Yashwant Mahajan.

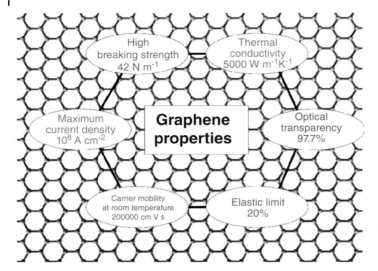

Figure 30.1 A small selection of the many reported properties of graphene.

surpassing the other 2D films studied within the same paper [4]. This amazingly large carrier mobility triggered further research into the physical properties of graphene, which are summarized in Figure 30.1. Furthermore, the large specific surface area of graphene ($2630 \, \text{m}^2/\text{g}$) and ability to mass produce in very large quantities at a relatively low level of expense make graphene an extremely attractive material for energy storage and generation applications. This chapter explores fundamental elements that make graphene potentially useful as a material for energy storage and generation devices and also discovers how researchers have designed, implemented, improved, and reported graphene-based devices. This chapter assumes a basic knowledge of the electrochemistry of carbon nanomaterials.

30.2
Graphene: Why All the Fuss?

One could quite easily be forgiven for asking the question in the subtitle. To the uninitiated, graphene is just a very thin piece of graphite, and therefore in macroscopic terms should not act any differently to the bulk material. This could not be further from reality – 2D materials exhibit quantum confinement unlike bulk materials and therefore the unique properties reported above can be exploited *only* under finite conditions. Graphene, a single graphite sheet, should behave as a 2D quantum well, and therefore its properties cannot be grossly assumed to be the same as graphite [5]. It is the observed differences in behavior between the single sheet and the bulk material which make graphene so fascinating to

Table 30.1 Comparison of the properties of several graphene-derived materials.

Material	Young's modulus (TPa)	Tensile strength (GPa)	Resistivity ($\mu\Omega$ cm)	Electron mobility (cm^2/V/s)	Optical conductance (%)	Specific surface area (m^2/g)
Polycrystalline graphite [44]	0.032	0.069	590	15 000	Opaque	90
Graphene [45]	1	130	1	200 000	97.7	2630
CNTs [46]	0.2–1	13–126	10	100 000	N/A	1315

researchers. In order to put graphene into context, it is useful to compare it to other carbonaceous materials such as carbon nanotubes (CNTs) and fullerenes; graphene is the building block of these graphitic materials [6]. The general properties of graphites, graphenes, and CNTs are listed in Table 30.1. Graphene offers significant advantages over CNTs and graphite in almost every possible way, whether that be the Young's modulus, electron mobility, or surface area. The very high surface area coupled with the electron mobility, in addition to its impressive mechanical strength, makes graphene an ideal material to study for energy storage and generation applications such as fuel cells, photovoltaic cells (PVCs), or supercapacitors.

Graphene oxide (GO) is a material often misquoted as graphene despite its remarkably different properties. GO does not exhibit the extreme hydrophobicity of graphene, and therefore can be suspended in water fairly comfortably compared to graphenes that prefer organic solvents. The electronic properties of GO are also completely different from graphene. Consequently, the electrochemical profiles related to graphene oxides are significantly different than observed for graphene working electrodes [7]. This fundamental understanding is useful for energy generation devices as it infers that graphene oxide might allow more energy efficient electronic transitions than graphene if used as an interfacial material.

30.3
Graphene and Graphene Oxide in Energy Storage Devices

With the increasing consumer demand for electronic products, the global electricity demand increases as reported by the International Energy Agency [8]. Presently, electronic devices have much improved energy efficiencies compared to primitive ones, but the number of units that are produced means that the energy demand is ever increasing. It is therefore useful to design more efficient energy storage devices in order to combat increasing demands in energy consumption. Graphene is ideally placed to be a material for energy storage devices because it exhibits an immensely large specific surface area of 2630 m^2/g (see Table 30.1), allowing a highly charged electric double layer to be formed

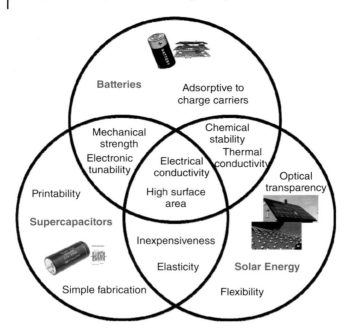

Figure 30.2 Venn diagram representing graphene's properties and their potential uses in several applications. The diagram is not exhaustive nor are the individual elements exclusive to these applications.

provided the entire surface can be accessed by the electrolyte. Figure 30.2 summarizes the useful properties of graphene that can be exploited for several energy devices. Additionally, graphene exhibits a high electrical conductivity and chemical stability; consequently, graphene-based devices may depreciate very slowly compared to devices fabricated from other nanomaterials. Evidence also suggests that GO can be chemically manipulated to exhibit very high surface areas that may be useful for supercapacitor applications [9]. Readers are referred to the reviews provided for further information on graphene-based energy storage devices [10]. This section discusses some applications of graphene and GO in energy storage devices such as supercapacitors, batteries, PVCs, fuel cells, and electrolyzers.

30.3.1
Supercapacitors

Supercapacitors are static electric double-layer devices that store an extremely large amount of charge, the value of which is normally commensurate with the surface area of the material used to fabricate the device [10b]. They are used in applications where a large amount of charge is required in a short span of time. Supercapacitors have several applications such as electric vehicles for acceleration or for the opening of fuselage doors in airplanes, because they deliver the charge requirements for the operation of high-energy demand electronic parts. Supercapacitors are also useful for effective power management of smaller devices such as cameras. They also exhibit shorter charge/discharge times and are generally robust enough to compete with hundreds of charge/discharge cycles.

Fabrication of graphene-based supercapacitors is not as simple as it may appear, however, and many efforts to achieve the theoretical capacitance value (in excess of 550 F/g) have failed [10a]. This can be attributed to several factors, such as the stacking and reorganization of graphene sheets reducing the effective surface area of the material [11]. There have been several examples of pseudocapacitance, a phenomenon that gives the impression of higher capacitance due to electron transfer reactions within the supercapacitor, that has seen specific capacitances in excess of 550 F/g, but that suffers from shorter lifetimes due to inefficiencies in the electron transfer reactions after several uses [11]. Therefore research is directed toward exploiting both the surface area of graphene and improving the efficiency of electron transfer reactions within the device, which combines capacitative and pseudocapacitative effects to produce very large specific capacitances. This is normally achieved by incorporating graphenes within metal oxides [10a], such as cobalt, zinc, ruthenium, or tin, though polymers such as polyaniline [12] have also been considered as useful materials for supercapacitors. Metal oxides such as cobalt can promote restacking of graphene sheets; however, care must be taken when fabricating such materials.

Since 2015 there have been several advances in graphene supercapacitor research. Currently, there is a focus toward the paradoxical "3D graphene" networks – the idea being a large porous network covered with graphene or GO to expand the effective surface area of a material. There are significant problems with decorating materials with graphenes, such as a high contact resistance for reduced graphene oxides (rGO). Furthermore, the 3D architecture must remain intact after fabrication, while maintaining this is not so easy. However, copper foam can be fabricated electrochemically and covered with GO that can subsequently be electrochemically reduced on the 3D foam surface, fabricating a 3D network of rGO [11]. The copper structure is also used as the current collector for the supercapacitor. Asymmetric supercapacitors are more commonly researched within the past few years as they normally exhibit higher energy densities than symmetric supercapacitors. An asymmetric supercapacitor is a capacitor that uses two different materials fabricated on opposing conducting plates to build the supercapacitor. Graphene/graphene symmetric capacitors have shown energy densities of 3.6 Wh/kg, whereas MnO_2/graphene asymmetric supercapacitors exhibit an energy density of 25.2 Wh/kg [13], representing a near ninefold increase in energy density. Graphene supercapacitors can also be fabricated through inkjet printing of graphene oxides onto tin foils and undergo subsequent thermal reduction of the graphene oxide to graphene. One perceived advantage of this method is that thermally reduced graphenes are highly defective [3,14], therefore will provide a balance between high surface area for double-layer capacitance and improved electron transport efficiency for

Table 30.2 Applications of graphene in supercapacitors.

Graphene	Specific capacitance (F/g)	Current density (A/g)	Comments	Ref.
Graphene/polyaniline	261	1 mA/cm^2	Highly flexible material	[16]
Graphene/polyaniline	909	1 mA/cm^2	*In situ* synthesis yields improved specific capacitance	[17a]
Nitrogen-doped graphene	220	0.8	Paper shows different specific capacitances, depending upon a measurement method	[17b]
GO reduced by elemental copper on Ti foil	148	1	Likely an impure graphene and specific capacitance suffers as a result	[17c]
Graphene/MnO$_2$/carbon nanotubes/Ni	251	1	High conductivity and low resistance	[17d]
Graphene/NiO	425	2	21% loss after 2000 cycles	[17e]
3D graphene	339	0.25	Experiments performed under alkaline conditions	[17f]

pseudocapacitance. Unfortunately, though the energy densities were shown to be higher, the specific capacitance of the printed graphene supercapacitors ranges from 48 to 132 F/g, which is considerably lower than that observed for other graphene-based supercapacitors [15]. The elasticity of graphene has also been exploited to design a supercapacitor that can be stretched by incorporating graphene into polyaniline, while maintaining a good specific capacitance of 261 F/g. Such devices may provide a solution for unusual circuit designs where the supercapacitor may need to be bent or stretched into place [16]. Table 30.2 summarizes some of the latest graphene-based supercapacitors [16,17], and further reading on porous graphene materials for supercapacitor design is provided for interested readers [18]. It is worth noting that the generic method used for the measurement of capacitance is galvanostatic charge/discharge, which leads to dramatically different results depending on user interpretation. Recent work has demonstrated that supercapacitors can be measured more accurately with minor rectification of the circuitry [19].

30.3.2
Batteries

Graphene and graphene oxide are the subject of many research papers focusing on the manufacturing of batteries, and in particular lithium-ion batteries. Similarly to supercapacitors, research on batteries is growing every year as there is more demand for smaller batteries that store more charge in order to improve the overall battery lifetime of consumer electronic devices. Batteries exhibit much slower discharge times than supercapacitors and provide lower energy densities over a much longer period of time than a supercapacitor, though the principle of using graphene as electrode materials within batteries is similar, in that the increased surface area allows improved adsorption of charge compared to graphite. Early reports on graphene-based batteries focused on improving the storage of lithium ions within nanostructures, because standalone metal oxide nanostructures such as tin oxide suffer from stress cracking after repeated charging/discharging, which effectively disconnect the charge storing material from the current collectors and reduce battery performance [20]. Therefore, graphene was explored to improve the lifetime of the battery through prevention of stress cracking over several charge/discharge sycles. The flexibility of graphene was hoped to be advantageous and take some of the stress off the tin oxide during charge/discharge cycles. Such a design recorded a specific capacity of 570 mA h/g after 30 cycles, while tin oxide only exhibited 550 mA h/g and reduced to 60 mA h/g after 15 cycles.

The energy and power densities of batteries need to be much higher and be more stable over several hundred charge cycles for uses in high power devices such as electric vehicles. Many cathodic materials, such as silicon, sulfur, and graphite, have been applied to deliver various specific capacities but all experience different problems. Sulfur-based materials in particular suffer because they lack the required conductivity for batteries even though they exhibit a theoretical specific capacity of 1672 mA h/g and also tend to dissolve into electrolytes and experience swelling during discharge [21]. However, graphene can be incorporated into sulfur-based cathodes by wrapping sulfur particles in poly(ethylene glycol) (PEG) and graphene. The coating therefore protects sulfur from dissolution (PEG) and enhances the conductivity (graphene) of the material while exploiting the specific capacity of sulfur. The fabrication procedure is illustrated in Figure 30.3. Such a procedure allows specific capacities of 600 mA h/g for over 100 charge/discharge cycles.

Three-dimensional graphene foams are also studied for use in lithium-ion batteries. Fe_3O_4 is considered a useful anode material for lithium-ion batteries due to its relative abundance, stability, high theoretical specific capacity, inexpensiveness, and nontoxicity. However, Fe_3O_4 suffers from volume expansion and low conductivity similar to sulfur. A different approach is used by Luo *et al.* who grafted Fe_3O_4 nanoparticles into 3D graphene foams instead of building graphene around the nanoparticles [22]. In this approach, therefore, the nanoparticles are wired to current collectors using graphene as a flexible substrate. This approach achieved a specific capacity of 785 mA h/g for over 500 cycles, perhaps due to the stability of the 3D graphene framework. Other research includes sandwiching silicon nanowires between graphene oxide sheets achieving specific capacities of 1600 mA h/g [23], graphene-coated lithium iron phosphate achieving specific capacities of 208 mA h/g using less defective electrochemically exfoliated graphene [24], and MoS_2/graphene electrodes have

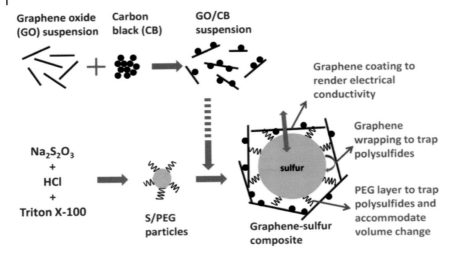

Figure 30.3 Preparation of sulfur-based cathodes for lithium-ion batteries.

been designed for sodium-ion batteries with a specific capacity of 230 mA h/g
[25]. Further applications are listed in Table 30.3 [25,26] as there are too many
to discuss within this text. Interested readers are referred to a review for information on graphene used in batteries for further strategies [27].

Table 30.3 Applications of graphene toward batteries.

Graphene	Specific capacity (mA h/g)	Current density (mA/g)	Comments	Ref.
Graphene/ MnO	870	100	Synthesized from Hummers-type GO. Small initial losses after charge cycling but stable over 50 cycles	[26a]
GO/MoS$_2$ paper	230	25	Over 90% capacity remained after 15 cycles	[25]
GO/Si nanoparticles	2300	120	Very high specific capacitance with a high retention (87%) after 152 cycles	[26b]
Graphene/Sn nanoparticles	1022	200	*In situ* preparation method using chemi-cal vapor deposition. 96% retention after 100 cycles	[26c]
rGO/SnS$_2$	630	200	Long cycle life of 500 cycles, retaining 84% of the charge capacity	[26d]
Graphene/ NiCo$_2$O$_4$	806	100	Figure quoted after 55 cycles, with a capacity retention of around 50%	[26e]
rGO/C/Ge	1384	200	96.2% capacity retention after "hundreds of cycles"	[26f]

30.3.3
Solar Energy

The conductivity and surface area of graphene also holds great potential in the field of PVCs. Traditional electrode materials for PVCs such as indium tin oxide (ITO) and fluorine tin oxide (FTO) have several disadvantages (as listed in Wang *et al.* [28]) and thus alternatives are widely investigated. Apparently, graphene's high optical transparency makes it an ideal current collector for PVCs, and its conductivity will theoretically allow efficient electron movement throughout PVC circuitry.

While graphene electrodes in PVCs may provide a good solution toward improved PVC efficiency, the fundamental electrochemistry of graphene is not conducive to effective electrode materials for dye-sensitized solar cells (DSSCs) because the majority of the surface area of a graphene sheet is electrochemically inactive [29]. Moreover, many graphene-based composites, such as the polystyrene graphene composite [30] reported by Stankovich *et al.*, were simply not conducting enough for PVC applications. This is perhaps why reports on graphene-based PVCs did not publish until 2008. Transparent and conductive graphene electrodes were reported in abundance thereafter, and were derived from solution-based graphenes fabricated through wet chemical methods such as the Hummers method. One of the earliest examples of graphene in a DSSC is illustrated in Figure 30.4; the device incorporated graphene as a current collector behind a TiO_2 layer that acts as an electron conduit between a ruthenium dye and the graphene [28]. The primitive device was reported to exhibit a high conductivity of 550 S/cm and an optical transparency of over 70% for wavelengths between 1000 and 3000 nm, properties that indeed qualify graphene as a highly attractive electrode material for DSSC applications. It is also an ultrasmooth

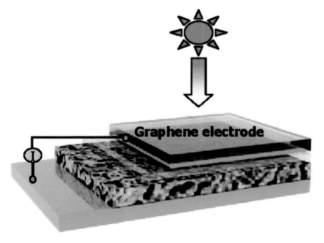

Figure 30.4 Schematic of the construction of the DSSC in Ref. [28]. The DSSC design (from top to bottom) is as follows: graphene, TiO_2, dye, and Au.

material that is reportedly beneficial to PVCs and DSSCs because high levels of surface roughness cause excessive short circuiting between the current collector and the semiconductor material. This approach yielded a graphene-based DSSC with a short-circuit current density of $1.01 \, mA/cm^2$ and an open-circuit potential of 0.7 V. Unfortunately, when the graphene was replaced with an FTO current collector, the short-circuit current density was $3.02 \, mA/cm^2$ and the open circuit potential was +0.76 V, but, on a positive note, the device could be fabricated easily and at a minimum of expense for future investigation.

There were also reports on graphene-based counterelectrodes for application within DSSCs, once again exploiting the transparency, conductivity, and flexibility of graphene. ITO electrodes were modified with composite films fabricated from poly(ethylenedioxythiophene) (PEDOT) and polystyrenesulfonate (PSS) with a ratio of 8:5 by weight. The graphene used in this example was stabilized in solution using 1-pyrenebutyrate as reported by Xu *et al.* [31]. Unfortunately, the prescribed method exhibited two-thirds of the efficiency of a standard platinum counterelectrode. As an author's perspective, we believe that the voltammetric profiles obtained for the redox couple I_2/I_3^- are very similar to platinum so it could be argued that the electrode is not graphene at all. Nevertheless, this report was primitive and proved the concept that these materials could potentially be incorporated into solar energy applications to avoid expensive materials such as platinum.

Other nanomaterials, such as CuS_2, were also used in graphene composites, but had problems with surface reactions when using a polysulfide electrolyte [32], particularly in the case of platinum current collectors. However by using GO, such surface reactions were avoided; reportedly 75% of the surface reactions were avoided, which improves the lifetime of the electrode. The GO/S_2 composite counterelectrode was incorporated into a DSSC design with an efficiency of 4.4%. Graphene with higher efficiency and PVCs can be designed by chemical doping of graphene with bis(trifluoromethanesulfonyl)amide [33]. The graphene composite is in contact with a SiO_2 wafer, effectively creating a metal–semiconductor junction or a Schottky junction. The addition of the graphene as a dopant increases the Schottky barrier height, allowing higher energy electronic transitions between the valence and conduction bands of the semiconductor and therefore more absorbed energy from photons is captured by the system. The introduction of the dopant improves the device efficiency from 1.9 to 8.6%.

Several other graphene-based PVC and DSSC technologies have been investigated and reported, but there are too many to discuss in this chapter. Table 30.4 lists some of the latest PVC and DSSC applications of graphene [34], and further applications can be found within the reviews provided [35].

30.4
Graphene and Graphene Oxide in Energy Generation Devices

Energy storage devices were not the only area that graphene was considered to be useful, and consequently there was intensive research in many areas,

Table 30.4 Applications of graphene in solar technologies.

Graphene	Efficiency (%)	Comments	Ref.
Graphene/zinc phosphide	1.9	Field effect solar cell	[34a]
B-doped graphene	9.2	Electrode used as a counterelectrode for the I^-/I^{3-} couple in a DSSC	[34b]
rGO/CdSe or CdTe	3.3	Moderately useful open-circuit potential	[34c]
Graphene/Pt	0.4	Poor conversion efficiency	[34d]
Graphene/CdSe	0.53	Flexible design but requires improved efficiency	[34e]
Graphene/Ag	8.0	Similar efficiency to ITO	[34f]

including energy generation devices. The increasing energy demand of society and rapid depletion of fossil fuels require more efficient and greener energy generation solutions than ever before, and, in particular, the electrodes designed within electrochemical devices must be highly efficient, durable, sometimes flexible, chemically inert or resistant to fouling, abundant and readily available, and inexpensive to the end user. Graphitic materials could therefore be considered important because they exhibit these properties in addition to being highly conducting. Graphene itself may prove to be problematic in some cases, because, as discussed previously, the fundamental structure of graphene is not particularly conducive to fast and efficient electron transfer reactions at the solid–liquid interface. Yet this has not restricted researchers using graphene in several types of fuel cells and electrolyzers, which are discussed in this section and are summarized in Figure 30.5. Fuel cells have been investigated intensively for many years as a potential replacement for fossil fuels, as they can be operated from materials such as glucose, methanol, ethanol, phosphoric acid, and so on [36]. The concept of a fuel cell is to generate energy from other sources than fossil fuels, and ideally utilize fuels that are in greater supply. This section discusses how graphenes have been incorporated into such devices and analyzes the role of graphene and how effectively it performs within such technology.

30.4.1
Fuel Cells

The improvement of fuel cells is an essential requirement for the future of transportation because of the finite oil supply the earth has to offer. Fuel cells are an alternative mode of engine-powering devices that use materials such as methanol as a fuel rather than the customary crude-oil-based fuel such as diesel. The general *modus operandi* of a fuel cell is illustrated in Figure 30.6 [36a]; a very basic description of a fuel cell's operation is provided in the figure caption. The points

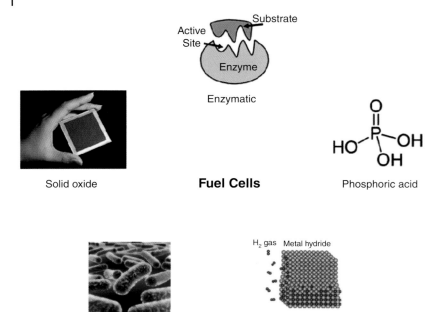

Figure 30.5 Nonexhaustive schematic of various fuel cells.

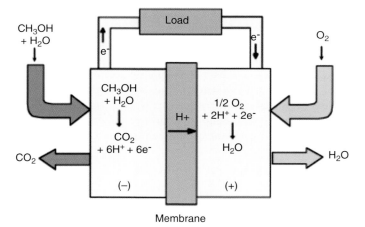

Figure 30.6 Generic fuel cell design. In this example, methanol is used as the fuel. Methanol is oxidized on a working anode in one side, producing protons and electrons. The electrons are conducted through the circuit, creating a current to power the load, while the protons diffuse through a membrane to the cathode region. The protons reduce oxygen to produce water [36a].

of interest from the perspective of a graphene are of course the electrodes, the membrane, and even the build quality.

In a proton exchange membrane fuel cell (PEMFC), the membrane is a key component because it allows an oxidizing species to travel from the anode to the cathode without mixing the electrolytes in the respective anodic and cathodic compartments. The transfer efficiency of the oxidizing species, namely H^+, is, therefore, the key mechanism in the efficiency of a PEMFC, while the stability of the membrane must be sufficiently high such that it can withstand high temperatures and numerous loads with little or no membrane deformation. The high mechanical and thermal stability of graphene could therefore be a very useful property for PEMFCs, provided that protons can efficiently permeate the structure, while being impermeable to other molecules. One of graphene's reportedly exciting properties was its impermeability to almost anything [36b], which is unfortunately not the case for PEMFCs as the proton exchange membranes (PEMs) must permit protons to be transferred across the membrane to the cathodic region, in order for oxygen to be reduced to water. The impermeability is a result of the relative strength of its σ bonds, which are reportedly the strongest known chemical bonds in nature [36b]. It should, therefore, come as no surprise that even protons struggle to permeate a *completely pristine* graphene sheet, with conflicting reports demonstrating a large range of barriers for permeation [36c,36d,37]. It is likely that the purity of the graphene utilized in each instance is related to the observed permeation barrier. In reality, graphene should not permeate H^+ at all if it is pure and defect free. There is a route around this, however, which is discussed later.

Consequently, early attempts to create fuel cell membranes were limited to nanocomposites fabricated from highly defective graphene oxides. GO is more attractive because it contains more proton-permitting channels and allows an increased absorption of water molecules. One such study incorporated GO into a Nafion-based nanocomposite membrane to yield a composite with an improved ionic exchange capacity and water uptake compared to GO, which was retained at temperatures up to 120 °C [38]. GO is also used in low-temperature PEMFCs, such as in the work reported by Cao *et al.* who utilize GO in a poly(ethylene oxide) membrane network [39].

More recently, fundamental graphene investigations have revealed some remarkable secrets, provided the graphene is applied correctly. Much like putting a pin prick in a balloon, such a small perforation is required in a pristine graphene sheet to allow H^+ ions *only* to permeate the sheet [36c]. If this is possible, the mechanical and thermal stability of graphene would provide an extremely strong platform for a good membrane material in PEMFCs as it would offer the selective permeability to H^+, while maintaining a good thermal stability. Indeed, a report by Achtyl *et al.* demonstrated that H^+ does indeed permeate pristine graphene sheets, but only through the defective regions naturally occurring upon a pristine graphene sheet [40]. Graphene, therefore, could act as a nanofiltering material for PEMFCs in the future, provided the graphene is pure enough and applied correctly inside the fuel cell device. Only then could its

Table 30.5 Applications of graphene in fuel cells.

Graphene	Max power density[a]	Voltage (V)	Comments	Ref.
rGO/Pd	1.6 W/g	0.23	Paper titles itself to demonstrate the properties of rGO/Pd, but rGO/Pt demonstrates a higher maximum power density (2.6 W/g)	[41a]
Multilayer graphene/Pt	60 W/g	0.65	Graphene exhibits poor mass transport capabilities compared to nanotubes and carbon black, leading to worse power densities	[41b]
Graphene/FeN	885 mW/m^2	Approx. 0.75 but not reported properly	Very low current at this power density	[41c]
Polydopamine-modified graphene	192.1 mW/cm^2	0.96	Membrane for PEM fuel cells improves cell power density by 38% compared to nongraphene version	[41d]
GO/Nafion	0.8 mW/cm^2	0.55	Low humidifying polymer membrane	[41e]

a) Units can vary as reported in the publications.

long-term properties such as cycle numbers and thermal stability be correctly tested in an applicable environment. Further applications of graphene to several fuel cells are presented in Table 30.5 [41].

30.4.2
Electrolyzers

Another energy source that is receiving increasing attention is hydrogen energy. Academics predict that hydrogen is a cleaner fuel because it does not release carbon upon combustion in air; rather it creates water as opposed to CO_2 and NO_x-type gases. Currently, the disadvantage of hydrogen from an economic standpoint is the production of hydrogen, and how it can be produced in a green manner that is carbon friendly. There is a current legislative drive in several nations such as the United Kingdom, Brazil, Canada, Iceland, Japan, and Russia, toward increasing the use of hydrogen fuel. The United Kingdom, for example, has started a consultation with the Department of Energy and Climate Change on how to define green hydrogen. Such a definition may pave the way for tax credits in the United Kingdom for hydrogen users.

Consequently, there is a foreseeable demand for low-energy electrolyzers, creating the maximum amount of "green" hydrogen at the bare minimum of expense and compromise to the environment. The important permutations to create low-energy electrolyzers for green hydrogen are the ease of procurement of a raw material to electrolyze and the electrode design for the electrolyzer. Obviously, it is the latter case where nanotechnology can have an influence. Academic studies have already demonstrated that graphene can be incorporated into electrode designs for electrolyzers, such as a urea electrolyzer reported by Wang *et al.* [42]. Their work focused on urea as it is a renewable hydrogen carrier with the potential to be used as an energy supplier. It is defined as renewable as it is produced as a waste product from the metabolism of proteins in the liver of mammals, including humans. Thus, it is conceivable that perhaps one day, from the economic standpoint, hydrogen could be driven by human excrement. Indeed, there is already a bus service running in the United Kingdom that is powered on human feces and household waste, but that runs on biomethane instead of biohydrogen. In the work of Wang *et al.*, the electrode design was based on an electrochemical fabrication of an rGO/Ni composite. Under alkaline conditions, the composite electrode catalyzes the electrochemical reaction of urea, forming hydrogen gas as a result as suggested in the reaction below.

$$CO(NH_2)_{2(aq)} + H_2O_{(l)} \rightarrow N_{2(g)} + 3H_{2(g)} + CO_{2(g)}$$

A close examination of their work reveals that this is a case of improved surface area available for the reaction to take place, therefore giving the impression of electrocatalysis and indicating that the rGO itself does not catalyze the reaction. This is not an issue, however, as the increase in surface area offered by the rGO/Ni composite allows more nickel to be exposed to the electrolyte; therefore, a highly increased passage of current is observed for the rGO/Ni composite when compared to both an Ni catalyst and rGO itself. Such a high passage of current is, therefore, indicative of enhanced electrolysis of urea. An additional benefit of using the composite material was its ability to avoid any effects of surface blocking from excessive redox product build up on the electrode surface when potential cycling is too high. Effectively, this means that the lifetime of this material is longer due to a lack of passivation of the surface, providing an additional advantage for this approach. The amperometric currents were also demonstrated to be twice as high as a separate nickel catalyst used for comparative purposes for the electrolysis of urea, indicating a more efficient electrolysis process.

Further application on the production of hydrogen from organic wastes is an academic report by Hou and coworkers, describing the effectiveness of MoS_2 combined with nitrogen-doped graphene aerogels as a hydrogen producer in microbial electrolytic cells (MECs) [38]. The developed material was rigorously tested using polarization techniques, impedance, and analysis of hydrogen

Table 30.6 Applications of graphene for hydrogen splitting of electrolytes for potential uses in electrolyzers.

Graphene	Hydrogen source	Comments	Ref.
CNT/graphene hybrid with MoS_2	H_2SO_4, H_2O	Several composites of molybdenum studied	[43a]
Graphene/cobalt composite	$NaBH_4$, H_2O	Require a more widely available input material	[43b]
Graphene/TiO_2	Na_2S, Na_2SO_3, H_2O	Photocatalytic evolution	[43c]
GO/CdS	MeOH, H_2O	Strong photostability	[43d]
GO/WS_2	H_2SO_4	Rigorously characterized	[43e]
rGO/WC	H_2SO_4	Cocatalytic support	[43f]

formation and Tafel analysis. It was observed that platinum on carbon (10% Pt on Vulcan XC-72) offered a more catalytic approach and thus a higher production of hydrogen in this application [38]. Tafel analysis further demonstrated that the current density increases logarithmically as a function of potential at a slower rate using platinum on the carbon catalyst compared to the MoS_2/graphene composite. However, the large amount of hydrogen produced by the graphene electrode ($0.16\,m^3\ H_2/m^3/d$) and the relative cost of the graphene compared to the platinum on the carbon catalyst makes this method an attractive approach for the electrolysis of organic wastes. Though graphene is a lesser researched material for electrolyzer electrodes, there are still many examples of electrode design in the academic literature, such as MoS_2-based materials. Table 30.6 lists other electrolyzer-based graphene electrodes [43].

30.4.3
Outlook

Much of the hype surrounding graphene has not yet been fully realized despite the extensive levels of graphene research in the decade following the original research articles from Geim and Novoselov. Despite this assessment, graphene-based electrodes for supercapacitors and batteries continue to produce countless research articles, commentaries, reviews, and columns reporting new breakthrough technologies that will solve all the world's energy problems. However, this is far from the reality. It is true that the specific capacitance of graphene-based supercapacitors is steadily increasing, and the specific capacity of graphene-based batteries is also increasing, but currently neither are at the stage of market implementation. One should expect many years of further research before such technology envisions a breakthrough significant enough to be implemented in an electronic device suitable for consumers.

Q1 References

1 Rogers, B., Pennathur, S., and Adams, J. (2008) *Nanotechnology: Understanding Small Systems*, CRC Press, Boca Raton, FL.

2 Novoselov, K.S., Geim, A.K., Morozov, S.V., Jiang, D., Zhang, Y., Dubonos, S.V., Grigorieva, I.V., and Firsov, A.A. (2004) *Supramol. Sci.*, **306**, 666–669.

3 Brownson, D.A.C., Kampouris, D.K., and Banks, C.E. (2012) *Chem. Soc. Rev.*, **41**, 6944–6976.

4 Novoselov, K.S., Jiang, D., Schedin, F., Booth, T.J., Khotkevich, V.V., Morozov, S.V., and Geim, A.K. (2005) *Proc. Natl. Acad. Sci. U. S. A.*, **102**, 10451–10453.

5 Berger, C., Song, Z., Li, X., Wu, X., Brown, N., Naud, C., Mayou, D., Li, T., Hass, J., Marchenkov, A.N., Conrad, E.H., First, P.N., and de Heer, W.A. (2006) *Supramol. Sci.*, **312**, 1191–1196.

6 Geim, A.K. and Novoselov, K.S. (2007) *Nat. Mater.*, **6**, 183–191.

7 Brownson, D.A.C., Lacombe, A.C., Gomez-Mingot, M., and Banks, C.E. (2012) *RSC Adv.*, **2**, 665–668.

8 International Energy Agency (2014) International Energy Agency.

9 Song, W., Ji, X., Deng, W., Chen, Q., Shen, C., and Banks, C.E. (2013) *Phys. Chem. Chem. Phys.*, **15**, 4799–4803.

10 (a) Brownson, D.A.C., Kampouris, D.K., and Banks, C.E. (2011) *J. Power Sources*, **196**, 4873–4885; (b) Zhang, L.L. and Zhao, X.S. (2009) *Chem. Soc. Rev.*, **38**, 2520–2531; (c) Stoller, M.D., Park, S., Zhu, Y., An, J., and Ruoff, R.S. (2008) *Nano Lett.*, **8**, 3498–3502; (d) Yin, S., Zhang, Y., Kong, J., Zou, C., Li, C.M., Lu, X., Ma, J., Boey, F.Y.C., and Chen, X. (2011) *ACS Nano*, **5**, 3831–3838; (e) Yang, X., Cheng, C., Wang, Y., Qiu, L., and Li, D. (2013) *Supramol. Sci.*, **341**, 534–537.

11 Dey, R.S., Hjuler, H.A., and Chi, Q. (2015) *J. Mater. Chem. A*, **3**, 6324–6329.

12 Yan, J., Wei, T., Fan, Z., Qian, W., Zhang, M., Shen, X., and Wei, F. (2010) *J. Power Sources*, **195**, 3041–3045.

13 Cao, J., Wang, Y., Zhou, Y., Ouyang, J.-H., Jia, D., and Guo, L. (2013) *J. Electroanal. Chem.*, **689**, 201–206.

14 Ambrosi, A., Bonanni, A., Sofer, Z., Cross, J.S., and Pumera, M. (2011) *Chem. A Eur. J.*, **17**, 10763–10770.

15 Le, L.T., Ervin, M.H., Qiu, H., Fuchs, B.E., and Lee, W.Y. (2011) *Electrochem. Commun.*, **13**, 355–358.

16 Xie, Y., Liu, Y., Zhao, Y., Tsang, Y.H., Lau, S.P., Huang, H., and Chai, Y. (2014) *J. Mater. Chem. A*, **2**, 9142–9149.

17 (a) Xiong, S., Shi, Y., Chu, J., Gong, M., Wu, B., and Wang, X. (2014) *Electrochim. Acta*, **127**, 139–145; (b) Sahu, V., Grover, S., Tulachan, B., Sharma, M., Srivastava, G., Roy, M., Saxena, M., Sethy, N., Bhargava, K., Philip, D., Kim, H., Singh, G., Singh, S.K., Das, M., and Sharma, R.K. (2015) *Electrochim. Acta*, **160**, 244–253; (c) Li, W. and Yang, Y. (2014) *J. Solid State Electrochem.*, **18**, 1621–1626; (d) Zhu, G., He, Z., Chen, J., Zhao, J., Feng, X., Ma, Y., Fan, Q., Wang, L., and Huang, W. (2014) *Nanoscale*, **6**, 1079–1085; (e) Wu, C., Deng, S., Wang, H., Sun, Y., Liu, J., and Yan, H. (2014) *ACS Appl. Mater. Interfaces*, **6**, 1106–1112; (f) Hu, J., Kang, Z., Li, F., and Huang, X. (2014) *Carbon*, **67**, 221–229.

18 Zhang, X., Zhang, H., Li, C., Wang, K., Sun, X., and Ma, Y. (2014) *RSC Adv.*, **4**, 45862–45884.

19 Kampouris, D.K., Ji, X., Randviir, E.P., and Banks, C.E. (2015) *RSC Adv.*, **5**, 12782–12791.

20 Paek, S.-M., Yoo, E., and Honma, I. (2009) *Nano Lett.*, **9**, 72–75.

21 Wang, H., Yang, Y., Liang, Y., Robinson, J.T., Li, Y., Jackson, A., Cui, Y., and Dai, H. (2011) *Nano Lett.*, **11**, 2644–2647.

22 Luo, J., Liu, J., Zeng, Z., Ng, C.F., Ma, L., Zhang, H., Lin, J., Shen, Z., and Fan, H.J. (2013) *Nano Lett.*, **13**, 6136–6143.

23 Wang, B., Li, X., Zhang, X., Luo, B., Jin, M., Liang, M., Dayeh, S.A., Picraux, S.T., and Zhi, L. (2013) *ACS Nano*, **7**, 1437–1445.

24 Hu, B. Lung.-Hao., Wu, F.-Y., Lin, C.-T., Khlobystov, A.N., and Li, L.-J. (2013) *Nat. Commun.*, **4**, 1687.

25 David, L., Bhandavat, R., and Singh, G. (2014) *ACS Nano*, **8**, 1759–1770.

26 (a) Gao, F., Qu, J.-y., Zhao, Z.-b., Dong, Y.-f., Yang, J., Dong, Q., and Qiu, J.-s. (2014) *New Carbon Mater.*, **29**, 316–321; (b) Chang, J., Huang, X., Zhou, G., Cui, S., Hallac, P.B., Jiang, J., Hurley, P.T., and Chen, J. (2014) *Adv. Mater.*, **26**, 758–764; (c) Qin, J., He, C., Zhao, N., Wang, Z., Shi, C., Liu, E.-Z., and Li, J. (2014) *ACS Nano*, **8**, 1728–1738; (d) Qu, B., Ma, C., Ji, G., Xu, C., Xu, J., Meng, Y.S., Wang, T., and Lee, J.Y. (2014) *Adv. Mater.*, **26**, 3854–3859; (e) Chen, Y., Zhu, J., Qu, B., Lu, B., and Xu, Z. (2014) *Nano Energy*, **3**, 88–94; (f) Yuan, F.-W. and Tuan, H.-Y. (2014) *Chem. Mater.*, **26**, 2172–2179.

27 Kucinskis, G., Bajars, G., and Kleperis, J. (2013) *J. Power Sources*, **240**, 66–79.

28 Wang, X., Zhi, L., and Müllen, K. (2008) *Nano Lett.*, **8**, 323–327.

29 Brownson, D.A.C., Munro, L.J., Kampouris, D.K., and Banks, C.E. (2011) *RSC Adv.*, **1**, 978–988.

30 Stankovich, S., Dikin, D.A., Dommett, G.H.B., Kohlhaas, K.M., Zimney, E.J., Stach, E.A., Piner, R.D., Nguyen, S.T., and Ruoff, R.S. (2006) *Nature*, **442**, 282–286.

31 Xu, Y., Bai, H., Lu, G., Li, C., and Shi, G. (2008) *J. Am. Chem. Soc.*, **130**, 5856–5857.

32 Radich, J.G., Dwyer, R., and Kamat, P.V. (2011) *J. Phys. Chem. Lett.*, **2**, 2453–2460.

33 Miao, X., Tongay, S., Petterson, M.K., Berke, K., Rinzler, A.G., Appleton, B.R., and Hebard, A.F. (2012) *Nano Lett.*, **12**, 2745–2750.

34 (a) Vazquez-Mena, O., Bosco, J.P., Ergen, O., Rasool, H.I., Fathalizadeh, A., Tosun, M., Crommie, M., Javey, A., Atwater, H.A., and Zettl, A. (2014) *Nano Lett.*, **14**, 4280–4285; (b) Jung, S.-M., Choi, I.T., Lim, K., Ko, J., Kim, J.C., Lee, J.-J., Ju, M.J., Kim, H.K., and Baek, J.-B. (2014) *Chem. Mater.*, **26**, 3586–3591; (c) Tong, S.W., Mishra, N., Su, C.L., Nalla, V., Wu, W., Ji, W., Zhang, J., Chan, Y., and Loh, K.P. (2014) *Adv. Funct. Mater.*, **24**, 1904–1910; (d) Dong, P., Zhu, Y., Zhang, J., Peng, C., Yan, Z., Li, L., Peng, Z., Ruan, G., Xiao, W., Lin, H., Tour, J.M., and Lou, J. (2014) *J. Phys. Chem. C*, **118**, 25863–25868; (e) Gao, Z., Jin, W., Li, Y., Song, Q., Wang, Y., Zhang, K., Wang, S., and Dai, L. (2015) *J. Mater. Chem. C*, **3**, 4511–4514; (f) Yusoff, A.R.b.M., Lee, S.J., Shneider, F.K., da Silva, W.J., and Jang, J. (2014) *Adv. Energ. Mater.*, **4**, 1301989.

35 (a) Bonaccorso, F., Colombo, L., Yu, G., Stoller, M., Tozzini, V., Ferrari, A.C., Ruoff, R.S., and Pellegrini, V. (2015) *Supramol. Sci.*, **347**, (b) Roy-Mayhew, J.D. and Aksay, I.A. (2014) *Chem. Rev.*, **114**, 6323–6348; (c) Yin, Z., Zhu, J., He, Q., Cao, X., Tan, C., Chen, H., Yan, Q., and Zhang, H. (2014) *Adv. Energ. Mater.*, **4**, 1300574.

36 (a) Bagotsky, V.S. (2012) *Fuel Cells*, John Wiley & Sons, Inc, pp. 99–106; (b) Lamy, C., Rousseau, S., Belgsir, E.M., Coutanceau, C., and Léger, J.M. (2004) *Electrochim. Acta*, **49**, 3901–3908; (c) Hamnett, A. (1997) *Catal. Today*, **38**, 445–457; (d) Rabaey, K., Lissens, G., Siciliano, S., and Verstraete, W. (2003) *Biotechnol. Lett.*, **25**, 1531–1535.

37 Berry, V. (2013) *Carbon*, **62**, 1–10.

38 Hou, Y., Zhang, B., Wen, Z., Cui, S., Guo, X., He, Z., and Chen, J. (2014) *J. Mater. Chem. A*, **2**, 13795–13800.

39 Cao, Y.-C., Xu, C., Wu, X., Wang, X., Xing, L., and Scott, K. (2011) *J. Power Sources*, **196**, 8377–8382.

40 Achtyl, J.L., Unocic, R.R., Xu, L., Cai, Y., Raju, M., Zhang, W., Sacci, R.L., Vlassiouk, I.V., Fulvio, P.F., Ganesh, P., Wesolowski, D.J., Dai, S., van Duin, A.C.T., Neurock, M., and Geiger, F.M. (2015) *Nat. Commun.*, **6**, 6539.

41 (a) Carrera-Cerritos, R., Baglio, V., Aricò, A.S., Ledesma-García, J., Sgroi, M.F., Pullini, D., Pruna, A.J., Mataix, D.B., Fuentes-Ramírez, R., and Arriaga, L.G. (2014) *Appl. Catal. B: Environ.*, **144**, 554–560; (b) Marinkas, A., Hempelmann, R., Heinzel, A., Peinecke, V., Radev, I., and Natter, H. (2015) *J. Power Sources*, **295**, 79–91; (c) Liu, Y., Jin, X.-J., Dionysiou, D.D., Liu, H., and Huang, Y.-M. (2015) *J. Power Sources*, **278**, 773–781; (d) He, Y., Wang, J., Zhang, H., Zhang, T., Zhang, B., Cao, S., and Liu, J. (2014) *J. Mater. Chem. A*, **2**, 9548–9558; (e) Lee, D.C., Yang, H.N., Park, S.H., and Kim, W.J. (2014) *J. Membr. Sci.*, **452**, 20–28.

42 Wang, D., Yan, W., Vijapur, S.H., and Botte, G.G. (2013) *Electrochim. Acta*, **89**, 732–736.

43 (a) Youn, D.H., Han, S., Kim, J.Y., Kim, J.Y., Park, H., Choi, S.H., and Lee, J.S. (2014) *ACS Nano*, **8**, 5164–5173; (b) Zhang, F., Hou, C., Zhang, Q., Wang, H., and Li, Y. (2012) *Mater. Chem. Phys.*, **135**, 826–831; (c) Zhang, X.-Y., Li, H.-P., Cui, X.-L., and Lin, Y. (2010) *J. Mater. Chem.*, **20**, 2801–2806; (d) Gao, P., Liu, J., Lee, S., Zhang, T., and Sun, D.D. (2012) *J. Mater. Chem.*, **22**, 2292–2298; (e) Yang, J., Voiry, D., Ahn, S.J., Kang, D., Kim, A.Y., Chhowalla, M., and Shin, H.S. (2013) *Angew. Chem. Int. Ed.*, **52**, 13751–13754; (f) Yan, Y., Xia, B., Qi, X., Wang, H., Xu, R., Wang, J.-y., Zhang, H., and Wang, X. (2013) *Chem. Commun.*, **49**, 4884–4886.

44 (a) Shornikova, O.N., Kogan, E.V., Sorokina, N.E., and Avdeev, V.V. (2009) *Russ. J. Phys. Chem.*, **83**, 1022–1025; (b) (2013) *Entegris*, Entegris Inc.

45 (a) Lee, C., Wei, X., Kysar, J.W., and Hone, J. (2008) *Supramol. Sci.*, **321**, 385–388; (b) Kuzmenko, A.B., Heumen, E.van., Carbone, F., and van der Marel, D. (2008) *Phys. Rev. Lett.*, **100**, 117401.

46 (a) Yu, M.-F., Lourie, O., Dyer, M.J., Moloni, K., Kelly, T.F., and Ruoff, R.S. (2000) *Supramol. Sci.*, **287**, 637–640; (b) Dürkop, T., Getty, S.A., Cobas, E., and Fuhrer, M.S. (2004) *Nano Lett.*, **4**, 35–39.

31

Inorganic Nanotubes and Fullerene-Like Nanoparticles at the Crossroad between Materials Science and Nanotechnology and Their Applications with Regard to Sustainability

Leela S. Panchakarla[1,2] and Reshef Tenne[1]

[1]*Weizmann Institute of Science, Department of Materials and Interfaces, 234 Herzl St., Rehovot 76100, Israel*
[2]*Indian Institute of Technology Bombay, Department of Chemistry, Powai Mumbai 400076, India*

31.1
Introduction

Fullerenes (zero-dimensional) and nanotubes (one-dimensional) are unique class of materials among a large variety of low dimensional materials. Following the breathtaking progress with carbon fullerenes and carbon nanotubes over the past three decades, this field largely expanded more than two decades ago by discovering their analogs from inorganic layered (2D) compounds. Indeed, a lot of research has been done in the field of inorganic nanotubes (INTs) and fullerene-like nanostructures (IFs) based on layered materials including WS_2 [1], MoS_2 [2,3], BN [4], and so on [5,6]. Nanotubes from inorganic compounds with a quasi-isotropic (3D) structure include SiO_2 [7], GaN [8], ZnO [9], and so on. However, this chapter focuses on recent developments in INTs and IFs from layered materials. Numerous layered materials have been synthesized in the fullerene/nanotubular forms since the first formation of inorganic WS_2 fullerene-like structures/nanotubes in 1992 [1]. Extensive reviews on this field can be found in the literature [5,10,11]. In this chapter, we restrict ourselves to recent developments in the synthesis and doping of MoS_2 and WS_2 fullerenes and nanotubes. Characterizing low dimensional structures at individual particle level by various techniques, especially transmission electron microscopy (TEM), are presented. Furthermore, IF/INT from WS_2 and MoS_2 show great promise for future energy saving and reducing the reliance on energy supply from fossil sources, in particular. The usage of these materials as efficient solid lubricants and for reinforcing polymer and other nanocomposites is discussed. These technologies were shown to reduce energy consumption by reducing friction and improving the mechanical, electrical, and thermal properties of composites and are destined for future large-scale applications. This chapter also surveys the

Nanotechnology for Energy Sustainability, First Edition. Edited by Baldev Raj, Marcel Van de Voorde, and Yashwant Mahajan.
© 2017 Wiley-VCH Verlag GmbH & Co. KGaA. Published 2017 by Wiley-VCH Verlag GmbH & Co. KGaA.

recent development of synthesis and structural characterization of fullerene-likes and nanotubular forms of ternary and quaternay misfit layered compounds.

Although nanotubular structures have long been known for many years, the sheer interest in this field began only after the synthesis of carbon fullerenes [12] and nanotubes [13] and immediately afterward the realization of WS_2 (MoS_2) nanotubes/fullerene-likes [1,2]. Pauling was the first scientist to study the mechanism of folding of inorganic layered materials with asymmetric structure (chlorites) [14]. Chlorites consist of two different crystal structures with different lattice periodicities and stack together alternately, thus creating substantial stress in the flat 2D structure. Pauling hypothesized that the asymmetry in the lattice along the *c*-axis would force the layers to fold in order to relax the stress, producing tubes and scrolls. Based on this logic, he concluded, therefore, that layered materials that are not asymmetric along the *c*-axis, like MoS_2 and $CdCl_2$, are not expected to form rolled structures by the above mechanism. However, irrespective of the above argument, layered materials behave differently when they are brought down to low dimensions. Nanoflakes of layered materials are unstable against folding due to the large surface energy related to the peripheral atoms and fold into rolled structures such as fullerenes and nanotubes. Folding of such nanosheets requires overcoming a large elastic energy, which demands huge activation energy [15]. Thus, most of the synthetic techniques to make nanotubes and fullerene-like structures from 2D compounds involve high temperatures.

31.2
Synthesis and Structural Characterization

Unlike carbon fullerenes and nanotubes, which consist purely of carbon atoms, fullerene-like structures and nanotubes from transition metal dichalcogenides (MX_2, where M = transition metal and X = S or Se) consist of a three atom thick layer where sixfold bonded metal atoms are sandwiched between two layers of threefold bonded chalcogenide atoms [5]. There exist strong covalent bonds within the layer and weak van der Walls forces between the layers. Thus, MX_2 IFs and INTs are structurally more complex and possess large internal strain compared to their carbon analogs, which is compensated by their larger diameter and multiwalled IFs and INTs structure [16]. The high temperatures ($>700\,°C$) required for their synthesis imply that the reactions are generally kinetically controlled and very difficult to maneuver, which also implies large size and shape distribution. INTs exhibit in many cases behavior entirely different from that of their carbon analogs. For example, due to quantum confinement of carriers (electron/hole), the bandgap of carbon nanotubes and quantum dots increases with decreasing diameter of tube/particle. On the contrary, the bandgap of INTs of WS_2 (MoS_2) decreases upon reducing the diameter of the nanotubes due to the curvature of the three atom thick S-W-S layer and consequently distortion of the metal–sulfur chemical bond [5]. Bulk $2H\text{-}WS_2$ (MoS_2) are indirect bandgap semiconductors, whereas theory predicts that zigzag $(0,n)$ nanotubes of these materials possess direct bandgap [5].

Studies on inorganic fullerene-likes (IFs) and nanotubes (NTs) from layered materials are ever expanding since their first realization in WS_2 in 1992 [1]. Heating thin tungsten film in the presence of hydrogen sulfide atmosphere yields polyhedral and cylindrical structures of WS_2 [1]. There have been several inorganic layered metal chalcogenides such as TiS_2 [17], SnS_2 [18], and TaS_2 [19]; metal halides such as CdI_2 [5], $NiCl_2$ [20,21], and so on; metal hydroxides such as $Co(OH)_2$ [22], $Ni(OH)_2$ [23], $Mg(OH)_2$ [23]; metal oxides such as V_2O_5 [24] and Cs_2O [25]; nitrides such as BN [4,26]; and metal phosphates [22] as well as ternary misfit layered compounds such as $LaS-CrS_2$, $SnS-SnS_2$, and so on [27], which were synthesized as IFs/INTs. There are several physical and chemical methods known today [11] to synthesize different IFs and NTs including laser ablation [21], solar ablation [28], shock wave [29], plasma ablation [15], microwave [30], reaction of WO_3 or MoO_3 in the presence of H_2S [31], decomposition of trichalcogenides [32], and chemical vapor transport of chalcogenides in the presence of transport agent such as iodine [33], hydrothermal/solvothermal [34]. Among many MX_2 fullerenes/nanotubes, where M = W, Mo, Ti, V, Ta, and so on, synthesized so far, only MoS_2 and WS_2 IFs/NTs could be synthesized in large quantities [35] (several hundreds of kilogram a day in case of IF-WS2 (industrial grade); 100 g/day of INT-WS_2 and about 1 g/day of IF-MoS_2). Large quantities of $Mo(W)S_2$ IFs/INTs are achieved by reacting nanoparticles of $Mo(W)O_x$ in a reducing atmosphere with H_2S gas using a fluidized bed reactor (FBR) [35].

In this case, the oxide nanoparticle reacts with H_2S and forms MS_2 shell within a fraction of a second engulffing the oxide core. The inert MS_2 shell, together with the dynamic ambient in the FBR, prevents aggregation of the nanoparticles. The oxide core is further reduced and converted to sulfide by slow diffusion-controlled reaction. Thus, after the first instant, the rate of the reaction depends on the diffusion rate of the sulfurizing agent and the final diameter of the IFs/INTs depends on the starting precursor oxide diameter [36]. The growth mechanism of WS_2 nanotubes by this method follows a different route [35]. It was proposed that the nonvolatile oxide mixture WO_{3-x} undergoes reduction in the presence of H_2 and produces volatile oxide phase; this phase is further reduced to a nonvolatile phase that serves as a nucleation center for the $WO_{2.72}$ nanowhisker growth. Gradual sulfurization of this oxide core creates a hollow multiwall WS_2 nanotube. Typical SEM and TEM images of WS_2 nanotubes produced by this process is shown in Figure 31.1 [35,37]. The reaction of volatile metal halides or carbonyles with H_2S gas at high temperatures also yields IFs and INTs [36,38]. Here, the nanoparticles grow from a small nuclei outward in a kinetically controlled reaction through the gas phase (unlike solid diffusion in the previous case). Different IF-NbS_2, IF-TiS_2, IF-WS_2, IF-MoS_2, and INT-TiS_2 were produced by this method [36]. Most of the synthetic methods are detailed in several papers and reviews [5,6,10,11,34,36]. Here, we focus on recent developments in high-temperature synthesis of IFs and NTs of MoS_2 and WS_2.

As we mentioned earlier, synthesizing inorganic nanotubes/fullerene-likes with small number of walls and with smaller diameters is much challenging due to high elastic energy of folding [15]. Figure 31.2 shows the energy per atom of

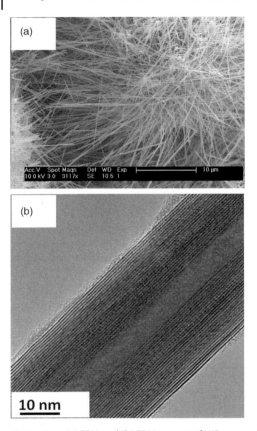

Figure 31.1 (a) SEM and (b) TEM images of WS_2 nanotubes.

MoS_2 nanotubes with respect to the number of atoms in the unit length for different number of layers, which are calculated by density functional tight-binding theory [15]. The energy of nanoribbons of MoS_2 with the same number of layers is compared. Both nanotubes and nanoribbons demonstrate that the energy per atom increases as the number of atoms decreases. Dangling bonds in the rim of the nanoribbons are responsible for increased energy per atom, whereas increased elastic energy of folding makes the nanotubes less stable at smaller diameters. Below a certain point, the energy of folding increases more steeply for nanotubes than the energy of the nanoribbons, which leads to a lesser stability of smaller diameter nanotubes compared to straight nanoribbons [15]. Thus, synthesizing smaller diameter INTs/IFs demands exotic reaction conditions.

Nonetheless, WS_2 nanotubes with smaller diameters (3–7 nm) with less number of walls (1–3) were realized by irradiation of multiwall WS_2 nanotubes by inductively coupled radiofrequency plasma [15]. Under these conditions, the electron temperature reaches 10^4 K. Interaction of the plasma with either point or line defect on the surface results in exfoliation of a few nanoslabs on top of

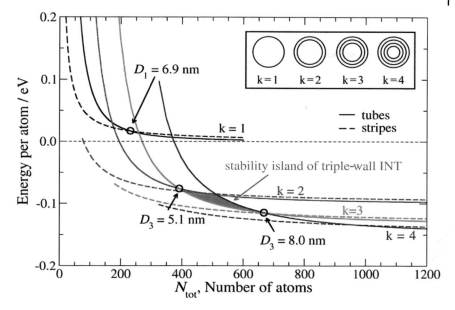

Figure 31.2 The calculated energy per atom for MoS$_2$ nanotubes and nanostripes with 1–4 walls as a function of the number of atoms in the tube unit cell – N_{tot}. In the range of ~390 $< N_{tot} <$ 670, triple layer nanotubes are more stable than nanotubes with $k=2$ and $k=4$ and corresponding to outer diameters of 5.1 nm $< D_3 <$ 8.0 nm. The diameter (D_k) represents the outer diameters of the nanotubes with k shells. Note that nanotubes with the same (outer) diameters but different numbers of shells have consequently a different (total) number of atoms. Thus, a single-wall tube with larger diameter may have less atoms than triple walled tubes with smaller diameter (adapted from Ref. [15]).

the multiwalled nanotubes, subsequently leading to unzipping of nanosheets from the outer surface. The elastic strain energy of rolled sheet on top of the nanotube is released when it gets exfoliated through an inverted umbrella effect [15]. This phenomenon can be compared to "Walden inversion" of nucleophilic attack in SN$_2$ reactions in organic chemistry. The result of the inversion are small daughter nanotubes on top of the multiwall WS$_2$ nanotubes. A schematic reaction mechanism of this effect is presented in Figure 31.3. TEM images in Figure 31.4 show small-diameter nanotubes with 1–3 walls produced by plasma treatment of multiwall WS$_2$ nanotubes. The nanotube yield was found to increase up to a certain degree both with reaction time and with the plasma power.

The smallest stable inorganic symmetric closed-cage structures synthesized were nanooctahedra [39,40]. These nanoclusters are produced under extreme conditions that are far from equilibrium. Laser ablation of MoS$_2$ and MoSe$_2$ yielded nanooctahedra 3–6 nm in size consisting of three–five layers [41]. Theoretical calculations predicted that the properties of these particles strongly

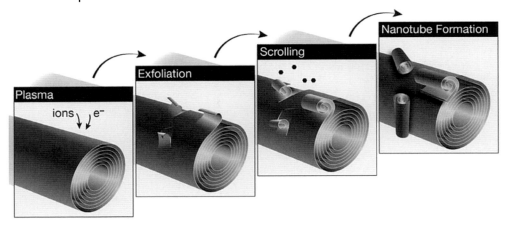

Figure 31.3 Proposed mechanism of the formation of daughter nanotubes by plasma treatment of the multiwall (mother) nanotubes (adapted from Ref. [15]).

depend on their shape [42]. Hollow clusters with similar composition are expected to show different behavior depending on their shape. The structure speculated by *ab initio* calculations combined with experimental TEM images revealed that 3D octahedra are formed by seaming of triangular MoS_{2-x} nanostructures [39]. And the structures are stabilized by the formation of S_2 pairs at the edges. Unlike bulk and nanotubular MoS_2 that are semiconductor MoS_2, octahedra particles exhibit metallic-like character [39,42]. The detailed analysis of the structure of nanooctahedral particles was done by aberration-corrected transmission electron microscope. Figure 31.5 shows atomic resolution phase-contrast TEM images of MoS_2 nanooctahedron. The structure at the apex fits well with density functional tight binding calculated model for an energetically stable structure of octahedral coordination of sulfur atoms.

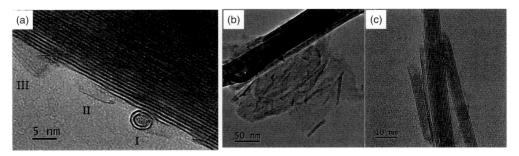

Figure 31.4 (a) HRTEM of a two-layer daughter nanoscroll viewed head-on along its axis (I) and attached to the large multiwall nanotube; an initial scrolling stage of exfoliated single layer (II) and three layers (III) before the formation of nanotube. (b) A large number of daughter nanotubes next to a treated multiwall nanotube. (c) A group of daughter nanotubes isolated from plasma-treated multiwall WS_2 nanotubes by sonication (adapted from Ref. [15]).

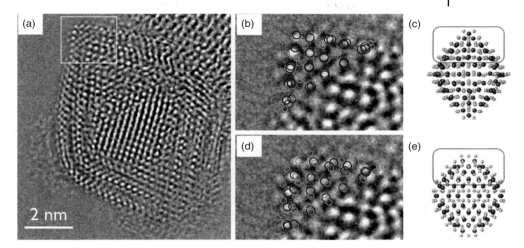

Figure 31.5 Atomic resolution image of a MoS$_2$ nanooctahedron taken in an image-side aberration-corrected FEI Titan 80–300 under NCSI conditions. Full structure (a). A magnified part of the tip of the octahedron revealing the sulfur atoms is shown in (b) and (d), with a superposition of the models presented in (c) and (e). The models in (c) and (e) correspond to two of the 15 hypothetical structures proposed in Ref. [42]. Mo atoms are displayed in red; S atoms are displayed in yellow. In Figure 31.5b, one of the most stable structures (no. 5 of Ref. [42]) coincides, whereas the less stable structure fails to match in Figure 31.5d (adapted from Ref. [39]).

Concentrated solar or artificial light (solar ablation) has also been used to synthesize small diameter nanoparticles of different materials including MoS$_2$, WS$_2$ IFs and INTs [28]. This technique also generates high temperatures ($\sim$2700 K) and provides conditions for far-from-equilibrium reactions. These high-energy techniques could drive highly exergonic reaction accessing less thermodynamically stable phases compared to traditional chemical synthesis routes. Few layers of relatively unstable 1T-MoS$_2$ phase were recently observed in IF-MoS$_2$ [43], which is synthesized by solar ablation (see Figure 31.6) [43]. This observation indicates that the large temperature gradients lead to rapid quenching of the nanoclusters that does not allow them to anneal out structural defects.

Most of the chemical methods for the synthesis of IFs or INTs involve handling dangerous chemical gases such as H$_2$S and H$_2$, which requires extra safety features while scaling up the process to an industrial level. Solid-state precursors like sulfur instead of H$_2$S and alkali metal tetrahydroborates (NaBH$_4$/LiBH$_4$) or tetrahydroaluminates (NaAlH$_4$/LiAlH$_4$) instead of hydrogen gas were proposed as possible alternative to address this problem. Heating a mixture of WO$_3$/MoO$_3$ nanopowders with sulfur and alkali metal tetrahydroborates in evacuated quartz ampoules to 900 °C yields IFs of WS$_2$/MoS$_2$ [44]. The above reaction mixture was also found to yield IFs of MoS$_2$ when it was subjected to solar/photothermal ablation [44]. In this case, a small amount of carbon is added to the ampoule to increase the solar light absorption. Here, carbon also serves as

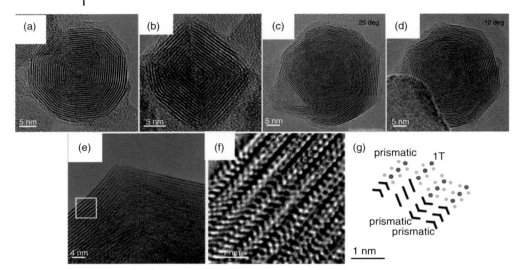

Figure 31.6 Typical MoS_2 closed nanoparticles in which a nanooctahedral core undergoes morphology change to quasi-spherical outer shells, obtained in the solar furnace. (a, b) Two hybrid nanoclusters of roughly the same diameter (about 26 nm) and number of layers (about 20). (c, d) TEM tilting experiment of an individual hybrid MoS_2 nanoparticle of diameter 43 nm and 28 layers, substantially larger than those in Figure 31.6a and b, and reinforcing the clarity of the transition from core octahedra to surrounding quasi-spherical shells. (e) High-resolution TEM images of a hybrid nanoparticle. The marked frame is enlarged in Figure 31.6f, showing atomic resolution of the MoS_2 layers. The atomic model is overlaid in red (Mo) and yellow (S) and detailed in Figure 31.6g with black lines as visual guides to their appearance in the TEM image. The chevron motif correlates with a prismatic coordinated MoS_2 layer, which as a bulk phase is semiconducting. The pattern of diagonal lines indicates the 1 T phase, which has been predicted to be metallic (adapted from Ref. [43]).

an additional reducing agent. In solar ablation, highly concentrated sunlight is focused on small areas of the quartz ampoule for short period of times. In these series of experiments, high temperatures ($\sim$1700 °C) and far from equilibrium conditions with large irradiation flux and temperature gradients were produced. The reaction mechanisms in these photothermal conditions are different from those of furnace annealing. The hydrogen gas produced in this reaction has a dual role. On the one hand, it reacts with sulfur generating thereby H_2S. Furthermore, the sublimed MoO_3 is reduced by hydrogen gas and subsequently reacts with the H_2S converting it to IF-MoS_2 nanoparticles. The size of the IFs is determined by the rate of the MoO_3 evaporation and the temperature gradients. However, in general no IF particles with sizes beyond 200 nm were observed.

Recently, microwave-assisted rapid synthesis was developed for IFs of MoS_2/WS_2 [30]. This procedure is single step and takes less time (typically 60 s). A mixture of $M(Co)_6$ (M = Mo or W), sulfur powder, and polypyrrole nanofibers (PPy-NFs) was subjected to heat by microwave irradiation (frequency = 2.45 GHz

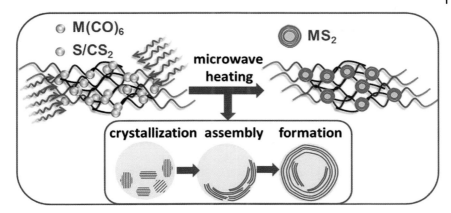

Figure 31.7 Schematic of the experimental process and formation of MS₂ IFs (adapted from Ref. [30]).

and power 1250 W) for 60 s to yield the respective IFs. Here, the PPy-NFs served as microwave absorbers and produced heat and transferred it to the reactants. The experimental process and formation of IFs are schematically shown in Figure 31.7. The microwave heating activated the reaction between metal carbonyl and sulfur powder and yielded small-size MS₂ nanoparticles (few nanometers). These freshly formed thermally activated MS₂ particles aggregated through rapid collisions between them and were stabilized by van der Walls interactions. In the later stage, the large surface energy of these aggregates promoted directional growth that form ring-like MS₂ [30]. Typical TEM images of IFs of MoS₂ produced by microwave irradiation method are shown in Figure 31.8.

Relatively mild reaction conditions have also been used to synthesize IFs of MoS₂. Heating of MO(Co)₆ with iodine at 450–650 °C under metal–organic chemical vapor deposition conditions yielded giant IF-MoS₂ "bubbles" with about a 5 nm think shell [45]. The reaction initially produced amorphous particles consisting of Mo, S, and iodine. Subsequent annealing of these amorphous particles to higher temperature under inert conditions yielded fullerene-like

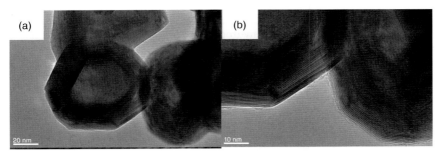

Figure 31.8 (a) and (b) HRTEM images of MoS₂ particles produced by microwave irradiation method (adapted from Ref. [30]).

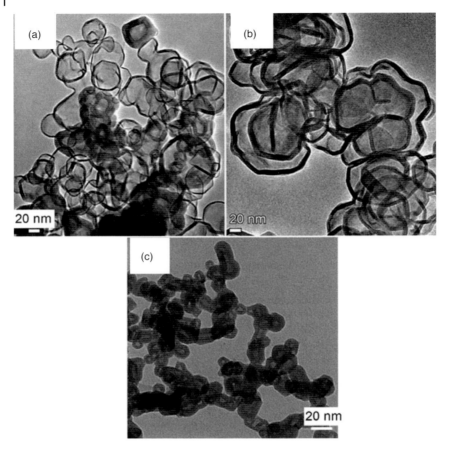

Figure 31.9 Overview TEM images of the product obtained after annealing the precursor particles obtained at (a) 350, (b) 450, and (c) 550 °C for 30 min at 850 °C under argon flow (adapted from Ref. [45]).

particles. The diameter of the particles was found to decrease with increasing reaction temperatures. TEM images of the MoS_2 IFs synthesized at different temperature are shown in Figure 31.9. Iodine released in the thermal decomposition of amorphous particles generated large voids in the nanoparticles. The increased iodine pressure inside the nanoparticles ruptured their walls. Subsequent annealing stitched the patches and healed the ruptured nanoparticles producing the large IFs.

Unlike WS_2 nanotubes, the synthesis of MoS_2 nanotubes is found to be quite challenging. Even though several procedures were proposed for the synthesis of MoS_2 nanotubes, an efficient way to make them in large quantities through sulfurizing the respective oxides has not been materialized so far. In the case of WS_2 nanotube synthesis, a nonvolatile phase of anisotropic nanowhiskers $W_{18}O_{49}$ serves as an intermediate. This phase shows a sublimation temperature larger than the sulfurization temperature (800–900 °C). The mechanism of the synthesis goes as follows: in the first step, the WO_3 nanoparticle precursors assume an anisotropic growth at high temperature to the nonvolatile long nanowhiskers of $W_{18}O_{49}$ phase. In the second step, the suboxide whiskers are converted to sulfide nanotubes under reducing conditions. On the other hand, MoO_3 vaporizes above 700 °C, which is therefore not a suitable starting material for growing the subsequent nanotubes. Thus, the lack of a stable anisotropic MoO_{3-x} phase limits the synthesis of MoS_2 nanotubes with high yields through this path. However, several strategies were applied to stabilize the molybdenum suboxide phase by substituting impurities of other elements. Recently, Pb-stabilized molebdnum suboxide nanowhiskers were synthesized by shock wave treatment [29]. Shock waves generate momentarily high temperatures (up to 10 000 °C for a few microseconds) and far-from equilibrium conditions, which helps inserting Pb into the elongated MoO_{3-x} nanowhiskers stabilizing them against vaporization and collapse during the subsequent high-temperature sulfurization process. Pellets containing MoS_2 and Pb powder were exposed to shock wave in the hypersonic shock tunnel. In this process, compressed gas at one end and low-pressure region at the other side of the piston is created. Motion of the heavy piston inside the compression tube adiabatically compresses helium to a high pressure and high temperature by the end of the stroke, rupturing the diaphragm and creating a shock wave in the tubes [29]. The pellet materials get uniform and instantaneous heating and quenching in this process. In the present experiment, Pb-stabilized oxide whiskers were formed on top of the pellet. The oxide whiskers were further successfully converted to sulfides, that is, INT-MoS_2 by reacting with H_2S at high temperatures in conventional furnaces.

Another high-temperature and nonequilibrium method such as ablation by a focused solar irradiation was also used for synthesizing MoS_2 nanotubes [46]. A mixture of MoS_2 and Pb powder was inserted into quartz ampoules and vacuum sealed. This reaction mixture was subjected to solar ablation to produce MoS_2 nanotubes (Figure 31.10). High-resolution TEM analysis indicated that the Pb atoms were incorporated into the MoS_2 lattice. Figure 31.11 shows the high-resolution aberration-corrected HAADF image of MoS_2 nanotubes synthesized in solar ablation. As high angle annular dark field images are atomic number sensitive, the presence of lead atoms could be detected inside the MoS_2 nanotubes, though in lesser amounts compared to their concentration in the oxide precursor. MoS_2 INTs growth in solar ablation was shown to happen in three steps [46]. In the first step, intense heating generated by focused solar beam rapidly evaporates MoS_2 and Pb powders. These vapors react rapidly with the surrounding gases, especially with oxygen and water vapor that are generated by outgassing from the quartz ampoule. Putatively, high temperature promotes the growth of Pb-stabilized crystalline MoO_{3-x} nanowhiskers. In the second step, MoS_2 layers are formed on the oxide whiskers surface by reacting with sulfur. In the third step, the oxide core is converted back to the sulfide, but now in the form of MoS_2 nanotubes, by reaction with the sulfur vapors. The schematic reaction mechanism is presented in Figure 31.12.

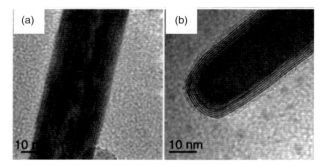

Figure 31.10 (a and b) High-magnification TEM images of MoS_2 nanotubes after exposure of MoS2 and Pb mixture for 10 min to focused solar irradiation (adapted with the permission from Ref. [46]).

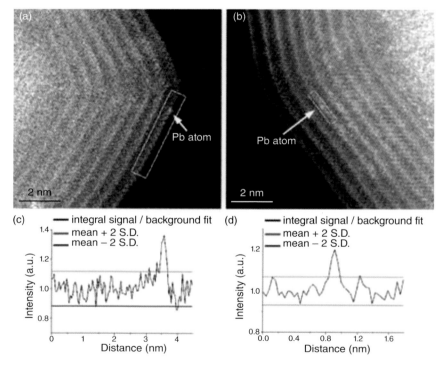

Figure 31.11 (a, b) High-resolution aberration-corrected HAADF image of a single MoS_2 nanotube. (c, d) Normalized line profile data taken from the marked layers in the HAADF images. The profile data are normalized to the signal intensity of MoS_2 layers lacking extraordinary bright dots. Signal enhancements (SE) significantly beyond background noise (horizontal lines) can be identified as individual Pb atom (adapted with the permission from Ref. [46]).

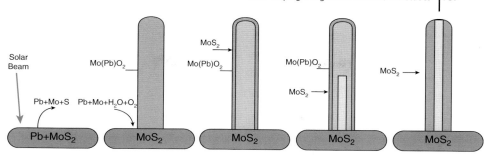

Figure 31.12 Schematic of the formation mechanism of MoS_2 nanotubes (adapted with the permission from Ref. [46]).

31.3
Doping Inorganic Fullerenes/Nanotubes

Chemical modifications are necessary to tune the chemical and physical properties of a material for various applications. Doping has been used as an effective way to tune the properties of semiconducting materials. Chemical doping brings significant changes in a material; for instance, boron and nitrogen doping in graphene or carbon nanotubes turns them into p-type and n-type materials, respectively [47]. Nitrogen doping in silicon is effectively used for increasing its mechanical strength by locking dislocations [48]. Doping inorganic fullerenes/nanotubes with minute amounts (<0.1 at%) impurities is very challenging. Most importantly, the impurities, such as rhenium or niobium or other atoms, must occupy substitutional sites in order for them to become electronically active. doping of IFs/INTs by Re atoms has been nevertheless achieved leading to significant electronic changes in their physicochemical behavior. This progress offers various potential applications [36,49–57]. Re-doping in IF MoS_2 nanoparticles was shown to improve its lubrication characteristics [57], catalytic properties in electrochemcial hydrogen evolution reaction [58], and improved kinetics in sodium ion batteries [55].

Doping of nanoparticles can be achieved also by different means including intercalation and functionalization [36]. However, substitutional dopants are less mobile in the lattice and possess extra advantage by showing higher resistance to external factors such as extreme pH, high ionic strength, and potentials [36]. This discussion focuses on the recent advances in achieving different ways of substitutional doping of IFs/INTs.

Theoretical calculations by Ivanovskaya *et al.* confirmed that nanotubes obtained by Nb substitution of Mo in MoS_2 (Nb_{Mo}) are thermodynamically more stable than pristine nanotubes. Furthermore, Nb substitutional doping (>3 at%) renders the nanotubes metallic irrespective of their chirality and diameter [59]. On the other hand, minute amount (doping) of Re in MoS_2 nanotubes turn them into n-type materials [60]. Nb doping (alloying) of MoS_2 IFs (> 5 at%) has been achieved by reacting the respective halides in H_2S environment [61]. Nb doping (alloying) of WS_2 nanotubes was achieved by heating Nb_2O_5-coated

$W_{18}O_{49}$ nanorods at $1100\,^\circ$C in H_2S [51]. Ti doping in MoS_2 nanotubes was attained by reacting Mo-Ti alloy in the presence of H_2S at high temperature ($950\,^\circ$C) [52]. Alloying W with Mo in nanotubes also been achieved [54]. Recently, Re-doped MoS_2 IFs/INTs were obtained by *in situ* and postannealing methods as detailed next [62]. Reaction between Re-doped MoO_{3-x} nanoparticles (doped nanoparticles are synthesized *in situ* by reacting the respective oxides) with H_2S gas in reducing atmosphere produced Re-doped MoS_2 fullerenes [62]. In this method, 20–7000 ppm (0.002–0.7 at%) Re doping could be achieved. Postannealing methods also used to dope Re into MoS_2 IFs/INTs. High-temperature heating of ReO_3 ($ReCl_3$) and MoS_2 IFs/INTs in evacuated quartz ampoules also yielded Re doping in IFs/INTs [62]. Figure 31.13a shows the SEM images of Re-doped MoS_2 IFs. A HRTEM image of Re-doped MoS_2 IF is shown as an inset with line profile revealing interlayer spacing of 0.624 nm. A typical TEM image of Re-doped MoS_2 nanotube synthesized by postannealing method is shown in Figure 31.13b. The low concentration of Re in the IFs/INTs could not be measured accurately by standard methods, such as EDS, XRD, Auger, and so on. The rhenium concentration was measured, accurately, by careful dissolution of the powder and using inductively coupled plasma mass spectrometry (ICP-MS). The local environment around the Re atom in doped MoS_2 was studied by X-ray absorption fine structure (XAFS) measurements. XAFS along with theoretical calculations helped understand the Re bonding in the doped IFs/INTs [62]. However, direct visualization of Re dopants in MoS_2 IF was achieved by high angle annular dark field imaging mode in scanning transmission electron microscopy (see Figure 31.14) [56]. Re-doped MoS_2 IFs were found to be less agglomerated compared to undoped MoS_2 IFs, due to the self-repulsion between the doped particles caused by charge trapping on the surface. This phenomenon helps use these particles in oil additives, which helps improve their dispersion, with favorable influence on the tribological behavior of these lubricants [57]. Nb doping of IF-MoS_2 at a level of 0.05 at% has been also demonstrated leading to substantial improvements in the hydrogen evolution reaction [63].

31.4
Applications

31.4.1
Tribological Properties

Energy saving is very important and one of the imminent global challenges of today. Most of the mechanical systems consume more than the required energy to compensate for the friction losses. Strategies for energy conservation via tribology have been an active field. Lowest friction attainable, durable, nonwettable, nontoxic, corrosion resistant, and low surface energy materials are needed for future lubricants. Effective lubrication can be achieved by solid lubricants

(a)

(b)

Figure 31.13 (a) HRSEM and HRTEM (inset) images of a typical Re(0.12 at%):IF-MoS$_2$ NP. The interlayer spacing as shown by the line profile of Re:IF-MoS$_2$ (0.627 nm) coincides with that of the undoped IF-MoS$_2$. (b) HRTEM image of a postsynthesis Redoped INT-WS$_2$ (adapted from Ref. [62]).

beyond oil and greases in certain applications and can support fluid lubricants in other applications. Soft materials such as graphite and MoS$_2$ have been used as solid lubricants or additives. MoS$_2$/WS$_2$ possess low interlayer shear modulus due to the weak interlayer interaction between the 2D layers, which allows them facile shearing. However, high reactivity of the edges of MoS$_2$ platelets to oxygen/moisture makes them less durable [36]. Thanks to IFs/INTs possessing none or fewer edges and exposing only their inert van der Walls surface outward makes these materials better alternatives for improved lubrication. In fact, WS$_2$ and MoS$_2$ fullerene-like particles were shown to exhibit excellent tribological properties superior to all known solid lubricants, mostly under high loads. This

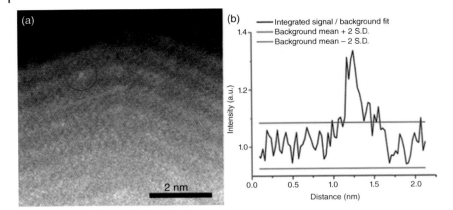

Figure 31.14 (a) High-magnification HAADF-HRSTEM of the outer surface of a Re:IF-MoS₂ nanoparticle (0.12 at% rhenium in the dry powder). The circle indicates a local enhancement on the HAADF scattering intensity. (b) Background corrected integrated intensity profile along the MoS₂ layer containing the local HAADF enhancement in Figure 31.14a. The gray and black dotted lines depict the ±2 standard deviation limits regarding the background intensity fluctuation. The peak at 1.25 nm along the line scan corresponds to the maximum intensity highlighted in Figure 31.14a, which can be assigned to a rhenium dopant atom as it is much higher than the background noise ripple level (adapted from Ref. [56]).

attribute of the IFs nanoparticles has been effectively used in various potential applications, such as lubricating additives [36,64].

The unique structure and bonding of INTs/IFs along with their very good mechanical properties and high-pressure resilience contribute substantially to their improved tribological properties. Various mechanical tests were performed by macroscopic or *in situ* electron microscopy on IFs/INTs revealing many of the unique mechanical properties and deformation and failure mechanism of these materials. The resilience of WS_2 nanotubes to shock waves was also tested. These experiments revealed that INT-WS_2 can withstand up to 21 GPa transient shear stress [65]. The significant stability under shock loading is indicative of their high tensile strength. Other studies found that WS_2 nanotubes can withstand high hydrostatic pressure (20 GPa) but got exfoliated when they were exposed to high uniaxial pressures [66]. Recent *in situ* TEM studies of individual WS_2 nanotubes revealed their anomalous tensile properties [67]. It was found that larger diameter nanotubes could bear much higher loadings due to the intershell "cross-linking" and higher defect densities, compared to smaller diameter nanotubes that are either free or have less defects. A WS_2 nanotube with 30 nm outer diameter was shown to sustain a tensile force of ~1053 nN with corresponding facture strength of 17.1 GPa, whereas a WS_2 nanotube with 50 nm outer diameter had shown ultimate tensile force of up to ~3987 nN with strength of ~19.6 GPa [67]. Previous mechanical measurements by AFM revealed that individual WS_2 nanotubes have Young's modulus of 171 GPa, interlayer shear modulus of 2 GPa, intralayer shear modulus of 77 GPa, and a radial modulus of 1.7–5.8 GPa [68]. Individual IF-WS_2 nanoparticles are shown to withstand a stress of 1–2.5 GPa under uniaxial loading before failure. Higher stiffness and strength was observed in IF-MoS_2 nanoparticles (up to 3.5 GPa) compared to IF-WS_2 nanoparticles. Due to their seamless hollow closed structures, flexible layered structure, and small size, IF-WS_2 and IF-MoS_2 nanoparticles are also able to withstand very high elastic stress. Recent review summarizes studies on the mechanical properties of individual IFs and INTs of WS_2 and MoS_2 by *in situ* electron microscopy measurements [68].

After intense research and development efforts, a large number of products based on (industrial grade) IFs as additives in different oils and greases and for machine working fluids were commercialized. The production of IF nanoparticle-formulated lubricants (oils and greases) has reached sales over 1000 metric tons per year [69]. Furthermore, being minerals that are mined for many decades, it is accepted that WS_2 (MoS_2) are environment-friendly and nontoxic materials. More recently, several tests have clearly indicated that IF/INT of these materials are nontoxic as well and are biocompatible [68]. Combined with their excellent tribological performance and their mechanical robustness, these findings offer these IF/INT several intriguing medical applications. Indeed, metallic coatings impregnated with IF nanoparticles were shown to be self-lubricating and could find medical applications especially as coatings of orthodontic archwires [70,71]. For instance, orthodontic stainless steel wires coated with IFs showed 50% reduction in friction compared to the uncoated ones. Reduction in the friction of NiTi orthodontic wire substrates coated with cobalt film impregnated with IF nanoparticles was also studied, while keeping the unique shape memory characteristic of the alloy intact. Self-lubricating coatings can also find applications in different fields including aerospace, machining, sports, and so on.

The highly crystalline order coupled with their chemical inertness and their quasi-spherical morphology are well documented attributes of the IFs and contribute to their enhanced lubrication behavior. These properties confer also high elasticity and robustness allowing the IF nanoparticles to roll as well as easily slide in suitable loading regimes. Theoretical and experimental investigations of the friction mechanism of individual inorganic fullerenes revealed that it consists of a superposition of three main mechanisms, that is, rolling, sliding, and exfoliation combined with material transfer to the underlying substrate (third body) [66,72–76]. These experiments concluded that under relatively low normal stress, that is, <0.5 GPa, rolling is an important lubrication mechanism. Under a higher stress, that is, <1 GPa, the average tolerance between the two matting surfaces is too small (<100 nm) to allow free rolling and consequently sliding of the IF nanoparticles becomes more relevant mechanism. Under higher pressures (>1 GPa), exfoliation of fullerenes becomes the dominant mechanism. These mechanisms are schematically displayed in Figure 31.15. The exfoliated

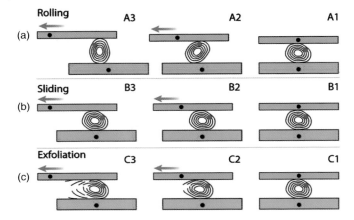

Figure 31.15 The three main friction mechanisms of multilayered IF NP discussed in the literature: rolling (a), sliding (b), and exfoliation (c). The bottom surface is stationary while the upper surface is moving to the left. The red mark is a point of reference, that is, gold nanoparticle in the present experiments (adapted from Ref. [72]).

molecular layers of WS_2/MoS_2 are deposited on the asperities of the shearing surfaces providing effective lubrication.

Notwithstanding their effective initial lubrication behavior, the gradual agglomeration of IF nanoparticles makes their oil suspensions less stable, thus shortening their shelf life. Some studies have been carried out to reduce the agglomerating behavior of IFs by surface functionalization. Interestingly, recent studies of Re-doped IF-MoS_2 nanoparticles found them to exhibit reduced agglomeration compared to undoped counter parts, thus producing more stable oil suspensions [57,58]. Tribological studies of oils formulated with 1 wt% of doped IFs in oils showed reduced friction coefficients coupled with negligible wear. Remarkably also, due to their effective lubrication, IF nanoparticles were shown to reduce the drag and viscosity of various lubricating media, for example, in medical gels [77]. The remarkable tribological and rheological properties of Re-doped IFs could be attributed also to their atomically even van der Waals surfaces. Furthermore, the doped IFs are negatively charged causing mutual repulsion between the particles. Thus, a number of applications based on doped IFs/INTs as alternative additives or lubricant additives are foreseen.

31.4.2
Nanocomposites

Sustainability can be achieved by making better and smatter materials by improving their electrical, mechanical, and thermal properties. However, it is not always possible to enhance the inherent property of a material to a desirable extent. Composite materials, on the other hand, show novel/enhanced properties that cannot be achieved by the individual component constituting it. There is a lot of research devoted to composite materials reinforced by using nanomaterials as a minority phase in recent years. Composite materials of CNTs, graphene, BN, and so on were shown to have enhanced electrical, mechanical, and storage capacities with some technologies already commercialized [73–75].

The good mechanical properties of INTs and IFs, as mentioned earlier [68], signify plenty of opportunities to use it as a minority phase in composite materials for improving their mechanical behavior and thermal stability. These materials find application in various fields including high-strength construction materials, shielding, surface transportation, and so on. Since IF-WS$_2$ and more recently INTs of WS$_2$ and MoS$_2$ became available in substantial amounts, various groups started studying their reinforcing effect in polymer nanocomposites in recent years. The first study concerning polymer nanocomposites formulated with IF/INT-WS$_2$ nanoparticles dates back to 2003/2004 [78,79].

A large number of studies were published dealing with the thermal stability and mechanical properties of thermoplastic and also thermosetting polymers in which minute amounts (usually less than 2 wt%) of IF-WS$_2$ nanoparticles were incorporated [80,81]. For example, improved thermal stability and mechanical behavior of nanocomposites based on polypropylene IF-WS$_2$ nanoparticles were reported [82–85]. Remarkable improvements in modulus, tensile strength, thermal stability, and thermal conductivity of a processable poly(ether ether ketone) (PEEK) was achieved by adding IF-WS$_2$ [86]. The addition of IF nanoparticles improved the flexural modulus and strength of carbon fiber (CF)-reinforced poly(phenylene sulfide) by 17 and 14%, respectively, without any loss in toughness while increasing the storage modulus [87]. In another study, significant enhancement in the peel and shear strength of epoxy resin contact with 0.5% IF was measured [88]. The fracture toughness of the epoxy was improved up to 70% by addition of alkyl silane-treated IF-WS$_2$ nanoparticles [89]. Reinforcing silica aerogels [90] and AZ31 Mg alloy [91] composites and concrete [92] was achieved by addition of small amounts of WS$_2$ nanotubes.

Improved electrical conductivity has also been observed for INT composites. Recent study found that adding 0.85 wt% INT-WS$_2$ increased the conductivity of PANI by two orders of magnitude compared to undoped PANI, which is inherently a good conductor by itself [93]. The improved conductivity was attributed to the efficient electron transfer at the INT–polymer interface (INT served as a donor). In another recent report, WS$_2$ nanotubes were mixed with poly(3-hexylthiophene) (P3HT) and CdSe quantum dots. The energetic matching between the WS$_2$ nanotubes and the quantum dots led to a vectorial electron transport from the photoexcited CdSe quantum dot through the P3HT and to the WS$_2$ nanotube, suggesting it to be a potentially useful system for the active layer of a hybrid solar cells [94].

31.5
Fullerenes and Nanotubular Structures from Misfit Layered Compounds

So far, we have concentrated on nanotubes and fullerene-likes of layered materials made from one type of molecular layer. On the other hand, a relatively new class of materials called misfit layered compounds (MLCs) consisting of two molecular slabs with different lattice periodicities at least along one direction were studied [27,95]. Recently, substantial amount of work was devoted to bulk misfit compounds, including synthesis and structure characterization, and were tested for potential thermoelectric properties [96–99]. High-efficiency thermoelectric properties demand material to have high Seebeck coefficient, large electrical conductivity coupled with a reduced thermal conductivity. In bulk materials, these transport properties are interdependent making it difficult to alter one property and keeping the other unchanged, which limits the performance of thermoelectric devices. These transport properties can be decoupled both by decreasing the dimensionality of the material and by creating superlattices [100,101]. Decreasing the dimensionality of a material leads in general to enhanced phonon scattering and hence to a reduced thermal conductivity, thus enhancing the thermoelectric properties [102–104]. On the other hand, in misfit layer compounds (MLCs), one of the layers helps conducting the electrons, whereas the second layer acts as a phonon scattering region, thus decreasing the lattice thermal conductivity of the misfit compound (in the $a–b$ plane) [102]. The structural advantage of misfit layer compounds as thermoelectric materials could be further enhanced by bringing it down to the nanoscale. However, synthesizing misfit compounds in low dimensional forms is challenging. Nonetheless, recent studies showed that MLCs could be synthesized in nanotubular forms [27,105,106]. Here, we describe the latest development in synthesizing nanotubular forms and characterization of these unique one-dimensional misfit compounds.

MLCs can be considered as alternating stacking of two layers along the c-axis. The general formula of (chalcogenide-based) MLCs can be considered as $[(MX)_{1+x}]_m[TX_2]_n$, where M = rare earths, Sn, Pb, Bi, and so on; T is Cr, V, Ta, Nb, and so on; and X is S and Se [95]. One of the two layers (MX) consists of two-atom thick distorted NaCl structure (pseudotetragonal), whereas the other layer is three-atom thick pseudohexagonal, in which the transition metal atom layer is sandwiched between two layers of chalcogen atoms. MLCs show strong intralayer bonding and the layers are stacked together by weak van der Walls forces and polar forces due to partial charge transfer between the layers. These compounds can also be called incommensurate composite crystals. In MLCs, each subsystem possesses its own unit cell, but the composite structure lacks three-dimensional periodicity [95]. The lattice parameters of the sublayers differ from each other at least in one direction leading to an incommensurate structural modulation. Superspace symmetry is used sometime to describe the symmetry of MLCs.

Stress generated due to the lattice mismatch between the layers in MLCs can be partially relieved by rolling the structure along the incommensurate lattice vector (usually a). Furthermore, it was recently shown that a driving force for the annihilation of the dangling bonds in the peripheral atoms drives the layered materials to roll into perfectly seamless nanotubular structures [27]. Chemical vapor transport (CVT) has been used to synthesize bulk three-dimensional structures of MLCs. Bernaerts *et al.* were the first to observe micrometer-sized scroll-like tubular structures of MLCs [107]. CVT of Pb, S, and Nb yielded tubular structures of $(PbS)_{1+x}(NbS_2)_n$. Gomez-Herrero *et al.* found micrometer-size scrolls and tubular structures of $(BiS)_{1+x}(NbS_2)_n$ by CVT of Bi, Nb, and S [108].

Nanotubular structures of misfit compounds were first observed by serendipity in 2003 [105]. Laser ablation of SnS_2 substrates yielded nanotubular structures of SnS-SnS_2 superstructures [105]. Nd:YAG laser pulses are believed to evaporate some of the sulfur from the SnS_2 target, which resulted in a spontaneous formation of the SnS-SnS_2 misfit nanotubular structures. However, the yield of the nanotubes was rather small and the method produces short nanotubes only. More recently, however, several techniques have been developed to systematically synthesize nanotubes from MLCs in high yields [27]. To that end, modified chemical/physical vapor transport techniques were used [106,109,110]. In one of these techniques, either elements or metal chalcogenides of the precursors were inserted into a quartz ampoule along with a transport agent (e.g., iodine or chlorine) and the ampoule was sealed under vacuum. Sometimes, an excess of chalcogen itself acted as a transport agent. The typical CVT reaction to synthesize nanotubes was carried out in a two-zone vertical/horizontal furnace in two steps. In the first step, the entire ampoule was heated to high temperature (or alternatively the ampoule was heated in a temperature gradient and the precursor was placed at low temperature) for few hours to allow the reaction to happen among the reactants. In the second step, the ampoule was heated with steep temperature gradient. The precursor was placed at the high-temperature end and the other end was placed at a lower temperature. The reaction generally happened for a few hours. Subsequently, the ampoule was removed rapidly from the furnace to cool down at room temperature. Selection of the temperatures and the gradient varied from one reaction mixture to another. It is important to note that small temperature gradients (less than $100\,°C$) were used for synthesizing bulk single-crystal misfit compounds. These gradients were appreciably higher (400–$700\,°C$) in the case of the nanotubes synthesis. Depending on the energetics of the reaction, the nanotubes accumulated either at the cold zone (exothermic reaction) or at the hot (endothermic reaction) zone of the ampoule. Radovsky *et al.* adopted this technique to produce long SnS-SnS_2 nanotubes with relatively good yields [106,111]. In this case, small amounts of bismuth and Sb_2S_3 were used in the reaction mixture. It is believed that Bi helps stabilize the orthorhombic SnS aiding the process of SnS-SnS_2 nanotube formation. There were no nanotubes observed in the absence of Bi, while Sb_2S_3 is believed to have acted as a cocatalyst to improve the yield of the nanotubes by helping scroll the structure. Figure 31.16 shows a typical scanning electron microscopy (SEM) images of SnS-SnS_2 misfit nanotubes synthesized by the CVT method. The scrolling process of the SnS-SnS_2 slabs and steps of the scrolling (marked with

Figure 31.16 (a) High-magnification backscattering electrons (BSE) SEM image illustrating the exfoliation/scrolling of SnS$_2$/SnS misfit layers into tubular (scroll) structures. Sulfides of bismuth appear as bright spots in the BSE contrast. (b)Secondary electrons (SE) image of tubule's agglomerate. Red arrows indicate nanosheets in the midst of a scrolling process. Blue arrows point to nanoscrolls exhibiting helical wound growth step that can be clearly seen in the inset and is marked by short white arrows. (c) SE image of macroscopic amounts of nanotubes, nanoscrolls, and several unscrolled nanosheets. (adapted from Ref. [106]).

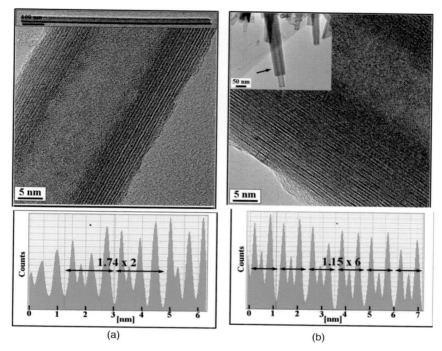

(a)   (b)

Figure 31.17 (a) TEM images of SnS/SnS$_2$ tubule with O-T-T . . . periodicity. Top: High and low (inset) magnification images. Bottom: Line profile obtained from the region enclosed in the rectangle in the upper image. (b) TEM images of SnS-SnS$_2$ tubule with O-T . . . periodicity. Top: High and low (inset) magnification images. Bottom: Line profile obtained from the region enclosed in the rectangle in the upper image (adapted with the permission from Ref. [109]).

white arrows) are shown in Figure 31.16b. The end product encompassed a mixture of nanotubes and nanoscrolls of distinct superstructure along with unscrolled nanosheets (see Figure 31.16c). There were two kinds of stacks commonly observed in between orthorhombic (O) SnS and trigonal (T) SnS$_2$ in the order O-T-O-T . . . with 1.15 nm lattice spacing or O-T-T, O-T-T . . . superstructure with 1.74 nm lattice periodicity along the c-axis. Typical TEM images of SnS-SnS$_2$ nanotubes with O-T-T... and O-T-O-T . . . superstructures are presented in Figure 31.17. The misfit nanotubes with O-T-O-T-T . . . with lattice spacing of 2.89 nm have also been occasionally observed [111].

The CVT method was further extended to produce a large variety of MLC nanotubes including PbS-NbS$_2$, PbS-TaS$_2$, SnS-TaS$_2$, BiS-TaS$_2$, SbS-TaS$_2$, LnS-TaS$_2$ (Ln = La, Ce, Nd, Ho, and Er), and LaSe-TaSe$_2$ [109,110,112]. Lanthanide-based misfit nanotubes were found at the hot zone of the ampoule in the two-zone furnace after the reaction, whereas Pb-, Bi-, and Sn-based misfit compounds were usually observed at the cold zone of the reaction. Figure 31.18a

Figure 31.18 SEM images of the tubular structures and common by-products from (a) LaS-TaS$_2$, (b) ErS-TaS$_2$ MLC. Scrolling steps are visible in the significant part of the tubular crystals (adapted from Ref. [112]).

and b shows the SEM images of LaS-TaS$_2$ and ErS-TaS$_2$, respectively. Due to the structural modulation, the rock salt lattice of LnS undergoes orthorhombic distortion. Also, the lattice parameters of selenide-based misfit compounds are slightly larger than that of the sulfur counterparts.

Relatively large quantities of SnS-SnS$_2$ misfit nanotubes up to 20 mg/batch have also been synthesized using horizontal/vertical flow rector [106]. In this case, a mixture of SnS$_2$ and Bi powder was placed in a quartz buret, which was placed in a horizontal quartz reactor. This mixture was heated in a single-zone furnace above 700 °C under argon flow. The product was found to accumulate at the low-temperature region. This reaction yielded simultaneously SnS-SnS$_2$ misfit nanotubes along with pure SnS$_2$ nanotubes. The main growth mechanism in this case is believed to be a Bi-catalyzed vapor–liquid–solid process [106].

The primary driving force for the formation of tubular structures from MLC is the misfit stress between the MX and the TX$_2$ layers. Upon scrolling, the strain energy associated with the misfit is reduced [27,106]. In addition, as a result of the presence of atoms with dangling bonds on their rims, nanoparticles of compounds with layered structures are known to be unstable in the planar form and fold into seamless hollow structures, such as nanotubes [27]. It is shown here that the combination of these two stimuli leads to the formation of new kinds of nanotubes from MLC compounds. The presence of dislocation-like defects on their surfaces suggests that the transition from scroll-like to concentric tubules is likely to occur by a mechanism that involves high-temperature dislocation migration.

MLC nanotubes based on hexagonal CrS$_2$ and VS$_2$ could not be easily obtained by the CVT method. Nevertheless, a thermal annealing technique was developed in recent years to synthesize a large variety of LnS-CrS$_2$ (Ln = La, Ce, Nd, Gd, and Tb) and LnS-VS$_2$ nanotubes [27,113,114]. The thermal annealing technique was found to produce large quantities of such nanotubes in a short time. In this procedure, a mixture of Ln(OH)$_3$ and Cr(OH)$_3$ was annealed at high temperature (above 800 °C) in the presence of H$_2$S and H$_2$ to yield the

(a)   (b)

(c)   (d)

Figure 31.19 SEM images of (a) LaS-CrS$_2$ (b) CeS-CrS$_2$, (c) GdS-CrS$_2$, and (d) TbS-CrS$_2$ nano-scrolls and nanotubes. Insets show corresponding high-resolution SEM images (adapted with the permission from Ref. [27]).

respective misfit nanotubes. The reaction temperature depended on the molecular weight of the Ln ion in the reaction mixture. For instance, LaS-CrS$_2$ nanotubes could be grown at 825 °C, whereas TbS-CrS$_2$ nanotubes required higher temperatures (>900 °C). Unlike NbS$_2$ and TaS$_2$, which adopt trigonal prismatic structure, CrS$_2$ and VS$_2$ adopt octahedral structure in misfit compounds. Figure 31.19 shows SEM images of as-synthesized (a) LaS-CrS$_2$, (b) CeS-CrS$_2$, (c) GdS-CrS$_2$, and (d) TbS-CrS$_2$ nanoscrolls and nanotubes. Selected area electron diffraction (SAED) is a powerful technique to characterize these structures at individual nanotube level, which is not possible with X-ray diffraction technique. Besides identifying the crystal structure, the SAED pattern provides additional information about the nanotube including its chirality, number of different layers, superstructure periodicity, and so on. High-resolution TEM images along with the corresponding SAED pattern of one of a LaS-CrS$_2$ nanoscroll is presented in Figure 31.20. Superstructure (alternative stacking of LaS and CrS$_2$ subunits) with 1.14 nm periodicity along the *c*-axis can be clearly seen in Figure 31.20a (see line profile at the bottom). The inset in Figure 31.20a shows the corresponding low-magnification TEM image of the same nanoscroll. The superstructure periodicity can also be confirmed by the SAED pattern in Figure 31.20b. The tube axis is shown in the SAED pattern using a green arrow and the basal reflection are marked with white arrows. Segmented blue and

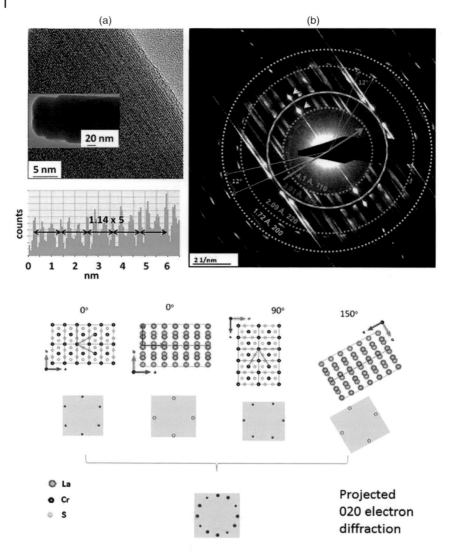

Figure 31.20 (a) HRTEM image of a LaS-CrS$_2$ nanoscroll, shown alongside a line profile (bottom) obtained from the region indicated by the green rectangle. A low-magnification TEM image of the nanoscroll is shown as an inset. (b) SAED pattern taken from the area shown in Figure 31.20a. The tube axis and basal reflections are indicated using green and white arrows, respectively. Spots corresponding to the same interplanar spacings are marked by segmented circles and measured values, with corresponding Miller indices indicated. Blue and orange colors correspond to the LaS and CrS$_2$ subsystems, respectively. The red color corresponds to the combined *d* spacings. The bottom panel shows the projected 020 reflection of the LaS and CrS$_2$ layers, for two layers each of LaS and CrS$_2$ within a single nanoscroll with corresponding projected 020 electron diffraction patterns. All the layers together produce the final reflections (bottom). There are 12 spots, of which 8 have higher intensities, which is similar to the observed 020 reflections in Figure 31.20b (adapted from Ref. [113]).

yellow circles indicate, respectively, the LaS and CrS_2 subsystem. The measured interplanar values and related Miller indices are shown, as well. In this specific nanoscroll, there are two kinds of LaS layers present with different rolling vectors. Thus, one observes eight pairs of each {110} and {220} LaS reflections with d-spacings of 4.1 and 2.09 Å, respectively, on a circle. These layers are observed to be chiral, which results in splitting of the diffraction spots. Each of the above spots is split by 12°. The chiral angle is 6°, which is half the azimuthal splitting. There are 12 sets of each {200} and {020} reflection present on a circle with corresponding d-spacings of 1.72 and 3.01 Å, respectively. The multiplicity factor of these plane is six, suggesting that there are two CrS_2 layers with different rolling vector present in the same nanoscroll. Also, these CrS_2 layers are found to be chiral with chiral angle 6°. The lattice parameters of both the LaS and the CrS_2 layers in the b-direction are the same; thus, they are commensurate along this direction. This is manifested in the fact that the {020} reflections of LaS are superimposed on top of the {020} reflections of CrS_2. A careful observation can lead one to conclude that the intensity of the 12 {020} spots are not the same and 8 of them are more intense. The superimposition of these spots can be understood by referring to the schematics in the bottom panel of Figure 31.20. Further details of the MLC nanotubes can be probed by aberration-corrected TEM. A typical high-resolution HAADF STEM image along with atomic-resolution energy-filtered TEM elemental maps of a CeS-CrS_2 nanotube is shown in Figure 31.21. The relative position and orientation between the layers can be readily seen from the image.

Analysis of the nanotubes synthesized with different times and temperatures can help proposing the growth mechanism of such nanotubes, which is presented in Figure 31.22. It is believed that the initial binary hydroxide mixtures were converted to oxide upon annealing. The ternary oxide compounds reacted subsequently with H_2S at high temperature and converted to the ternary sulfide [113]. The large differences in the densities of $LaCrO_3$ and LaS-CrS_2 led to huge volumetric changes upon conversion, which generated substantial stress. One way to remove the stress was by growing sheets or whisker vertically through pores/defects/crevices in the film. Simultaneously, these upgrowing nanosheets were curled by the misfit stress between the LaS and the CrS_2 layers. In this fashion, both nanotubes and nanoscrolls were formed. As the temperature increased, the diameter and/or length of the nanotubes increased, too. It has also been noticed that the nanotubes to nanoscrolls ratio in the reaction product depended on the symmetry of the synthesized material [27]. For instance, orthorhombic crystals of GdS-CrS_2 produced larger number of nanotubes compared to nanoscroll, whereas triclinic LaS-CrS_2 produced more nanoscrolls compared to nanotubes. Maybe, the more complex unit cell imposes additional boundary conditions to produce seamless closed nanotubes.

By using thermal annealing, nanotubes were synthesized in high yields. Quaternary nanotubular structures could also be synthesized by the thermal annealing method [113]. In this case partially, the Ln atom was substituted with another lanthanide atom or Sr(II) in the LnS-CrS_2 nanotubes. It is commonly

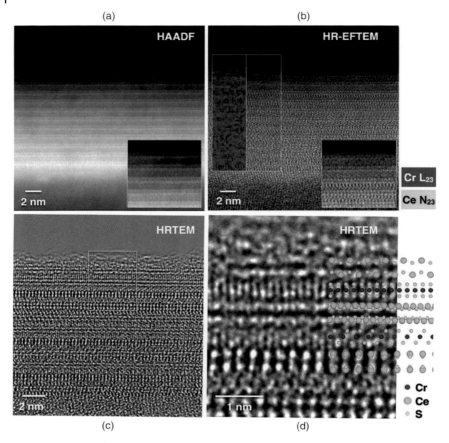

(a)   (b)

(c)   (d)

Figure 31.21 (a) High-resolution HAADF STEM image of a CeS-CrS$_2$ nanotube. The high-angle scattering signal is highly sensitive to atomic number, with CeS double layers appearing bright against sandwiched CrS$_2$ layers. The inset shows a magnified region of the outer shells. (b) Atomic-resolution energy-filtered TEM elemental maps taken using energy-selecting windows of 20 and 25 eV, corresponding to the Ce N$_{23}$ and Cr L$_{23}$ core-loss excitations, respectively. The modulation in the color-coded images obtained from the Ce and Cr signals shows the atomic layer modulation. Inset: a magnified region of the outer shells. (c) HRTEM images taken under optimized phase contrast conditions, with projections of the atomic helices corresponding to bright dots. (d) Magnified region of Figure 31.20c with superimposed crystallographic projections of CeS double layers and hexagonal CrS$_2$ layers (adapted from Ref. [113]).

observed that LnS-CrS$_2$ possess O-T-O-T . . . superstructure. Other superstructures such as O-T-O-T-T . . . or O-T-T-O-T-T . . . have not been observed so far in this family of nanotubes. CrS$_2$ is a metastable structure that requires one electron per molecular formula to stabilize the hexagonal structure. The lanthanides are more stable in the Ln(III) form and consequently the LnS provides the extra electron to CrS$_2$ and stabilize the misfit structure, imposing the

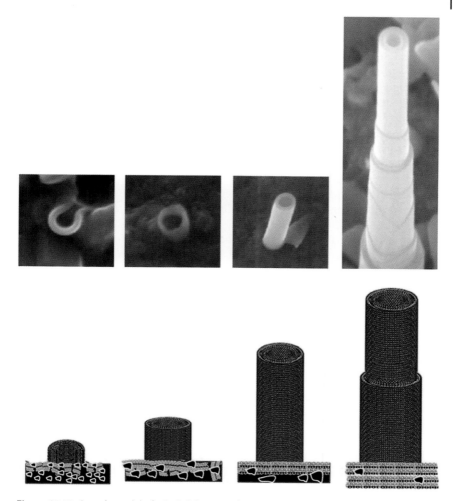

Figure 31.22 Growth model of a LnS-CrS$_2$ nanotube. Models along with SEM images in different stages of the reaction are shown to illustrate the growth of a nanotube (adapted from Ref. [113]).

O-T-O-T . . . order. On the other hand, O-T-O-T-T . . . superstructure was observed in 20 at% Gd- or 20 at% Ce-substituted LaS-CrS$_2$ misfit compounds. It was proposed that about 6 at% La vacancies are present in the lattice of the misfit compound (Las)$_{1.2}$CrS$_2$ (LaS-CrS$_2$) such that charge balance among the La^{3+}, Cr^{3+}, and S^{2-} in the lattice is maintained. It was hypothesized that these vacancies can be filled with substituent atoms such as Ce^{3+} or Gd^{3+}, which can donate that extra charge to CrS$_2$ extra layer and stabilize the O-T-O-T-T . . . structure. However, this hypothesis has to be confirmed experimentally. It is important to note that as the fraction of the TX$_2$ component increases in the misfit compound, the electrical conductivity of the system increases. Thorough electron microscopy analysis of Ce- and Gd-substituted $LaS-CrS_2$ nanotubes can be found in the literature [27,113].

Nanotubes of misfit-layered compounds were discussed thus far. However, fullerene-type structures were also reported very recently. Solar ablation of a mixture of PbS and SnS_2 powders yielded quasi-spherical fullerene-like PbS-SnS_2 superstructures with diameters ranging from 20 to 100 nm [115]. Unlike nanotubes that are folded along one-axis (usually the a-axis), fullerene-like nanoparticles get folded along two axes in the $a–b$ plane. The unique environment created in the solar furnace with temperatures exceeding 2500 °C, and large temperature gradients in the reaction zones help create these misfit fullerene-like particles. Misfit stress along with difference in the thermal expansion coefficient of PdS and SnS_2 subsystems induced the folding of the system along two axes at elevated temperatures.

In layered materials, van der Waals forces between the molecular layers keep the layers together. However, in misfit compounds, besides the van der Walls forces, polar forces between the sublayers also play an important role to stabilize the misfit superstructure. Charge transfer is believed to happen usually from the MX slab to the TX_2 layer. The amount of charge transfer depends on the subsystems that form the misfit compounds [27]. It is clear that CrS_2- and VS_2-based misfit compounds tend to show more charge transfer effect than NbS_2- and TaS_2-based compounds, because they are stable layered compounds on their own. It is also observed that LnS intercalates donate more charge than SnS and PbS do. The amount of charge transfer cannot be easily measured, experimentally. However, the effect of charge transfer can be observed both through the spacing between the MX and the TX_2 layers and through the shifts in the Raman bands. It is to be noted that, traditionally, the MX subsystem was thought to transfer charge to the TX_2 layer. However, it was recently realized that this is not always the case. In fact, theoretical calculations found that in the case of $SnS-SnS_2$ and $PbS-SnS_2$ cases, small but definite charge transfer happens from the SnS_2 system to SnS and PbS subsystems [116]. Furthermore, it was shown that in the case of the $LnS-TaS_2$ nanotubes, charge modulation along the a-axis is prevalent with some La atoms giving 0.1 charge, while others give more than 0.5 unit charge to the hexagonal TaS_2 layer [112]. Furthermore, the metal to metal charge transfer is partially compensated by opposite charge transfer from the sulfur atom of the TaS_2 to the sulfur atom of the LnS layer. This reversal charge transfer can be associated with the tendency of the lattice to have a reduced overall dipole and thereby minimize its free energy of formation.

Figure 31.23 shows plots of the ion size versus c-axis periodicities, which reveal that the interlayer spacing between the MS and the TX_2 layers is correlated with the ion size and the degree of charge transfer between layers. LnS intercalated (Ln = lanthanides) misfit compounds show lesser interlayer spacing due to higher degree of charge transfer between the layers, imposing thus strong interlayer interaction. CrS_2-and VS_2-based misfits also show stronger interaction. As the Ln ion size increases, the c-axis periodicity also increases that can

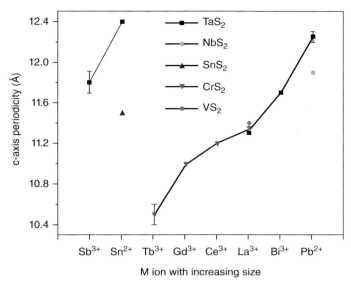

Figure 31.23 (a) *c*-axis periodicity values of tubular structures from different MX-TX$_2$ MLC with O-T stacking order. The *x*-axis of the graph represents different MX compounds with increasing ionic radius of the M atoms. (adapted with the permission from Ref. [27]).

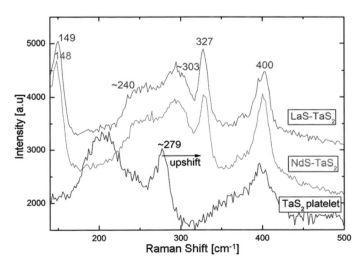

Figure 31.24 Raman spectra of single LaS-TaS$_2$ and NdS-TaS$_2$ tubular crystals. A spectrum recorded from a TaS$_2$ platelet is also shown for comparison (adapted from Ref. [112]).

be seen in Figure 31.23. Raman spectroscopy provides a strong indication for the charge transfer effects. These effects are very evident in layered materials such as carbon nanotubes and graphene [47]. These charge transfer effects were also found in misfit compounds [106,112]. Figure 31.24 shows the Raman spectra of single LaS-TaS$_2$ and NdS-TaS$_2$ tubular structures along with TaS$_2$ platelet. It is very clear from the spectra that the E$_{2g}$ mode, which is very sensitive to charge transfer, upshifts considerably due to charge transfer from LnS to TaS$_2$.

31.6
Conclusions

The fact that layered compounds suffer from inherent instability in the nano-range and form nanotubes and fullerene-like structures has been exploited to synthesize numerous new nanostructures with large prospects for exploitation to reduce mankind's dependence on depletable energy resources and help mitigate the large environmental risk associated with usage of fossil fuels. At the same time, this discovery provided rich research field, which led to new insights in the chemistry of materials and the elucidation of the physical properties of such nanostructures.

References

1 Tenne, R., Margulis, L., Genut, M., and Hodes, G. (1992) *Nature*, **360**, 444–446.

2 Margulis, L., Salitra, G., Tenne, R., and Talianker, M. (1993) *Nature*, **365**, 113–114.

3 Feldman, Y., Wasserman, E., Srolovitz, D.J., and Tenne, R. (1995) *Science*, **267**, 222–225.

4 Chopra, N.G., Luyken, R.J., Cherrey, K., Crespi, V.H., Cohen, M.L., Louie, S.G., and Zettl, A. (1995) *Science*, **269**, 966–967.

5 Tenne, R. (2006) *Nat. Nanotechol.*, **1**, 103–111.

6 Rao, C.N.R. and Govindaraj, A. (2005) *Nanotubes and Nanowires*, RSC Publishing, Cambridge, UK.

7 Fan, R., Wu, Y., Li, D., Yue, M., Majumdar, A., and Yang, P. (2003) *J. Am. Chem. Soc.*, **125**, 5254–5255.

8 Goldberger, J., He, R., Zhang, Y., Lee, S., Yan, H., Choi, H.-J., and Yang, P. (2003) *Nature*, **422**, 599–602.

9 Sun, Y., Fuge, G.M., Fox, N.A., Riley, D.J., and Ashfold, M.N.R. (2005) *Adv. Mater.*, **17**, 2477–2481.

10 Tenne, R. and Rao, C.N.R. (2004) *Phil. Trans. R. Soc. A*, **362**, 2099–2125.

11 Tenne, R., Rosentsveig, R., and Zak, A. (2013) *Phys. Status Solidi A*, **210**, 2253–2258.

12 Kroto, H.W., Heath, J.R., O'Brien, S.C., Curl, R.F., and Smalley, R.E. (1985) *Nature*, **318**, 162–163.

13 Iijima, S. (1991) *Nature*, **354**, 56–58.

14 Pauling, L. (1930) *Proc. Natl. Acad. Sci. USA*, **16**, 578–582.

15 Brüser, V., Popovitz-Biro, R., Albu-Yaron, A., Lorenz, T., Seifert, G., Tenne, R., and Zak, A. (2014) *Inorganics*, **2**, 177–190.

16 Seifert, G., Kohller, T., and Tenne, R. (2002) *J. Phys. Chem. B*, **106**, 2497–2501.

17 Chen, J., Tao, Z.-L., and Li, S.-L. (2003) *Angew. Chem.*, **115**, 2197–2201.

18 Yella, A., Mugnaioli, E., Panthöfer, M., Therese, H.A., Kolb, U., and Tremel, W. (2009) *Angew. Chem., Int. Ed.*, **48**, 6426–6430.

19 Schuffenhauer, C., Parkinson, B.A., Jin-Phillipp, N.Y., Joly-Pottuz, L., Martin,

J.-M., Popovitz-Biro, R., and Tenne, R. (2005) *Small*, **1**, 1100–1109.

20 Rosenfeld Hacohen, Y., Popovitz-Biro, R., Grunbaum, E., Prior, Y., and Tenne, R. (2002) *Adv. Mater.*, **14**, 1075–1078.

21 Rosenfeld Hacohen, Y., Popovitz-Biro, R., Prior, Y., Gemming, S., Seifert, G., and Tenne, R. (2003) *Phys. Chem. Chem. Phys.*, **5**, 1644–1651.

22 Ni, B., Liu, H., Wang, P.-p., He, J., and Wang, X. (2015) *Nat. Commun.*, **6**, 8756.

23 Zhuo, L., Ge, J., Cao, L., and Tang, B. (2009) *Cryst. Growth Des.*, **9**, 1–6.

24 Levi, R., Bar-Sadan, M., Albu-Yaron, A., Popovitz-Biro, R., Houben, L., Shahar, C., Enyashin, A., Seifert, G., Prior, Y., and Tenne, R. (2010) *J. Am. Chem. Soc.*, **132**, 11214–11222.

25 Albu-Yaron, A., Arad, T., Popovitz-Biro, R., Bar-Sadan, M., Prior, Y., Jansen, M., and Tenne, R. (2005) *Angew. Chem. Int. Ed.*, **44**, 4169–4172.

26 Kalay, S., Yilmaz, Z., Sen, O., Emanet, M., Kazanc, E., and Çulha, M. (2015) *Beilstein J. Nanotechnol.*, **6**, 84–102.

27 Panchakarla, L.S., Radovsky, G., Houben, L., Popovitz-Biro, R., Dunin-Borkowski, R.E., and Tenne, R. (2014) *J. Phys. Chem. Lett.*, **5**, 3724–3736.

28 Levy, M., Albu-Yaron, A., Tenne, R., Feuermann, D., Katz, E.A., Babai, D., and Gordon, J.M. (2010) *Israel J. Chem.*, **50**, 417–425.

29 Brontvein, O., Jayaram, V., Reddy, K.P.J., Gordon, J.M., and Tenne, R. (2014) *Z. Anorg. Allg. Chem.*, **640**, 1152–1158.

30 Liu, Z., Zhang, L., Wang, R., Poyraz, S., Cook, J., Bozack, M.J., Das, S., Zhang, X., and Hu, L. (2016) *Sci. Rep.*, **6**, 22503.

31 Rothschild, A., Sloan, J., and Tenne, R. (2000) *J. Am. Chem. Soc.*, **122**, 5169–5179.

32 Nath, M. and Rao, C.N.R. (2001) *Chem. Commun.*, 2236–2237.

33 Remskar, M., Mrzel, A., Skraba, Z., Jesih, A., Ceh, M., Demšar, J., Stadelmann, P., Lévy, F., and Mihailovic, D. (2001) *Science*, **292**, 479–481.

34 Rao, C.N.R. and Govindaraj, A. (2009) *Adv. Mater.*, **21**, 4208–4233.

35 Zak, A., Ecker, L.S., Efrati, R., Drangai, L., Fleischer, N., and Tenne, R. (2011) *Sens. Transducers*, **12**, 1–10.

36 Levi, R., Bar-Sadan, M., and Tenne, R. (2012) *Springer Handbook of Nanomaterials* (ed. R. Vajtai), Springer, New York.

37 Zak, A., Sallacan-Ecker, L., Margolin, A., Genut, M., and Tenne, R. (2009) *Nano*, **04**, 91–98.

38 Schuffenhauer, C., Popovitz-Biro, R., and Tenne, R. (2002) *J. Mater. Chem.*, **12**, 1587–1591.

39 Sadan, M.B., Houben, L., Enyashin, A.N., Seifert, G., and Tenne, R. (2008) *Proc. Natl. Acad. Sci. USA*, **105**, 15643–15648.

40 Parilla, P.A., Dillon, A.C., Jones, K.M., Riker, G., Schulz, D.L., Ginley, D.S., and Heben, M.J. (1999) *Nature*, **397**, 114–114.

41 Parilla, P.A., Dillon, A.C., Parkinson, B.A., Jones, K.M., Alleman, J., Riker, G., Ginley, D.S., and Heben, M.J. (2004) *J. Phys. Chem. B*, **108**, 6197–6207.

42 Bar-Sadan, M., Enyashin, A.N., Gemming, S., Popovitz-Biro, R., Hong, S.Y., Prior, Y., Tenne, R., and Seifert, G. (2006) *J. Phys. Chem. B*, **110**, 25399–25410.

43 Albu-Yaron, A., Levy, M., Tenne, R., Popovitz-Biro, R., Weidenbach, M., Bar-Sadan, M., Houben, L., Enyashin, A.N., Seifert, G., Feuermann, D., Katz, E.A., and Gordon, J.M. (2011) *Angew. Chem. Int. Ed.*, **50**, 1810–1814.

44 Wiesel, I., Arbel, H., Albu-Yaron, A., Popovitz-Biro, R., Gordon, J.M., Feuermann, D., and Tenne, R. (2010) *Nano Res.*, **2**, 416–424.

45 Yella, A., Panthöfer, M., Kappl, M., and Tremel, W. (2010) *Angew. Chem. Int. Ed.*, **49**, 2575–2580.

46 Brontvein, O., Stroppa, D.G., Popovitz-Biro, R., Albu-Yaron, A., Levy, M., Feuerman, D., Houben, L., Tenne, R., and Gordon, J.M. (2012) *J. Am. Chem. Soc.*, **134**, 16379–16386.

47 Panchakarla, L.S., Govindaraj, A., and Rao, C.N.R. (2010) *Inorg. Chimica Acta*, **363**, 4163–4174.

48 Sumino, K. and Imai, M. (1983) *Philos. Mag. A*, **47**, 753–766.

49 Friend, R.H. and Yoffe, A.D. (1987) *Adv. Phys.*, **36**, 1–94.

50 Deepak, F.L., Popovitz-Biro, R., Feldman, Y., Cohen, H., Enyashin, A., Seifert, G.,

and Tenne, R. (2008) *Chem. Asian. J.*, **3**, 1568–1574.

51 Zhu, Y.Q., Hsu, W.K., Firth, S., Terrones, M., Clark, R.J.H., Kroto, H.W., and Walton, D.R.M. (2001) *Chem. Phys. Lett.*, **342**, 15–21.

52 Hsu, W.K., Zhu, Y.Q., Yao, N., Firth, S., Clark, R.J.H., Kroto, H.W., and Walton, D.R.M. (2001) *Adv. Func. Mater.*, **11**, 69–74.

53 Nath, M., Mukhopadhyay, K., and Rao, C.N.R. (2002) *Chem. Phys. Lett.*, **352**, 163–168.

54 Tannous, J., Dassenoy, F., Bruhács, A., and Tremel, W. (2009) *Tribol. Lett.*, **37**, 83–92.

55 Woo, S.H., Yadgarov, L., Rosentsveig, R., Park, Y., Song, D., Tenne, R., and Hong, S.Y. (2015) *Israel J. Chem.*, **55**, 599–603.

56 Yadgarov, L., Stroppa, D.G., Rosentsveig, R., Ron, R., Enyashin, A.N., Houben, L., and Tenne, R. (2012) *Z. Anorg. Allg. Chem.*, **638**, 2610–2616.

57 Rapoport, L., Moshkovich, A., Perfilyev, V., Laikhtman, A., Lapsker, I., Yadgarov, L., Rosentsveig, R., and Tenne, R. (2011) *Tribol. Lett.*, **45**, 257–264.

58 Chhetri, M., Gupta, U., Yadgarov, L., Rosentsveig, R., Tenne, R., and Rao, C.N.R. (2015) *Dalton Trans.*, **44**, 16399–16404.

59 Ivanovskaya, V.V., Heine, T., Gemming, S., and Seifert, G. (2006) *Phys. Status Solidi B*, **243**, 1757–1764.

60 Tiong, K.K., Liao, P.C., Ho, C.H., and Huang, Y.S. (1999) *J. Cryst. Growth*, **205**, 543–547.

61 Deepak, F.L., Cohen, H., Cohen, S., Feldman, Y., Popovitz-Biro, R., Azulay, D., Millo, O., and Tenne, R. (2007) *J. Am. Chem. Soc.*, **129**, 12549–12562.

62 Yadgarov, L., Rosentsveig, R., Leitus, G., Albu-Yaron, A., Moshkovich, A., Perfilyev, V., Vasic, R., Frenkel, A.I., Enyashin, A.N., Seifert, G., Rapoport, L., and Tenne, R. (2012) *Angew. Chem. Int. Ed.*, **51**, 1148–1151.

63 Chhetri, M., Gupta, U., Yadgarov, L., Rosentsveig, R., Tenne, R., and Rao, C.N.R., (2016) *ChemElectroChem*. doi: 10.1002/celc.201600291.

64 Rosentsveig, R., Gorodnev, A., Feuerstein, N., Friedman, H., Zak, A., Fleischer, N.,

Tannous, J., Dassenoy, F., and Tenne, R. (2009) *Tribol. Lett.*, **36**, 175–182.

65 Zhu, Y.Q., Sekine, T., Brigatti, K.S., Firth, S., Tenne, R., Rosentsveig, R., Kroto, H.W., and Walton, D.R.M. (2003) *J. Am. Chem. Soc.*, **125**, 1329–1333.

66 Joly-Pottuz, L., Martin, J.M., Dassenoy, F., Belin, M., Montagnac, G., Reynard, B., and Fleischer, N. (2006) *J. Appl. Phys.*, **99**, 023524.

67 Tang, D.-M., Wei, X., Wang, M.-S., Kawamoto, N., Bando, Y., Zhi, C., Mitome, M., Zak, A., Tenne, R., and Golberg, D. (2013) *Nano Lett.*, **13**, 1034–1040.

68 Kaplan-Ashiri, I. and Tenne, R. (2015) *JOM*, **68**, 151–167.

69 Zak, A., Sallacan-Ecker, L., Margolin, A., Feldman, Y., Popovitz-Biro, R., Albu-Yaron, A., Genut, M., and Tenne, R. (2010) *Fuller. Nanotub. Car. N.*, **19**, 18–26.

70 Adini, A.R., Feldman, Y., Cohen, S.R., Rapoport, L., Moshkovich, A., Redlich, M., Moshonov, J., Shay, B., and Tenne, R. (2011) *J. Mater. Res.*, **26**, 1234–1242.

71 Samorodnitzky-Naveh, G.R., Redlich, M., Katz, A., Adini, A.R., Gorodnev, A., Rapoport, L., Moshkovich, A., Cohen, S.R., Rosentsveig, R., Moshonov, J., Shay, B., and Tenne, R. (2010) *Int. J. Nano Biomater.*, **3**, 140–152.

72 Tevet, O., Von-Huth, P., Popovitz-Biro, R., Rosentsveig, R., Wagner, H.D., and Tenne, R. (2011) *Proc. Natl. Acad. Sci. USA*, **108**, 19901–19906.

73 Lahouij, I., Dassenoy, F., Vacher, B., and Martin, J.-M. (2011) *Tribol. Lett.*, **45**, 131–141.

74 Lahouij, I., Bucholz, E.W., Vacher, B., Sinnott, S.B., Martin, J.M., and Dassenoy, F. (2012) *Nanotechnology*, **23**, 375701.

75 Rabaso, P., Ville, F., Dassenoy, F., Diaby, M., Afanasiev, P., Cavoret, J., Vacher, B., and Le Mogne, T. (2014) *Wear*, **320**, 161–178.

76 Tannous, J., Dassenoy, F., Lahouij, I., Mogne, T., Vacher, B., Bruhács, A., and Tremel, W. (2010) *Tribol. Lett.*, **41**, 55–64.

77 Sedova, A., Ron, R., Goldbart, O., Elianov, O., Yadgarov, L., Kampf, N., Rosentsveig, R., Shumalinsky, D., Lobik, L., Shay, B.,

Moshonov, J., Wagner, H.D., and Tenne, R. (2015) *Nanomater. Energy*, **4**, 30–38.

78 Zhang, W., Ge, S., Wang, Y., Rafailovich, M.H., Dhez, O., Winesett, D.A., Ade, H., Shafi, K.V.P.M., Ulman, A., Popovitz-Biro, R., Tenne, R., and Sokolov, J. (2003) *Polymer*, **44**, 2109–2115.

79 Rapoport, L., Nepomnyashchy, O., Verdyan, A., Popovitz-Biro, R., Volovik, Y., Ittah, B., and Tenne, R. (2004) *Adv. Eng. Mater.*, **6**, 44–48.

80 Naffakh, M. and Díez-Pascual, A. (2014) *Inorganics*, **2**, 291–312.

81 Naffakh, M., Díez-Pascual, A.M., Marco, C., Ellis, G.J., and Gómez-Fatou, M.A. (2013) *Prog. Polym. Sci.*, **38**, 1163–1231.

82 Naffakh, M., Martín, Z., Fanegas, N., Marco, C., Gómez, M.A., and Jiménez, I. (2007) *J. Polym. Sci. B*, **45**, 2309–2321.

83 Naffakh, M., Marco, C., Gómez, M.A., and Jiménez, I. (2008) *J. Phys. Chem. B*, **112**, 14819–14828.

84 Naffakh, M., Marco, C., Gómez, M.A., Gómez-Herrero, J., and Jiménez, I. (2009) *J. Phys. Chem. B*, **113**, 10104–10111.

85 Naffakh, M., Diez-Pascual, A.M., Remskar, M., and Marco, C. (2012) *J. Mater. Chem.*, **22**, 17002–17010.

86 Naffakh, M., Díez-Pascual, A.M., Marco, C., Gómez, M.A., and Jiménez, I. (2010) *J. Phys. Chem. B*, **114**, 11444–11453.

87 Díez-Pascual, A.M. and Naffakh, M. (2012) *Polymer*, **53**, 2369–2378.

88 Shneider, M., Dodiuk, H., Kenig, S., and Tenne, R. (2010) *J. Adhes. Sci. Technol.*, **24**, 1083–1095.

89 Shneider, M., Dodiuk, H., Tenne, R., and Kenig, S. (2013) *Polym. Eng. Sci.*, **53**, 2624–2632.

90 Sedova, A., Bar, G., Goldbart, O., Ron, R., Achrai, B., Kaplan-Ashiri, I., Brumfeld, V., Zak, A., Gvishi, R., Wagner, H.D., and Tenne, R. (2015) *J. Supercrit. Fluids*, **106**, 9–15.

91 Huang, S.-J., Ho, C.-H., Feldman, Y., and Tenne, R. (2016) *J. Alloys Compd.*, **654**, 15–22.

92 Nadiv, R., Shtein, M., Peled, A., and Regev, O. (2015) *Constr. Build. Mater.*, **98**, 112–118.

93 Voldman, A., Zbaida, D., Cohen, H., Leitus, G., and Tenne, R. (2013) *Macromol. Chem. Phys.*, **214**, 2007–2015.

94 Bruno, A., Borriello, C., Haque, S.A., Minarini, C., and Di Luccio, T. (2014) *Phys. Chem. Chem. Phys.*, **16**, 17998–18003.

95 Wiegers, G.A. (1996) *Prog. Solid State Chem.*, **24**, 1–139.

96 Shikano, M. and Funahashi, R. (2003) *Appl. Phys. Lett.*, **82**, 1851–1853.

97 Liu, C.-J., Huang, L.-C., and Wang, J.-S. (2006) *Appl. Phys. Lett.*, **89**, 204102.

98 Putri, E., Wan, C., Wang, Y., Norimatsu, W., Kusunoki, M., and Koumoto, K. (2012) *Scripta Mater.*, **66**, 895–898.

99 Jood, P., Ohta, M., Nishiate, H., Yamamoto, A., Lebedev, O.I., Berthebaud, D., Suekuni, K., and Kunii, M. (2014) *Chem. Mater.*, **26**, 2684–2692.

100 Hicks, L.D. and Dresselhaus, M.S. (1993) *Phys. Rev. B*, **47**, 16631–16634.

101 Snyder, G.J. and Toberer, E.S. (2008) *Nat. Mater.*, **7**, 105–114.

102 Tritt, T.M. and Subramanian, M.A. (2006) *MRS Bull.*, **31**, 188–198.

103 Boukai, A.I., Bunimovich, Y., Tahir-Kheli, J., Yu, J.-K., and Goddard Iii, W.A., and Heath, J.R. (2008) *Nature*, **451**, 168–171.

104 Hochbaum, A.I., Chen, R., Delgado, R.D., Liang, W., Garnett, E.C., Najarian, M., Majumdar, A., and Yang, P. (2008) *Nature*, **451**, 163–167.

105 Hong, S.Y., Popovitz-Biro, R., Prior, Y., and Tenne, R. (2003) *J. Am. Chem. Soc.*, **125**, 10470–10474.

106 Radovsky, G., Popovitz-Biro, R., Staiger, M., Gartsman, K., Thomsen, C., Lorenz, T., Seifert, G., and Tenne, R. (2011) *Angew. Chem. Int. Ed.*, **50**, 12316–12320.

107 Bernaerts, D., Amelinckx, S., Van Tendeloo, G., and Van Landuyt, J. (1997) *J. Cryst. Growth*, **172**, 433–439.

108 Gomez-Herrero, A., Landa-Canovas, A.R., and Otero-Diaz, L.C. (2000) *Micron*, **31**, 587–595.

109 Radovsky, G., Popovitz-Biro, R., Stroppa, D.G., Houben, L., and Tenne, R. (2014) *Acc. Chem. Res.*, **47**, 406–416.

110 Radovsky, G., Popovitz-Biro, R., and Tenne, R. (2014) *Chem. Mater.*, **26**, 3757–3770.

111 Radovsky, G., Popovitz-Biro, R., and Tenne, R. (2012) *Chem. Mater.*, **24**, 3004–3015.

112 Radovsky, G., Popovitz-Biro, R., Lorenz, T., Joswig, J.-O., Seifert, G., Houben, L., Dunin-Borkowski, R.E., and Tenne, R. (2016) *J. Mater. Chem. C*, **4**, 89–98.

113 Panchakarla, L.S., Popovitz-Biro, R., Houben, L., Dunin-Borkowski, R.E., and Tenne, R. (2014) *Angew. Chem. Int. Ed.*, **53**, 6920–6924.

114 Panchakarla, L.S., Lajaunie, L., Tenne, R., and Arenal, R. (2016) *J. Phys. Chem. C*, **120**, 15600–15607.

115 Brontvein, O., Albu-Yaron, A., Levy, M., Feuerman, D., Popovitz-Biro, R., Tenne, R., Enyashin, A., and Gordon, J.M. (2015) *ACS Nano*, **9**, 7831–7839.

116 Lorenz, T., Joswig, J.-O., and Seifert, G. (2014) *Semicond. Sci. Technol.*, **29**, 064006.

32

Nanotechnology, Energy, and Fractals Nature

Vojislav V. Mitić,[1,2] Ljubiša M. Kocić,[1] Steven Tidrow,[3] and Hans-Jörg Fecht[4]

[1]*University of Niš, Univerzitetski trg 2, 18000 Niš, Serbia*
[2]*Institute of Technical Sciences of Serbian Academy of Sciences, Knez Mihailova 35, 11000 Belgrade, Serbia*
[3]*Alfred University, 1 Saxon Drive, Alfred, NY 14802, USA*
[4]*University of Ulm, Albert-Einstein-Allee 47, D-89081 Ulm, Germany*

32.1
Introduction

Even before the beginning of known civilization(s), humans have utilized matter or material(s) to improve their standard of living and quality of life. For instance, charcoal was used by prehistoric humans to create drawings on cave walls that simultaneously convey knowledge, material science, engineering, history, and art in an interesting example of the early interdisciplinary ability of humans. At present, one of the greatest challenges, if not the greatest challenge, for material scientists and engineers is to accelerate the discovery and development of materials that can be utilized to address the myriad of challenges that civilization faces with regard to water, energy, environment, medicine, affordability, economy, and so on, while simultaneously sustaining or continuing to increase the standard of living and quality of life of the growing population. As the most important international energy agencies anticipate, the global energy demands in the middle of the twenty-first century should be doubled. A certain part of this energy will be provided by the renewable and alternative energy sources (water, wind, solar, bio, etc.) and other innovative energy applications with increasing trends. From all of these points of view, nanotechnology would contribute to the sphere of energy growth.

Material models and methods that a priori accurately predict material parameters as a function of temperature are needed and can be used to reduce the time and cost of discovering and developing affordable materials with properties that both outperform materials in current technologies and provide novel properties upon which new technologies can evolve. Such material models can be used to more effectively overcome the myriad of challenges that society faces.

Nanotechnology for Energy Sustainability, First Edition. Edited by Baldev Raj, Marcel Van de Voorde, and Yashwant Mahajan.
© 2017 Wiley-VCH Verlag GmbH & Co. KGaA. Published 2017 by Wiley-VCH Verlag GmbH & Co. KGaA.

Many of the properties of solid-phase materials are governed by the arrangement of the atoms. The unit cell is the smallest group of atoms that can generate the entire crystal by translation in 3D. Physical properties derived/computed on the unit cell can easily be extrapolated to entire crystal, except for the perturbations introduced by defects. Similarly, most of the physical and mechanical behavior of practical engineering components and structures can be analyzed taking Euclidian geometric dimensions. However, unlike the orderly unit cells at the atomic scale and the Euclidian surfaces at macroscale, the surface of nanostructures is characterized by randomness and is difficult to characterize quantitatively. In order to analyze these nanostructures, many authors proposed a fractal approach. The notable trend in the recent literature is that a wide range of disordered systems, for example, linear and branched polymers, biopolymers, epoxy resins, and percolation clusters can be characterized by the fractal nature over a microscopic correlation length. It is favorable to the fact that energy transformations are permitted on a small scale [1–3].

This chapter will discuss the role of fractal geometry and analysis in energetic questions from the nanotechnology point of view.

32.2
Short Introduction to Fractals

This chapter, which discusses the role of fractal geometry and analysis with regard to material and material properties, is geared toward the science and development of technology(ies), including nanotechnology(ies), which can be used to address projected trends of energy demand. At least three items are essential with regard to addressing energy: (i) identification of the location(s) and quantity(ies) of freely stored energy reservoirs, (ii) *energy harvesting*, and (iii) *short- and long-term energy storage*. Each of these three items has *fractals* as a common point or intersection. Energy, although not usually viewed as having the properties of a physical object, does have a mapping within fractals as though energy has the property of being a physical object. In other words, fractals, a crucial construct within modern physics, can be used to enhance the cultivation of energy.

Fractals are geometric objects having broken, fragmented, wrinkled, or amorphous form or being highly irregular in some other way. Euclid's standard geometry fails to describe such objects so that they are subjects of *fractal geometry*. The term fractal, first used by Mandelbrot [4] as a neologism derived from the Latin adjective *fractus*, means *fragmented, irregular*.

The *topological dimension* D_T describes common, intuitive dimension(s) in Euclidian geometry with $D_T = 1$ for curves, 2 for surfaces, 3 for solids, and so on. Note that Euclidian dimensionality as described using integers is strictly smaller than the Hausdorff dimension D_H, which is a natural extension of D_T. Unlike the topological dimension, D_T, the Hausdorff or fractal dimension D_H is typically a noninteger for a fractal object and maps into Euclidean space when $D_H = D_T$. For

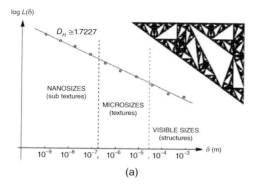

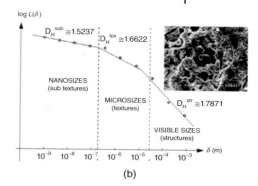

Figure 32.1 The Richardson–Mandelbrot diagram for (a) mathematical fractal (Conway triangle, $D_H \approx 1.7227$) and (b) real fractal – the SE microphotograph of BaTiO$_3$ ceramics doped by 0.1 wt% of holmium (Ho$_2$O$_3$). Fractal dimensions differ in different ranges of magnitude, as expected from real fractals (prefractals).

example, a calm liquid surface has $D_T = D_H = 2$, that is, the superficial layer of liquid molecules can be approximated as a mathematical plane. Any disturbance, for example, by heating, will make the surface geometry more complicated, with $2 < D_H \leq 3$. The upper limit, $D_H = 3$, can occur through evaporation of all liquid particles, transforming the planar layer into a three-dimensional space.

It may be noted that fractal dimension is not a unique descriptor. It is likely that fractal structures or patterns can have the same D but are dramatically different. There are many other fractal descriptors (e.g., *spectral dimension D_S* [5], see in the next section) based on different qualities of materials such as textures, contours, shapes [6], and so on.

Many fractals possess strict geometric self-similarity, they are invariant upon transformation of similarity (*homothety*), such as *Cantor set, Koch snowflake, Sierpinski triangle, Conway triangle* (Figure 32.1a), *Menger sponge,* and so on. But many more are made of parts that resemble the whole in some other way. Sometimes, the resemblance may be only approximate or statistical (*Mandelbrot set* and *Julia sets*). Unlike these ideal (mathematical) fractals, *physical (real) fractals* are approximately close to their ideal counterparts; they are called *prefractals* (or *proto-fractals*) and typical examples include trees, forests, earth relief, microparticles, pores labyrinth in porous materials, and so on.

The Hausdorff distance is given by

$$d_h(A, \quad B) = \max\{\max\{d(y, A), y \in B\}, \max\{d(x, B), x \in A\}\}, \qquad (32.1)$$

where $d(s, A) = \min\{d(s, t), t \in A\}$ is the distance of a point s to the set A, and $d(s, t)$ is the Euclid distance between two points (see [7]). The Hausdorff distance is used to compare the closeness of a prefractal B to that of an ideal fractal A.

Having in mind the notion of Hausdorff distance d_h, it is not difficult to understand that *self-similarity* that is equivalent to *scale invariance* is the direct consequence of *contractive (or dilation) symmetry* [8]. Namely, the usual meaning of *being symmetric* understands that for a symmetric object S, there exists transformation T, so that $T(S) = S$, provided that T has unit norm, $\|T\| = 1$. This means that T does not change the size of the object, simply transforms the object inside itself. On the other hand, for fractal object S, there exists transformation T that satisfies

$$\|T\| < 1, \text{ and } T(S) = S. \tag{32.2}$$

In this case, T is called transformation of dilation symmetry.

Fractal objects can be classified by their fractal dimension, such as fractal dots (usually called *fractal dust*) with $D_T = 0$ and $0 < D_H < 1$ (e.g., Cantor set, $D_H \approx 0.6309$), fractal lines $D_T = 1$ and $1 < D_H < 2$ (e.g., Koch snowflake, $D_H \approx 1.2619$ [1]), fractal surface $D_T = 2$ and $2 < D_H < 3$ (e.g., Menger sponge, $D_H \approx 2.7268$), and so on [7,9].

32.3
Nanosizes and Fractals

The most important fractal object property is that it preserves its "degree of fractalness," the property expressed by (Hausdorff) fractal dimension, in all scales of magnitude. This means that if some feature exists in the scale of millimeters, the same type of feature will be observable down to the nanoscale. Since there are no mathematical fractals in the real world, this scaling feature fails at or below the element level or elementary particle size due to the discrete nature of matter. Practically, if some feature exists in the scale of millimeters, the same will be observable in nanosizes. It is illustrated (Figure 32.1) by using the concept of evaluating fractal dimension with the *box counting* method [1,2,4].

The box counting method is the most popular algorithm for estimating fractal dimension, as illustrated using the Richardson–Mandelbrot diagram. The object or image of a chosen object, in this case a quadratic triangle, is overlaid with a regular continuously connected grid of boxes with each box side having length modulus δ, square box. As the box size is decreased through decreasing the modulus, δ_n, starting with $\delta_1 = 1$, the number of boxes covering or partially covering the object are counted. For simplicity, assume that the object or its image is all in black and white, like the Conway triangle (also, *pinwheel fractal*) [10], shown in Figure 32.1a, with the black color representing the object or image of the object. If at least one black point lies within the box, it is counted as containing the object. As δ_n decreases, a larger number of boxes will cover the object or object's image. The number of boxes covering the object or object's image becomes $N(\delta_n) = N_n$ for modulus $\delta = \delta_n$.

The set of pairs $\{(\delta_n, N_n), n = 1, \ldots, m\}$ is connected with (Hausdorff) fractal dimension D_H, through the relation

$$N_n = K\delta_n^{-D_H}, \tag{32.3}$$

where K is a constant. Taking the logarithm of the equation, with the logarithm base being arbitrary, yields

$$\ln N_n = \ln K - D_{\mathrm{H}} \ln \delta_n, \tag{32.4}$$

which is a linear function in the $(\ln \delta_n, \quad \ln N_n)$ – coordinate system. Within such a coordinate system, the data points, $\{(\ln \delta_n, \ln N_n), n = 1, \ldots, m\}$, should be a linear function having a line slope coefficient $-D_{\mathrm{H}}$. Further, as $n \to \infty$ or the size of boxes goes toward zero, Eq. (32.4) yields the Hausdorff fractal dimension to be as follows [1,4]:

$$D_{\mathrm{H}} = \lim_{n \to \infty} \frac{\ln N_n}{\ln(1/\delta_n)}. \tag{32.5}$$

Thus, the box counting method, a simple straightforward method that can be automated through computer algorithm, is well suited for the determination of fractal dimensionality.

Figure 32.1 shows two typical examples. Figure 32.1a shows a genuine fractal (Conway triangle), from which construction follows that $N_n = 2^{4n}$ and $\delta_n = 5^{-n}$, wherefrom Eq. (32.5) gives. Note that within the Richardson–Mandelbrot diagram (Figure 32.1a), the data points (δ_n, N_n) follow a linear curve, which occurs whenever an object has ideal fractal or Euclidean dimensionality. Further, note that the linear trend of the data points extends over the entire metric scale, from macro- to micro- to nanoscale, which from Eq. (32.2) illustrates dilation symmetry. If a natural object or product of technology shows some extent of scale invariance, it means that there exists a contractive symmetry transformation that more or less closely fulfills relation (32.2). The degree of closeness depends on the shape of the Richardson–Mandelbrot diagram for a real fractal, like Figure 32.1b, which does not possess a continuous straight line like that of an ideal fractal, like Figure 32.1a. As noted by Kaye [11], real diagrams are typically piecewise linear, globally making a concave graph, possessing subranges of linearity. For the case of Figure 32.1b, there are three linear subranges that roughly correspond to *visual sizes* $> 4 \cdot 10^{-5}$ m, the smallest dimension a human eye can distinguish, *microsizes*, from $4 \cdot 10^{-5}$m down to a few hundreds of nanometers, and *nanosizes*, all the way down to angstroms.

According to Kaye [11], the real material configuration corresponding (roughly) to visual sizes, representing *microscale*, is called *structure* and represents *texture*. Accordingly, the range of *nanoscale* expresses the material's *subtexture*.

From the discussions above, there is an important conclusion: Every real (physical) fractal or *prefractal* transfers veraciously its morphology across scales from millimeter to micrometer range or from micrometer to nanometer range, although the fractal dimensionality may change slightly in a concave manner as the modulus δ for box counting approaches zero. In other words, the degree of fractality, the closeness to being an ideal fractal, influences regularity in morphology distribution throughout the wider range. In terms of fitting diagrams, the closer the box counting fit of diagram (Figure 32.1b) is to the straight line, the greater the preservation of fractal properties. Such linearity of the

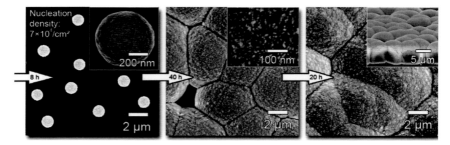

Figure 32.2 Nanocrystalline diamond nucleation and early-stage growth on an untreated silicon substrate [12,13].

dimensionality is of high practical importance, especially in the domain of nano-technologies, since many operative fractal techniques at the nanoscale use methods similar to fractal-building algorithms. A striking example is provided from the investigation of the formation of diamond films, Figure 32.2 [12,13].

Within 3 years of the pioneering book of Mandelbrot on fractals [4], Messier and Yehoda [14] reported about fractal morphology of vapor-deposited thin films and stated: "Such morphological structures are found to have a strong similarity in shape and form over six orders of magnitude in film thickness." Examining the pyrolytic graphite deposits, these authors noticed the so-called "cauliflower" morphology of the surface of many films. With regard to the process morphology produced by vapor deposition, Serov [15] has pointed out that "various kinds of orderliness are observed, dome-shaped structures can be built in chains, form linear and cross-linked structures, and can form precise helicoids structures."

In general, diamond films produced by conventional chemical vapor deposition (CVD) consist of micrometer-size crystallites having a columnar growth structure. Due to rather chaotic processes, such films suffer from high surface roughness (>>100 nm) and low fracture toughness. Through improved methods, nanocrystalline chemical depositions (NCDs) and ultrananocrystalline chemical depositions (UNCDs), with film surface roughness of nanometric dimensions, grain sizes of 5–100 nm for NCD and 3–5 nm for UNCD, show superior mechanical strength and excellent tribological properties [16]. Although terminology like "*nanofractal*" and "*microfractal*" is used in the literature, since fractals scale from micro- to nanoscale, a more appropriate term or phrase to use in place of "nanofractal" and "microfractal" is "nanogeometry."

In addition to the topological fractal (Hausdorff) dimension D_H, which incorporates the Euclid dimension D_T when D_H takes on an integer value, the spectral dimension [17], D_S, is an important parameter for characterizing the fractal environment. The spectral dimension, D_S, is defined as the return probability of the random walk, that is, the probability of being at the origin at time t or after n steps, $P(0, t) \propto t^{-(1/2)D_S}$ and is related to D_H through the relation:

$$\frac{2D_H}{1 + D_H} \leq D_S \leq D_H.$$

The spectral dimension describes the dynamical rate of spreading material through a fractal structure over time. Each moving molecule or other particle has a certain number of closest neighbors, which determines the rate at which the moving material diffuses. This kind of diffusion, called *anomalous diffusion* is characterized by a nonlinear relationship to time, where the mean squared displacement $\langle x^2(t) \rangle$ deviates from the linear time dependence and obeys the power law rule:

$$\langle x^2(t) \rangle \propto t^{\alpha}.$$

For $\alpha > 1$, the phenomenon is called *superdiffusion* and *subdiffusion*, provided $\alpha < 1$. Superdiffusion can be the result of active cellular transport processes. From the energetic point of view, the especially important case is diffusion through porous media.

Fractals have dilatation symmetry, that is, they are invariant under scaling. This means that an emergence of energy on a fractal-shaped physical object can be considered in any scale from subnanometer to distances of light years [4,8]. On the other hand, time as a quantity conjugate to energy has translation symmetry on all feasible scales. Moreover, the principle of energy conservation follows from symmetry under time translations, which is the consequence of the *first Noether theorem* published in German by Emmy Noether [18] (see additional Ref. [71] for English translation).

This attribute implies the conservation of energy law is consistent with recent notable trends of fractal characterization over microscopic correlation lengths of a wide range of disordered systems, for example, linear and branched polymers, biopolymers, epoxy resins, and percolation clusters. It is favorable to the fact that energy transformations are permitted on a small scale [1–3].

As far as the practical side of determining fractal dimension is concerned, it might be useful to mention our experience in fractal microelectronic, especially specific intergranular relations within grain surface coatings. A practical question for fractals is what is the measurement range in which fractality can be identified? By generalization, the minimum information needed ranges over at least three orders of magnitude, for example, on the level of $1–100\,\mu m$ with a more acceptable model providing six to seven orders of magnitude. Since our typical grain size is on average about $50\,\mu m$, we need magnifications that reveal roughly 50 grains, $3000\,\mu m$, $300\,\mu m$, $30\,\mu m$ (roughly 1 grain), $3\,\mu m$, $300\,nm$, and finally $30\,nm$. Such scaling investigations will be enough for fractal dimension D_H to be estimated with an error of 3–5%. The error can be reduced down to 1% by combining successive magnification with some geometric actions, for example, by SEM picture rotations for some angle and by procedure repeating. The other way is to analyze several details, at the same time (for early attempts see [19–21]). Ultimately, the accuracy in fractal dimension estimating strongly depends on the proper choice of the scaling range. For example, if one needs to determine D_H for an oak leaf, it is pointless to observe the oak from the distance of $5\,km$, since the oak canopy will look like a point to the naked eye. Also, magnification of the leaf surface to the nanometer level will not reveal any information about the leaf as a whole.

32.4
Energy and Fractals

World's need for energy has imposed the whole spectra of technological challenges that further transform to scientific tasks. This chapter discusses the role of fractal geometry and analysis in energetic questions. In fact, some of the early fractal applications have been used as a tool in energy research, applying on diverse energy technologies, ranging from photovoltaics [22] to fuel cells [23] and carbon capture [24].

Three items are essential regarding energetic questions: (i) free energy stock location, (ii) energy harvesting, and (iii) short- and long-term energy storage.

All three items have their specific common points with fractals. Also, the concept of energy as a property of physical objects shares some features characteristic to fractal objects. In other words, fractal, as a crucial concept of modern theoretical physics, is closely connected with the process of "cultivating" the "wild" energy.

Let us list what do fractals and energy have in common.

i) Both fractals and energy are properties; while energy is a physical object's property, fractality is a geometrical one;

ii) As well as energy, fractality is omnipresent, an unlimited entity concerning space and time. There is a strong belief and many evidences that the whole Universe is permeated both by fractal structures and by energetic fields;

iii) Both entities, energy and fractality, exist in micro- and macroworld. Any fractal is dividable down to nanoscale, so is the energy field;

iv) Energy is closely connected with geometry structures and especially with nanogeometry. Some energetic extremal problems as a solution have smooth, symmetric Euclidean objects, another have fractals;

v) If an energetic situation can be described by fractals, then the potential energy corresponds to fractals constant in time, while kinetic energy needs time-dependent fractals;

vi) The energy spectrum and energy spectral density of some time series (signals), by the rule, have a fractal structure.

Let discuss the above items. In (i), the term "geometry," as is the custom in plain language, means "shape" rather than the "science of geometry". In this sense, geometry is a property. And it is more present in everyday life than we are usually aware of. Just note that all our senses often convey information on the quality of some matter by interacting with some geometry or shape. The touch feeling of smooth or rough surface, olfactory, or taste data generate according to geometry of particles. With fair precision, our eye can distinguish metal from fur or wood from feather.

Connected with (ii), much misunderstanding has been caused by the concept "fractals everywhere" being chosen by Barnsley [7] as the title of his famous book. This question was treated in the above discussion connected with Figure 32.1 and by separating "ideal (or mathematical) fractal" from "real, physical fractal." The main issue here is that any real fractal such as a ball of crumpled paper or a neuron cell is closer to ideal fractal than to ideal Euclidean object, measuring with Hausdorff distance (1).

The practical value of the item (iii) is that the fractal objects' interaction and energy is possible at any reasonable scale of magnitude, including nanoscale. This is a consequence of fractal independence on scales.

The relation energy geometry and energy nanogeometry (iv), the question of central importance for this text, is scientifically very both challenging and involving at the same time. Just consider the two opposed energy problems. (a) Surface tension, which measures the energy needed to create a surface, makes the soap film surface in free space to take a spherical form. In this way, the film deforms to minimize its surface and thus its energy. (b) In order to supply live tissues uniformly, there exist in nature various types of drain capillaries, veins and vessels. They serve to distribute oxygenated blood from arteries to the tissues of the body and to feed deoxygenated blood from the tissues back into the veins.

Another symmetric branching structure is shown in Figure 32.3. It is the fourth iteration of the so-called *Sierpinski quadratic curve*, having the L-system code [25]:

$$\delta \quad = 90°$$
$$\text{axiom} = F + XF + F + XF$$
$$\text{rule} \quad = X \rightarrow XF - F + F - XF + F + XF - F + F - X.$$

It is immediately clear that this curve, having many branches, can serve as a paradigm for an energy collector or energy emitter. The principle that causes

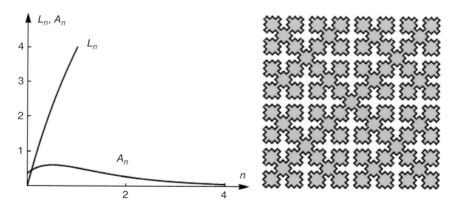

Figure 32.3 The Sierpinski square curve ($D_H = 2$, when $n \rightarrow \infty$) displays typical behavior of fractal object. While one dimension (length L_n) increases unlimitedly, another one (area it covers, A_n) decreases to zero.

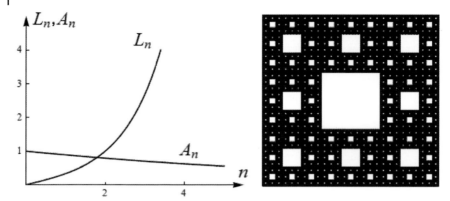

Figure 32.4 The Sierpinski carpet ($D_H \approx 1.8928$, when $n \to \infty$). While the length L_n (the sum of circumferences of all white squares) increases to infinity, the black area, A_n decreases to zero.

such branched geometry is simple. The extensive length of the circumference embraces a very small area. Figure 32.3 shows how the length quickly increases to infinity with the number of iterations, while the area comprised tends to zero.

Sometimes, the branching property is not visible at a glance. Being fractal by definition, branching can be functional in the way that it represents the essence of constructing a fractal object that is representable by a tree type graph of dynamical addresses for each point. Such objects are displayed in Figures 32.4 and 32.5. The first one, known as *Sierpinski carpet* [4] (Figure 32.4) has similar property to Sierpinski quadratic curve, that is, the circumference unlimitedly increases while the area vanishes with increasing n.

The 3D extension of the Sierpinski carpet is the Menger sponge [4] (Figure 32.5). Here, the area increases rapidly with n, making the body of the sponge thinner until it goes to 0.

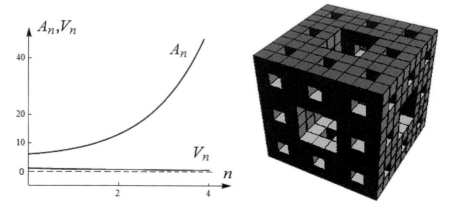

Figure 32.5 The Menger sponge fractal ($D_H \approx 2.7268$ when $n \to \infty$). In this case, the area A_n increases unlimitedly, while the volume V_n decreases to zero.

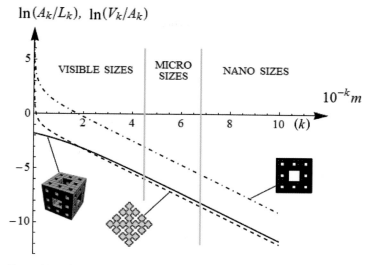

Figure 32.6 The ratios area versus length A_n/L_n for Sierpinski quadratic curve (Figure 32.2) and Sierpinski carpet (Figure 32.3) as well as the ratio V_n/A_n for Menger sponge (Figure 32.4) in logarithmic scale and with $n = k \log_3 10$.

Both Sierpinski carpet and the Menger sponge are nice examples of electrodes in electrolytic batteries. On the other hand, the Sierpinski square curve is used as a model of broadband antennas.

The ratio A_k/L_k for Sierpinski square curve and Sierpinski carpet, or V_k/A_k for Menger sponge, in logarithmic scale is shown in Figure 32.6. Note that all ratios decrease steadily and exponentially toward zero.

Another moment is also of interest. If some flow (fluid, flow of particles, electrons, etc.) uses the winding path, it spends more energy as if it goes straight. It can be expressed by evaluating the energy of the piecewise linear curve:

$$E_{fc} = \sum_{i=1}^{n-2} |\Delta^2 v_i|^2,$$

where v_i is the vector containing the two- or three-dimensional vertices data of the curve and $\Delta^2 v_i$ is the second forward difference $\Delta^2 v_i = v_{i+2} - 2 v_{i+1} + v_i$. This "geometrical energy" directly correlates to mechanical potential energy that can be released under some circumstances. On the other hand, E_{fc} reflects the "capacity" for energy storage or energy exchange of a prefractal object. Figure 32.7a visualizes this energy for increasing iterations degree n of prefractals of Sierpinski quadratic curve and the Koch snowflake.

One of the cleanest, most sustainable ways to generate power is wind energy. It does not produce any of the toxic and heat-trapping emissions that contribute to global warming (aside from the embodied energy and emissions associated with manufacturing). The fluctuation of wind speed affects energy conversion of

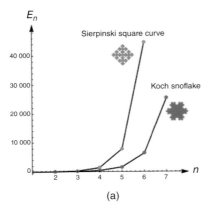

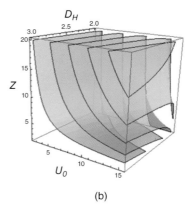

(a)   (b)

Figure 32.7 (a) Energy of fractal curves as function of iteration number *n*. The higher fractal dimension ensures higher energy. (b) The wind profile logarithmic law with fractal correction (32.6) showing the average wind speed as function of friction velocity u_0, height z, and fractal dimension of the surface D_H [26,27].

wind energy systems since wind power is proportional to the cube of wind speed. A little change of the wind speed might cause extreme instability in wind energy systems. Therefore, an accurate prediction of wind speed is essential in order to improve performances of wind energy systems. It is well known from the air fluid experiments that the turbulent stress acts like a viscous stress, that is, it is directly proportional to the gradient of mean horizontal wind velocity $\bar{u}$, $\tau'_{zx} = \eta' \, \partial\bar{u}/\partial z$, where η' is the coefficient of turbulent viscosity.

Taking account of the friction velocity $u_0 = \sqrt{\tau_\omega/\rho}$, representing the *shear stress* term τ_ω for the fluid of density ρ, (κ is the Karman's constant), one gets the so-called wind profile logarithmic law $\bar{u}(z) = (u_0/\kappa)\ln(z/z_0)$, and z_0 is known as roughness length (or roughness height), which is the height (above the surface) where the wind has zero speed. Having known the fractal dimension D_H of the surface (thus, $2 < D_H < 3$) over which the wind blows, the following fractal correction is reasonable (Figure 32.7b) [26,27].

$$\bar{u}(z) = \frac{u_0}{\kappa} \ln \frac{z}{z_0} (3 - D_H). \tag{32.6}$$

A similar rule holds for liquid fluids flowing through channels or tubes. Some analogy with rheological behavior of nanofluids can be established. A nanofluid is a base fluid containing colloidal suspension of nanoparticles. If μ_b denotes the viscosity of the base fluid, μ_n the viscosity of the nanofluid, ϕ volume fraction of nanoparticle in base mixture, and ϕ_m the maximum volume fraction of nanoparticles and depends on the aspect ratio of the nanotubes, then $\mu_n = \mu_b(1 - \phi/\phi_m)^{-2}$.

Applying the microrheological model for fractal aggregation in shear flow, the application of the fractal concept can be used to predict the viscosity of nanofluid at high shear rate. Considering that the geometry of the aggregates can be

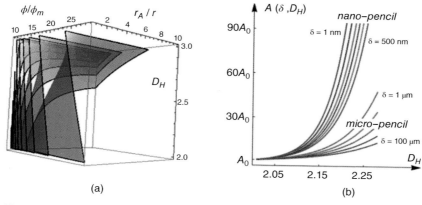

Figure 32.8 (a) Illustration of Eq. (32.7) describing viscosity of the nanofluid. (b) The size of the contact area $A(\delta, DH_f)$ versus fractal dimension for δ from micro- and nanoranges (relation (32.8)).

described as fractal-like structure with a fractal dimension D_H, the fractal correction similar to Eq. (32.6) is possible [28]:

$$\mu_n = \mu_b \left(1 - \frac{\phi}{\phi_m} \left(\frac{r_A}{r} \right)^{3-D_H} \right)^{-2}, \tag{32.7}$$

where r_A and r are the aggregates and primary nanoparticle radii, respectively. The formula (32.7) is illustrated in Figure 32.8, where the level surfaces of μ_n are displayed in function of ratios ϕ/ϕ_m and r_A/r.

The issue of solar energy is specific. Since the solar radiation is uniformly distributed over a medium-sized surface (say 10^3 m $\times 10^3$ m), it is natural to have as many receptors as possible. For example, a team of physicists and chemists from the University of Oregon is exploring ways to use fractals to improve solar energy conversion [29]. The team hopes a better understanding of fractal branching patterns can help optimize collection of sunlight, thus improving the efficiency of solar panels. They proposed to use fractal analysis to predict nanoscale patterns that will provide efficient, novel solar structures. They planned to synthesize new solar materials with fractal geometries predicted to have properties mimicking the light-harvesting fractal patterns found in nature.

32.5
Toward Fractal Nanoelectronics

The world's huge interest in the development of renewable and new-generation energy sources, especially the new battery systems (Li-ion technologies), has a research and development focus. Simultaneously, in the development of electric car technologies, battery systems are playing an increasing role from the viewpoint of how many times in electrochemical reactions recharging can be performed. Beside the electrochemical processes and rechargeable speed, the storage capacity has special significance, which is enriched by the fractal nature of the microcapacitors network spread throughout the bulk of some electronic ceramics with high permittivity (dielectric constant). In the series of papers [30–38], some fundamental electrochemical laws have been revised from the point of view of fractal nature to gain corrections of the relevant formulas. In this way, the new frontiers in the direction of electrochemical fractal micro-electronics are established.

The wide investigations of perovskite ferroelectric and dielectric materials, which as it is well known heavily contribute to microelectronic devices, bring to the fore the whole spectrum of methods that fall in the group of nanotechnological methods. Being in the massive usage, PTC thermistors, varistors, and piezo-electric and optoelectronic elements owe their superb electronic characteristics to doping selected additives, since the additives impact the host material by the action of ions. As semiconductors, they are part of integrated circuits and dynamic random access memories, as thin-film perovskite light collectors with a sandwich structure, and are used in solar cells. The high-performances multi-layer ceramic capacitors based on perovskite ceramics can be used to store energy. The dielectric properties of such ceramics can be modified by using various types of donor or acceptor dopants.

The dielectric properties of $BaTiO_3$ ceramics can be a particularly good example that represents many perovskites. It can be modified by using various types of donor or acceptor dopants. Substitution of the barium or titanium ion with small concentrations of ions with a similar radius could lead to structure and microstructure changes and, furthermore, modify the dielectric properties. Also, rare earth ions have been used as doping elements to improve the dielectric properties of $BaTiO_3$-based materials. The electrical properties strongly depend on the site preference of rare earth ions in $BaTiO_3$ sublattices. The ions with large radius and low valence, such as Nd^{3+} and La^{3+}, predominantly occupy the Ba^{2+} sites, giving donor doped behavior, while ions with small radius, such as Yb^{3+}, prefer the Ti^{2+} sites, conferring acceptor doped behavior. The rare earth ions with intermediate range of ionic radius such as Ho^{3+} and Dy^{3+} can simultaneously occupy both Ba and Ti sites; they can act as either donor or acceptor.

Perovskite materials are important for both electrolyte and electrodes of solid oxide fuel cells and alkaline fuel cells for their potential as low-cost electrode catalyst materials. Perovskites, with formula ABO_3, can be controlled by doping with aliovalent cations to introduce mixed valency/oxide ion vacancies. The numeric value of the so-called *Goldschmidt's tolerance factor* (GFT, see the next section) helps classify the crystal structure of perovskites that can be of an ideal cubic structure, then have distortion from cubic symmetry, or gets octahedral face sharing to make hexagonal perovskites thereby losing the conducting property. This is a typical example of modeling electric properties of material using nanostructure.

Using the Mandelbrot's formula that gives a relationship between the length L of a fractal "space filling" curve of fractal dimension D_H that "fills" the two-dimensional area of size A, that is, [4],

$$L^{-D_H} \propto \sqrt{A},$$

in combination with Richardson law of variable yardstick, upon which a fractal curve length depends on the measurement precision, that is, on the measure yardstick length δ

$$L = \delta^{1-D_H},$$

gives A as a function of D_H

$$A(D_H) = K \, \delta^{2((1/D_H)-1)} (K \text{ is constant}).$$

The physical meaning of the last equation is the functional dependence of the capacitor electrode area size with fractal dimension D_H that can be measured using different algorithms, as it is explained above.

Such considerations did help in gaining the approximate value of neck cross-section area size A in Coble two-grain contact model, where real cross section is approximated by a, b semi-axes elliptic disk as follows:

$$A_{Coble} = 2 \, ab \, \pi \delta^{2((1/D_H)-1)}.$$

From the last two formulas, it is clear that the surface area increases when δ gets smaller; theoretically, for $1 < D_H$, $A \to +\infty$, when $\delta \to 0_+$.

In addition, both the area of ceramic surface (as a cluster of grains) and the area of a single grain surface, depend on the unit of measure δ and local fractal dimension D_H and is given by

$$A(\delta, D_H) \sim \delta^{2-D_H}. \tag{32.8}$$

The relationship (32.8) is illustrated by the graphs in Figure 32.9. The contact area size increases with more precise measuring (smaller δ), which is evident for all fractal dimensions >2. For the nanorange of δ (1 nm $< \delta <$ 100 nm), the area size rapidly increases. Even for "smooth" fractal surfaces, that is, surfaces with D_H close to 2 (as it is the case with BaTiO$_3$ ceramic grains surfaces having $2.079 < D_H < 2.095$), the area size A duplicates its value if the unit of measure δ decreases for the factor 2.87389×10^{-4}.

While the fractal dimension of BaTiO$_3$ ceramic grain's surface is modest, the dimension of the specimen surface is much higher, as Figure 32.8b shows. The ordinate A_0 is the area size of the ideally flat contact surface. The fractal dimension is calculated using max-gray level box counting method of the SEM, which yields $D_H = 1.7531$. For other specimens, similar values are obtained in the range from 1.7529 to 1.8025.

These results led to the revision of the formula for parallel plate capacitor with plates area size A and a separation d ($d \ll A^{1/2}$)

$$C = \varepsilon_r \varepsilon_0 \frac{A}{d}, \tag{32.9}$$

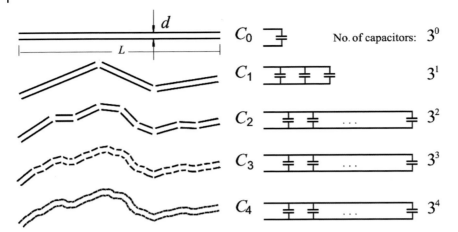

Figure 32.9 (a, b) The constructive way is used to explain fractal character of an intergranular microcapacitor grain surface [20].

where ε_r and ε_0 are relative and vacuum permittivity of $BaTiO_3$ ceramics grains' contact zone, respectively. Namely, the intergranular microcapacitor, formed in the contact zone of two ceramics grains, is *not* a parallel plate capacitor. It is a fractal capacitor that may be thought of as being the product of a iterative process described by Figure 32.9a. The top part of the figure shows a parallel plate capacitor as described above. It corresponds to a capacitor (denoted by C_0) with adjacent grains' perfectly flat contact surfaces, which do not exist in reality. On the contrary, the contact surface is rough and uneven, so that the following fractal model will be a good approximation. Suppose that the flat parallel geometry of C_0 (Figure 32.9a) is replaced by three flat capacitors "Z"-shaped configuration connected in parallel, forming the unique capacitor C_1.

In the next stage, each of the three linear segments is replaced by another smaller "Z" made of three smaller flat capacitors. By continuing this procedure, the sequence of more and more segmented chains of subcapacitors C_0, C_1, C_2, ... , is obtained (Figure 32.9b). It is suitable to take each smaller "Z" to be a shrunk *affine image* of the bigger "Z" standing in the above line. In this manner, the separating distance would be shrunk as well for some factor (although it is not shown in the figure for clarity reasons). This factor (called vertical scaling factor) makes the ratio $d/\sqrt{A}$ smaller for each iteration, allowing capacity to increase. Also, the sum of the lengths in each iteration increases due to wiggling form of the higher stage capacitor. In all, the capacity of C_n will be substantially bigger than C_0. This increment is characterized by the ratio $\bar{\alpha} = \lim_{n \to \infty} C_n/C_0$, wherefrom the "fractal capacitor" capacity, corrected by the factor $\bar{\alpha}$ has the following value:

$$\overline{C}_f = \lim_{n \to \infty} C_n = \bar{\alpha}\, \varepsilon_r \varepsilon_0 \frac{A}{d} = \bar{\alpha}\left(\varepsilon_r \varepsilon_0 \frac{A}{d}\right) = \bar{\alpha}\, C, \tag{32.10}$$

where C is the capacity of the capacitor having ideally flat plates. Since $\overline{C}_f > C$, it follows that $\overline{a} > 1$. The relation (32.10) is called fractal correction of capacity [35]. Another kind of fractal correction follows.

Regarding the perovskite ceramics, a hypothesis is stated [35,36] that fractality introduces some corrective factors. So, the morphology of grains, their fractal dimension in unity with intergranular contact pattern, introduces a "grain surface" factor, denoted by α_S obeying normalizing condition $0 < \alpha_S < 1$. The next parameter α_P (also, $0 < \alpha_P < 1$), represents the "negative space" of pores in the ceramics body. And finally, the real dynamics of the Fermi gas impose a necessity of movement factor inclusion α_M ($0 < \alpha_M < 1$), which describes dynamics of 3D Brownian motion. According to the hypothesis, the working temperature of ceramics must be influenced by these three fractality factors, making "correction" of "theoretic" temperature as $T_f = \underline{\alpha}\,T$, where $\underline{\alpha}$ is the function of α_S, α_P, and α_M, that is, $\underline{\alpha} = \phi(\alpha_S, \alpha_P, \alpha_M)$. The argument for this expectation hides the fact that geometrically irregular motion of a huge number of particles consumes extra energy from the system. In other words, fractality of the system represented by three factors α_S, α_P, and α_M should increase overall energy of the system, and this increment must be subtracted from the input energy, which is, in fact, an input thermal energy denoted by T. In other words, $T_f = T - \Delta T$ and by setting $\underline{\alpha} = T/T_f = 1 - \Delta T/T$, it yields $0 < \underline{\alpha} < 1$ [35,36]. If, in the Curie–Weiss law (T_c is Curie temperature) $\varepsilon_r = C_{cw}/(T - T_c)$, T is replaced with $\underline{\alpha}T$, the modified relative permittivity is $\underline{\varepsilon}_r = C_{cw}/(\underline{\alpha}T - T_c)$ [35], which brings us to another fractal modification of (32.9),

$$\underline{C}_f = \underline{\varepsilon}_r \varepsilon_0 \frac{A}{d} = \frac{\varepsilon_0 C_{cw}}{\underline{\alpha}T - T_c} \frac{A}{d}. \qquad (32.11)$$

32.6
The Goldschmidt's Tolerance Factor, Clausius–Mossotti Relation, Curie, and Curie–Weiss Law Bridge to Fractal Nanoelectronics Contribution

For nearly a century, the Goldschmidt's tolerance factor or formalism, a semi-empirical correlation relation that is based on the ionic radii construct, has led to the discovery and development of numerous perovskites that are based on the "ideal" cubic Pm$\overline{3}$m structure having chemical formula $\underline{ABC}_3$, where the $\underline{A}$ site and $\underline{B}$ site atom(s) are 12-fold and 6-fold coordinated with the $\underline{C}$ site atom, respectively. GTF, which heretofore continues to be used extensively within the scientific community, may be expressed as follows:

$$T = \frac{R_A + R_B}{\sqrt{2}(R_B + R_C)},$$

where R_A, R_B, and R_C are the "effective" ionic radii of the $\underline{A}$ site, $\underline{B}$ site, and $\underline{C}$ site atom(s), respectively, where T ranges from roughly 0.77 to about 1.05 with the "ideal" cubic perovskite forming when T is 1.00. For $T > 1$, the material is often associated with high dielectric, ferroelectric, and antiferroelectric material properties [39]. For $T < 1$, the material is often associated with low-symmetry material [39]. GTF has been used in conjunction with various sets [40–53] as well as other sets of ionic radii with the revised crystal and ion radii of Shannon [53], which include coordination dependence, typically used by the large fraction of the scientific community. Using the ionic radii construct, the lattice parameter of the perovskite can be estimated using the C site–B site–C site (C–B–C) interaction, through the relationship

$$a_1 = 2(R_B + R_C),$$

and using the A–C–A interaction, through the relationship

$$a_2 = \frac{2}{\sqrt{2}}(R_A + R_C),$$

with the average unit cell lattice parameter written as follows:

$$\bar{a} = \frac{1}{2}(a_1 + a_2) = (R_B + R_C) + \frac{1}{\sqrt{2}}(R_A + R_C),$$

where R_A, R_B, and R_C are previously defined. While over the past century, GTF has led to the discovery and development of numerous perovskite materials, it has recently been reinvestigated [45] due to the following reasons: (a) GTF, a correlation relation, is not based on physical principles; (b) increasing numbers of materials lie outside GTF crystal symmetry predictions; (c) GTF lattice parameter predictions are both systematically faulty, through the average of two numbers, and are typically imprecise independent of radii sets [40–53] used; and (d) GTF provides very limited information about the a priori prediction of a significant number of other material properties. Further, the reported room temperature radii sets [40–53] do not adequately address the variation in temperature for lattice parameter, volume, and other material properties. Recently, an extensible new "simple" material model [54–58] (NSMM) has been introduced based on fundamentals of geometry, through physical, planar, and volume constraints [54]. The five planar constraints [54], two of which are identical with GTF [59], are

$$a_{B-C} = 2(R_B + R_C)$$

for C–B–C interaction;

$$a_{A-C} = \frac{2}{\sqrt{2}}(R_A + R_C)$$

for A–C–A interaction;

$$a_{A-A} = 2(R_A)$$

for A–A interaction;

$$a_{C-C} = \frac{4}{\sqrt{2}}(R_C)$$

for C–C interaction; and

$$a_{A-B} = \frac{2}{\sqrt{3}}(R_A + R_B)$$

for A–B interaction, with the largest value defining the lattice parameter of the "ideal" cubic, Pm$\bar{3}$m, perovskite. In addition, the volume constraint [54] is

$$V_m \geq \frac{4\pi}{3}\left(R_A^3 + R_B^3 + 3R_C^3\right),$$

where V_m, the unit cell molar volume, must be greater than or equal to the total volume of the A site, B site, and C site atoms comprising the unit cell. Further packing factor restrictions, roughly 0.52–0.76, can be placed upon the volume constraints [54].

Through transformation into orthonormal ionic radii space [54], NSMM has been utilized to investigate "simple" and "simply mixed" "ideal" Pm$\bar{3}$m perovskites [55,56]. As specifically discussed, GTF may be mapped into ionic radii space, an orthonormal space coordinate system or basis set, by rearranging the GTF expression to have the mathematical form of a plane [54]:

$$Ax + By + Cz = 0,$$

that is,

$$R_A - T\sqrt{2}R_B - \left(T\sqrt{2} - 1\right)R_C = 0,$$

where $A = 1$, $x = R_A$, $B = -T\sqrt{2}$, $y = R_B$, $C = -\left(T\sqrt{2} - 1\right)$, and $z = R_C$, and through applying volume restrictions. NSMM [54], compared to GTF [57], a correlation relation and the "gold" standard for development of new perovskite materials for around the past century, shows why GTF has worked well, Figure 32.10a [54], yet why there are significant mounting discrepancies [48] between the experiment and the correlation relation. Based on the fundamental physical constructs, including an equivalent classical and quantum mechanical linear expansion coefficient of ions within a temperature-dependent isotropic lattice, "effective" temperature-dependent ionic radii were self-consistently determined using temperature-dependent Pm$\bar{3}m$ perovskite lattice parameters, through correlation based on fundamental geometric relations as contained within the NSMM [54–56]. As the number of boundary conditions increase, the ionic radii solution space will become more constrained and ultimately should yield fundamental genome properties of the "effective" temperature-dependent ionic radii. The reported temperature-dependent ionic radii well mimic "simple" "ideal" Pm$\bar{3}$m Perovskites and reasonably well mimic their solid solutions [55,56], with Figure 32.10b [55] representing but one case, one of the worst if not the worst case fit, of numerous possible perovskite solid solution examples. At present, the model produces room-temperature lattice parameters that have significantly improved fidelity compared to that of Goldschmidt's tolerance formalism [55,56,59] using room-temperature Shannon ionic radii data [53] as well

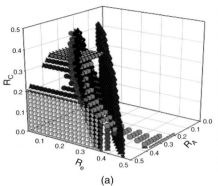

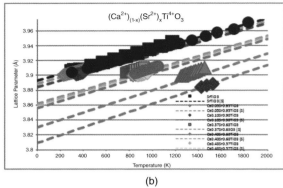

(a)   (b)

Figure 32.10 (a) NSMM constraints, compared with Goldschmidt's tolerance factor planes ($T = 1.07$ (red), 1.0 (black), and 0.77 (blue)), model why materials far from $T = 1.00$ may be cubic, while others with $T \cong 1.00$ may be non-cubic. HRPM lie near the A–C plane (blue with openings) and low symmetry materials lie near the B–C plane (green). (b) Temperature-dependent ionic radii, within the planar and volume constrained ionic radii model, are used to model the lattice parameter of "simple" Pm $\bar{3}$m perovskites and their solid solutions to significantly greater fidelity at room temperature than Shannon radii with Goldschmidt's formalism and are able to extend such fidelity over large temperature range.

as maintains such improved fidelity from low to high temperatures, about 100 K to near about the melting temperature of the material, respectively [55,56].

Since NSMM has been shown to provide significantly improved fidelity of material lattice parameter(s) and volume, NSMM values have been utilized within the *Clausius–Mossotti relation* (CMR) [57,60–62] to "self-consistently" determine the "effective" temperature-dependent polarizability of ions [57]. Importantly, note that CMR is contained directly in Maxwell's conductivity equation [63]:

$$\vec{J} = \sigma \vec{E},$$

where $\vec{J}$ is the current density, σ is the conductivity, and $\vec{E}$ is the electric field, and, further ties NSMM to fundamental equations [54,57,58] with conductivity related through a mathematical formula to complex frequency-dependent permittivity. CMR [57,60–62] can be written as follows:

$$N\alpha_T = 3\left[\frac{n^2 - 1}{n^2 + 2}\right],$$

where N and α_T are the number of particles and total atomic polarizability per unit cell volume, respectively, and n is the index of the refraction of the material; or in terms of permittivity as follows:

$$\frac{\varepsilon}{\varepsilon_0} = \varepsilon_r = \frac{3\,V_m + 8\,\pi\,\alpha_T}{3\,V_m - 4\,\pi\,\alpha_T},$$

where ε is the permittivity, ε_r is the relative permittivity, ε_0 is the permittivity of free space, V_m is the molar volume, and α_T is previously defined. In addition, the dielectric "catastrophe" [64],

$$3\,V_m = 4\,\pi\,\alpha_T,$$

a singularity condition in the dielectric constant, which mathematically results in a nonphysical infinite energy density [64,65]

$$\mu = \frac{1}{2}\varepsilon E^2,$$

due to any nonzero electric field, $\vec{E}$, within the molar volume, may be circumvented through a structural phase transition whereby the ion(s) in the molar volume rearrange to, at least in part, "effectively" store or release some of the additional energy differences that are net gained or released through other energy transfer processes [57], thereby circumventing the "apparent" approaching discontinuity in energy. Thus, due to atomic displacements or crystal distortions, energy may be gained or released from the structure through field changes, change in "effective" coordination with an associated change in polarization, which changes the space distribution of average equal potential energy surfaces within the structure that are related to the band structure. Within the CMR, the total polarizability of ions in the molar volume, V_m, follow the "oxide" additivity rule [60,66], a superposition construct, which states that the total polarizability of atoms or related energy in the molar volume is related to the sum of the individual polarizabilities of the ions or sum of total energy associated with the ions within the molar volume. Using CMR [57,60–62] within the developing NSMM, it has been shown that the structural phase transition temperatures of the "simple" "ideal" $Pm\overline{3}m$ perovskites and their solid solutions can be reasonably well modeled [57,58] and that the solid solutions show a non-Vegard's law behavior, Figure 32.11a [54], which is in closer

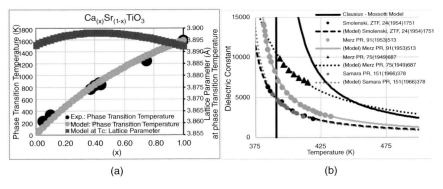

(a)   (b)

Figure 32.11 (a) The structural phase transition temperatures of the solid solution $Ca_{(x)}Sr_{(1-x)}TiO_3$ is but one of the numerous solid solutions modeled using temperature-dependent ion properties, radii, and polarizability, within CMR [46]. (b) Variations of temperature-dependent relative permittivity of $BaTiO_3$ as reported by several groups and even within groups are simulated using the modified CMR, using a depolarization factor and scaling factor to account for crystallite size and porosity, respectively.

agreement with the experiment than the use of Vegard's law [67–70], linear extrapolation between experimental end members. Further, it has recently been shown [58] that CMR, which is contained within Maxwell's fundamental conduction equation (MAS), incorporates both the Curie law [64] (CL) and Curie–Weiss law [64] (CWL). Hence, CMR, in addition to being directly linked to Maxwell's conduction equation, as integrated within the NSMM, directly links genomic-like temperature-dependent ion properties, radii and polarizability, with temperature-dependent ε_r, Curie constant, and Curie temperature, T_c, of materials, without empirical fitting to each material data set as required of both CL and CWL [64]. Through slight modification of CMR [58], using a depolarization term to account for variation of average crystallite size and a scaling factor to account for sample porosity, the variation in ε_r of BaTiO$_3$ has been reasonably well modeled, Figure 32.11b [58], especially near T_c, where variation from "ideal" is pronounced due to differing processing conditions as reported by several groups and even within the same group(s).

Although NSMM has thus far primarily been applied to "simple" and "simply mixed" perovskites [54–57], NSMM has already been shown to (a) replace GTF [59] or formalism, a correlation relation, through the use of fundamental geometric considerations [45]; (b) replace correlated static room temperature-, coordination-dependent ionic and anion radii of Shannon [40] with higher fidelity correlated coordination- and temperature-dependent radii that significantly extend the temperature range of high-fidelity modeling [55,56]; and (c) replace correlated room-temperature ion polarizability of Shannon [60] with higher fidelity correlated temperature- and coordination-dependent ion polarizabilities that extend the temperature range of high-fidelity modeling [57]. Through the inclusion of the Clausius–Mossotti relation [60–62], NSMM has also been shown to identify polarization-induced structural phase transition temperatures as well as model temperature-dependent ε_r of materials above the Pm$\overline{3}$m phase transition temperature [57,58], including variation in crystallite size and porosity, at least for BaTiO$_3$, through a slightly modified CMR, which includes a simple scaling factor for porosity and a depolarization term associated with the crystallite surface energy due to crystallite size. Thus, through the integration of CMR within NSMM, NSMM directly links genomic-like temperature-dependent ion properties, radii and polarizability, with temperature-dependent ε_r, Curie constant, and Curie temperature, T_c, of materials, without the empirical fitting to each material data set as required of both CL and CWL [64].

Of particular importance, NSMM provides a method for a priori determining the fabric of space and, therefore, time, through providing relatively high precision, the crystal symmetry, starting position(s) of atom(s), ion(s) and anion(s), and their temperature-dependent evolution within the structure. It is this property of atomic position that can be used to link simplistic general guidelines followed by experimentalists with first principles, which are typically based on computationally intensive algorithms, for discovery and development of new materials that have a wide range of material properties that can be a priori determined as a function of temperature using a combination of modeling techniques.

Through the development of models that can be used to fuse experimentalists, through known computationally intensive algorithms, with fundamental techniques of first-principle theorists, combinatorialists, data miners, informaticians, and others, the rate of a priori predicting, discovering, and developing new materials with a priori accurate, high fidelity prediction of material parameters as a function of temperature can be used to reduce the time and cost of discovering and developing affordable materials with properties that both outperform materials in current technologies and provide novel properties upon which new technologies can evolve. Such material models can be used to more effectively overcome the myriad of challenges that society faces with regard to water, energy, environment, medicine, affordability, economy, and so on, while simultaneously sustaining or continuing to increase the standard of living and quality of life of the growing population.

32.7
Summary

This chapter has brought together some fundamental constructs with regard to fractals related to energy, its storage, and nanostructures as illustrated schematically within the quaternary diagram, Figure 32.12. Fundamentally important fractal parameters (1) the Hausdorff or fractal dimension D_H, (2) Euclidean or topological dimension D_T, and (3) spectral dimension D_S that describes dynamics of fractal structures and their interrelations were discussed. The three dimensions represent vertices of a three-dimensional triangle that can be used to determine the fractal nature of the structure, object, and so on. One of the key

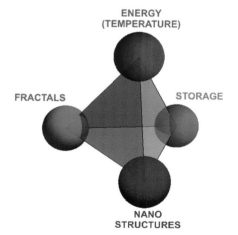

Figure 32.12 The tetrahedral diagram showing connectedness of four key components that binding together contribute to the relation energy nanoscale, which lies in the basis of this chapter.

properties of fractals is their independence of the scale. This brings us to the valuable conclusion that properties of fractals are valid on any scale, macro, micro, or nano. Fractality, as a scale-independent morphology, provides significant opportunity for energy storage. Through dilatation symmetry, fractals allow the mapping of energy in any scale from subnano to distances in light years. On the other hand, time as a quantity conjugate to energy has translation symmetry on all feasible scales. As a consequence, the law of conservation of energy can be applied on all feasible scales. From the viewpoint of scaling, the relation between large and small in fractal analysis is very important. An ideal fractal can be magnified endlessly but natural morphologies cannot. From six common features of the two entities, energy–fractal relationship, with some examples provided, one may conclude that the "classic" attitude that extremal solutions always lead to Euclidean and simple forms may lead to solutions that are contrary to fractal conclusions. For example, the soap film problem results in a sphere that represents the smallest surface containing the biggest volume. To the contrary, the smallest volume with largest surface size question inevitably has fractal solutions as Figures 32.3–32.5 show. An interesting parallel is found concerning the wind energy equations and the behavior of nanofluids. Fractal relationships with regard to electronic parameters and nanoscale morphology were discussed. As a characteristic case, the integral capacity of electronic ceramics was treated. The crucial element is again the length–surface relation that in the fractal environment reaches extremal values, Figure 32.8b and Figure 32.9. Specific microstructure of sintered materials, especially perovskite ceramics with fractal structured configuration of grains and pores, constitutes a favorable environment for very complex electronic phenomenology. Section 32.5 offers some theoretical fundaments to connect complex morphology with the expressed emergence of capacitive microcomponents and high-permittivity constants. On the other hand, there is possibility to include three sources of fractality: grains' surfaces, pores, and the vapor transport flow combined with electron gas during sintering process, which in combination with Curie–Weiss law may describe necessity of three corrective fractal factors α_S, α_P, and α_M, joining in the uniqe fractal influence of nanotexture through the functional dependence $\underline{\alpha} = \phi(\alpha_S, \alpha_P, \alpha_M)$ that influences the relative permittivity modification from ε_r to slightly bigger $\underline{\varepsilon_r}$.

In addition, developing a new simple material model (NSMM) was reviewed that links genomic-like properties of coordination and temperature-dependent ionic properties, radii and polarizability, with temperature-dependent material properties: (a) lattice parameter and volume, (b) crystal structure, (c) dielectric constant, (d) polarization-induced phase transition temperature, and (e) temperature- and frequency-dependent conductivity, through the complex permittivity as related to Maxwell's conduction equation. The latter three properties are linked to the Clausius–Mossotti relation that has been shown to incorporate both the Curie and Curie–Weiss laws. Most importantly, NSMM provides a method for a priori determining the fabric of space and therefore time and energy, by providing to relatively high precision the crystal symmetry, starting position(s) of atom(s), ion(s), and anion(s), and their temperature-dependent evolution within the structure. Although the relationships have not been linked directly, through fundamental formulas, to the microstructure, a preliminary modified CMR has been shown to emulate microstructural effects indicating that CMR may be directly related to fractal formalisms.

Further, as alluded to within the chapter, fractals can be used to discuss and describe a wide range of constructs, including but not limited to structures, contact surfaces, energy, flow of energy, and so on over large length- and timescales. The existence of the fractal nature of ceramic materials has been shown for a wide range of phenomena such as ferroelectric, ferromagnetic, PTC, piezo-, and optoelectronic properties, as well as electrochemical thermodynamic and fluid dynamics parameters. Thin-film layers and chemical vapor deposition techniques naturally produce fractal structures that under appropriate growth conditions provide new, desirable properties. Thus, fractals have generalized use for the discussion of a wide range of physical phenomena. While development of complex fractals to the best of our knowledge has not yet been developed, the physical phenomenon is often mapped into an appropriate reference frame, often a complex coordinate space, where mathematical solutions are easier to find and describe the phenomenon. Once the solutions in the reference frame are determined, the solutions can then be mapped back into real space and time, through mapping functions that often involve complex conjugates. The Clausius–Mossotti relation as related to Maxwell's conductivity equation through the complex permittivity serves as an example. Thus, the development of complex fractals may need to be considered in the future use of fractals to describe physical phenomena.

Acknowledgment

The authors wish to express their appreciation to anonymous reviewers for valuable suggestions.

References

1 Plonka, A. (2001) *Dispersive Kinetics*, Kluwer Academic Publishers, Dordrecht.

2 Gouyet, J.-F. (1996) *Physics and Fractal Structures*, Springer.

3 Abete, T. (2006) A statistical mechanical approach to the study of gels and colloidal systems. PhD thesis, UniversitadegliStudi di Napoli Federico II, Napoli, Italy.

4 Mandelbrot, B. (1982) *The Fractal Geometry of Nature*, W.H. Freeman, San Francisco.

5 Durhuus, B., Jonsson, T., and Wheater, J. (2006) On the spectral dimension of random trees. Fourth Colloquium on Mathematics and Computer Science DMTCS Proc. AG, pp. 183–192.

6 Florindo, J.B. and Bruno, O.M. (2012) Fractal descriptors based on Fourier spectrum applied to texture analysis. *Phys. A*, **391** (20), 4909–4922.

7 Barnsley, M. (1993) *Fractals Everywhere*, 2nd edn, Academic Press, San Diego.

8 Even, U., Rademann, K., Jortner, J., Manor, N., and Reisfeld, R. (1984) Direct electronic energy transfer on fractals. *J. Lumin.*, **31–32**, 634–638.

9 Falconer, K. (1990) *Fractal Geometry: Mathematical Foundations and Applications*, John Wiley & Sons, Ltd, Chichester.

10 Radin, C. and Conway, J. (1995) *Quaquaversal Tiling and Rotations*, preprint, Princeton University Press.

11 Kaye, B.K. (1989) *A Random Walk Through Fractal Dimensions*, John Wiley & Sons, Inc, New York.

12 Wiora, M., Brühne, K., and Fecht, H.-J. (2014) Synthesis of nanodiamond, in *Carbon-Based Nanomaterials and Hybrids, Synthesis, Properties, and Commercial Applications* (eds H.-J. Fecht, K. Brühne, and P. Gluche), Pan Stanford Publ. Pte. Ltd., pp. 220.

13 Wiora, M. (2013) Characterization of nanocrystalline diamond coatings for micro-mechanical applications. Doctoral dissertation, Ulm University, Germany.

14 Messier, R. and Yehoda, J.E. (1985) Geometry of thin film morphology. *J. Appl. Phys.*, **58** (10), 3739–3746.

15 Serov, I.N. *et al.* (2005) Generating and investigating of fractal nano-scale thin films (in Russian). TPKMM-2005, 4th Moscow International Conference, Moscow.

16 Fecht, H.-J., Brühne., K., and Gluche, P. (eds) (2014) *Carbon-Based Nanomaterials and Hybrids, Synthesis, Properties, and Commercial Applications*, Pan Stanford Publ. Pte. Ltd., pp. 5–48.

17 Durhuus, B., Jonsson, T., and Wheater, J. (2006) On the spectral dimension of random trees. Fourth Colloquium on Mathematics and Computer Science DMTCS Proc. AG, pp. 183–192.

18 Noether, E. (1918) Invariante varlationsprobleme. *Nachr. Ges. Wiss. Goettingen*, **2**, 235–257.

19 Mitić, V.V., Kocić, Lj.M., and Ristić, M.M. (1997) The fractals and BaTiO$_3$-ceramics sintering. *Key Eng. Mater.*, **136**, 1060–1063.

20 Mitic, V.V., Kocic, L.M., Miljkovic, M., and Petkovic, I. (1998) Fractals and BaTiO$_3$-ceramic microstructure analysis. *Microchim. Acta*, **15**, 365–369.

21 Petkovic, I., Mitic, V.V., and Kocic, Lj. (1998) Contribution to BaTiOr ceramics structure analysis by using fractals. *Folia. Anat.*, **26** (Suppl I), 67–69.

22 Bazilian, M. *et al.* (2013) Re-considering the economics of photovoltaic power. *Renew. Energ.*, **53**, 329–338.

23 HikosakaBehling, N. (2012) *Fuel Cells: Current Technology Challenges and Future Research Needs*, 1st edn, Elsevier Academic Press..

24 Stephens, J.C. (2006) Growing interest in carbon capture and storage (CCS) for climate change mitigation. *SSPP*, **2** (2), 4–13.

25 Peitgen, H.-O., Jürgens, H., and Saupe, D. (1992) *Chaos and Fractals: New Frontiers of Science*, Springer-Verlag.

26 Petković, D., Mitić, V.V., and Kocić, L.j. (2016) Adaptive neuro-fuzzy optimization of wind farm project investment under wake effect, in *Proceedings of the III Advanced Ceramics and Applications Conference, Belgrade (Serbia), 29th Sep.–1st Oct., 2014* (eds W.E. Lee, Rainer Gadow, Vojislav Mitic, and Nina Obradovic), Atlantis Press, pp. 265–282.

27 Mitić, V.V., Petković, D., and Kocić, Lj. (2016) TRIZ creativity approach to the design of an innovative wind turbine system, in *Proceedings of the III Advanced Ceramics and Applications Conference, Belgrade (Serbia), 29th Sep.–1st Oct., 2014* (eds W.E. Lee, Rainer Gadow, Vojislav Mitic, and Nina Obradovic), Atlantis Press, pp. 283–306.

28 Estellé, P., Halelfadl, S., Doner, N., and Maré, T. (2013) Shear history effect on the viscosity of carbon nanotubes water-based nanofluid. *Curr. Nanosci.*, **9** (2), 225–230.

29 Darren, W. Johnson and Taylor, Richard P. (2013) Role of Fractal patterns on new materials for solar energy applications: inorganic clusters, films and fractal geometry simulations, Research Corporation for Science Advancement Scialog Conference, Biosphere 2, Oct. 15–18, 2013 (invited). (http://www.fractal.org/Fractal-Research-and-Products/FractalSolarCells.pdf).

30 Mitić, V.V., Fecht, H.-J., and Kocić, L.j. (2015) Material science and energy fractal nature new frontiers. *Contemp. Mater.*

(Renewable Energy Sources), **VI – 2,** 33–206.

31 Paunovic, V., Zivkovic, L.j., and Mitic, V. (2010) Influence of rare-earth additives (La, Sm and Dy) on the microstructure and dielectric properties of doped $BaTiO_3$ ceramics. *Sci. Sintering,* **42,** 69–79.

32 Mitic, V.V., Nikolic, Z., Pavlovic, V.B., Paunovic, V., Miljkovic, M., Jordovic, B., and Zivkovic, L.j. (2010) Influence of rare-earth dopants on barium titanate ceramics microstructure and corresponding electrical properties. *J. Am. Ceram. Soc.,* **93** (1), 132–137.

33 Paunovic, V., Mitic, V.V., Prijic, Z., and Zivkovic, L.j. (2014) Microstructure and dielectric properties of Dy/Mn doped $BaTiO_3$ ceramics. *Ceramics Int.,* **40,** 4277–4284.

34 Mitić, V.V., Kocić, L.j., Paunović, V., and Pavlović, V. (2014) Fractal corrections of BaTiO3-ceramic sintering parameters. *Sci. Sintering,* **46** (2), 149–156.

35 Mitić, V.V., Paunović, V., and Kocić, Lj. (2014) Dielectric properties of $BaTiO_3$ ceramics and Curie–Weiss and modified Curie–Weiss affected by fractal morphology, in Advanced Processing and Manufacturing Technologies for Nanostructured and Multifunctional Materials (eds T. Ohji, M. Singh, and S. Mathur) Ceramic Engineering and Science Proceedings, vol. 35 (6), pp. 123–133.

36 Mitić, V.V., Paunović, V., Kocić, Lj., Jankovic, S., and Litovski, V. (2014) BaTiO3-ceramics microstructures new fractal frontiers, CIMTEC 2014, 13th Ceramics Congress, CJ-1.L14, Montecatini, Italy, June 9–13.

37 Mitić, V.V., Paunović, V., and Kocić, L.j. (2015) Fractal approach to BaTiO3-ceramics micro-impedances. *Ceramics Int.,* **41** (5), 6566–6574.

38 Mitić, V.V., Kocić, L.j., Paunović, V., Bastić, F., and Sirmić, D. (2015) The fractal nature materials microstructure influence on electrochemical energy sources. *Sci. Sintering,* **47** (2), 195–204.

39 Galasso, F.S. (1969) *Structure, Properties, and Preparation of Perovskite-Type Compounds,* Pergamon Press, London.

40 Bragg, W.L. (1920) The arrangement of atoms in crystals. *Phil. Mag.,* **40** (236), 169–189.

41 Landé, A. (1920) Über die Größe der Atome. *Z. Phys.,* **1** (3), 191–197.

42 Wasastjerna, J.A. (1923) Comment phys-math. *Helsingf,* **1,** 1.

43 Goldschmidt, V.M., Barth, T.F.W., Lunde, G., and Zachariasen, W.H. (1926) Geochemische Verteilungsgesetze: VII: Die Gesetze der Krystallochemie. *Skr. Norske Vidensk Akad. 1 Mat.-Nat.* **Kl 2** (1), 107.

44 Bragg, W.L. and West, J. (1927) The structure of certain silicates. *Proc. Roy. Soc. A,* **114** (768), 450–473.

45 Pauling, L. (1927) The sizes of ions and the structure of ionic crystals. *J. Am. Chem. Soc.,* **49** (1), 765–790.

46 Pauling, L. (1928) The sizes of ions and their influence on the properties of salt-like compounds. *Z. Kristallogr.,* **67** (3/4), 377–404.

47 Zachariasen, W.H. (1931) A set of empirical crystal radii for ions with inert gas configuration. *Z. Kristallogr.,* **80** (3/4), 137–153.

48 Arhens, L.H. (1952) The use of ionization potentials. Part 1: ionic radii of the elements. *Geochim. Cosmochim. Acta.,* **2** (3), 155–159.

49 Slater, J.C. (1964) Atomic radii in crystals. *J. Chem. Phys.,* **41** (10), 3199.

50 Fumi, F.G. and Tosi, M.P. (1964) Ionic sizes and born repulsive parameters in the NaCl-type alkali halides – I: the Huggins–Mayer and Pauling forms. *J. Phys. Chem. Solids,* **25** (1), 31–43.

51 Bohr, N. (1913) On the constitution of atoms and molecules: part I. *Philos. Mag. S-6,* **26** (151), 1–24.

52 Shannon, R.D. and Prewitt, C.T. (1969) Effective ionic radii in oxides and fluorides. *Acta. Cryst.,* **B25,** 925–945.

53 Shannon, R.D. (1976) Revised effective ionic radii and systematic studies of interatomic distances in halides and chalcogenides. *Acta. Cryst.,* **A32,** 751–767.

54 Tidrow, S.C. (2014) Mapping comparison of Goldschmidt's tolerance factor with Perovskite structural conditions. *Ferroelectrics,* **470** (1), 13–27.

55 Miller, V.L. and Tidrow, S.C. (2013) Perovskites: temperature and coordination

dependent ionic radii. *Integr. Ferroelectr.*, **148** (1), 1–16.

56 Miller, V.L. and Tidrow, S.C. (2015) Perovskites: "Effective" temperature and coordination dependence of 38 ionic radii. *Integr. Ferroelectr.*, **166** (1), 30–47.

57 Tidrow, S.C. (2015) Perovskites: some polarization induced structural phase transitions using "effective" temperature and coordination dependent radii and polarizabilities of ions. *Integr. Ferroelectr.*, **166** (1), 206–224.

58 Tidrow, S.C. (2016) Linking curie constant and phase transition temperature with fundamental ion properties. *Integr. Ferroelectr.*, **174** (1), 15–25.

59 Goldschmidt, V.M. (1926) Geochemische Verteilungsgesetze der Elemente. *Skr. Norske Vidensk Akad., Mat.-Naturv.*, **KlNo. 2**, 8.

60 Shannon, R.D. (1993) Dielectric polarizabilities of ions in oxides and fluorides. *J. Appl. Phys.*, **73** (1), 348–366.

61 Mossotti, O.F. (1850) *Mem. Mathem fisica Modena*, **24**, 49.

62 Clausius, R. (1879) Die mechanische U'grmetheorie. *Die Mechanische U'grmetheorie*, **2**, 62–64.

63 Maxwell, J.C. (1865) A dynamical theory of the electromagnetic field. *Philos. Trans. R. Soc. Lond.*, **155**, 459–512.

64 Kittel, C. (1974) *Introduction to Solid State Physics*, 5th edn, John Wiley & Sons, Chapters 13–15.

65 Halliday, D. and Resnick, R. (1974) *Fundamentals of Physics*, John Wiley & Sons, Inc. (revised edition).

66 Heydweiller, A. (1920) Dichte, dielektricitätskonstante und

refraktion fester salze. *Z. Phys.*, **3** (5), 308–317.

67 Denton, A.R. and Ashcroft, N.W. (1991) Vegard's law. *Phys. Rev. A*, **43** (6), 3161–3164.

68 Vegard, L. (1921) The constitution of the mixed crystals and the filling of space of the atoms. *Z. Phys.*, **5** (1), 17–26.

69 Vegard, L. and Dale, H. (1928) Tests on mixed crystals and alloys. *Z. Kristallogr.*, **67** (1), 148–162.

70 Vegard, L. (1928) Die Röntgenstrahlen im Dienste der Erforschung der Materie. *Z. Kristallogr.*, **67** (2), 239–259.

Additional References

71 Travel, M.A. (1971) Invariant variation problem. *Transport Theor. Stat.*, **1** (3), 183–207.

Further Reading

Russ, J.C. (1994) *Fractal Surfaces*, Springer.

Kiinkenberg, B. (1994) A review of methods used to determine the fractal dimension of linear features. *Math. Geol.*, **26** (1), 23–46.

Rao, C.B. and Raj, B. (2001) Estimation of Minkowski dimension using neighborhood operations. *Math. Geol.*, **33**, 369–376.

Shaotong, F. and Dianronga, H. (2005) Experimental determination of the fractal dimension of the ordinary fractal patterns with optical fractional Fourier transform. *Proc. SPIE*, **5642**, 543–548.

33

Magnesium Based Nanocomposites for Cleaner Transport

Manoj Gupta and Sankaranarayanan Seetharaman

National University of Singapore, Department of Mechanical Engineering, 9 Engineering Drive 1, Singapore 117 576

33.1
Introduction

Lightweight materials are important for energy efficient and cleaner transportation. Since it takes less energy to accelerate a lighter vehicle, light-weighting of automotive and aerospace components can directly influence the fuel consumption for better fuel efficiency and energy savings [1]. For example, in a V6 cylinder car, replacement of a cast iron engine block by Mg has reduced the weight from 86 kg to 30 kg, which in turn promoted extensive fuel savings [2]. Similarly, BMW also reduced the fuel consumption of its model cars by 30% (from 1990 to 2007) using the advanced lightweight engine concept [3] (see Figure 33.1).

When compared to other lightweight structural metals such as aluminum and titanium, Mg with a density (ρ_{Mg}: 1.74 g/cm^3) that is only slightly higher than that of plastics provides excellent weight savings. For this reason, it is of particular interest for structural weight reduction in automotive and aerospace applications. Besides being light, Mg and its alloys exhibit an excellent spectrum of influential properties such as damping characteristics, castability, machinability, weldability, and resistance to electromagnetic radiation. Recent research works also identified Mg alloys as potential medical implants owing to their biocompatibility [4,5].

Despite considerable efforts made thus far, the adoption of magnesium alloys that follow precipitation strengthening methods has been substantially hindered by their relatively low strengths, modulus, and ductility at room temperature, for use in critical engineering applications. This is primarily attributed to the fact that the dominant secondary phase ($Mg_{17}Al_{12}$) becomes thermally unstable at temperatures >120 °C [7]. Other than traditional precipitation control, unconventional processing methods are also used in recent years to influence the microstructural characteristics for mechanical

Nanotechnology for Energy Sustainability, First Edition. Edited by Baldev Raj, Marcel Van de Voorde, and Yashwant Mahajan.

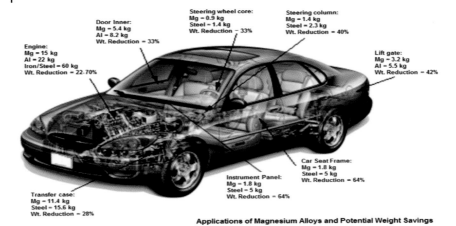

Figure 33.1 Automobile components made of magnesium alloys [6].

properties enhancement of Mg alloys. For example, rapid solidification/ powder metallurgy method has been used recently to fabricate an Mg–Zn–Y alloy with ~600 MPa yield strength. The achieved mechanical strength enhancement was due to the presence and uniform distribution of long-period ordered structures [8]. Using equal channel angular extrusion (ECAE) method, Yoshida *et al.* [9] developed ultrafine-grained AZ61 alloy with average grain size 0.5 μm for high-strength applications. Similar unconventional methods have been used by various other researchers [10,11]. While it is generally perceived that the high concentration of grain boundaries in these materials restricts the dislocations motion and promotes strength improvement, available literature also suggests the suppression of deformation twinning propensity due to grain refinement as a strengthening mechanism in addition to dislocation strengthening. While remarkable properties have been achieved using such untraditional fabrication methods, the processing difficulties limit their potential applications (see Figure 33.2).

On the other hand, composite making methodologies with the flexibility of constituent (matrix and reinforcement) selection also attract extensive R&D efforts on tailor-made magnesium matrix composites with superior properties (particularly, the Young's modulus, dimensional stability, and high-temperature performance) for demanding applications [13–16]. Especially, those magnesium nanocomposites reinforced with nanoscale ceramic reinforcement are receiving stupendous attention as the dispersion of hard, nanoscale phases into Mg matrix seems to be a viable economic option to enhance the strength of Mg alloys. Unlike micrometer scale reinforcement addition, the efficient dispersion of nanoscale reinforcement improves the strength of Mg without any adverse effects on ductility.

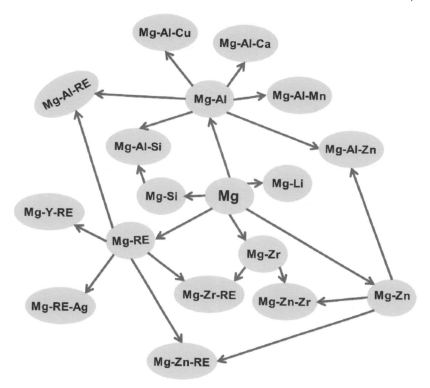

Figure 33.2 Development of Mg alloys [12].

33.2
Fabrication of Magnesium-based Nanocomposites

A variety of processing techniques are used to produce magnesium-based metal matrix composites that can be broadly classified into (i) solid-phase processing and (ii) liquid-phase processing [7,13,14]. The major significance in selecting the processing method is the capability of the process to produce materials with homogeneously distributed reinforcement particulates, which is essential to achieve superior mechanical properties. Figure 33.3 shows in detail of various fabrication methods that are capable of producing magnesium-based nanocomposites. The benefits and limitations of each methodology are listed in Table 33.1. From the literature, it can be seen that the processing methods such as (i) mechanical alloying and densification [17], (ii) microwave-assisted rapid sintering-based powder metallurgy [18], (iii) ultrasonic cavitation [19], and (iv) disintegrated melt deposition [20] are widely used. Since the processing methodologies have been extensively reported elsewhere [21,22], their individual benefits and limitations alone are tabulated here (see Table 33.1).

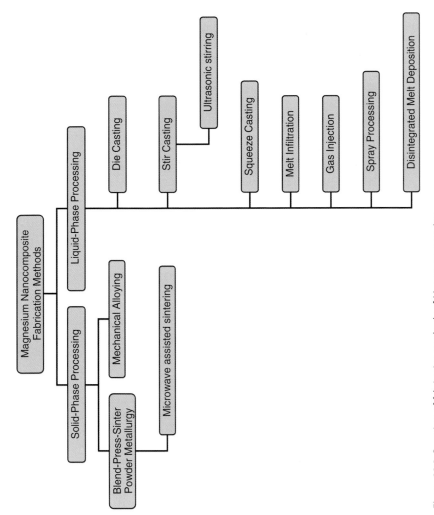

Figure 33.3 Overview of fabrication methods of Mg nanocomposites.

Table 33.1 Benefits and limitations of different synthesis technologies of Mg nanocomposites.

Fabrication Technique	Benefits	Limitations
Solid-Phase Processing		
Blend–press–sinter powder metallurgy	• Capable of using different types and volume fractions of reinforcement • Lesser interfacial reaction	• Complex parts cannot be fabricated • Ineffective dispersion of reinforcement • Difficulty in removal of binder materials
Mechanical alloying	• Effective dispersion of reinforcement • High-strength, equilibrium, and nonequilibrium alloys can be developed • Strengthening due to high dislocation density	• Danger in handling of highly reactive powders • Not suitable for bulk production • Inability to produce complex parts
Microwave sintering	• Rapid process • Energy-efficient technology	• Reinforcement segregation at larger volume fraction
Liquid-Phase Processing		
Pressure die casting	• Capable of making complex parts • Economical for large-quantity production • Effective reinforcement dispersion due to high sheer speed during processing	• Brittle interfacial reaction products
Stir casting	• Able to produce parts in large quantities • Economical process • Capable of using a larger reinforcement quantity (up to 30%)	• Reinforcement clustering • Undesirable brittle interfacial reaction products • Increase in viscosity of molten metal slurry • Damage to reinforcement due to high pressure • Casting defects such as porosity
Ultrasonic stirring	• Improved reinforcement dispersion	• Undesirable interfacial reaction
Squeeze casting	• Capable of using larger reinforcement quantity than stir casting (up to 40–50%) • Capable of making complex parts • Effective reinforcement dispersion	• Undesirable brittle interfacial reaction products • Reinforcement clustering
Melt infiltration	• Capable of using larger reinforcement quantity • Parts with complex geometry can be produced	• Damage to reinforcement due to larger injection pressure
Gas injection	• Reasonable uniform distribution of reinforcement for rapid cooling rates • Economical process	• Process efficiency not completely established

(continued)

Table 33.1 (*Continued*)

Fabrication Technique	Benefits	Limitations
Spray processing	• High solidification rate result in finer microstructure	• Larger residual porosity • Expensive process due to gas and residual materials cost • Limitation to produce complex and intricate shapes
Disintegrated melt deposition	• Combined benefits of stir casting and spray processing methods • Flexibility of using different types and volume fractions of reinforcement • Effective reinforcement distribution and lesser reinforcement segregation issues • Finer grain structure due to faster cooling rates	• Usage of gases

33.3
Mechanical Properties and Corrosion

33.3.1
Tensile and Compressive Properties

The available open literature indicates that the ceramic particles such as SiC, B_4C, TiC, Al_2O_3, Y_2O_3, ZrO_2, TiO_2, Bi_2O_3, Si_3N_4, TiN, BN, AlN, TiB_2, and carbon nanotubes (CNTs) are used as nanoreinforcement for Mg and Mg alloys [20–48]. Regardless of the processing methods, the inherent high-temperature stability, high strength, and high modulus of these finely dispersed ceramic phases (only a small volume percentage) offered distinct advantages (Figure 33.4) through a combination of different strengthening mechanisms such as grain refinement strengthening, Orowan strengthening, geometrically necessary dislocation (GND)-assisted strengthening effects, and so on without adversely affecting the ductility [49,50].

During solidification and recrystallization, the reinforcing phases would act as grain refiners by grain boundary pinning and grain nucleation mechanism for grain size strengthening. The presence of nanosize reinforcement is expected to hinder the movement of dislocations to contribute toward Orowan strengthening. Further, the strain and CTE mismatch between the matrix and the nanosize particles will also contribute toward an increase in dislocations around the interfacial region between the matrix and the particles, thus strengthening the material.

With regard to ductility enhancement, the grain refinement and the activation of nonbasal slip systems by means of texture modification are identified as the

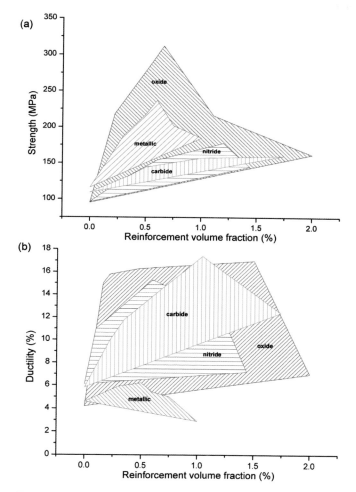

Figure 33.4 Tensile properties of Mg nanocomposites.

responsible mechanisms. Among different reinforcement studied so far, nano-Y_2O_3 and nano-B4C were found to be promising for strengthening and ductilization, respectively. Comparing the methods, often the DMD technique produced magnesium nanocomposites with significant improvement in 0.2% yield strength and ductility, whereas the PM-processed nanocomposites displayed a more modest improvement in strength as seen in the case of Mg/Al_2O_3 nanocomposites. While a maximum improvement of 80% in 0.2% yield strength (0.2% YS) and 200% in ductility were obtained for DMD-synthesized composites, the PM-processed Mg/Al_2O_3 nanocomposites, on the other hand, exhibited 33% increment in 0.2% YS and 27% in UTS [34].

Available literature also highlights the incorporation of nanoscale metallic particles such as copper [18], nickel [35,51], titanium [52], and aluminum [53] particles to improve the mechanical performance of Mg. In particular, the addition of nanosized Cu has led to an increase in 0.2% YS by ∼104%. However, the ductility was reduced with the increasing addition of Cu due to the formation of Mg_2Cu intermetallic phase.

The strength enhancement due to nanoparticle addition was also reported under compression loads (Figure 33.5). The failure strain was either unaffected or showed slight decrement except for the case of ZK60 alloy nanocomposites, wherein the addition of nanoscale Al_2O_3, Si3N4, and CNT reinforcement resulted ∼46, ∼52, and ∼69% improvement, respectively [20–48]. Significant

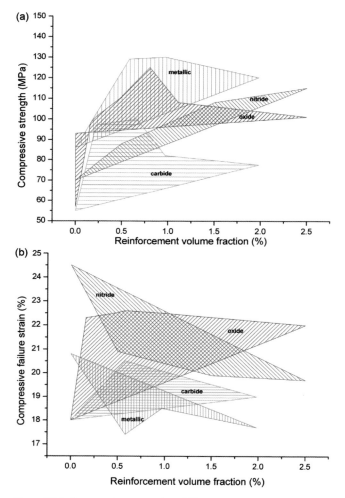

Figure 33.5 Compressive properties of Mg nanocomposites.

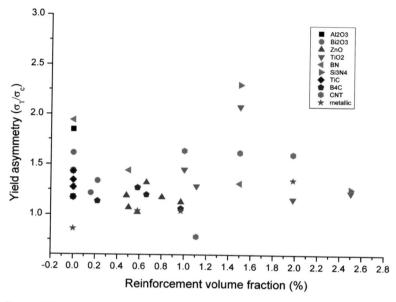

Figure 33.6 Tension–compression yield asymmetry of Mg nanocomposites.

improvement in ductility was also reported for nano-TiO_2 reinforcement addition to Mg [45].

The improvement in mechanical behavior was also evident at elevated temperatures up to 150 °C [54–56].

Further, it should also be noted that the nanocomposites display different yield strengths under tension and compression loading and the ratio of tension–compression yield asymmetry is summarized as shown in Figure 33.6. While it is common for Mg-based alloys and composites to display a large degree of anisotropy, the incorporation of reinforcement in nanoscale either aggravates or propitiates the tension–compression asymmetry based on the type of reinforcement and matrix material. In general, the yield strength under tension will be 1.5–2 times greater than that of the compressive yield strength when the hot extruded Mg materials are tested along the extrusion direction. This difference in yield strengths under tension and compression loading (tension–compression yield symmetry) can be attributed to the strong influence of preferred orientation on the deformation behavior of Mg (with basal planes aligned parallel to the extrusion direction). While the twinning process with relatively low critical resolved shear stress (CRSS) governs the compressive deformation and results in lower yield strength, the inability of twinning and the relatively larger CRSS value of basal/nonbasal slips result in larger tensile yield strength. Based on the mechanical properties reported earlier, it can be understood that the nanoscale reinforcement such as B4C, BN, ZnO, and CNT influence the crystallographic texture changes to positively affect the mechanical characteristics. The same has been verified using electron backscattered diffraction methods that clearly indicates the weakening of the fiber texture to facilitate slip transition by non-basal cross-slip activation to possibly enhance the dislocation activity for basal slip for larger uniform deformation [40,41].

33.3.2
Dynamic Mechanical Properties

The feasibility evaluation of magnesium nanocomposites' usage in vehicles, aircraft, and armor applications requires a thorough assessment of the dynamic mechanical performance of these nanocomposites. However, the available literature suggests that only limited studies have been carried out so far [57–61]. Guo *et al.* [57] and Chen *et al.* [58] investigated the influence of nano-Al_2O_3 reinforcement on the dynamic mechanical performance of Mg-6Al and AZ31 alloys developed using DMD method. Results indicated that the addition of nanoparticles increased the material strength by a constant value for both low and high rate loading (Figure 33.7). However, their strain hardening behavior remained unchanged. In order to understand the deformation mechanism in detail, the texture configurations of Mg nanocomposites were studied before and after testing (both quasi-static and dynamic tension tests) using X-ray diffraction method [59]. It revealed (i) change in initial texture and (ii) texture retention during tensile tests due to nanoparticle addition as the fundamental reason for enhancement of dislocation movement and increased ductility of the nanocomposites. Similar observations have been reported by Xiao *et al.* [60]. On the other hand, hot deformation experiments conducted on Mg/ZnO nanocomposites clearly identify the dynamic recrystallization (DRX) and dynamic recovery (DRY) zones between 375–400 °C and 300–350 °C, respectively, at a strain rate of 0.01 s^{-1}, and a highly localized flow along the maximum shear stress plane at higher strain rates due to the inadequate time and low thermal conductivity [61].

33.3.3
Creep Properties

Conventionally, Mg alloys exhibit poor creep resistance due to the presence of eutectic components (beta phase plus alpha Mg) along the grain boundaries and dendrites. Hence, it is extremely necessary to improve the creep properties of Mg alloys in order to extend their applications in automotive industry that requires adequate creep resistance at temperatures about 150–200 °C. Ferkel and Mordike [23] investigated the creep properties of SiC nanoparticle-strengthened magnesium. In this study, the tests were performed at 200 and 300 °C, with stresses of 35 and 45 MPa, respectively. The results (Figure 33.8) indicated minimum creep rate, that is, better creep resistance due to a dislocation movement-controlled deformation rather than the diffusion creep. It was also found that the small amount of nano-SiC has led to similar or better creep resistance to that of the commercial creep-resistant magnesium alloy such as WE43, WE54, and

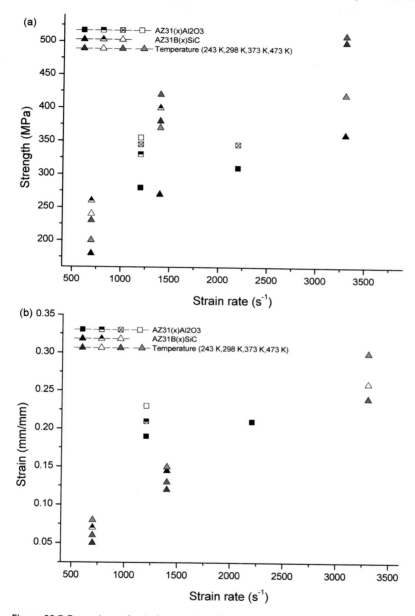

Figure 33.7 Dynamic mechanical properties of Mg nanocomposites.

QE22. Dieringa [62] also reported enhancement in creep resistance due to the addition of nanoscale reinforcement. The correlation of true stress exponents revealed dislocation climb and pipe diffusion mechanisms as the responsible rate-controlling deformation process during creep [63,64].

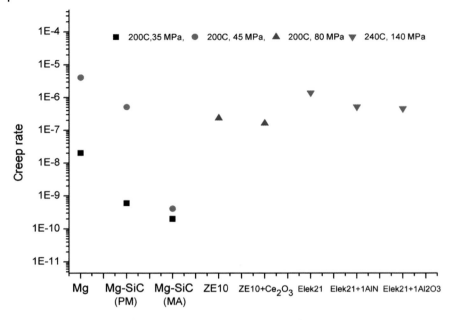

Figure 33.8 Creep rates of Mg nanocomposites reported in the literature.

33.3.4
Fatigue Properties

The fatigue properties of magnesium nanocomposites are important when cyclic loads are present in the engineering application. During the past two decades, only few detailed studies have been conducted with the primary objective of quantifying and rationalizing the intrinsic influence of nanosized reinforcement on the cyclic fatigue life and fracture behavior of magnesium alloys. Srivatsan *et al.* [65–67] investigated the high cycle fatigue response of CNT and nano-Al_2O_3 particle-reinforced AZ31 alloy nanocomposites. In these studies, it has been observed that the addition of nanoscale reinforcements improved the endurance strength of AZ31 alloys under tension–tension ($R = 0.1$) and fully reversed ($R = -1$) fatigue loads (tested at 106 cycles) as shown in Figure 33.9.

33.3.5
Corrosion Properties

Mg materials exhibit poor corrosion resistance due to nonprotective nature of oxide film. Hence, it becomes essential to evaluate the corrosion characteristics

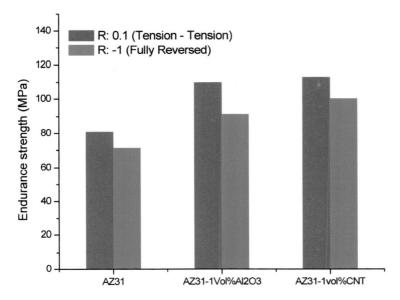

Figure 33.9 Endurance strength of AZ31 alloys nanocomposites reported by Srivatsan *et al.* [65–67].

of Mg nanocomposites before being put into real-time applications. Simple hydrogen evolution (immersion) tests and electrochemical measurements such as polarization and electrochemical impedance spectroscopy methods are often used for this purpose. Kukreja *et al.* [68] investigated the corrosion nature of nanoalumina-reinforced AZ31B nanocomposites and reported that the average corrosion rate for $AZ31B/Al_2O_3$ nanocomposite was approximately one-third that for monolithic AZ31B alloy. In a similar research work by Aung *et al.* [69], it was identified that the CNT-reinforced Mg nanocomposites suffered from higher corrosion rates due to microgalvanic action mechanism. With regard to dry corrosion, it is extremely important to understand their oxidation characteristics considering the elevated temperature exposure of magnesium alloys during secondary processing (heat treatment, rolling, forming, welding, etc.) and end service conditions. It is generally perceived that with an increase in temperature, the magnesium oxide film becomes increasingly nonprotective (porous) leading to higher corrosion rate. Alloying with rare earth elements is one of the methods used to improve the elevated temperature properties of magnesium [21,70]. Ye and Liu [21] suggested that the incorporation of thermally stable reinforcements will also help improve the high-temperature behavior of magnesium. For the first time, Nguyen *et al.* [71] investigated the role of nanosized reinforcement in determining oxidation characteristics of magnesium alloys using the thermogravimetric method under ambient atmospheric condition. The increasing amount

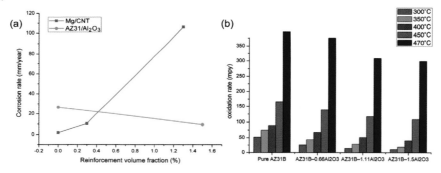

Figure 33.10 Results of corrosion (a) and oxidation studies (b) conducted on Mg nanocomposites [68,69,71].

of nano-Al_2O_3 addition has improved the oxidation resistance (retardation in oxidation) of AZ31B alloy below 400 °C, which was attributed to the fairly protective oxide layer (Figure 33.10).

33.4
Engineering Properties

33.4.1
Wear

Improving the wear resistance of soft magnesium alloys is one of the primary objectives of composites development as the incorporation of hard ceramic reinforcement would enhance the indentation resistance of matrix material. However, the delamination of micrometer-size particles promotes crack nucleation and propagation at the particle matrix interface to result in severe wear compared to unreinforced materials. On the other hand, the use of low-volume fraction nanosized reinforcement has been an attractive option as it would aid in overcoming the delamination issues to improve the wear resistance of Mg alloys. Dry sliding wear tests have been carried out by different research groups on nanoparticle-reinforced magnesium composites. Lim *et al.* [72] investigated the wear characteristics of DMD-processed Mg composites containing various amounts (up to 1.11 vol%) of nanosized alumina particulates. The volumetric wear rates calculated using weight loss data indicated a consistent improvement in wear resistance with increasing amounts of nanoalumina reinforcement. Similar results were also reported for commercial AZ31B and its Ca-modified counterparts [73]. Microscopic investigations conducted in these studies identify abrasive wear as the responsible mechanism at lower sliding speed that eventually changed into adhesive

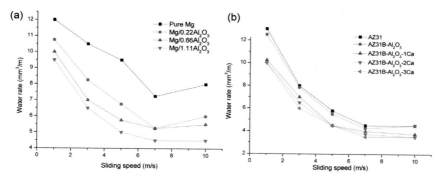

Figure 33.11 Wear characteristics of Mg nanocomposites [72,73].

wear for higher sliding speeds. However, it was also confirmed that the delamination wear was not evident in Mg nanocomposites (Figure 33.11).

33.4.2
Machinability

As magnesium nanocomposites are envisaged for high-performance applications, it is significantly important that suitable machining methods are developed to facilitate their industrial development. Available literature suggests that only very limited studies have been carried out so far in this aspect [74–76]. However, it is believed that the metal removal rate for nanocomposites could be low due to the "shielding" effect caused by nanoceramic particle reinforcement. In addition, the machining cost could be inevitably high due to severe tool wear. Further, it was also reported by some researchers that the unconventional machining processes such as EDM or laser could not be used for machining Mg nanocomposites as the surface finish would be relatively poor and the microstructure was often changed due to the high heat generated during these machining processes. Hence, an optimized approach should be identified for a cost-effective and highly productive machining process. With this objective, a few research works have been published in recent years.

Li *et al.* [74,75] assessed the machinability of Mg nanocomposites by micromilling method and reported that the cutting force increases with the increase in the spindle speed, the feed-rate, and/or the volume fraction. Recently, Teng *et al.* [76] also studied the micromachinability of Mg nanocomposites reinforced with 1.98 vol% nanosized titanium/titanium diboride particles. From the experimental results, it was deduced that the spindle speed and depth of cut have significant influence on surface roughness. Among the nanocomposites, Mg/Ti nanocomposite derived larger cutting force compared to that of Mg/TiB$_2$. However, Mg/TiB$_2$ exhibited better machinability in terms of surface morphology.

33.5
Potential Applications in Transport Industries

With the advent of novel high-performance magnesium nanocomposites, their targeted applications in critical engineering sectors are also gaining importance simultaneously. While recent years have seen extensive utilization of magnesium and its alloys in transportation and electronics industries, there exists enormous application potential for the lightweight magnesium nanocomposites that exhibit superior mechanical performance.

In aerospace industry, weight reduction remains one of the critical objectives to address the increasing need for better fuel efficiency. However, the use of magnesium and its alloys have been largely controlled by the aviation regulators owing to the dangers surrounding the metal's flammability. Recently, Federal Aviation Administration (FAA) has cleared the use of high-performance Mg alloys in seat components, provided they meet the required flammability performance [77]. The potential applications of magnesium alloys and nanocomposites in aerospace industry hence include the following [7]:

  i) Rudder petal assembly
 ii) Gear box – transmission housings
iii) Thrust reversers
 iv) Seat frame
  v) Seat arm support
 vi) Floor and galley components

The weight saving attribute of magnesium-based materials are especially attractive for their potential use in automobiles for reduced fuel energy consumption and emission. The current applications include engine block, crankcase, engine valve covers, gearbox housing, instrument panels, steering components, radiator supports, door and lift gate inners, seat structures (for car, bus, and train), rear subframe and fairing support bracket (for motorcycle), roof components, steering wheel armature, and shifter support bracket. Magnesium alloys and composites are used in the pan of steering wheel frame, seats, and instrument panels of Ford Chryster, Toyota, Lexus, Mercedes, and Audi vehicles. Mg alloys are also used in the air intake system and engine casing of BMW six-cylinder and eight-cylinder engines, the oil pump body of McLaren Motors (FL-V12), the cam shaft drive chain case in Porshe AG (911 series) [20].

Military aircraft such as the Eurofighters Typhoon and Tornado and F16 Fighting Falcon fighter jet also benefit from the lightweight characteristics of Mg alloys for transmission castings. Other military applications include ballistic protection and personnel protection (helmet and body armor). Magnesium alloys and composite materials also witness a remarkable incline in the utilization for sports, electronics, and marine applications [20].

Recently, magnesium is also receiving considerable attention for use as biocompatible, biodegradable, and potential osteoconductive material for implant materials, microclips, and suture wires. With performance, at par or superior to the commercial alloys, magnesium nanocomposites are also envisaged for use in applications such as those mentioned above [78].

33.6
Challenges

There exist stringent norms and regulations for magnesium alloys to be used in aerospace and automotive applications. It is primarily due to the perceived flammability concerns of magnesium and its alloys. Further, the corrosion attributes of Mg alloys also equally alarm their utilization in critical engineering applications. Specifically, for magnesium nanocomposites, the key challenge is uniform dispersion of nanoscale reinforcement. Further, the incorporation of relatively stable reinforcement also affects the corrosion attributes of Mg alloys. Generally, most magnesium alloys exhibit higher bare corrosion rate compared to other structural metals such as aluminum, steel, and so on. Further, the low protection performance of surface treatments also adversely affects the corrosion resistance to result in higher corrosion level of magnesium aerospace components relative to aluminum ones. However, with the advancement in magnesium processing and surface modification technologies such as GA1 ALGAN 2M and composite coating2 Gardobond® X4729 from Chemetall GmbH and Magoxide® and Magpass® from AHC Oberflächentechnik, high-purity magnesium materials with relatively better corrosion resistance have been developed. Other major concerns are the lack of adequate strength and thermal softening-associated strength deterioration at elevated temperature. Recently, Magnesium Elektron Ltd. (UK) developed new high-strength alloys Elektron 21 and Elektron 675 that have mechanical properties comparable to aerospace aluminum structural alloys. With effective bonding and efficient load transfer characteristics, magnesium nanocomposites are also envisaged for use in applications that require greater specific mechanical properties and energy absorption capability [79]. The overall improvement in absolute and specific mechanical properties that can be realized through nanocomposite technology hold promise of cutting down on the overall material cost, fuel consumption, carbon footprint, and energy required for driving even a battery-operated vehicle.

33.7
Conclusions

To summarize, there has been a renewed interest in nanoparticles reinforced magnesium composites due to the increasing availability of different nanoscale ceramic/metallic particles and sophisticated fabrication methods. It has been shown that the advanced solid-state and liquid-state processing techniques such as mechanical alloying, microwave sintering-assisted powder metallurgy, disintegrated melt deposition, and ultrasonic-assisted stir casting are reasonably effective in dispersing the nanoscale reinforcements in Mg matrix. Owing to their inherent capabilities, the incorporation of nanosized ceramic/metallic particles provided enhanced mechanical and engineering properties compared to conventional Mg composites reinforced with micrometer-sized particles. It was shown that the addition of nanoparticles simultaneously/significantly increased the strength and/or ductility of Mg under tensile and compressive loads. The incorporation of nanoparticle reinforcement had also improved the fatigue and creep response of Mg. While the carbon nanotubes aggravated the wet corrosion aspects of pure Mg, the addition of nano-Al_2O_3 improved the corrosion resistance of AZ31 alloy under both dry and wet conditions. Other engineering properties such as wear and machinability of Mg nanocomposites were also improved by nanoreinforcement addition. While these published properties of Mg-based nanocomposites encourage their use in critical engineering applications, further extensive studies are underway covering their overall engineering aspects for effective technology transfer.

References

1 Watarai, H. (2007) Trend of research and development of magnesium alloys: reducing the weight of structural materials in motor vehicles. *Sci. Technol. Trends*, 85–97.

2 Tharumarajah, A. and Koltun, P. (2007) Is there an environmental advantage of using magnesium components for light-weighting cars? *J. Clean. Prod.*, **15** (11), 1007–1013.

3 Mattes, W.G. (2007) The BMW approach to Tier2 Bin5. 13th DEER Conference, 13–16th August, Detroit, pp. 1–29.

4 Denkena, B., Witte, F., Podolsky, C., and Lucas, A. (2005) Degradable implants made of magnesium alloys. 5th Euspen International Conference, Montpellier, France, pp. 1–4.

5 Gu, X., Zheng, Y., Cheng, Y., Zhong, S., and Xi, T. (2009) *In vitro* corrosion and biocompatibility of binary magnesium alloys. *Biomaterials*, **30**, 484–498.

6 Kulekci, M.K. (2008) Magnesium and its alloys applications in automobile industry. *Int. J. Adv. Manuf. Tech.*, **39**, 851–865.

7 Kainer, K.U. (2003) *Magnesium Alloys and Technologies*, Wiley-VCH Verlag GmbH.

8 Inoue, A., Kawamura, Y., Matsushita, M., Hayashi, K., and Koike, J. (2001) Novel hexagonal structure and ultrahigh strength of magnesium solid solution in the Mg–

Zn–Y system. *J. Mater. Res.*, **16**, 1894–1900.

9 Yoshida, Y., Arai, K., Itoh, S., Kamado, S., and Kojima, Y. (2005) Realization of high strength and high ductility for AZ61 magnesium alloy by severe warm working. *Sci. Technol. Adv. Mater.*, **6** (2), 185–194.

10 Ma, A., Jiang, J., Saito, N., Shigematsu, I., Yuan, Y., Yang, D., and Nishida, Y. (2009) Improving both strength and ductility of a Mg alloy through a large number of ECAP passes. *Mater. Sci. Eng. A*, **513–514**, 122–127.

11 Yamashita, A., Horita, Z., and Langdon, T.G. (2001) Improving the mechanical properties of magnesium and a magnesium alloy through severe plastic deformation. *Mater. Sci. Eng. A*, **300** (1–2), 142–147.

12 Mordike, B.L. and Ebert, T. (2001) Magnesium: properties–applications– potential. *Mater. Sci. Eng. A*, **302**, 37–45.

13 Kainer, K.U. (2006) *Basics of Metal Matrix Composites*, Wiley-VCH Verlag GmbH.

14 Clyne, T.W. and Withers, P.J. (1995) *An Introduction to Metal Matrix Composites*, Cambridge University Press.

15 Mohamed, F.A., Lavernia, E.J., and Ibrahim, I.A. (1991) Particulate reinforced metal matrix composites: a review. *J. Mater. Sci.*, **26** (2), 1137–1156.

16 Evans, A., Marchi, C.S., and Mortensen, A. (2003) *Metal Matrix Composites in Industry: An Introduction and a Survey*, Springer.

17 Suryanarayana, C. (2001) Mechanical alloying and milling. *Prog. Mater. Sci.*, **46** (1), 1–184.

18 Wong, W. and Gupta, M. (2007) Development of Mg/Cu nanocomposites using microwave assisted rapid sintering. *Compos. Sci. Technol.*, **67** (7), 1541–1552.

19 Cao, G., Konishi, H., and Li, X. (2008) Recent developments on ultrasonic cavitation based solidification processing of bulk magnesium nanocomposites. *Int. J. Metal Casting*, **2** (1), 57–65.

20 Gupta, Manoj. and Sharon, NaiMuiLing. (2010) *Magnesium, Magnesium Alloys and Magnesium Composites*, John Wiley & Sons.

21 Ye, H.Z. and Liu, X.Y. (2004) Review of recent studies in magnesium matrix composites. *J. Mater. Sci.*, **9**, 6153–6171.

22 Dieringa, H. (2010) Properties of magnesium alloys reinforced with nanoparticles and carbon nanotubes: a review. *J. Mater. Sci.*, **46** (2), 289–306.

23 Ferkel, H. and Mordike, B. (2001) Magnesium strengthened by SiC nanoparticles. *Mater. Sci. Eng. A*, **298** (1–2), 193–199.

24 Thein, M.A., Lu, L., and Lai, M.O. (2006) Effect of milling and reinforcement on mechanical properties of nanostructured magnesium composite. *J. Mater. Process. Technol.*, **209**, 4439–4443.

25 Goh, C.S., Wei, J., Lee, L.C., and Gupta, M. (2007) Properties and deformation behaviour of Mg–Y$_2$O$_3$ nanocomposites. *Acta Mater.*, **55** (15), 5115–5121.

26 Goh, C.S., Wei, J., Lee, LC., and Gupta, M. (2006) Simultaneous enhancement in strength and ductility by reinforcing magnesium with carbon nanotubes. *Mater. Sci. Eng. A*, **423**, 153–156.

27 Paramsothy, M., Chan, J., Kwok, R., and Gupta, M. (2012) Nanoparticle interactions with the magnesium alloy matrix during physical deformation: tougher nanocomposites. *Mater. Chem. Phys.*, **137**, 472–482.

28 Md., ErshadulAlam., Han, S., Nguyen, Q.B., Hamouda, A.M.S., and Gupta, M.

(2011) Development of new magnesium based alloys and their nanocomposites. *J. Alloys Compd.*, **509**, 8522–8529.

29 Wong, W.L.E. and Gupta, M. (2006) Simultaneously improving strength and ductility of magnesium using nano-size SiC particulates and microwaves. *Adv. Eng. Mater.*, **8** (8), 735–740.

30 Wong, W., Karthik, S., and Gupta, M. (2005) Development of hybrid Mg/Al$_2$O$_3$ composites with improved properties using microwave assisted rapid sintering route. *J. Mater. Sci.*, **40** (13), 3395–3402.

31 Tun, K.S. and Gupta, M. (2007) Improving mechanical properties of magnesium using nano yttria reinforcement and microwave assisted powder metallurgy method. *Compos. Sci. Technol.*, **67** (13), 2657–2664.

32 Tun, K.S. and Gupta, M. (2008) Effect of heating rate during hybrid microwave sintering on the tensile properties of magnesium and Mg/Y$_2$O$_3$ nanocomposite. *J. Alloys Compd.*, **466** (1), 140–145.

33 Tun, K.S. and Gupta, M. (2008) Effect of extrusion ratio on microstructure and mechanical properties of microwave-sintered magnesium and Mg/Y$_2$O$_3$ nanocomposite. *J. Mater. Sci.*, **43** (13), 4503–4511.

34 Hassan, S.F. (2006) Creation of new magnesium-based material using different types of reinforcements. PhD thesis, National University of Singapore.

35 Tun, K.S. (2009) Development and characterization of new magnesium based nanocomposites. PhD thesis, National University of Singapore.

36 Nguyen, Q.B. (2009) Development of nanocomposites based on magnesium alloys system AZ31B. PhD thesis, National University of Singapore.

37 Tun, K.S., Jayaramanavar, P., Nguyen, Q.B., Chan, J., Kwok, R., and Gupta, M. (2012) Investigation into tensile and compressive responses of Mg-ZnO composites. *Mater. Sci. Technol.*, **28** (5), 582–588.

38 Sankaranarayanan, S., Nayak, U.P., Sabat, R.K., Suwas, S., Almajid, A., and Gupta, M. (2014) Nano-ZnO particle addition to monolithic magnesium for enhanced tensile and compressive response. *J. Alloys Compd.*, **615**, 211–219.

39 Seetharaman, S., Subramanian, J., Tun, K.S., Hamouda, A.M.S., and Gupta, M. (2013) Synthesis and characterization of nano boron nitride reinforced magnesium composites produced by the microwave sintering method. *Mech. Compos. Mater.*, **6** (5), 1940–1955.

40 Sankaranarayanan, S., Sabat, R.K., Jayalakshmi, S., Suwas, S., Almajid, A., and Gupta, M. (2015) Mg/BN nanocomposites: nano-BN addition for enhanced room temperature tensile and compressive response. *J. Compos. Mater.*, **49** (24), 3045–3055.

41 Sankaranarayanan, S., Sabat, R.K., Jayalakshmi, S., Suwas, S., and Gupta, M. (2014) Effect of nanoscale boron carbide particle addition on the microstructural evolution and mechanical response of pure magnesium. *Mater. Des.*, **56**, 428–436.

42 Habibi, M.K., Hamaouda, A.M.S., and Gupta, M. (2013) Hybridizing boron carbide (B4C) particles with aluminum (Al) to enhance the mechanical response of magnesium based nano-composites. *J. Alloys Compd.*, **550**, 83–93.

43 Sankaranarayanan, S., Habibi, M., Jayalakshmi, S., Jia Ai, K., Almajid, A., and Gupta, M. (2015) Nano-AlN particle reinforced Mg composites: microstructural and mechanical properties. *Mater. Sci. Technol.* **31** (9), 1122–1131.

44 Habibi, M.K., Hamaouda, A.M.S., and Gupta, M. (2013) Using hierarchical composite approach to improve mechanical response of Mg and Mg–Bi₂O₃ nano-composites. *Mater. Des.*, **49**, 627–637.

45 Meenashisundaram, G.K., Nai, M.H., Almajid, A., and Gupta, M. (2015) Development of high performance Mg–TiO₂ nanocomposites targeting for biomedical/structural applications. *Mater. Des.*, **65**, 104–114.

46 Meenashisundaram, G.K., Sankaranarayanan, S., and Gupta, M. (2014) Enhancing overall tensile and compressive response of pure Mg using nano-TiB₂ particulates. *Mater. Charact.*, **94**, 178–188.

47 Meenashisundaram, G.K., Nai, M.H., Almajid, A., and Gupta, M. (2016) Reinforcing low-volume fraction nano-TiN particulates to monolithical, pure Mg for enhanced tensile and compressive response. *Materials*, **9** (3), 1–21.

48 Meenashisundaram, G.K. and Gupta, M. (2015) Synthesis and characterization of high performance low volume fraction TiC reinforced Mg nanocomposites targeting biocompatible/structural applications. *Mater. Sci. Eng. A*, **627**, 306–315.

49 Callister, W.D. and Rethwisch, DavidG. (2011) *Materials Science and Engineering: An Introduction*, John Wiley & Sons.

50 Dieter, G.E. and Bacon, David. (1990) *Mechanical Metallurgy*, McGraw-Hill Publishers.

51 Jayalakshmi, S., Sankaranarayanan, S., Sahu, S., and Gupta, M. (2013) Effect of nano-nickel particle addition on the microstructure and tensile response of pure Mg. Twenty-Second International Conference on Processing and Fabrication of Advanced Materials, Singapore, December 2013.

52 Zhong, X., Wong, W., and Gupta, M. (2007) Enhancing strength and ductility of magnesium by integrating it with aluminum nanoparticles. *Acta Mater.*, **55** (18), 6338–6344.

53 Meenashisundaram, G.K. and Gupta, M. (2014) Low volume fraction nano-titanium particulates for improving the mechanical response of pure magnesium. *J. Alloys Compd.*, **593**, 176–183.

54 Hassan, S.F. (2008) High-temperature tensile properties of Mg/Al₂O₃ nanocomposite. *Mater. Sci. Eng. A*, **486**, 56–62.

55 Hassan, S.F., Paramsothy, M., Patel, F., and Gupta, M. (2012) High temperature tensile response of nano-Al₂O₃ reinforced AZ31 nanocomposites. *Mater. Sci. Eng. A*, **558**, 278–284.

56 Mallick, A., Tun, K.S., Vedantam, S., and Gupta, M. (2010) Mechanical characteristics of pure Mg and a Mg/Y₂O₃ nanocomposite in the 25–250°C temperature range. *J. Mater. Sci.*, **45** (11), 3058–3066.

57 Guo, Y.B., Shim, V.P.W., and Tan, B.W.F. (2012) Dynamic tensile properties of magnesium nanocomposite. *Mater. Sci. Forum*, **706–709**, 780–785.

58 Chen, Y., Guo, Y.B., Gupta, M., and Shin, V.P.W. (2013) Dynamic tensile response of magnesium nanocomposites and the effect of nanoparticles. *Mater. Sci. Eng. A*, **582**, 359–367.

59 Habibi, M.K., Pouriayevali, H., Hamouda, A.M.S., and Gupta, M. (2012) Differentiating the mechanical response of hybridized Mg nano-composites as a function of strain rate. *Mater. Sci. Eng. A*, **545**, 51–60.

60 Xiao, J., Shu, D.W., and Wang, X.J. (2014) Effect of strain rate and temperature on the mechanical behavior of magnesium nanocomposites. *Int. J. Mech. Sci.*, **89**, 381–390.

61 Selvam, B., Marimuthu, P., Narayanasamy, R., Senthilkumar, V., Tun, K.S., and Gupta, M. (2015) Effect of temperature and strain rate on compressive response of extruded magnesium nano-composite. *J. Magn. Alloys*, **3** (3), 224–230.

62 Dieringa, H., Hort, N., and Kainer. K.U. (2012) Ultrasonic stirring as a production process for nanoparticle reinforced magnesium alloys and the compression creep response of ze10 reinforced with ceria nanoparticles. 15th European Conference on Composite Materials, Venice, Italy, 24–28 June 2012.

63 Sillekens, W.H., Jarvis, D.J., Vorozhtsov, A., Bojarevics, V., Badini, C.F., Pavese, M., Terzi, S., Salvo, L., Katsarou, L., and Dieringa, H. (2014) The ExoMet project: EU/ESA research on high-performance light-metal alloys and nanocomposites. *Metall. Mater. Trans. A*, **45** (8), 3349–3361.

64 Kumar, Harish. and Chaudhari, G.P. (2014) Creep behavior of AS41 alloy matrix nano-composites. *Mater. Sci. Eng. A*, **607**, 435–444.

65 Srivatsan, T.S., Godbole, C., Paramsothy, M., and Gupta, M. (2012) The role of aluminum oxide particulate reinforcements on cyclic fatigue and final fracture behavior of a novel magnesium alloy. *Mater. Sci. Eng. A*, **532**, 196–211.

66 Srivatsan, T.S., Godbole, C., Paramsothy, M., and Gupta, M. (2012) Influence of nano-sized carbon nanotube reinforcements on tensile deformation, cyclic fatigue, and final fracture behavior

of a magnesium alloy. *J. Mater. Sci.*, **47** (8), 3621–3638.

67 Srivatsan, T.S., Godbole, C., Quick, T., Paramsothy, M., and Gupta, M. (2013) Mechanical behavior of a magnesium alloy nanocomposite under conditions of static tension and dynamic fatigue. *J. Mater. Eng. Perform.*, **22** (2), 439–453.

68 Kukreja, M., Balasubramaniam, R., Nguyen, Q.B., and Gupta, M. (2009) Enhancing corrosion resistance of Mg alloy AZ31B in NaCl solution using alumina reinforcement at nanolength scale. *Corros. Eng. Sci. Technol.*, **44** (5), 381–383.

69 Aung, N.N., Zhou, W., Goh, C.S., Nai, S.M.L., and Wei, J. (2010) Effect of carbon nanotubes on corrosion of Mg–CNT composites. *Corros. Sci.*, **52** (5), 1551–1553.

70 Friedrich, H.E. and Mordike, B.L. (2006) *Magnesium Technology: Metallurgy, Design Data*, Springer-Verlag, Berlin.

71 Nguyen, Q.B., Gupta, M., and Srivatsan, T.S. (2009) On the role of nano-alumina particulate reinforcements in enhancing the oxidation resistance of magnesium alloy. *Mater. Sci. Eng. A*, **500**, 233–237.

72 Lim, C.Y.H., Leo, D.K., Ang, J.J.S., and Gupta, M. (2005) Wear of magnesium composites reinforced with nano-sized alumina particulates. *Wear*, **259**, 620–625.

73 Shanthi, M., Nguyen, Q.B., and Gupta, M. (2010) Sliding wear behaviour of calcium containing AZ31B/Al_2O_3 nanocomposites. *Wear*, **269** (5–6), 473–479.

74 Li, J., Liu, J., Liu, J., Ji, Y., and Xu, C. (2013) Experimental investigation on the machinability of SiC nano-particles reinforced magnesium nanocomposites during micro-milling processes. *Int. J. Manuf. Res.*, **8**, 64–84.

75 Liu, J., Li, J., and Xu, C. (2013) Cutting force prediction on micromilling magnesium metal matrix composites with nanoreinforcements. *J. Micro Nano-Manuf.*, **1** (1), 0110101–01101010.

76 Teng, X., Huo, D., Wong, W.L.E., and Gupta, M. (2015) Experiment based investigation into micro machinability of Mg based metal matrix composites (MMCs) with nano-sized reinforcements.

Proceedings of the 21st International
Conference on Automation and
Computing, University of Strathclyde,
Glasgow, UK, pp. 1–6.

77 Magnesium ELektron (September 21,
2015) *Magnesium ELektron News*. http://
www.magnesium-elektron.com/news-tags/
aircraft-seats (accessed 10 December
2015).

78 Staiger, M.P., Pietak, A.M., Huadmai, J.,
and Dias, G. (2006) Magnesium and its
alloys as orthopedic biomaterials: a review.
Biomaterials, **27** (9), 1728–1734.

79 Ostrovsky, Y. (2007) Present state and
future of magnesium application in
aerospace Industry. International
Conference: "New Challenges in
Aeronautics," Moscow.

34
Nanocomposites: A Gaze through Their Applications in Transport Industry

Kottan Renganayagalu Ravi,[1] Jayakrishnan Nampoothiri,[1] and Baldev Raj[2]

[1]*PSG Institute of Advanced Studies, Structural Nanomaterials laboratory, Coimbatore, 641004, Tamil Nadu, India*
[2]*National Institute of Advanced Studies, Bangalore 560 012, Karnataka, India*

34.1
Introduction

Improved fuel efficiency, reduced emissions, enhanced performance with stringent safety rules, and requirement of low maintenance along with increased style and comfort options are the few challenges confronted by the transport industry [1–3]. Engineering of new materials and manufacturing processes plays an important role in the fight against these problems. The main objective of a new technological development in materials engineering is to meet competitive pressures, cost, durability, performance, and recyclability.

Nanocomposites are a new class of materials exhibiting excellent thermal and mechanical properties, apposite to replace conventional high-density metal parts in transport industry [4]. The use of nanocomposites in automotive components is expected to improve production rates, environmental and thermal stability, promote recycling, and reduce weight. At present, several polymer nanocomposites were considered potential material for structurally noncritical parts such as the front and rear fascia, cowl, vent grills, valve/timing covers, and truck beds [5,6]. This could result in several billion kilograms of weight reduction per year. In addition, the evolution of polymer nanocomposites improved the esthetics of transport carriages and their performance [7].

Although polymer/polymer nanocomposite materials hold edge over metallic materials in structurally noncritical automobile components, still the development of lightweight metallic materials such as Al and Mg alloys are major focus for structural and high-temperature withstanding applications [8,9]. For the past three decades, discontinuously reinforced light alloy metal matrix composites (MMCs) have been proposed as potential material for automobile applications due to their high specific modulus, fatigue, and

Nanotechnology for Energy Sustainability, First Edition. Edited by Baldev Raj, Marcel Van de Voorde, and Yashwant Mahajan.

wear properties [10–12]. However, some of their mechanical properties still fall short of the requirements of extensive use in large-scale engineering practices. In this situation, metal matrix nanocomposites (MMNCs) have emerged as a new class of materials that can result in a desired combination of mechanical properties such as high ductility and high specific strength [13,14].

Ceramics are potential contestants for various transport engineering applications such as high-temperature structures, electrical insulators, wear resistance parts, and various coatings, due to their high hardness, chemical inertness, and high electrical and thermal insulating properties [15,16]. Though, a widespread application of ceramics in engineering arena was limited because of its low fracture toughness. Incorporation of ductile metallic materials or another ceramic phase with the concerned ceramic particles can make the ceramics suitable for engineering applications and are known as ceramic matrix composites. Current developments in nanotechnology have made it possible to tailor ceramic materials with added functionalities [17,18]. Selected nanoceramics with distinct physical–mechanical properties with specific geometrical aspects have been used to reinforce monolithic ceramics. It can improve mechanical and other functional properties. In this perspective, latest developments in the fabrication technology and mechanical properties and potential applications of polymer, metal, and ceramic nanocomposites are discussed in this chapter.

34.2
Polymer Matrix Nanocomposites in Transport Sector

Polymer nanocomposites (PMNCs) are a mixture of materials in which at least one dimension of the reinforcement is less than 100 nm [18]. Low density, eco-friendly, and superior mechanical properties with high production rate, good surface finish with improved corrosion resistance, and better working temperature are the major prerequisites for developing PMNCs. These materials contain nanosized fillers such as nanoparticles, nanofibres, nanotubes, and nanoclays engulfed either in thermoset or thermoplastic polymer material [6]. Polymer nanocomposites are considered potential alternatives to conventional polymer composites in which a twofold increase in tensile modulus and strength can be achieved without decreasing impact resistance. In addition, the heat distortion temperature of the nanocomposites can be increased up to 100 °C, which further extends the use of the PMNCs to high-temperature applications [19,20]. Besides their improved mechanical and physical properties, PMNCs can be readily extruded or molded into near-net shape, which simplifies their manufacturing process [21]. Since high degrees of stiffness and strength can be realized using a small fraction of high-density inorganic material, PMNCs are much lighter compared to conventional polymer composites. It has been reported that a significant use of PMNCs could

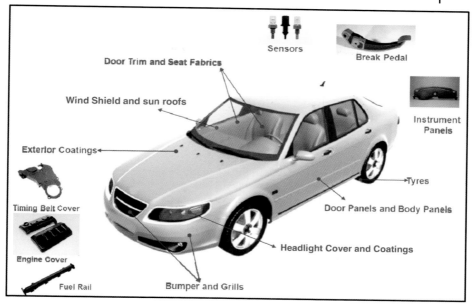

Figure 34.1 Potential automobile applications of PMNCs.

reduce the weight of the vehicle by 20–40% (Figure 34.1), consequently increasing the fuel efficiency about 20% [6,22].

As already mentioned the major prerequisites for developing PMNCs are low density, eco-friendly, and superior mechanical properties with high production volume, good surface finish with improved corrosion resistance, and better working temperature are [23]. The following section discusses about PMNCs and illustrates some successes and remaining challenges from the perspective of recent efforts.

34.2.1
Common Matrices and Reinforcement Materials Used in PMNCs

The selection of matrix and reinforcements largely depends on the end user applications and the property requirements. Among the additives and fillers, nanosized clay [24] is the most widely accepted one, mainly due to significant improvement in mechanical properties of PMNCs (Table 34.1). Apart from that, carbon nanofibers, carbon nanotubes (CNTs), graphite, graphene, polyhedral oligomeric silsesquioxane (POSS), and organosilanes are also being used as reinforcement in PMNCs. Recently, various nanoceramic oxides such as TiO_2, ZnO, and SiO_2 and metallic nanoparticles such as silver, gold, and Cu nanoparticles are also being employed as reinforcement material [6,25,26].

Table 34.1 Comparison of mechanical properties of nanoclay-reinforced PMNCs with PMCs [6].

Properties	PMCs	PMNCs with 2–5% Nanoclay
Tensile strength	1×	>1.5–2×
Flexural modulus	1×	>4–5×
Notch strength (Izod)	1×	>0.5–1×
HDT	1×	>1.5–2×
LCTE	1×	<1.5–2×
Specific gravity	1×	<0.5–1×
Shrinkage and warpage	1×	<0.5–1x

The inherent superior properties, such as good thermal, electrical, flexural, and mechanical properties, made thermoplastics (polyolefin, polyamide, polypropylene sulfide, polyetheretherketone, polyethylene terephthalate, and polycarbonates) attractive matrix materials for the fabrication of PMNCs [27,28]. In addition, thermosets, such as epoxy, conductive polymers, and thermoplastic elastomers such as butadiene–styrene diblock copolymer are also being used as matrices in transport industries [21].

34.2.2
Manufacturing Methods for PMNCs

Melt intercalation, template synthesis, exfoliation adsorption, and *in situ* polymerization intercalation are the four major routes available for synthesis of PMNCs (Table 34.2). Depending on the synthesis route, the microstructure of PMNCs is categorized into unitercalated, intercalated, and/or flocculated or exfoliated structures [5,29–32].

In order to improve the dispersion of the nanofiller in the polymer matrix to obtain desired properties of final composites, researchers have investigated different modifications in the above-mentioned traditional methods. For example, *in situ* polymerization processes are customized to enable redox or catalytic chain transfer or even photoinduced polymerization. In addition to the aforementioned traditional processing methods, microwave-induced synthesis, one-pot synthesis, template-directed synthesis, electrochemical synthesis, electro spinning synthesis, self-assembly synthesis, and intermatrix synthesis are emerging as new methods for fabrication of PMNCs [33–35].

34.2.3
Potential Applications of PMNCs in Transport Industry

Potential applications of PMNCs in aerospace and automobile industries are summarized in Tables 34.3 and 34.4 [6,18,33,36].

Table 34.2 Major Synthesis routes for fabrication of PMNCs with brief description.

Method	Process Description	Remarks
Melt intercalation	In this process, polymer matrix is annealed at elevated temperature, thereafter filler is added, and finally it undergoes kneading for improving the distribution	Advantages: Eco-friendly – no solvent usage Scalable and compatible with industrial processes such as extrusion and injection molding Limitations: High temperature used in this process can damage surface modification of fillers
Exfoliation adsorption	Filler such as layered silicate is first mixed and dispersed in solvent before mixing it with the proper polymer solution. The polymer chains then intercalate and displace the solvent within the silicate interlayers. Finally, on the removal of the solvent, a multilayer structure is formed as the sheets reassemble trapping the polymer chains	Limitations: Extensive usage of solvents – not eco-friendly Applications Water-soluble polymers and intercalated nanocomposites are produced using this method
In situ polymerization	This process comprises the inflammation of the filler in liquid monomer or monomer solution as the low molecular weight monomer seeps in between the interlayers causing the swelling. Polymerization starts either by using heat, radiation, and initiator diffusion or by using an organic initiator or catalyst fixed through cationic exchange. The monomers then polymerize in between the interlayers forming intercalated or exfoliated nanocomposites	Advantages: Better exfoliation can be achieved compared to melt and exfoliation adsorption methods
Template synthesis	This method involves the formation of the inorganic filler in an aqueous solution or gel containing the polymer and the filler precursors. The matrix polymer serves as a nucleating agent and enhances the growth of the filler crystals. As those crystals grow, the polymer is confined within the layers and thus forms the nanocomposite	Applications This process is mainly used for the synthesis of double-layer hydroxide-based nanocomposite Limitations This process is not fully developed for the synthesis of layered silicates. High temperature used during synthesis degrades the polymer and the resulting aggregation tendency of the growing inorganic crystals

Table 34.3 Potential applications of PMNCs in aerospace industry.

Aerospace Industry

Properties Required	PMNC Type	Potential Candidate
Permeability and outgassing	Incorporation of impermeable, high-aspect-ratio silicate, or graphite flake in resin	Cryogenic tanks, optical benches, interferometers, and antenna truss structures
Oxidation resistance	Incorporate high-temperature, oxidation-resistant fillers,silicate, CNTs, POSS, and so on To form passivating layers or slow oxidative erosion in resin	Thermal protection systems, atomic oxygen resistance, and space structures
Electrostatic dissipation	Addition of high-aspect-ratio conductive particles such as CNTs,, graphite flake, and metals, as percolated networks in resin between conductive fibers	Charge-dissipating adhesives, coatings, and gap fillers
Electromagnetic interference	Create films of highly percolated networks of conductive nanofillers (nickel nanostrand veil, SWNT buckypaper, etc.) that can absorb and dissipate broadband frequencies	Bus compartment enclosures and electronic enclosures
Lightning strike	Addition of conductive nanofillers (nickel nanostrands, CNTs, etc.) as highly percolated coatings, appliqués, resins, or veils that can carry large currents with controlled mode failure	Composite aircraft exteriors
Thermal conductivity	Incorporate highly thermally conductive particles (CNTs, metals, etc.) into resin and optimize structure for heat transfer along continuous path to heat sink	Thermally conductive adhesives, gaskets, radiators, doublers, electronics boards, and solid-state laser heat removal
Thermal protection systems	Use thermally conductive and insulating nanofillers within resin to help larger structural components to direct heat away from protected systems or improve mechanical properties at high temperature	Aircraft brakes, reentry vehicles, and missiles
Coefficient of thermal expansion (CTE)	Incorporate nanofillers with low expansion coefficients and good matrix bonding such as functionalized CNTs, CNFs, and silicate, into resin, or as fiber sizing to reduce CTE mismatch with fiber by composite effect and restriction of polymer motion	Adhesives, space apertures with improved thermal cycling durability
Toughness	Incorporate nanoparticles such as CNTs, layered silicate, and silica into resin to increase energy dissipation on failure through deformation, pullout, crack bridging, and so on, at needed plies	Membrane structures, damage-tolerant structures
Modulus	Incorporate high-modulus nanoparticles such as continuous CNT yarns/sheets as reinforcement or grow reinforcements between plies to increase out-of-plane modulus	Stable, precision structures

Compression strength	Incorporate high-strength nanoparticles such as functionalized CNTs into the resin	Propulsion tanks, fittings
Interfacial shear stress	Grow high-strength nanoparticles such as CNTs from fiber to tailor the interfacial properties as a smart sizing	High-temperature composites, vehicle health monitoring
Interlaminar shear strength	Incorporate nanofilled resins with increased toughness at mid-plies via coating or prepregging	Tubular structures

Table 34.4 Potential applications of PMNCs in automobile industry.

Major Property Requirement	PMNC Type	Potential Candidate
Abrasion Resistance	Soot (carbon black), silica, and organo-silane are incorporated in rubber to enhance abrasive properties. Addition of CNT and CNF can improve tensile strength	Tyres
High strength/ modulus	Nylon—nanoclay composites	Timing belt cover, engine cover, fuel line, fuel hoses, and fuel valves
Lightweight with improved stiffness	Polyolefin reinforced with organoclays, nanosilicates, carbon fibers, and CNTs	Door frames, seat backs, step assists, heavy-duty electrical enclosures, sail panel, box rails, fascia, grills, hood louvers, instrument panels, side trims, body panels, and fenders
Magnetic properties	Polymers loaded with magnetic nanoparticles (nano-Fe_3O_4)	Sensors
Electrical conductivity	Polymer electrolyte fuel cells with CNT, carbon black, and grapheme	Fuel cells
Wear and frictional resistance	Mineral oils and other lubricants with nanodiamonds, fullerenes, and CuO nanoparticles	Lubricants in bearing surfaces and other parts of engine
Heat transfer properties	Nanoparticles such as Al_2O_3, CuO, TiO_2, ZnO, and MgO and CNTs in water or ethylene glycol	Coolants
Antimicrobial/antibacterial and antiodor properties	Fabric with nanostructured antimicrobial agents such as silver, titanium oxide, gold, copper, and zinc oxide and chitosan nanoparticles, silver-based nanostructured materials, titania nanotubes, CNTs, nanoclay, and gallium	Automotive fabrics
Flame retardancy	Fabrics with layered nanosilicates (clay) and CNTs	

34.3
Lightweight High-strength Metal Matrix Nanocomposites

For the lightweight high-strength automobile component application, low density metals such as Al, Mg, and Ti were considered as the matrix constituent for the fabrication of MMNCs [8,37–45]. Among the matrices, the most investigated one is Al and its alloys. The main reason for the acceptance of "aluminum" matrices is that it is not only suitable for transport industry but also adaptable for other structural applications with flexible processing capability [46]. Apart from aluminum, significant number of work has been reported on Mg-based MMNCs [47]. On the other hand, nanodispersoids exist in different varieties according to the applications. The major classification of reinforcements is like oxides, carbides, hydrides, and borides. Among them, alumina (Al_2O_3) [48–51] and yttria (Y_2O_3) [52–54] are common oxides, while Si_3N_3 and AlN [55] are popularly used nitrides. Carbides such as B_4C, SiC, and TiC [56–61] are nanocarbides, which researchers consider as reinforcement particles. The well-established boride refiner for aluminum, that is, TiB_2 [62–66] is one of the best reinforcements of interest. Moreover, CNTs [67], graphene [68,69], and graphite [70,71] particles are also employed as the reinforcement. Recently, *in situ* generated intermetallic particles such as Al_3Ti [72,73], NiAl [74,75], and Mg_2Si [76,77] had also been successfully used as reinforcement particles.

34.3.1
Manufacturing of Metal Matrix Nanocomposites

A brief description of various processing methods available for manufacturing of MMNCs is given in Table 34.5 along with its advantages and limitations [74,78–89]. The major classification of processing methods of MMNCs is solid-state processing, liquid-state processing, and semi-solid processing. Solid-state processing comprises either simple blending of matrix and reinforcement powders or reactive milling of powders to form *in situ* nanostructure materials. Liquid-state processing is a viable large-scale processing method. In this process, either *ex situ* nanoparticles or precursors for *in situ* generation of nanosized reinforcement particles are mixed with molten metal using conventional or nonconventional stirring method. Semisolid processing is an emerging method for the manufacturing of MMNCs, which combines advantages of both solid-state and liquid-state processing [90–93].

The choice of a manufacturing method for fabrication of transport sector components is made on the basis of its industrial scalability. Among the various manufacturing routes, liquid- and semisolid state processing of MMNCs satisfy this criterion. In the following section, the major issues associated with processes relevant for manufacturing of components in transport

Table 34.5 Potential processing methods for manufacturing of MMNCs for transport application.

Process		Description	Remarks	Potential Automobile and Aerospace Components
Solid-state processing	Powder metallurgy	Simple blending of matrix and reinforcement particles using a mechanical mixer	Advantages: Uniform dispersion of reinforcement. Large fraction of reinforcement can be added. Good interfacial bonding. Nanostructured matrix grains can be achieved; further improve the strength of nanocomposites Limitations: Complex and lengthy process. Not scalable to industrial level production. Contamination from milling media. Complex consolidation and machining process is required to obtain near-net shape components	Engine mountings, Driveline Shift forks, drive shafts, and gears Wheels and power line cases Transmission housing Aircraft electrical and electronic parts Electronic parts and packaging; antenna parts
	Mechanical alloying	High-energy ball milling of powders of matrix and reinforcement together. Nanocomposites are obtained by repeated cold welding and fracture of powder particles		
	Cryo milling	Mechanical alloying at low temperature (in the presence of liquid nitrogen)		
	Friction stir processing	Uses frictional torsion force to induce severe plastic deformation and disperse or generates nanoparticle with fine grained matrices	Advantages: Simple and hassle free process. Uniform dispersion of nanoparticles can be achieved. Limitations: Not scalable to industrial level production. Suitable only for simple geometries	
	Equal-channel angular pressing process	Uses shear force for inducing severe plastic deformation and associated generation or dispersion of nanoparticles in ultrafine grained matrices		

(continued)

Table 34.5 (Continued)

Process		Description	Remarks	Potential Automobile and Aerospace Components
Liquid-state processing	Stir casting	*Ex situ* particles will be added to molten metal with the help of vortex formed by a specially designed stirrer	Advantages: Simple and easy process. Scalable to large-scale industrial Limitations: Possibility of rejection of particles. Agglomeration of particles. Air entrapment. Only a small fraction of reinforcement can be added. Possibility of contamination from stirrer materials. Poor wettability of reinforcement particles. Lack of fluidity	Engine components like piston, cylinder block/liner, crank case and cylinder head cover, intake manifold, engine mountings, connecting rod. Suspension and driveline components such as shift forks Drive shafts, gears, and wheels Power line cases Transmission housing Air craft structures; mid fuselage, landing gear, and other structures. Hubble Telescope antennae waveguide/boom
	Ultrasonic-assisted fabrication	Uses cavitation implosion phenomenon to improve wettability; dispersion and generation of *in situ* nano particles	Advantages: Simple and easy scalable process. Uniform dispersion of particles with reduced porosity. Good interfacial bonding. Increases the kinetics of *in situ* nanoparticle formation Limitations: Possibility of rejection of particles. Agglomeration of particles. Only small fraction of reinforcement can be added. Contamination from sonotrode dissolution	
Semisolid processing	Combo casting or rheocasting	It uses the thixotropic property of materials. Application of shear force in semisolid condition of a metal makes it to behave like a Newtonian fluid. Addition of nanoparticles under this condition can improve dispersion	Advantages: Scalable to large-scale industrial application. Uniform dispersion of reinforcement particles with reduced porosity. It solves problems associated with the lack of fluidity. Near-net shape component can be manufactured. Combines the advantage of both solid- and liquid-state processing. Limitations: This process is not adaptable for all matrix alloys, Certain solidification criteria should be followed by matrix materials	

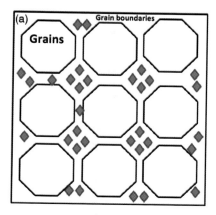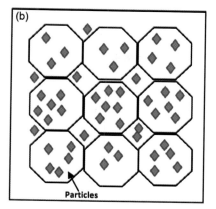

Figure 34.2 Possible distribution of the matrix and nano-reinforcement particles in the composites.

industry, that is, liquid-state and semisolid-state processing, are discussed along with possible solutions.

i) Dispersion of nanoparticles:

Because of the low wettability of the nanosized particles, the nanoparticles would agglomerate in the grain boundaries (Figure 34.2a), not to be homogeneously dispersed throughout the matrix [67], and the optimal exploitation of the strengthening potential is lost. Figure 34.2b shows that the reinforcement particles are uniformly distributed in the matrix and it is expected to possess best mechanical properties [68]. Adoption of unconventional stirring by ultrasonic processing is one of the simple and scalable approaches to obtain uniform dispersion of nanoparticles.

ii) Wettability and achievement of good interfacial bond:

The high surface energy and low wettability of ceramic nanoparticles leads to the formation of clusters of nanoparticles. The agglomeration of particles does not allow the formation of proper physical–chemical bond between matrix and reinforcement and it is not effective in hindering the movement of dislocations; thus, it reduces the strengthening capability of nanoparticles [69]. Several unconventional production methods have been studied by researchers in order to overcome the wettability issue; ultrasonic-assisted stirring is one among the best processes available for improving the wettability of nanoparticles with molten metal.

iii) Thermal stability of nanoparticles:

MMCs produced by conventional *ex situ* methods are not thermodynamically stable, as interfacial reaction occurs between matrix alloy and reinforcement particles [94]. The kinetics of interfacial reaction expected to increase many fold when reinforcement particles are in the nanorange. These interfacial reactions are well known to have several undesirable effects on the mechanical and corrosion properties of composites. The formation of thermodynamically stable particle by *in situ* reaction is one of the possible approaches to overcome this problem.

iv) Limited fluidity:

It is difficult to fabricate complex-shaped MMNC components through the conventional casting process due to lack of fluidity [95]. The manufacturing of near-net shape components through the semisolid process is not affected by the fluidity of liquid metal and hence this process is suitable for near-net shape thin-wall castings of MMNCs.

34.3.2
Mechanical Property of MMNCs

A comparison of mechanical properties, especially tensile strength and elongation, of MMC and MMNC is given in Table 34.6 [39,96–102]. The addition of micrometer-size reinforcement particle improves the yield and tensile strength of MMCs with substantial reduction in ductility. Studies on MMNCs emphasize

Table 34.6 Comparison of mechanical properties between MMC and MMNC.

Matrix Material	Reinforcement			Yield Strength (MPa)	Ultimate Tensile Strength (MPa)	Elongation (%)	Hardness (GPa)
	Type	Fraction%	Size				
Al-6061	—	—	—	80	140	10	588..8
Al-6061	TiB$_2$	12	1–2 µm	—	173.6	7	931.7
Al-6061	SiC	10	40–45 µm	—	168.8	4	686.5
Al-6061	MWCNT	1	20–30 nm	110.2	189.49	9	764.9
Al	—	—	—	50	120	20	—
Al	SiC	10	10 µm	110	180	15	—
Al	Al$_2$O$_3$	2	25 nm	135	210	18	—
Al-2014	—	—	—	96			441.3
Al-2014	TiB$_2$	2	1–2 µm	119			549.2
Al-2014	TiB$_2$	2	20–30 nm	198			666.9
A356	—	—	—	75	135	5.5	585
A356	SiC	15	40 µm	90	150	2	
A356	SiC	2	20–30 nm	160	300	3.8	765
AZ91	—	—	—	90	130	7.5	—
AZ91	SiC	10	7 µm	135	152	0.8	—
AZ91	SiC	1	30 nm	150	220	5.5	—
AZ91	CNT	1	50 nm	150	210	7.5	—
AZ91	CNT and SiC	0.7%CNT +0.3%SiC	SIC 30 nm, CNT 50 nm	200	300	9	—

Table 34.7 Potential applications of MMNCs in ground transport industry.

Major Properties Required	MMNC Type	Candidates
High strength and stiffness	Al with nanosized TiB_2, Al_2O_3, CNT, and other *in situ* ceramics and borides	Connecting rod, brake rotors, and calipers; wheels, power line, and transmission parts
Lightweight and high-energy damping capacity	Al/Gr, Al/Fly ash, Al MMNC foams and Mg/Mg_2Si MMNCs	Crumple zones, frame members, struts, seat shell, steering column, steering wheel, bonnet inner part, door panels and tail/boot lid
High-wear resistance	Al with nanosized SiC, TiB_2, graphite, and MoS_2 CNT	Piston, cylinder block, cylinder liner, cam shaft, valve tappets, rocker arm, bearings gears, shift forks, and drive shafts
High-temperature properties	Al with nanosized TiB_2, CNT, and *in situ* ceramics	Crank case, cylinder head, piston and cylinder liners, intake manifold, and wrist pin
High-thermal conductivity	Al-based MMNCs with Gr, CNT, and nitrides	Cylinder liners, break unit parts, turbo and super charges, and electronic parts
Self-lubricating	Al-based MMNCs with Gr, MoS_2, TiB_2, and other solid lubricants	Bearing journals, cylinder liners, pistons, CV joints, and gear and power drives
Self-cleaning	Superhydrophobic and anticorrosion nanocoatings, UV preventive nanocoating	Coolant pumps and water jackets, windshield, exposed metallic components. Underbody coatings and for exterior beatification

that the same or higher level of strength can be obtained without an appreciable decrease in ductility by adding a very small fraction of nanoparticles.

34.3.3
Potential Applications of MMNCs in Automobile Sector

The uniqueness and tailoring ability of MMNCs have led numerous applications in transport industry. The potential usage of MMNCs in ground transport industry is listed in Table 34.7 [10,12–14,37,103].

34.3.4
Metal Matrix Nanocomposites for Aerospace Applications

Aerospace diligence is one of the leading adopters of advanced composite materials, particularly MMNCs. The industry demands for lightweight materials with high strength to weight ratio along with the ability to fabricate complex shapes, which is important for space applications. In fact, composites constitute an average of 25% of aircraft by weight since their introduction. Some of the potential uses of MMNCs in aerospace industry are listed in Table 34.8 [104].

Table 34.8 Potential candidates of MMNCs in aerospace industry.

Major Properties Required	MMNC Type	Candidates
High-specific strength	Al/SiC, *in situ* ceramics	Structural parts such as mid-fuselage, landing gear, and actuators
High-temperature properties	Graphite–aluminum	Hubble Telescope antennae waveguide/boom
Lightweight	Al/Gr, Al/SiC, Al/Al$_2$O$_3$ and Mg/Gr, Mg/SiC, and MMNC-based foams	Frame members, struts, seat shell, and other body parts
Good thermal and electrical property	Al/SiC, Al/Gr, and Al/CNT	Electronic parts and packaging; antenna parts

34.3.5
Metallic Foam-based Lightweight Structure

In the transportation design, the requirement of weight reduction with improved performance is often in conflict with the necessities of increased safety. Aluminum foams are ultralightweight material (density 0.3–1 g/cm^3) with the capability of damping noise and vibration [105]; they also have excellent impact energy absorbing capability [106]. Possessing these combinations of properties makes aluminum foam an interesting choice of material for meeting the above-mentioned contradicting design requirement [107]. Potential applications of metal foam in the automotive industry are listed in Table 34.9 [105].

The low mechanical strength of aluminum foams limits its range of applications. In order to broaden the range of their applications, the nanoscale reinforcements, such as alumina (Al$_2$O$_3$), silicon carbide (SiC), ceramic fibers, carbon nanotubes, and so on, are added to improve the mechanical properties.

Table 34.9 Potential aluminum foam automotive components.

Major Properties Required	Candidates
Weight reduction	Casting core element, floating body element, and constructions with high stiffness at low weight
Energy absorber	Crash absorber for mobile and stationary transportation means, safety pads for lifting and conveying systems, protective covers, and blast mitigation
Acoustic, electromagnetic, and fire insulation	Casings cover plates, filling material, absorption under different conditions (high temperatures, moisture, powder, vibrations, etc.)
	Heat resistant, nonflammable constructions

Since a much smaller volume fraction of nanoparticles is required to obtain desired mechanical properties, weight contributions from the particle are negligible. Apart from improving the mechanical property, nanoparticles are found to be more effective in stabilizing metallic foams compared to micrometer-size particles [108,109].

34.4
Ceramic Matrix Nanocomposites in Transport Industry

Ceramic nanocomposites can be classified into four categories: intragranular, intergranular, hybrid, and nano/nanocomposites. Figure 34.3 depicts the differences between four types [15]. Intragranular, intergranular, and hybrid type are the ones in which one of the phases (most likely the matrix phase) may not be in the nanoregime. Nano/nanotype CMNCs consisting of both phases are around 100 nm or less [15].

34.4.1
Manufacturing Methods for CMNCs

The synthesis of CMNCs requires special processing routes. In fact, fabrication of ceramic nanocomposites is complicated and existing conventional methods associated with incorporation of micrometer-size reinforcement phases need to be modified carefully according to the adaptability of

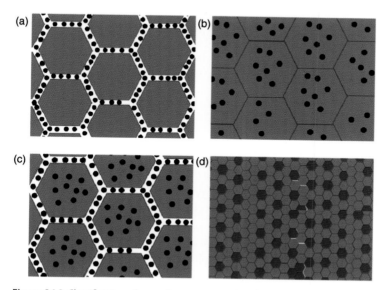

Figure 34.3 Classification of ceramic nanocomposites: (a) intertype, (b) intratype, (c) hybrid type, and (d) nano/nanotype.

nanometric particles. The basic principle behind the successful process is that the specific process could enable the crystalline phases to nucleate but suppress the growth of nuclei. Further consolidation of a fine-sized powder mixture can make the components [15].

Table 34.10 [16,110] summarizes the different processes available for synthesis of ceramic nanocomposites. The major methods used for the fabrication of CMNCs are plasma synthesis, CVD, sputtering, sol–gel technique, combustion synthesis, intercalation, organometallic pyrolysis, and mechanical milling. All the above-mentioned methods are used to generate powder composites, and consolidation of nanopowder is one of the important processes involved in the manufacturing of near-net shape CMNC components. Pressureless sintering, reaction sintering, spark plasma sintering, hot pressing, and HIPing are the major consolidation techniques for the fabrication of CMNC components [15,111].

34.4.2
Mechanical Properties of CMNCs

Incorporation of nanoparticles into ceramic matrix has an impact on densification, retarding grain growth, and elimination of porosity; consequently, it impacts the mechanical properties. A simple comparison of mechanical properties of CMNCs with CMCs is given in Table 34.11. It can be seen that addition of silicon carbide nanopaticles to Al_2O_3 ceramic matrix shows 40–60% and 30–40% improvement in fracture strength and toughness, respectively, compared to that of Al_2O_3/SiC microcomposites. Still, the biggest improvement in the mechanical property of CNMCs is observed while adding carbon-derived nanomaterials, particularly CNTs and graphene. Regarding CNTs, their incorporation has a tremendous impact on almost all properties of ceramic materials. For example, the results showed up to 250% increase in fracture toughness for addition of CNTs [17,112,113].

34.4.3
Potential CMNC Entrants in Transport Industry

CMCs can be fabricated as strong as metals, but they are much lighter and can withstand much higher temperatures. These advantages led to their application in automotive and aerospace engineering. Developments in sensor and its usage in automobiles made CMNCs familiar to ground transport industry. They improve safety features along with performance. CMNCs also serve as an advanced solution in aircraft construction. They act as strengthening elements for the structures such as frames or stringers or as the outer layer for honeycomb-type structures used at fuselage and wings. Thermal protection for turbo engines is made of zirconia-based nanocomposites. The nanocarbon–carbon composites are used in missiles, space shuttles, and reentry vehicles. It is also used as a brake disk material and

Table 34.10 Major synthesis routes for fabrication of CMNC powders with brief description.

Materials	Composite Type	Process	Process Description
$LaCeO_{2-x}$, $CuCeO_{2-x}$, and ZrO_2/Al_2O_3	Oxide/oxide	Plasma phase synthesis	The presence of plasma helps conduct electric energy to enhance kinetics of the reaction. Equilibrium plasma mode – synthesis of composite powders. Non-equilibrium plasma mode – thin-film coatings and sintering of CMNCs.
SiC/TiC/C, SiC/TiSi$_2$, SiC/TiC, and SiC/C	Nonoxide/ nonoxide	Chemical vapor deposition	Multicomponent gas reactions make nanocomposites. The main advantage of CVD is its easein process controls. Hydrocarbon-associated contamination can affect the quality of product
TiO_2/Al_2O_3, mullite/ ZrO_2, and mullite/ TiO_2	Oxide/oxide	Sol–gel technique	Conversion of a homogeneous solution into a stable phase and known as sol–gel transition. The gel is a flexible solid filling with same volume as that of solution, which following the drying process shrinks and transforms to the desired phase. The major advantage of the process is that high melting point materials can be incorporated into a noncrystalline xerogel matrix at reasonably low temperatures. Although this process is fairly viable, raw materials can be expensive
AlN/BN	Nonoxide/ nonoxide		
Al_2O_3/SiC	Oxide/nonoxide		
SiO2/BaSO$_4$ V_2O_5/SiO_2, Mg_2SiO_4, Al_2O_3/SiO_2, and ThSiO4	Glass-ceramics		
Au_2O_3/SiO_2	Oxide/oxide,	Sputtering	Cold evaporation method that uses pure metals, alloys, and compounds to deposit a film of the material on a substrate. The usage of reactive gases such as oxygen, nitrogen, and hydrogen along with inert gas can lead to the production of oxides or nitride thin films by *in situ* reaction
Ni$_3$N/AlN	Nitride/nitride		
TiO_2/SiO_2	Glass-ceramics	Intercalation	Cations are exchanged chemically followed by calcination to get CMNCs
SiC/Al_2O_3, $Si_3N_4/$ ZrO_2, TiN/Al_2O_3, and TiC/Al_2O_3	Nonoxide/oxide	Organometallic pyrolysis	Pyrolyzing of organometallic precursors to produce ceramic materials. Suitable for producing ceramic fibers, coatings, or reactive amorphous powders
TiB_2/TiN, ZnS/GaP, and TiN/AlN	Nonoxide/ nonoxide		

(continued)

Table 34.10 (Continued)

Materials	Composite Type	Process	Process Description
Cu/NiO_2	Oxide/metal	Combustion synthesis/spray conversion	Ceramic powders are formed by igniting the precursor solutions. An intermediate amorphous solid forms from the liquid precursor and the local rearrangement of the ions without species interdiffusion leads to the formation of crystalline phases
$Cu-NiO-NiFe_2O_4$	Oxide/oxide		
Al_2O_3/ZrO_2			
FeO/SiO_2	Glass-ceramics		
Ti/TiB_2	Nonoxide/metal		
WC-Co	Nonoxide/metal		
Al_2O_3/Fe, Cr, $Al_2O_3/$ Fe-Cr, and Fe/α -Al_2O_3	Oxide metal	Mechanical alloying	Ball milling of powders of matrix and reinforcement together to form nanocomposites

brake lining for military and civil aircraft. When combined with nanoaddi-tive-integrated ceramic matrix, nanocomposites represent a unique solution for the radomes of the hypersonic airplanes. Table 34.12 summarizes the usage of CMNCs in ground transport and aerospace industries [15,114].

Table 34.11 Mechanical properties of CMNCs and CMCs.

System	Fracture Strength (MPa)	Fracture Toughness (MPa/m^2)	Hardness (GPa)
Al_2O_3/SiC_p microcomposite	283	2.4	18.5–19.5
Al_2O_3/SiC_p nanocomposite	549–646	4.5	19.1–21.2
Si_3N_4/SiC_p microcomposite	700	4–5.3	16.5–17
Si_3N_4/SiC_p nanocomposite	1300	6–7.5	14–18.8
Al_2O_3/CNT	650	5.7	15.9
MgO/SiC	340–700	1.7–4	—
Al_2O_3/ZrO_2	—	8–10	4
SiC/CNT	321	3.8	—
ZrB_2-SiC/CNT	616	4.6	—
3Y-TZP/Mo	2100	11.5	—
$SiC-Si_3N_4$	640	6.6	—
Mulite/CNT	512	3.3	—
Graphene/Al_2O_3	600	—	15–18

Table 34.12 Potential applications of CMNCs in automobile and aerospace industries.

Key Properties	Ceramic Matrix	Ceramic Reinforcement	Parts/Components
Wear resistance, toughness, and good electrical and thermal properties	Al_2O_3	CNT/graphene	Cutting tools, electric contacts, armor plates, and corrosion-resistant pipes
Mechanical, chemical, and thermal properties	Si_3N_4	CNT/graphene	Gas turbine, air craft engine components, and bearings
Electrical, piezoelectric, and magnetorestrictor properties	$BaTiO_3$	CNT/graphene	Sensors and transducers
Good mechanical properties with excellent fracture toughness and high-temperature stability. Large energy bandgap and high breakdown electrical field	ZrO_2	CNT/graphene	Fuel cells and oxygen sensors
Electrical properties	TiN and FeN	CNT/graphene	Capacitors and electronic conductors
High electrical and optical properties	Mulite	CNT/graphene	Sensors
Wear resistance and thermal stability	Carbon	SiC	Break disks, rotors, and clutch assemblies
High flexural strength, fracture toughness, and good thermal stability	SiC and TiB_2	Si_3N_4 and SiC	Gas turbine parts such as turbine rotor, back plate, orifice liner, and extension liner
Improved mechanical properties and good thermal stability	SiC	SiC	Combustion chambers, vanes, and burners

34.5
Nanocomposite Coating

Coating plays an important role in improving the efficiency and lifespan of the automobile or aerospace components by reducing frictional loss, providing thermal insulation, and reducing corrosion and erosion. Nanocomposites are used in producing functional coatings for self-cleaning, scratch-resistant, anticorrosion, and ultralow friction coating applications [67]. Various types of functional nanocomposite coatings used in transport applications are summarized in Table 34.13 [69,115–120].

34.6
Challenges and Opportunities for Nanocomposites

Nanosized reinforcements are now being used to improve the performance of composites in aerospace and automotive industries. Among the nanocomposites,

Table 34.13 Functional nanocomposite coatings used in transport industry.

Major Property Requirement	Coating Type	Potential Candidate
Scratch resistance	Incorporate nanosized organosilanes, nanoalumina, silica, zirconia, titania, and so on to clear coat matrix to enhance scratch strength	Polymer glass parts such as headlight covers, wind shield, and sun roofs
Anticorrosive nature	Incorporating nanoparticles into electrodeposition coating formulation is another approach to improve the anticorrosion performance of a car body. Nanoparticles such as nano-SiO_2, nano-TiO_2, nanoclay, nanocarbon tubes, and so on are used to improve electrocoating properties	Exterior parts and underbody of an automobile
Weather resistance	Nanoparticles such as zinc oxide, iron oxide, cerium oxide, titanium oxide, and silica have been incorporated into polymeric coatings to enhance their resistance against sunlight	Exterior surfaces of an automobile
Self-cleaning coating	Ultrathin layer aluminum oxide coatings	Mirrors and headlights used in cars
Photocatalytic, self-cleaning and superhydrofilic coatings	Utilizing the photocatalytic activity of TiO_2 to create germicidal and self-cleaning surfaces	Antifogg glass in automobile
Low thermal conductivity coating for thermal protection	Nanocrystalline YSZ coating using plasma spraying process	Thermal protection of turbine blades and hypersonic vehicles

PMNCs have already found several interesting applications in transport industry. Experts in PMNCs had already identified that along with clays, the new generation of nanoparticles such as CNTs, magnetic nanoparticles, and graphene could drive the market dynamics. Studies on PMNCs suggest that the research world should concentrate more on developing simple hassle free, industrially viable methods for fabrication of PMNCs since the requirement of PMNCs in transport sector is growing very rapidly in both interior and exterior of vehicles.

Though significant opportunities have been identified for MMNCs, a number of barriers must be overcome to ensure widespread adoption of these materials in high production volume transport industry. These challenges can be classified as either technological or infrastructural issues. In the case of some MMNC components, final product costs may be higher than those of the components made of conventional materials because of the higher costs associated with raw materials, primary and secondary processing such as shape fabrication, and machining. Exploration of new technological and infrastructural development such as selection of inexpensive raw materials, near-net shape forming technology, rapid and inexpensive machining processes, and so on could reduce the costs and may lead to quick acceptance of these materials.

Ceramic nanocomposites have already been identified as potential candidates in the transport industry due to their widespread application as high-temperature material, high-sensitive sensors, and energy-efficient solar cells, as well as in various manufacturing stages of transport crafts. The optical, electrical, and magnetic properties of these kinds of materials will form the basis for their widespread and industrial applications in the very near future. The budding technologies such as additive manufacturing and 3D printing can make the components familiar to the world of transport industry.

References

1 Cole, G.S. and Sherman, A.M. (1995) Light weight materials for automotive applications. *Mater. Charact.*, **35** (1), 3–9.

2 Eliezer, D., Aghion, E., and Froes, F.H. (1998) Magnesium science, technology and applications. *Adv. Perform. Mater.*, **5** (3), 201–212.

3 Jambor, A. and Beyer, M. (1997) New cars: new materials. *Mater. Des.*, **18**, 203–209.

4 He, F., Han, Q., and Jackson, M.J. (2008) Nanoparticulate reinforced metal matrix nanocomposites: a review. *Int. J. Nanoparticles*, **1** (4), 301–309.

5 Hussain, F., Hojjati, M., Okamoto, M., and Gorga, R.E. (2006) Review article: Polymer-matrix Nanocomposites, Processing, Manufacturing, and Application: An Overview. *J. Compos. Mater.* **40** (17), 1511–1575.

6 Vivek, P., Yashwant (2011) Polymer nanocomposites drive opportunities in the automotive sector. *Nanotechnol. Insights*, **2**(4), 17–24.

7 Zhao, W., Li, M., and Peng, H.X. (2010) Functionalized MWNT-doped thermoplastic polyurethane nanocomposites for aerospace coating applications. *Macromol. Mater. Eng.*, **295** (9), 838–845.

8 Alam, M.E., Han, S., Nguyen, Q.B. *et al.* (2011) Development of new magnesium based alloys and their nanocomposites. *J. Alloys Compd.*, **509** (34), 8522–8529.

9 Borgonovo, C., Apelian, D., and Makhlouf, M.M. (2011) Aluminum nanocomposites for elevated temperature applications. *JOM*, **63** (2), 57–64.

10 Allison, J.E. and Cole, G.S. (1993) Metal-matrix composites in the automotive industry: opportunities and challenges. *JOM*, **45** (1), 19–24.

11 Rohatgi, P. (1991) Cast aluminum-matrix composites for automotive applications. *JOM*, **43** (4), 10–15.

12 Chawla, N. and Chawla, K.K. (2006) Metal-matrix composites in ground transportation. *JOM*, **58** (11), 67–70.

13 Casati, R. and Vedani, M. (2014) Metal matrix composites reinforced by nano-particles: a review. *Metals (Basel)*, **4** (1), 65–83.

14 Borgonovo, C. and Apelian, D. (2011) Manufacture of aluminum nanocomposites: a critical review. *Mater. Sci. Forum*, **678**, 1–22.

15 Bhaduri, S. (1998) Recent developments in ceramic nanocomposites. *JOM*, **50** (1), 44–51.

16 Silvestre, J., Silvestre, N., and De Brito, J. (2015) An overview on the improvement of mechanical properties of ceramics nanocomposites. *J. Nanomater.* doi: 10.1155/2015/106494.

17 Palmero, P. (2015) Structural ceramic nanocomposites: a review of properties and powders, synthesis, methods. *Nanomaterils*, **5**, 656–696.

18 Henrique, P., Camargo, C., Satyanarayana, K.G., and Wypych, F.

(2009) Nanocomposites: synthesis, structure, properties and new application opportunities. *Mat. Res.*, **12** (1), 1–39.

19 Njuguna, J. and Pielichowski, K. (2003) Polymer nanocomposites for aerospace applications: properties. *Adv. Eng. Mater.*, **5** (11), 769–778.

20 Njuguna, J. and Pielichowski, K. (2004) Polymer nanocomposites for aerospace applications: characterization. *Adv. Eng. Mater.*, **6** (4), 204–210.

21 Njuguna, J. and Pielichowski, K. (2004) Polymer nanocomposites for aerospace applications: fabrication. *Adv. Eng. Mater.*, **6** (4), 193–203.

22 Garcøs, B.J.M., Moll, D.J., Bicerano, J. et al. (2000) Polymeric nanocomposites for automotive applications. *Adv. Mater.*, **12** (23), 1835–1839.

23 Giannelis, E.P. (1998) Polymer-layered silicate nanocomposites: synthesis, properties and applications. *Appl. Organomet. Chem.*, **12** (10–11), 675–680.

24 Somwangthanaroj, A., Tantiviwattanawongsa, M., and Tanthapanichakoon, W. (2012) Mechanical and gas barrier properties of nylon 6/clay nanocomposite blown films. *Eng. J.*, **16** (2), 93–105.

25 Park, C., Wise, K.E., Kang, J.H. et al. (2008) Multifunctional nanotube polymer nanocomposites for aerospace applications: adhesion between SWCNT and polymer matrix. https://ntrs.nasa.gov/search.jsp, 1–3.

26 Tjong, S.C. (2006) Structural and mechanical properties of polymer nanocomposites. *Mater. Sci. Eng. R Rep.*, **53** (3–4), 73–197.

27 Paul, D.R. and Robeson, L.M. (2008) Polymer nanotechnology: nanocomposites. *Polymer (Guildf).*, **49** (15), 3187–3204.

28 Njuguna, J., Silva, F., and Sachse, S. (2011) Nanocomposites for vehicle structural applications. doi: 10.5772/23261.

29 Mittal, V. (2010) Polymer nanocomposites: synthesis, microstructure, and properties, in *Optimization of Polymer Nanocomposite Properties*, pp. 1–20.

30 Mittal, V. (2015) Synthesis of polymer nanocomposites. *Polymer (Guildf).*, 1–26.

31 Huang, H., Shu, D., Fu, Y. et al. (2014) Synchrotron radiation X-ray imaging of cavitation bubbles in Al-Cu alloy melt. *Ultrason. Sonochem.*, **21** (4), 1275–1278.

32 Thostenson, E.T., Li, C., and Chou, T.W. (2005) Nanocomposites in context. *Compos. Sci. Technol.*, **65** (3–4), 491–516.

33 Pandey, J.K., Reddy, K.R., Mohanty, A.K., and Misra, M. (2014) *Handbook of Polymernanocomposites. Processing, Performance and Application*, Springer.

34 Privalko, V.P., Shantalii, T.A., and Privalko, E.G., Polyimides reinforced by a sol–gel derived organosilicon nanophase: synthesis and structure–property relationships. **6**, doi: 10.1007/0-387-26213-X_4.

35 Fukushima, Y. and Inagaki, S. (1987) Synthesis of an intercalated compound of montmorillonite and 6-polyamide. *J. Incl. Phenom.*, **5** (4), 473–482.

36 Baur, J. and Silverman, E. (2007) Challenges and Opportunities in Multifunctional Nanocomposite Structures for Aerospace Applications, *MRS Bulletin*. **32** (4), 328–334.

37 Muley, A.V., Aravindan, S., and Singh, I.P. (2015) Nano and hybrid aluminum based metal matrix composites: an overview. *Manuf. Rev.* doi: org/10.1051/mfreview/2015018.

38 Yang, X., Zou, T., Shi, C. et al. (2016) Effect of carbon nanotube (CNT) content on the properties of *in-situ* synthesis CNT reinforced Al composites. *Mater. Sci. Eng. A.*, **660**,11–18.

39 Yang, Y., Lan, J., and Li, X. (2004) Study on bulk aluminum matrix nano-composite fabricated by ultrasonic dispersion of nano-sized SiC particles in molten aluminum alloy. *Mater. Sci. Eng. A*, **380** (1), 378–383.

40 Zhou, D.S., Tang, J., Qiu, F. et al. (2014) Effects of nano-TiCp on the microstructures and tensile properties of TiCp/Al-Cu composites. *Mater. Charact.*, **94**, 80–85.

41 Lim, C.Y.H., Leo, D.K., Ang, J.J.S., and Gupta, M. (2005) Wear of magnesium composites reinforced with nano-sized

alumina particulates. *Wear*, **259** (1–6), 620–625.

42 Meenashisundaram, G., Nai, M., Almajid, A., and Gupta, M. (2016) Reinforcing low-volume fraction nano-tin particulates to monolithical, pure Mg for enhanced tensile and compressive response. *Materials (Basel)*, **9** (3), 134.

43 Wong, W.L.E. and Gupta, M. (2007) Development of Mg/Cu nanocomposites using microwave assisted rapid sintering. *Compos. Sci. Technol.*, **67** (7–8), 1541–1552.

44 Meenashisundaram, G.K., Seetharaman, S., and Gupta, M. (2014) Enhancing overall tensile and compressive response of pure Mg using nano-TiB2 particulates. *Mater. Charact.*, **94**, 178–188.

45 Kondoh, K., Threrujirapapong, T., Imai, H. *et al.* (2008) CNTs/TiC reinforced titanium matrix nanocomposites via powder metallurgy and its microstructural and mechanical properties. *J. Nanomater.*, **2008** (1). doi: org/10.1155/2008/127538.

46 Kaufman, J.G. and Rooy, E.L. (2004) *Aluminum Alloy Castings: Properties, Processes, and Applications*, ASM International.

47 Gupta, M. and Wong, W.L.E. (2015) Magnesium-based nanocomposites: lightweight materials of the future. *Mater. Charact.*, **105**, 30–46.

48 Poirier, D., Drew, R.A.L., Trudeau, M.L., and Gauvin, R. (2010) Fabrication and properties of mechanically milled alumina/aluminum nanocomposites. *Mater. Sci. Eng. A*, **527** (29–30), 7605–7614.

49 Hesabi, Z.R., Hafizpour, H.R., and Simchi, A. (2007) An investigation on the compressibility of aluminum/nano-alumina composite powder prepared by blending and mechanical milling. *Mater. Sci. Eng. A*, **454–455**, 89–98.

50 Nguyen, V.S., Rouxel, D., Hadji, R. *et al.* (2011) Effect of ultrasonication and dispersion stability on the cluster size of alumina nanoscale particles in aqueous solutions. *Ultrason. Sonochem.*, **18** (1), 382–388.

51 Alam, M.E., Han, S., Nguyen, Q.B. *et al.* (2011) Development of new magnesium based alloys and their nanocomposites. *J. Alloys Compd.*, **509** (34), 8522–8529.

52 Tun, K.S. and Gupta, M. (2007) Improving mechanical properties of magnesium using nano-yttria reinforcement and microwave assisted powder metallurgy method. *Compos. Sci. Technol.*, **67** (13), 2657–2664.

53 Hassan, S.F., Tun, K.S., Gasem, Z.M. *et al.* (2015) Effect of hybrid reinforcement on the high temperature tensile behavior of magnesium nanocomposite. *Int. J. Mater. Res.*, **106** (12), 1298–1302.

54 Hassan, S.F. and Gupta, M. (2007) Development of nano-Y2O3 containing magnesium nanocomposites using solidification processing. *J. Alloys Compd.*, **429** (1–2), 176–183.

55 Cao, G., Choi, H., Oportus, J. *et al.* (2008) Study on tensile properties and microstructure of cast AZ91D/AlN nanocomposites. *Mater. Sci. Eng. A*, **494** (1–2), 127–131.

56 Conzone, S.D., Blumenthal, W.R., and Vainer, J.R. (1995) Fracture toughness of TiB$_2$ and B4C using the single-edge precracked beam, indentation strength, chevron notched beam, and indentation strength methods. *J. Am. Ceram. Soc.*, **78** (8), 2187–2192.

57 Moradi, M.R., Moloodi, A., and Habibolahzadeh, A. (2015) Fabrication of nano-composite Al-B4C foam via powder metallurgy-space holder technique. *Procedia Mater. Sci.*, **11** (2000), 553–559.

58 Gupta, M., Lu, L., Lai, M.O., and Ang, S.E. (1995) Effects of type of processing on the microstructural features and mechanical properties of AlCu/SiC metal matrix composites. *Mater. Des.*, **16** (2), 75–81.

59 Zhou, Y., Zhang, H., and Qian, B. (2007) Friction and wear properties of the co-deposited Ni-SiC nanocomposite coating. *Appl. Surf. Sci.*, **253** (20), 8335–8339.

60 Song, M.S., Zhang, M.X., Zhang, S.G. *et al.* (2008) *In situ* fabrication of TiC particulates locally reinforced aluminum matrix composites by self-propagating reaction during casting. *Mater. Sci. Eng. A*, **473**, 166–171.

61 Campus, D. (2005) Production of TiC reinforced-aluminum composites with the addition of elemental carbon. *Mater. Lett.*, **59**, 3795–3800.

62 Lu, L., Lai, M.O., and Chen, F.L. (1997) *In situ* preparation of TiB_2 reinforced Al base composite. *Adv. Compos. Mater.*, **6** (4), 299–308.

63 Mandal, A., Maiti, R., Chakraborty, M., and Murty, B.S. (2004) Effect of TiB_2 particles on aging response of Al-4Cu alloy. *Mater. Sci. Eng. A*, **386** (1–2), 296–300.

64 Pramod, S.L., Rao, A.K.P., Murty, B.S., and Bakshi, S.R. (2015) Effect of Sc addition on the microstructure and wear properties of A356 alloy and A356–TiB_2 in situ composite. *Materials and Design*, **78**, 85–94.

65 Nampoothiri, J., Raj, B., and Ravi, K.R. (2015) Role of ultrasonic treatment on microstructural evolution in A356/TiB 2 in-situ composite. *Trans. Indian Inst. Met.*, **68** (6), 1101–1106.

66 Chakraborty, M., Mandal, A., Kumar, G.S.V. *et al.* (2012) Recent developments in aluminium alloy reinforced titanium diboride *in-situ* composites. *Ind. Foundry J.*, **58** (11), 29–34.

67 Bakshi, S.R., Lahiri, D., and Agarwal, A. (2010) Carbon nanotube reinforced metal matrix composites: a review. *Int. Mater. Rev.*, **55** (1), 41–64.

68 Sharma, N., Kumar, N., Dhara, S. *et al.* (2012) Tribological properties of ultra nanocrystalline diamond film-effect of sliding counterbodies. *Tribol. Int.*, **53**, 167–178.

69 Bartolucci, S.F., Paras, J., Rafiee, M.A. *et al.* (2011) Graphene–aluminum nanocomposites. *Mater. Sci. Eng. A.*, **528**, 7933–7937.

70 Saravanan, M., Pillai, R.M., Ravi, K.R. *et al.* (2007) Development of ultrafine grain aluminium-graphite metal matrix composite by equal channel angular pressing. *Compos. Sci. Technol.*, **67** (6), 1275–1279.

71 Roshini, P.C., Nagasivamuni, B., Raj, B., and Ravi, K.R. (2015) Ultrasonic-assisted synthesis of graphite-reinforced Al matrix nanocomposites. *J.. Mater. Eng. Perform.* doi: 10.1007/s11665-015-1491-4.

72 Liu, Z., Han, Q., and Li, J. (2011) Ultrasound assisted in situ technique for the synthesis of particulate reinforced aluminum matrix composites. *Composites B Eng.*, **42** (7), 2080–2084.

73 Liu, Z.W., Wang, X.M., Han, Q.Y., and Li, J.G. (2013) Effect of ultrasonic vibration on direct reaction between solid Ti powders and liquid Al. *Metall. Mater. Trans. A*, **45** (2), 543–546.

74 Udhayabanu, V., Ravi, K.R., and Murty, B.S. (2011) Development of in situ NiAl-Al_2O_3 nanocomposite by reactive milling and spark plasma sintering. *J. Alloys Compd.*, **509** (Suppl 1), S223–S228.

75 Udhayabanu, V., Ravi, K.R., and Murty, B.S. (2013) Ultrafine-grained, high-strength NiAl with Al_2O_3 and Al_4C_3 nanosized particles dispersed via mechanical alloying in toluene with spark plasma sintering. *Mater. Sci. Eng. A*, **585**, 379–386.

76 Zhang, J., Zhao, Y., Xu, X., and Liu, X. (2013) Effect of ultrasonic on morphology of primary Mg 2 Si in *in-situ* Mg 2 Si/Al composite. *Trans. Nonferrous Met. Soc. China*, **23**, 2852–2856.

77 Zhang, S.L., Zhao, Y.T., and Chen, G. (2010) *In situ* (Mg2Si+MgO)/Mg composites fabricated from AZ91-Al 2 (SiO3)3 with assistance of high-energy ultrasonic field. *Trans. Nonferrous Met. Soc. China (Engl. ed.)*, **20** (11), 2096–2099.

78 Suryanarayana, C., Ivanov, E., and Boldyrev, V. (2001) The science and technology of mechanical alloying. *Mater. Sci. Eng. A*, **304–306**, 151–158.

79 Suryanarayana, C. and Al-Aqeeli, N. (2012) Mechanically alloyed nanocomposites. *Prog. Mater. Sci.*, **58** (4), 383–502.

80 Goh, C.S., Wei, J., Lee, L.C., and Gupta, M. (2006) Development of novel carbon nanotube reinforced magnesium nanocomposites using the powder metallurgy technique. *Nanotechnology*, **17** (1), 7–12.

81 Tang, F., Hagiwara, M., and Schoenung, J.M. (2005) Microstructure and tensile properties of bulk nanostructured Al-5083/SiCp composites prepared by cryomilling. *Mater. Sci. Eng. A*, **407**, 306–314.

82 Tjong, S.C. (2007) Novel nanoparticle-reinforced metal matrix composites with enhanced mechanical properties. *Adv. Eng. Mater.*, **9** (8), 639–652.

83 Surappa, M.K. (2003) Aluminium matrix composites: challenges and opportunities. *Sadhana*, **28** (1–2), 319–334.

84 Dehghan Hamedan, A. and Shahmiri, M. (2012) Production of A356-1wt% SiC nanocomposite by the modified stir casting method. *Mater. Sci. Eng. A*, **556**, 921–926.

85 Cao, G., Kobliska, J., Konishi, H., and Li, X. (2008) Tensile properties and microstructure of SiC nanoparticle-reinforced Mg-4Zn alloy fabricated by ultrasonic cavitation-based solidification processing. *Metall. Mater. Trans. A*, **39** (4), 880–886.

86 Nampoothiri, J., Raj, B., and Ravi, K.R. (2015) Effect of ultrasonic treatment on microstructure and mechanical property of *in-situ* Al/2TiB$_2$ particulate composites. *Mater. Sci. Forum*, **830–831**, 463–466.

87 Alhawari, K.S., Omar, M.Z., Ghazali, M.J. et al. (2013) Wear properties of A356/Al$_2$O$_3$ metal matrix composites produced by semisolid processing. *Procedia Eng.*, **68**, 186–192.

88 Kumar, S.D. and Chakraborty, A.M.M. (2015) Effect of thixoforming on the microstructure and tensile properties of A356 alloy and A356-5TiB 2 *in-situ* composite. *Trans. Indian Inst. Met.*, **68** (2), 123–130.

89 Flemings, M.C. and Martinez, R.A. (2006) Principles of microstructural formation in semi-solid metal processing. *Solid State Phenom.*, **116–117**, 1–8.

90 Rajan, T.P.D., Pillai, R.M., and Pai, B.C. (1998) Reinforcement coatings and interfaces in aluminium metal matrix composites. *J. Mater. Sci.*, **33**, 3491–3503.

91 Zhou, D., Qiu, F., Wang, H., and Jiang, Q. (2014) Manufacture of nano-sized particle-reinforced metal matrix composites: a review. *Acta Metall. Sin.*, **27**, 798–805.

92 Han, Y., Dai, Y., Shu, D. et al. (2006) First-principles calculations on the stability of AlTiB$_2$ interface. *Appl. Phys. Lett.*, **89** (14), 2004–2007.

93 Mutale, C.T., Krafick, W.J., and Weirauch, D.A. (2010) Direct observation of wetting and spreading of molten aluminum on TiB$_2$ in the presence of a molten flux from the aluminum melting point up to 1033 K (760 °C). *Metall. Mater. Trans. B*, **41** (6), 1368–1374.

94 Pramod, S.L., Bakshi, S.R., and Murty, B.S. (2015) Aluminum-based cast *in situ* composites: a review. *J. Mater. Eng. Perform.*, **24** (6), 2185–2207.

95 Ravi, K.R., Pillai, R.M., Amaranathan, K.R. et al. (2008) Fluidity of aluminum alloys and composites: a review. *J. Alloys Compd.*, **456**, 201–210.

96 Nam, D.H., Kim, Y.K., Cha, S.I., and Hong, S.H. (2012) Effect of CNTs on precipitation hardening behavior of CNT/Al-Cu composites. *Carbon*, **50** (13), 4809–4814.

97 Gazawi, A., Gabbitas, B., Zhang, D. et al. (2015) Microstructure and mechanical properties of ultrafine structured Al-4wt %Cu-(2.5–10) vol. %SiC nanocomposites produced by powder consolidation using powder compact extrusion. *J. Res. Nanotechnol.*, **2015**, 1–15.

98 Ramesh, C.S., Pramod, S., and Keshavamurthy, R. (2011) A study on microstructure and mechanical properties of Al 6061–TiB$_2$ *in-situ* composites. *Mater. Sci. Eng. A*, **528** (12), 4125–4132.

99 Mula, S., Padhi, P., Panigrahi, S.C. et al. (2009) On structure and mechanical properties of ultrasonically cast Al-2% Al$_2$O$_3$ nanocomposite. *Mater. Res. Bull.*, **44** (5), 1154–1160.

100 Nampoothiri, J., Harini, R.S., Nayak, S.K. et al. (2016) Post *in-situ* reaction ultrasonic treatment for generation of Al e 4. 4Cu/TiB 2 nanocomposite: a route to enhance the strength of metal matrix nanocomposites. *J. Alloys Compd.*, **683**, 370–378.

101 Nie, K.B., Wang, X.J., Wu, K. et al. (2012) Development of SiCp/AZ91 magnesium matrix nanocomposites using ultrasonic vibration. *Mater. Sci. Eng. A*, **540**, 123–129.

102 Mazahery, A. and Shabani, M.O. (2012) Characterization of cast A356 alloy

reinforced with nano SiC composites. *Trans. Nonferrous Met. Soc. China (Engl. ed.)*, **22** (2), 275–280.

103 Rohatgi, P.K., Schultz, B., and Matters, M. (2015) Lightweight metal matrix nanocomposites: stretching the boundaries of metals synthesis, processing and properties of metal matrix. Material Matters, **2** (4), 16–21.

104 Siochi, E.J. and Harrison, J.S. (2015) Structural nanocomposites for aerospace applications. *MRS Bull.*, **40** (10), 829–835.

105 Davies, G.J. and Zhen, S. (1983) Review metallic foams: their production, properties and applications. *J. Mater. Sci.*, **18**, 1899–1911.

106 Banhart, J. (2001) Manufacture, characterisation and application of cellular metals and metal foams. *Prog. Mater. Sci.*, **46**, 559–632.

107 Garc, F., Mukherjee, M., Sol, E. *et al.* (2010) Metal foams: towards microcellular materials. *International Journal of Materials Research*, **101** (9), 1134–1139.

108 Wang, J., Yang, X., Zhang, M. *et al.* (2015) A novel approach to obtain *in-situ* growth carbon nanotube reinforced aluminum foams with enhanced properties. *Mater. Lett.*, **161**, 763–766.

109 Du, Y., Li, A.B., Zhang, X.X. *et al.* (2015) Enhancement of the mechanical strength of aluminum foams by SiC nanoparticles. *Mater. Lett.*, **148**, 79–81.

110 Belekar, R.M., Sawadh, P.S., and Mahadule, R.K. (2014) Synthesis and structural properties of Al_2O_3-ZrO_2 nano composite prepared via solution combustion synthesis. *Impact Int. J. Res. Eng. Technol.*, **2** (3), 142–152.

111 Cirakoglu, M., Bhaduri, S., and Bhaduri, S. (2002) Combustion synthesis processing of functionally graded materials in the Ti-B binary system. *J. Alloys Compd.*, **347** (1–2), 259–265.

112 Ahmad, I., Yazdani, B., and Zhu, Y. (2015) Recent advances on carbon nanotubes and graphene reinforced ceramics nanocomposites. *Nanomaterials*, **5** (1), 90–114.

113 Ohnabe, H., Masaki, S., Onozuka, M. *et al.* (1999) Potential application of ceramic matrix composites to aero-engine components. Composites: Part A, **30**, 489–496.

114 Kaya, H. (1999) The application of ceramic-matrix composites to the automotive ceramic gas turbine. *Compos. Sci. Technol.*, **59** (6), 861–872.

115 Shchukin, D.G. and Mçhwald, H. (2007) Self-repairing coatings containing active nanoreservoirs. *Small*, **3** (6), 926–943.

116 Nobel, M.L., Picken, S.J., and Mendes, E. (2007) Waterborne nanocomposite resins for automotive coating applications. **58**, 96–104.

117 Mohseni, M. and Ramezanzadeh, B. (2012) The role of nanotechnology in automotive industries. doi: 10.5772/49939.

118 Yari, H., Moradian, S., Ramazanzade, B. *et al.* (2009) The effect of basecoat pigmentation on mechanical properties of an automotive basecoat/clearcoat system during weathering. *Polym. Degrad. Stab.*, **94** (8), 1281–1289.

119 Yari, H., Mohseni, M., and Ramezanzadeh, B. (2008) Comparisons of weathering performance of two automotive refinish coatings: a case study. *J. Appl. Polym. Sci.*, doi: 10.1002/app.29341.

120 Keshri, A.K. and Agarwal, A. (2011) Wear behavior of plasma-sprayed carbon nanotube-reinforced aluminum oxide coating in marine and high-temperature environments. *J. Therm. Spray Technol.*, **20** (6), 1217–1230.

35

Semiconducting Nanowires in Photovoltaic and Thermoelectric Energy Generation

Guglielmo Vastola, Gang Zhang

Institute of High Performance Computing, 1 Fusionopolis Way, 16–16 Connexis, Singapore 138632, Singapore

35.1
Introduction

The quest for clean, abundant, and renewable energy represents one of the main challenges on humanity's path going forward. In this regard, devices such as polycrystalline silicon solar cells have reached sufficient industrial maturity to grant large-scale deployment and integration into existing national electricity grids, especially in countries such as Germany, Italy, and Japan [1] and, therefore, are among the best candidates to help solve our energy needs in an economically viable fashion. At the same time, however, commercially available solutions require a significant amount of the raw material to be manufactured. Indeed, each solar cell has usually a size of 15 cm × 15 cm and is about 200 μm thick [2]. Therefore, a power-plant class installation of, for example, 1 MW peak power does indeed require 2.5 tons of refined silicon, considering the standard 60-cell solar modules. As a result, it is clear that reducing the amount of silicon required for solar energy conversion offers significant benefits in terms of both an efficient use of natural resources and a reduced carbon footprint from material extraction, transport, and purification [3]. Moreover, it would clearly contribute to a reduction in module price, since wafer cost accounts for about 50% of the final cost of the solar module [4].

In an effort to reduce the material needed to build a semiconductor solar cell, thin-film devices have been a hot topic of research since the pioneering works of the late 1970s [4], where continuous progress is made and the current record of efficiency is 10.7% for a cell that is only 1.8 μm thick [5]. At the same time, significant efforts are ongoing to further reduce the amount of silicon used. However, as the size of the device shrinks further, the standard lithographic top-down approaches become less and less suitable, and traditional manufacturing becomes either impossible or very expensive.

Nanotechnology for Energy Sustainability, First Edition. Edited by Baldev Raj, Marcel Van de Voorde, and Yashwant Mahajan.

It is, therefore, natural to seek in nanotechnology the answer to further device miniaturization, in other words, to embrace the idea of device sizes on the order of the nanometer (10^{-9} m). Here, the advantages are twofold, namely, the reduced amount of material usage because of the small size and the emergence of new phenomena that are hindered at larger, conventional length scales. For example, the material in conventional cells has to be of high quality to allow long minority carrier diffusion length to be collected on macroscopic-size metallic contacts. On the other hand, nanotechnology-enabled solar cells allow much shorter minority carrier diffusion length, effectively allowing a lower quality, therefore lower cost, of silicon to be employed for cell manufacturing.

Among the vast ecosystem of nanostructures, nanowires (NWs) have shown significant versatility for solar cell applications, ranging from single-NW to NW-bundle solar cells, and therefore will be the main focus of this chapter. In particular, silicon nanowires (SiNWs) will be covered. This choice is justified by the importance of this material given its immediate compatibility with existing silicon-based microelectronic industry technology. A nanowire is a nanoscale structure that resembles the shape of a wire. However, its typical width is on the order of tens to hundreds of nanometers and its typical length is usually on the order of hundreds of micrometers. In the course of this chapter, we will review the fundamentals of nanowire synthesis. Later, we will address the properties of single-NW solar cells and then cover the case of NW-bundle cells. Globally, this chapter will report on the role of SiNWs in solar energy conversion application and provide a description of the current state of the art in this field.

35.2
Fabrication of Silicon and Silicon–Germanium Nanowires

In this Section, we will address the main techniques for silicon and silicon–germanium nanowire fabrication. The first technique, the vapor–liquid–solid (VLS) method, is a bottom-up approach particularly suited to achieve excellent control of NW length, width, and shape for low-volume NW production. On the other hand, the metal-assisted chemical etching (MACE) and deep reactive ion etching (DRIE) techniques are top-down approaches that can achieve significant production volumes and NW number density over the substrate, at the expense of a relatively lower control over each NW length and cross-section profile.

35.2.1
Vapor–Liquid–Solid Synthesis of Silicon Nanowires

The behavior of matter at the nanoscale offers many fascinating effects that become hindered at the human scale. One of these effects is self-assembly [6]. In self-assembly, a nanoscale material spontaneously forms three-dimensional, well-defined structures that usually have a size on the order of few to hundreds of nanometers. One example of this phenomenon is the spontaneous formation of three-dimensional quantum dots when germanium is deposited over a silicon substrate [7]. In self-assembly, synthesis occurs because it represents the path of energy minimization for the system. For example, for the case of germanium quantum dots, nucleation occurs because it allows the relief of the strain energy arising from the lattice mismatch between Ge and Si.

The self-assembly vapor–liquid–solid [8] technique allows the bottom-up growth of a NW. This technique achieves synthesis by using a catalyst, which in the case of typical silicon NW growth is a metal. The use of many metals have been extensively investigated, including copper [9], platinum [10], aluminum [11]; however, the most commonly studied catalyst is gold [12]. Therefore, in the following, we will focus on gold-catalyzed VLS growth of SiNWs.

Typical VLS growth proceeds as follows. Starting from a Si substrate, usually exposing the (1 1 1) facet, typically two monolayers of gold are deposited onto the substrate using a Knudsen cell. Then, the growth chamber is heated at above 800 °C, where the gold partially dewets the silicon substrate and rearranges in liquid droplets over a thinned Au thin film. At this point, silicon gas precursor silane (SiH_3) or disilane (Si_2H_6) is injected into the chamber together with He at a 20% gas precursor–80% He ratio at a pressure of $5 \cdot 10^{-25}$ Torr. Consequently, the gas precursor dissociates at the surface of the gold droplet and silicon is incorporated inside it. This effect can be understood by looking at the gold/silicon phase diagram, Figure 35.1a. From the figure we can see the equilibrium composition of silicon diluted in liquid gold/silicon alloy. As more silicon becomes trapped inside the gold droplet, its composition increases. Ultimately, supersaturation is reached and the excess silicon precipitates as solid by epitaxial attachment onto the silicon substrate (homoepitaxy). In other words, VLS is a three-step process, composed of (i) dissociation of gas precursor at the liquid/vapor interface with consequent incorporation inside the melt droplet, (ii) diffusion of silicon from the liquid/vapor to the liquid/solid interface, and (iii) precipitation of Si and homoepitaxy at the liquid/solid interface [12]. After the initial stages of NW growth out of the substrate, the Si composition inside the droplet reaches a steady state, where new silicon is continuously incorporated from the vapor phase and at the same time continuously precipitates at the gold/silicon interface. This way, the NWs generally grow along {1 1 1} directions and have hexagonal cross sections bounded by {2 1 1} sidewalls [12]. Postgrowth scanning electron microscope (SEM) images of typical silicon nanowires are shown in Figure 35.1b, where the gold nanodroplets are visible in bright color on top of the newly synthesized SiNWs. As a reference, NWs of diameters as small as 3 nm [13] up to 300 nm [12,14,15] can the synthesized using VLS. A reasonable growth rate is obtained using chamber temperatures in the range of $500 - 650$ °C. Under these conditions, growth typically occurs in 3–6 h. Interestingly, the growth rate is independent of wire diameter [16]. This finding is one important consequence of the fact that the rate limiting factor to NW growth is the gas precursor dissociation rate at the liquid/vapor interface [14]. In fact, by replacing silane with disilane, the growth rate at 400 °C was shown to be 31 μm/

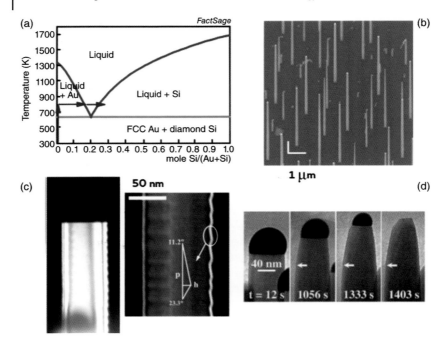

Figure 35.1 (a) Phase diagram of the Au–Si system, where the red arrow shows the path followed to initiate NW growth, (b) typical sample of Si nanowires grown by gold catalyst VLS, (c) TEM image of *in situ* NW, where the sawtooth faceting of the sidewalls is visible, and (d) TEM image of *in situ* late stages of growth of Si nanowires where tapering is evidenced. (Figure 35.1a is reproduced with permission from Ref. [12]. Copyright 2010 IOP Publishing. Figure 35.1b is reproduced with permission from Ref. [18]. Copyright 2006 the Nature Publishing Group. Figure 35.1c is reproduced with permission from Ref. [19]. Copyright 2005 the American Physical Society. Figure 35.1d is reproduced with permission from Ref. [16]. Copyright 2006 the American Physical Society.)

min, that is, 130 times faster than using silane [17], while producing wires of 20 − 80 nm diameter and length of millimeters [17].

A useful tool for investigating VLS NW growth is *in situ* transmission electron microscopy (TEM) [12]. In fact, a TEM apparatus can be mounted in conjunction with a chemical vapor deposition (CVD) growth chamber, ultimately allowing the recording of images and videos of real-time VLS self-assembly. Given that there are few and known risks arising from the influence of the TEM beam on the NW growth [12], this technique is a good choice to reveal details of growth that are hindered in postgrown analysis. For example, the shape of the gold droplet is particularly interesting. With reference to Figure 35.1c, the liquid droplet shape is governed by the surface tension of liquid gold, the liquid/solid interface energy, and, possibly, the lattice strain energy [20]. These forces must be in equilibrium at the triple junction at the NW/liquid rim. Consequently, this requirement will dictate the wetting angle of the droplet with respect to the NW sidewalls, ultimately determining its shape [19–21]. At the same time, another uniqueness shown by *in situ* TEM is the sawtooth faceting of the NW sidewalls, shown in Figure 35.1c. Such morphological peculiarities represent useful opportunities to test our models and understanding of the precise dynamics of nanowire growth [19].

One important aspect to consider when studying NWs for solar cell applications is the incorporation of gold into Si as the VLS growth proceeds. This is because the metal creates deep levels in the band structure of the semiconductor that act as local recombination centers for the photogenerated carriers. In particular, two main mechanisms become apparent as the NWs grow, namely, gold progressively incorporates into the silicon nanostructure and it also diffuses among concurrently growing NWs. Incorporation of gold into Si is governed by the solubility of the two elements in one another. For example, the concentration of Au in SiNW is typically in the range of 10^{16}–10^{20} atoms/cm^3, at growth temperatures below 500 °C [22]. This range of concentrations, in turn, gives a minority carrier lifetime of 2–4 ns [23]. This is indeed the reason why authors have explored alternative catalysts to gold, including Cu, Al, and Pt [9–11], to further limit metal incorporation into the NW. In general, however, the actual concentration depends on growth conditions [18]. As a result of gold incorporation, the catalyst droplet progressively shrinks and ultimately disappears [16]. At this point, growth terminates, as shown in Figure 35.1d. Notice that the fixed contact angle of the droplet with the NW rim induces tapering as the droplet volume decreases, a constant contact angle is maintained by reducing the contact area between droplet and Si, effectively reducing the NW diameter. As shown in Figure 35.1d, the latest stages of growth are characterized by the droplet shape change (flattening), highlighting size-effects of ultrasmall NW as well as nucleation of higher-angle facets [21].

Equally, gold diffuses along the NW sidewalls and is exchanged among different NWs. The net effect of this diffusion is growth instability where certain NWs grow in diameter while others shrink [18]. This instability is called Ostwald ripening and is driven by the fact that gold incorporation in a larger droplet is energetically favorable compared to incorporation into a smaller one, due to the different geometrical surface-to-volume ratio between the two [24]. Because this effect poses a challenge toward the growth of NW bundles with uniform size, methods to suppress or control Ostwald ripening have been recently investigated. In this regard, it has been found that oxidation of the NW sidewalls lowers the diffusion coefficient of gold, therefore acting as a barrier toward Ostwald ripening of SiNW arrays [25].

Taken together, the results discussed above show that VLS is a robust technique to synthesize NW with a high degree of control over the NW size, length, and shape. Therefore, it is the technique of choice for the manufacturing of single-NW solar cells. At the same time, it is also a low-volume and low-number density technique where the growth of highly packed NWs is difficult. A complementary approach, which allows larger volume productions while compromising on the control over NW morphology, is the chemical etching technique and is discussed in section 35.2.3.

35.2.2
Synthesis of Silicon–Germanium Superlattice Nanowires

The vapor–liquid–solid technique that we have introduced in the previous section offers indeed greater flexibility in terms of materials suitable for VLS growth. In fact, VLS allows the growth of monolithic nanowires that include multiple semiconductors [26], including silicon–germanium superlattice NWs. Here, NWs are composed of alternate segments of Si and Ge, with controllable degree of alloying [27]. Two main configurations are possible, namely, axial heterojunction NWs, where the materials are deposited along the length of wires [28], and core–shell heterojunction NWs, where one material realizes the core and the other is deposited as shell [29]. In the following, focus will be on the growth of axial Si–Ge NW because of their relevance as thermoelectric devices [30].

Typical VLS growth of SiGe axial superlattices proceeds as follows. The gas precursor disilane is flowed in the growth chamber at pressure $1 \cdot 10^{-5}$ Torr and temperature 560 °C to grow the Si portion. Then, the source gas is switched to digermane while keeping the chamber at growth temperature. This way, a Ge segment is grown on top of the Si one, and growth can then be either continued with digermane to obtain a single Si/Ge interface or the switching can be repeated again, to stack multiple Si-Ge and Ge-Si segments into a true axial superlattice.

Besides pure Ge/Si growth, deposition of Si/SiGe prealloyed heterojunctions is also possible. In this case, the VLS growth is altered by typically using a laser to ablate a solid target of the alloying specie (e.g., Ge) [27,28]. When the laser is switched on, the ablated atoms add to the vapor of the gas precursor; therefore, both species are incorporated into the catalyst and ultimately deposited in the solid. When the laser is turned off and as the partial pressure of the ablated material decreases, deposition of the pure specie can start again [27]. Figure 35.2a shows a typical Si/SiGe sample grown with this technique, where the atom count that shows the axial modulation of composition is shown in Figure 35.2c.

In the light of device application, the quality of the Si/Ge interface plays a pivotal role. Specifically, precursor switching during growth does not immediately translate into a change in deposited material because of the so-called "reservoir effect" of the liquid catalyst [31]. In fact, since the advancing of the solid/liquid interface is driven by supersaturation within the liquid alloyed droplet, the droplet is still saturated with one specie at the moment when the second is introduced in the chamber. As a result, under baseline growth conditions, the interface is diffuse with a typical interface width of 5–10 nm [32]. However, specific studies of the interface quality have identified growth conditions that achieve atomically sharp interfaces [31]. As shown in Figure 35.2b, atomically sharp interfaces are attainable by lowering the growth temperature such that the catalyst is solid, therefore, by using the vapor–solid–solid growth technique. As a result of the lowered kinetics of specie migration, the reservoir effect can be

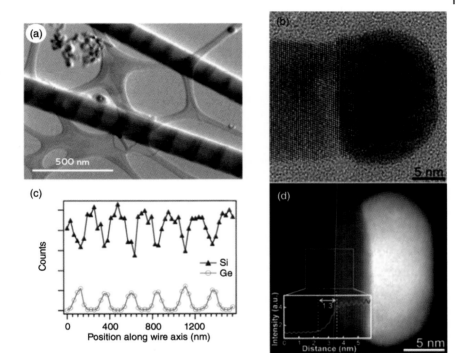

Figure 35.2 Silicon/silicon–germanium axial superlattice as obtained by VLS growth and laser ablation of *in situ* germanium for alloying, where the atomic specie count is shown in (c). (b) TEM of abrupt Si–Ge interface, obtained by combining VLS with vapor–solid–solid growth (see text for details). (d) Shows the atomic count across the interface as measured by high-angle annular dark-field scanning. (Figure 35.2a and c reproduced with permission from Ref. [27]. Copyright 2002 the American Chemical Society. Figure 35.2b and d reproduced with permission from Ref. [31]. Copyright 2009 the American Association for the Advancement of Science.)

avoided at the expense of a lower growth rate of the wires, and the composition change is abrupt at the interface (Figure 35.2d). Indeed, modeling has confirmed the role of kinetics, including the growth rate, in controlling a sharp or a diffuse interface [33].

35.2.3
Top-down Approaches

High packing density SiNW bundles can be obtained using the simple, low-cost, low-temperature technique of metal-assisted chemical etching (MACE) [34–36]. In this top-down technique, the starting point is a standard Si thin film. Silver nanoparticles are deposited on top of the film using an aqueous solution of hydrofluoric acid (HF) at concentrations around $5\,M$ [37] containing $AgNO_3$ in sealed vessels. This step is done at 20 °C for 1 min. The solution selectively etches the Si below the Ag nanoparticles. As a result, the sample emerges as a dense array of SiNWs, where the NWs are the remnants of the initial Si thin film, where the Ag nanoparticles were not present. After etching, the sample is cleaned from the Ag nanoparticles by using NH_3 and deionized water. A typical MACE sample is shown in Figure 35.3.

Control of the etching direction is achieved by choosing the substrate surface and etching conditions [35]. Using Si(1 1 1) substrates and high $[HF]/[H_2O_2]$ ratios, the etching direction is $\langle 1 0 0 \rangle$. At the same time, if the ratio $[HF]/[H_2O_2]$ is low, etching proceeds along the $\langle 1 1 1 \rangle$ direction. On the other hand, etching of Si(1 0 0) samples shows the opposite trend, that is, etching along the $\langle 1 0 0 \rangle$ directions at low $[HF]/[H_2O_2]$ ratios as well as etching along $\langle 1 1 1 \rangle$ directions at high $[HF]/[H_2O_2]$ values. Further, small spacing (high NW packing density) results in preferential etching along the $\langle 1 0 0 \rangle$ directions, while etching proceeds perpendicular to the film free surface at larger spacing (lower NW packing density). Finally, both straight and zigzag NW shapes can be obtained. This results from the control of process temperature; in particular, at temperatures below 25 °C, the etching is straight. At the same time, increasing the temperature to 55 °C or increasing the Ag concentration results in NW with zigzag morphology [35].

A similar approach to MACE, which does not make use of metal catalyst, was demonstrated by P. Yang and coworkers to be equally useful to produce solar cell-grade SiNWs [39,40]. In this technique, the metal deposition is replaced by dip-coating of the initial Si substrate using a silica bead solution. In this step, silica self-assembles on the Si surface with the shape of closely packed nanoparticles with a diameter on the order of 50–100 nm [39]. Later, deep reactive ion etching (DRIE) is performed, where the silica protects the silicon underneath it from the acid. After immersion in HF to remove the silica beads, the sample emerges as a set of closely packed vertical NWs, as shown in Figure 35.3b.

Figure 35.3 (a) Representative sample attainable using the MACE technique. (b) Typical sample that can be obtained using the DRIE technique. (Figure 35.3a is reproduced with permission from Ref. [34]. Copyright 2005 John Wiley & Sons, Inc. Figure 35.3b is reproduced with permission from Ref. [39]. Copyright 2010 the American Chemical Society.

(a)

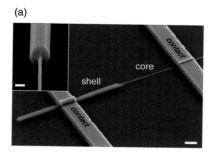

(b)

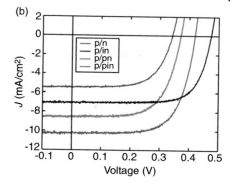

Figure 35.4 (a) SEM image of a representative core–shell single NW silicon solar cell. Notice the denudation of the core on the right side of the device, to allow carrier extraction. (b) $J - V$ curves of various combinations of p-, i-, and n- core–shell Si nanowire solar cells. (reproduced with permission from Ref. [38]. Copyright 2012 the National Academy of Sciences.)

Finally, it is important to note the relevance of chemical etching in achieving NW optimized for solar cell applications, in particular, NWs with tapered tip. In fact, tapered tip NWs further improve optical absorption because they offer a graded effective refractive index from the tip to the root [37]. Indeed, tapered SiNWs can be obtained by exposing a MACE sample to KOH. Because the etching rate is inversely proportional to the number of atomic bonds, the top corner atoms of the NWs are etched first, effectively resulting in a tapered morphology.

35.3
Nanowire-based Photovoltaics

In Section 35.2, the details of the growth of individual NW and of NW bundles have been discussed. Interestingly, these studies not only shed more light on the fundamental processes governing the NW growth and bundle synthesis, but also provided a platform where these nanostructures become the building blocks for device fabrication.

In order to engineer the NWs for photovoltaic applications, the first step is the insertion of dopants to create the proper n- or p- type band structure. Doping of VLS-grown SiNWs is performed by adding dopant gas precursors in the chamber during growth. Specifically, diborane (B_2H_6) and phosphine (PH_3) are utilized for p- and n-type doping, respectively. This way, p–n junction, as well as p-i-n junctions, can be obtained. In p-i-n junctions, a layer of silicon is left undoped and serves as an active region for exciton formation [41]. Introduction of the i-region reduces the saturation current and thus a large component of the leakage current in the diode. A lower leakage current, in turn, gives a larger V_{oc}

due to reduced shunt losses at the junction [41]. Doping can be achieved either along the longitudinal or along the radial direction of the NWs. If doping is along the longitudinal direction, p- and n-type modulation is achieved through switching of the doping gas precursors during growth. On the other hand, radial doping profiles are obtained by VLS growth of the core using one dopant. Then, CVD growth of the shell is commenced, and at this stage, gas precursors of the other dopant is used [42]. In particular, the shells are typically grown at higher temperature and lower precursor pressure to quench further axial elongation [12]. Notice that, in this method, the shell is polycrystalline, with grain size on the order of 20–80 nm [42]. Globally, the core/shell NWs have a diameter on the order of 300 nm [42]. On the other hand, doping of MACE-grown SiNW bundles is achieved by using an initially p- or n-type doped Si substrate, prior to etching. After the etching step, the sample is exposed to doping gas precursors, for example, $POCl_3$, achieving a core–shell doping distribution [34]. Interestingly, excessive sample exposure to doping of the other specie results in complete transformation of, for example, n-type NWs into p-type, effectively generating the p–n junction within the Si substrate [40]. As it will be discussed in Section 35.3.2, this configuration also shows very interesting properties for solar cells applications. In both cases, however, the photocurrent density J_{sc} tends to be larger than that in conventional solar cells, showing the advantage of the nanoscale geometry and reflecting its wider absorption of the solar spectrum [42].

Dopant concentration profiles within the NWs is an important factor that governs overall device quality. In fact, because of the particular thermodynamics of each doping species compared to that of silicon, the dopants may deposit primarily onto the surface (segregation) or they may effectively diffuse inside the bulk material. The particular doping profile depends on a variety of factors, including the NW growth rate and the peculiar thermodynamics of the dopant–silicon system [33]. For example, modeling has shown that a sharp or a diffuse doping profile can be tailored by controlling the growth conditions [33]. At the same time, experiments have shown that doping profiles also depend on the NW diameter [43]. In particular, NWs with diameter smaller than 22 nm show surface segregation of dopants, while NWs with diameter larger than this value show a uniform dopant distribution in the NW bulk.

At the same time, achieving high-quality metal contacts between the device and the load is another important factor to avoid degraded device characteristics. In fact, the Schottky metal/semiconductor interface is necessarily accompanied by pinning of the Fermi level between the two materials. Therefore, attention must be paid to ensure a low defect density at this interface [40].

Before reviewing the characteristics of SiNW-based solar cells, it is important to recall the performance of conventional, planar polycrystalline Si solar cells. Here, reference values are a short-circuit current density of J_{sc} of 42.7 mA/cm^2, fill factor (FF) of 82%, open-circuit voltage V_{oc} of 0.7 V, and efficiency of about 25% [44]. In the following two sections, solar cells based on individual NW and on SiNW bundles are discussed.

35.3.1
Single-Nanowire Solar Cells

Through successful doping and contact metalization for carrier collection, a single SiNW can be engineered into a photovoltaic device, as demonstrated by Lieber and coworkers [45]. In particular, both axial and core/shell NW solar cells can be manufactured.

In axial devices, the p-i-n junction is modulated along the NW length. Carrier separation occurs in the i-region, and built-in electric field provides exciton separation. Carriers flow in the n- and p-doped segments of the wire, ultimately being collected at the metal contacts. Axial devices have been measured with V_{oc} of 0.29 V and I_{sc} of 3.5 mA/cm^2 [41], with a fill factor of 54% [45] and overall apparent efficiency of 0.5% [41]. The efficiency is measured under 1-sun illumination (100 mW/cm^2) and under air mass (AM) 1.5G. The temperature dependence of V_{oc} has a slope of −2.97 mV/K. The quality of the p-i-n cell is directly proportional to the length of the i-segment, resulting in lower shunt losses at the junction [45].

Different from axial devices, radial devices have the p-i-n junction extending along the length of the wires [42]. This way, carrier separation occurs along the radial direction. Interestingly, the carrier collection distance is comparable to the minority carrier diffusion length [46], which represents a clear advantage over the axial configuration [47]. For this reason, the radial configuration minimizes bulk recombination of photocarriers.

Using a p-core with i- and n-type shells, NW devices typically show V_{oc} of 0.26 V, short-circuit current density J_{sc} of 23.9 mA/cm^2, fill factor of 55%, and efficiency of 3.4% [42]. Notice the significant increase in J_{sc} compared to axial devices, which shows both the increased absorption of solar spectrum and the reduced carrier recombination by the radial configuration. The maximum power at 1-sun illumination is about 72 pW per NW device. Here, the slope of the V_{oc} is −1.9 mV/K, similar to that of single-crystalline solar cells [48].

In an attempt to further improve device performance, synthesis of core/multishell NW was attempted where the core is a p-type wire and the shell itself is a p-n or a p-i-n multistructure [38]. Such device yields a larger V_{oc} of 0.5 V, has J_{sc} on the order of $10 − 12 \, mA/cm^2$, and has an efficiency of 6%, while the fill factor is 73%.

Because the NW size is comparable to the wavelength of the incident light, optical phenomena influence light absorption. This is a typical nanotechnological feature, since such effects cannot be exploited in conventional solar cells. Indeed, the NW absorption spectra show distinctive peaks originating from light resonances inside the NW, which acts as an optical cavity [49]. Moreover, the spectra changes as a function of NW diameter, further confirming the resonant geometric origin of the peaks. In particular, NW with larger diameters can support a higher number of optical modes, which indeed results in a higher absorption and ultimately supports the larger J_{sc} with respect to NW of smaller diameters. Overall, this distinctive feature shows that light absorption can be selectively tuned to specific spectral ranges by controlling the NW size and shape. Further simulation work [50] confirmed a direct correlation between the leaky-mode resonances and photocurrent density J_{sc}, where the absorption peaks are influenced by the NW diameter. These findings were later confirmed by Sandhu *et al.* [51].

Investigation of the role of Si crystallinity in device performance shows that the use of amorphous Si as the shell has a positive effect on the absorption efficiency [52]. Indeed, the simulations reported in Ref. [52] show that the a-Si (shell)/c-Si(core) NW has a greater absorption efficiency over the entire spectra than the pure c-Si NW, except for wavelength smaller than 450 nm. The increased absorption is similar to the one observed in nanocrystalline shell NW [53]. However, it is important to note that the increased absorption by nanocrystalline or amorphous Si comes at the expense of a reduced V_{oc}. A smaller V_{oc}, in fact, originates because the grain boundaries act as recombination centers [53]. Indeed, the overall conversion efficiency primarily originates from the high J_{sc} characteristic of nc-Si NWs. Finally, while the shell can sustain a relatively large defect density, the depletion region is the most critical region to defects. Here, defect density must be as low as possible to avoid exciton relaxation immediately after formation [40].

35.3.2
Silicon Nanowire Bundle Solar Cells

The SiNW bundles described in Section 35.2.3 can serve as template to obtain solar cell devices that can reach the size of centimeters [40]. Starting from a SiNW bundle sample, the device is obtained by evaporating metallic Al onto the rear surface of the sample, followed by sintering at 800 °C to remove the Schottky junction at the rear. Later, metallic Ag is also evaporated on the rear surface of the sample. Such solar cell showed an efficiency of 9.31%, V_{oc} of 0.54 V, and fill factor of 65.1% using a monocrystalline Si substrate. As a matter of comparison, efficency decreased to 4.7% using a polycrystalline Si substrate. By carefully optimizing the processing, research at GE and Caltech suggests that an efficiency as high as 20% can be attained using such SiNW-bundle cells [54].

Based on the dopant exposure time, three main cell configurations are possible. In fact, exposure time can be tuned to achieve an array of identical radial p–n NW. Depending on the exposure technique used, an axial doping profile can also be obtained [55]. However, if dopant exposure is prolonged, the entire NW can become n-type, with the p–n junction located within the silicon substrate. In this configuration, the NW array acts purely as an efficient light-trapping device [34].

Optical behavior of SiNW bundles is peculiar and very useful for solar cell applications. In fact, SiNW bundles have distinct absorption features, including a large surface/volume ratio of the sample due to high packing density, enhanced absorption due to cavity modes, and a possible nonuniform refractive index along the NW depth (achieved through NW tapering), reducing the Fresnel

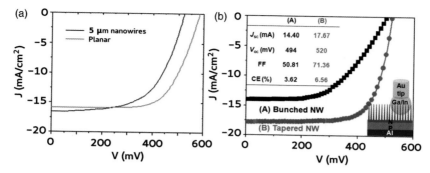

Figure 35.5 $J - V$ curve of a SiNW bundle solar cell compared to a planar device. (b) Improved performance obtained by tapering the NW, as shown in the $J - V$ device characteristic compared to a nontapered reference sample. (Figure 35.5a is reproduced with permission from Ref. [39]. Copyright 2010 the American Chemical Society. Figure 35.5b is reproduced with permission from Ref. [37]. Copyright 2010 The Optical Society.)

reflection over a broad wavelength at various angles of incidence [34,37,56,57]. Specifically, Garnett and Yang showed that the multiple internal reflections increase the path length of incident light up to 73 times [39].

The performance of an 8 μm NW sample compared to a planar reference sample with the same thickness is shown in Figure 35.5a. Under AM1.5G illumination, the NW cell had an efficiency of 4.83% or about 20% higher than a planar cell with the same thickness [15]. However, the efficiency of a 20 μm NW sample was 5.3%, about 35% lower than the corresponding planar cell. These results can be understood by considering both the mechanisms of enhanced light trapping and the surface recombination effects acting on the SiNW device. In fact, at shorter NW lengths, the effect of increased light trapping is seen dominating. On the other hand, longer wires show a larger surface recombination due to a large surface/volume ratio, ultimately degradating the overall efficiency compared to the reference flat device.

In the context of light trapping within the NW array, simulations have proven useful in linking the NW array absorption spectra to the specific geometrical features of the array such as each wire diameter and spacing [58]. In particular, by calculating the detailed balance between the light transmitted, reflected, and absorbed by the NW array as a function of geometric parameters, it was shown that more light absorption compared to that of a Si thin film of the same thickness is possible when the SiNW array periodicity is between 250 and 1200 nm. Moreover, simulations show that introducing a small degree of disorder in the NW array increases light absorption. In particular, a disorder of 10–20% from the uniform distribution shows optimal absorption enhancement [47].

Regarding specifically the optical properties attainable using SiNW solar cells, it is worth recalling Yablonovitch and Cody [59] who, using statistical mechanics and geometrical optics approaches, first derived the limit of a perfectly implemented randomized (i.e., Lambertian) light-trapping device. Both methods showed a maximum light intensity enhancement within the medium of $2n^2$ compared to the incident beam, where n is the refractive index of the medium. This corresponds to a path length enhancement of $4n^2$ (about 50 for silicon, and depending on the actual wavelength) when angle averaging is considered. The only assumptions in this derivation are that the medium does not absorb light (a good approximation for the very weak absorption limit) and that the light is fully randomized once it enters the structure. Under these two assumptions, a random array of nanowires may still exhibit the $4n^2$ limit, but a periodic array should be able to exceed it. Indeed, the light-trapping effect on SiNW arrays is beyond the Lambertian limit of $2n^2$, indicating that the array shows the characteristics of a photonic crystal [60]. The work reported in Ref. [44] shows that the best nanowire arrays have a path length enhancement factor of up to 73, exceeding the randomized scattering limit ($2n^2$, or about 25 in the specific case of Ref. [44]). As a direct consequence of the absorption enhancement, J_{sc} was reported as high as $180\,\text{mA/cm}^2$ [44].

Tapered SiNW bundles showed similar performance improvement because of the multiple light scattering at the NW tips [61]. Here, V_{oc} was $0.52\,\text{mV}$, J_{sc} was $17.6\,\text{mA/cm}^2$, fill factor of 71.36%, and conversion efficiency of 6.5% under 1.5 AM illumination. The total reflectance of tapered NW sample was less than 1% [37], compared to polished Si wafer that has reflectance of the order of 40%.

It is interesting to note that even amorphous Si (a-Si) NW cells have shown useful properties [56]. This is somewhat surprising, given the poor transport properties of a-Si (in particular, a carrier diffusion length of 300 nm [62]). At the same time, the absorption depth in a-Si is about two orders of magnitude smaller than in c-Si, therefore, allowing much thinner devices. However, the peculiar geometrical configuration of radial NW array solar cells enable to retain the advantage of a-Si, namely, the increased absorption, while not suffering from the small carrier diffusion length, because carrier are collected radially, effectively exploiting the small size of the NW. For a-Si:H nanowire core–shell devices, the improvement in short circuit current density is very significant. In fact, a J_{sc} of $9.3\,\text{mA/}cm^2$ was obtained, which was more than 150% larger than that of the control device. However, the dark current also increased because of the increased surface area. Because of that, open-circuit voltage is reduced to 0.46 V. The increased surface area and the surface damage caused by reactive ion etching also caused significant increase in series resistance, which reduces the fill factor to 0.26. Overall, the conversion efficiency was about the same for both thin-film and nanowire devices at about 1.2%.

If surface etching is performed for a longer period of time, SiNW tapering results in a true nanocone shape [56]. Here, the nanocone geometry, which showed superior performance in c-Si NW arrays, can also be employed for a-Si NW. In particular, a-Si NW arrays show an improved absorption of more than 90% of light, including angles of incidence below 60°. This value is significantly larger than that for c-Si NW arrays, which is 70%, and that of Si thin film, which is 45% [56].

Finally, it is worth mentioning that the versatility of SiNW array solar cells is demonstrated also in the capability to be integrated with glass substrates. In fact, a thin Si layer can be deposited on glass by electron beam evaporation and crystallized using laser sintering. From the bulk film, NW array is then fabricated by conventional MACE processing [63]. Metal contacts are achieved through Au tips at both n- and p-type sides of the device. Such solar cell shows V_{oc} of 0.45 V, J_{sc} of 40 mA/cm², and an overall efficiency of 4.4% at AM 1.5 G.

35.4
Introduction of Thermoelectric Effects

In addition to the application in solar PV, silicon nanostructures also have been applied in thermoelectric energy harvesting and cooling. Today, approximately 80% of the world's energy is generated by heat engines that use fossil fuel. However, the typical efficiency of these heat engines is only 30% [64]. This means huge heat energy is lost to the environment as waste heat. Hence, it is important to enhance the utilization efficiency of heat engines by converting the waste heat back to electric energy. Thermoelectric modules are able to convert heat directly into electricity, which can lead to an increase in the utilization efficiency and a reduction in the usage of fossil fuels and also carbon emission.

The typical thermoelectric device that underlies heat-to-electricity conversion is shown in Figure 35.6a: a junction is formed between two different types of conducting materials, with one being p-type (holes), while the other being n-type (electrons) [65]. When an electric current I is passing through the junction, both electrons and holes move away from the junction and carry heat energy Q, which is called Peltier effect. The ratio of Q to I is defined as the Peltier coefficient. On the other hand, as shown in Figure 35.6b, a temperature difference ΔT at the junction causes the carriers to flow away from the junction, leading to an open-circuit electromotive force V and thus forming an electrical generator. This is called Seebeck effect and is defined as the Seebeck coefficient [66].

Using thermoelectric cooler as an example, next we introduce the figure of merit, Z, to evaluate the performance efficiency of the thermoelectric device. Upon electrical current flowing through the cooler, electrons will absorb heat

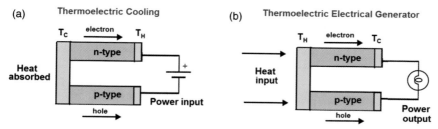

Figure 35.6 (a) Peltier effect (thermoelectric cooling). (b) Seebeck effect (thermoelectric generator). (Reproduced with permission from Ref. [65]. Copyright 2015 the Elsevier Publishing Group.)

energy from lattice at one end and transport it to another end, creating a cold end with temperature T_C and a hot end with temperature T_H. As a result of the Peltier effect, the rate of heat pumping at the cold junction is ST_CI, where S is the Seebeck coefficient of the material and I is the electrical current. The cooling effect is opposed by heat conducted from the hot end and Joule heat in the system. If half the overall Joule heat travels to each of the ends, and the heat transfer between the device and the surrounding air is ignored, the rate of absorption of heat from the source is given by [65]

$$Q_{ab} = ST_CI - \frac{1}{2}I^2R - K(T_H - T_C), \tag{35.1}$$

where K and R are the thermal conductance and electrical resistance of the system, respectively. The potential difference applied to the thermoelectric device is used to overcome the electrical resistance of the junction and balance the Seebeck voltage, which results from the temperature difference between the two ends. Thus, the input power is

$$P = S\Delta TI + I^2R. \tag{35.2}$$

Then, the coefficient of performance (COP) of the thermoelectric cooler is

$$COP = \frac{\text{Heat absorbed}}{\text{Electric power input}} = \frac{ST_CI - \frac{1}{2}I^2R - K(T_H - T_C)}{S\Delta TI + I^2R}. \tag{35.3}$$

The maximum temperature difference is $(T_H - T_C)_{max} = (1/2)S^2\sigma\kappa T_C^2$, where σ is the electronic conductivity and κ is the thermal conductivity of the material. Thus, the concept of figure of merit for a material is employed and given by $Z = (S^2\sigma/\kappa)$, with the maximum coefficient of performance being as follows:

$$COP_{max} = \frac{T_C\left[(1 + ZT_0)^{\frac{1}{2}} - \frac{T_H}{T_C}\right]}{(T_H - T_C)\left[(1 + ZT_0)^{\frac{1}{2}} + 1\right]}. \tag{35.4}$$

Here, T_0 is the average temperature. The figure of merit Z determines both the maximum temperature difference that can be achieved and the maximum coefficient of performance.

The figure of merit Z can also be used to evaluate the efficiency of energy generator. As shown in Figure 35.7, for thermoelectric energy generator with a fixed cold-end temperature of 300 K, the energy conversion efficiency increases with the temperature of the hot end and the figure of merit, Z. Frequently, the nondimensionalized figure of merit ZT is also used to describe the performance of thermoelectric materials [66].

In order to make a material competitive for thermoelectric generation, the ZT of the material must be at least higher than three. However, as shown in

Conversion efficiency %

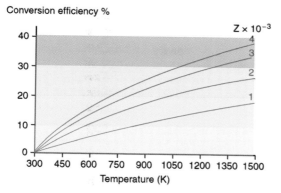

Figure 35.7 Conversion efficiency as a function of temperature and thermocouple material figure of merit. (Reproduced with permission from Ref. [65]. Copyright 2015 the Elsevier Publishing Group.)

Figure 35.8, the road to achieving ZT larger than three is not smooth. In fact, during 1970–2000, there was almost no progress in this aspect [67]. Although the thermoelectric figure of merit ZT of $CoSb_3$-based skutterudite system is higher than 1.5 in both p-type and n-type samples (not shown in Figure 35.8), there is still space to improve the figure of merit. The ideal case is to reduce the

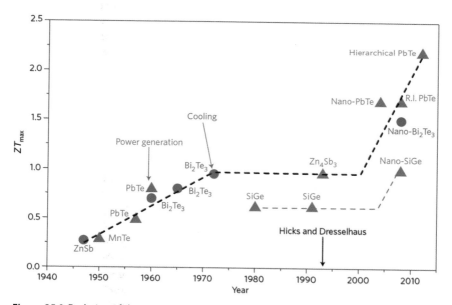

Figure 35.8 Evolution of the maximum ZT over time. Materials for thermoelectric cooling are shown as blue dots and for thermoelectric power generation as red triangles. (Reproduced with permission from Ref. [67]. Copyright 2013 the Nature Publishing Group.)

thermal conductivity without affecting the electrical conductivity. Moreover, an ultralow thermal conductivity is also required to prevent the backflow of heat from the hot end to cool end. Therefore, a reduction in thermal conductivity is crucial for thermoelectric applications. Recent theoretic and experimental studies [68] have demonstrated the possibility of achieving high ZT through using low-dimensional nanomaterials. In nanomaterials, the electrical conductivity and electron contributions to Seebeck coefficient are similar to those of bulk materials, but there is a remarkable reduction in thermal conductivity due to the strong boundary inelastic scattering of phonons, indicating that the electrical and thermal conductivities can be decoupled. This has led to the idea of using low-dimensional nanomaterials as a promising candidate to achieve high ZT values.

35.5
Thermal Conductivity of Silicon Nanowires

In nonmetallic materials, phonon, the vibrational mode of lattice, is the dominant heat carrier. In the past two decades, rapid developments in synthesis and processing of nanoscale materials and structures have created a great demand for understanding the thermal transport in nanomaterials.

Silicon nanowires have attracted considerable attention due to their excellent electrical and mechanical properties. Owing to their high surface-to-volume ratio, the strong phonon surface scattering in NWs can result in a low thermal conductivity compared to their bulk counterpart. For example, experiments [68,69] presented direct evidence that an approximately 100-fold reduction in thermal conductivity over bulk Si can be achieved in SiNWs, while the electrical conductivity and electron contributions to Seebeck coefficient are still similar to those of bulk silicon. Li *et al.* [70] reported experimentally the significant reduction of thermal conductivity in SiNWs compared to the thermal conductivity in bulk silicon. As shown in Figure 35.9, it is clear that as the NW diameter decreases, the thermal conductivity is reduced. More interestingly, the temperature dependence of thermal conductivity of SiNWs is qualitatively different from that in bulk silicon. In bulk silicon, thermal conductivity $\kappa \simeq T^3$ at low temperatures and $\kappa \simeq 1/T$ at high temperatures, with its peak value at about 25 K. However, the temperature corresponding to the peak value of the thermal conductivity shifts to a higher value with decreasing NW diameter. For the 22 nm NW, its thermal conductivity does not show a peak within the tested temperature range, demonstrating that the phonon boundary scattering dominates over the phonon–phonon Umklapp scattering in SiNWs.

On the other hand, at the low temperature range (from 20 to 60 K), the thermal conductivity curves for 115 and 56 nm NWs fit Debye T^3 law quite well, suggesting that the phonon spectrum of thick NWs is similar to that of their bulk counterpart. However, for the smaller diameter NWs, 37 and 22 nm, the

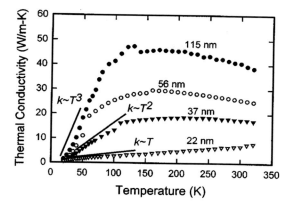

Figure 35.9 Measured thermal conductivity of different diameter Si nanowires. (Reproduced with permission from Ref. [70]. Copyright 2003 the American Institute of Physics.)

deviation from Debye T^3 law is apparent. This means that in thin nanowires, the phonon dispersion relation is altered due to the quantum confinement effect [71]. These remarkable results convincingly have raised the exciting prospect that SiNWs can be applied as high performance nanoscale thermoelectric materials.

Experimental observations showed that the surface of SiNWs was essentially rough, with root mean square ranging from 3 Å to 5 nm. The effect of surface roughness on phonon transport in SiNWs was investigated theoretically [72]. For a SiNW with diameter of 56 nm, the thermal conductivity decreases from about 40 to 2 W/mK with increasing surface roughness from 0.2 to 4 nm, as shown in Figure 35.10. In comparison to high frequency

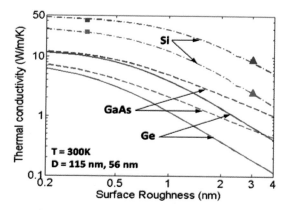

Figure 35.10 Effect of roughness root mean square on thermal conductivity of Si, Ge, and GaAs NWs with diameters of 115 nm (upper curve) and 56 nm (lower curve). (Reproduced with permission from Ref. [72]. Copyright 2010 the American Chemical Society.)

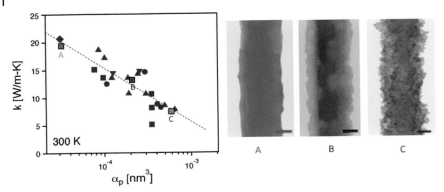

Figure 35.11 Thermal conductivity as a function of roughness factor α_p at 300 K. As α_p increases, the wires become rougher, with wavelengths in the 1–100 nm range and the thermal conductivity drops significantly. (Reproduced with permission from Ref. [73]. Copyright 2012 the American Chemical Society.)

phonons, low frequency phonons experience little thermal resistance from surface roughness [72].

A quantitative analysis of the surface roughness effect on thermal conductivity was explored experimentally in vapor–liquid–solid grown SiNWs with controllable roughness [73]. As shown in Figure 35.11, the thermal conductivity is correlated well with the power spectra of surface roughness and follows a power law in the 1–100 nm length range. Hence, the coherent phonon scattering off the surface is critical to phonon transport in rough NWs.

35.6
Thermoelectric Property of Silicon Nanowires

By using the density functional derived tight-binding method, we [74] have studied the size effect on thermoelectric power factor in silicon nanowires. Figure 35.12a and b shows the size effects on electrical conductivity σ and Seebeck coefficient S with different electron concentration. We can find Seebeck coefficient S decreases remarkably with increasing size. With the transverse dimension increases, the sharp DOS peaks widen and reduce S.

The dependence of figure of merit ZT on carrier concentration is shown in Figure 35.13. Increasing carrier concentration has two effects on ZT. On the one hand, the increase in carrier concentration will increase the electrical conductivity. On the other hand, the increase in carrier concentration will suppress the Seebeck coefficient S. These two effects compete with each

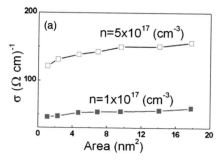 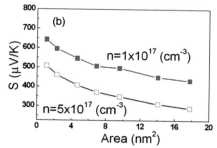

Figure 35.12 (a, b) Electrical conductivity and Seebeck coefficient versus cross-sectional area with different carrier concentration. (c) Maximum power factor versus cross-sectional area. (d) ZT versus carrier concentration. Details can be found in Ref. [74]. (Reproduced with permission from the American Institute of Physics.)

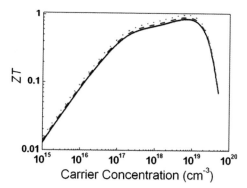

Figure 35.13 ZT versus carrier concentration. Details can be found in Ref. [74]. (Reproduced with permission from the American Institute of Physics.)

other. Therefore, ZT increases with carrier concentration increases, and there is an optimal carrier concentration yielding the maximum attainable value of ZT. Above this carrier concentration, ZT decreases with increasing carrier concentration.

35.7
Thermoelectric Property of Silicon–Germanium Nanowires

Besides SiNW, it has also been demonstrated that $Si_{1-x}Ge_x$ NW is a promising candidate for high-performance thermoelectric applications. Figure 35.14 shows the dependence of ZT in $Si_{1-x}Ge_x$ NWs on Ge content x [75]. The value of ZT of

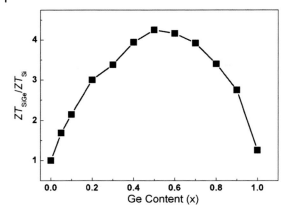

Figure 35.14 The dependence of ZT on Ge content x. Details can be found in Ref. [75]. (Reproduced with permission from the American Institute of Physics.)

$Si_{1-x}Ge_x$ NW increases with Ge content, reaches a maximum, and then decreases. At low Ge content, the small ratio of Ge atoms can induce large increase in ZT. For instance, in the case of $Si_{0.8}Ge_{0.2}$ NW, namely, 20% Ge, ZT is about three times that of pure SiNWs. And with 50% Ge atoms ($Si_{0.5}Ge_{0.5}$ NW), the ZT can be as high as 4.3 times that of pure SiNWs. Combine with the experimentally measured ZT of n-type SiNW that is about 0.6–1.0, it is exciting to see that we may obtain high ZT value of about $2.5 - 4.0$ in n-type $Si_{1-x}Ge_x$ NWs.

Figure 35.15 shows the ZT value of $Si_{0.5}Ge_{0.5}$ superlattice NWs with a different periodic length. As shown in this figure, the maximum value of ZT (ZT_{Max}) first increases with periodic length, reaches the maximum value at the periodic length L of 0.54 nm, and then decreases with the further increase in the periodic

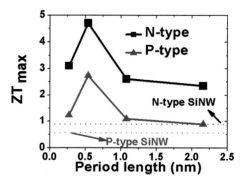

Figure 35.15 Dependence of ZT_{Max} on periodic length for p-type and n-type $Si_{0.5}Ge_{0.5}$ superlattice NWs. Details can be found in Ref. [30]. (Reproduced with permission from the American Institute of Physics.)

length. The achievable ZT_{Max} for n-type NWs is about twice its p-type counterparts. It is worth mentioning that in the real application of thermoelectric devices, the performance depends on the match of the n-type and p-type legs. Under the same periodic length, the device performance should be low, because the ZT_{Max} for n-type NWs is about twice its p-type counterparts. Therefore, in the thermoelectric design, n-type leg and p-type leg with different periodic lengths should be adopted; for instance, periodic length for p-type NW is 0.54 nm, but that for n-typed NW is 1.0 nm. Moreover, the values of ZT_{Max} for the superlattice NWs are much larger than those of pristine SiNWs because of their low thermal conductivity. The maximum value of ZT_{Max} for n-type superlattice NWs is 4.7 at the period length of 0.54 nm, which is a fivefold increase compared to the equivalent pristine SiNWs ($ZT_{\text{Max}} = 0.94$). The maximum value of ZT_{Max} for p-type NWs is 2.74 with the same periodic length, which is 4.6 times larger than that of p-type SiNWs ($ZT_{\text{Max}} \simeq 0.6$). The optimal carrier concentration yielding peak ZT on the same order of magnitude as the maximum achievable free carrier concentration for SiNWs demonstrates that superlattice SiGe nanowires are promising thermoelectric materials to achieve high ZT [30].

35.8
Thermoelectric Property of Other Nanowires

In addition to SiNWs, other nanowires, such as Bi_2Te_3 nanowires, are also under the spotlight. Recently, Bi_2Te_3 nanowires prepared by physical and chemical synthesis techniques have been demonstrated [76]. Bulk Bi_2Te_3 is a representative thermoelectric material, and forming nanostructures can lead to further enhancement in ZT. Recently, a peak ZT of 1.4 at 100 K was achieved in nanostructured bismuth antimony telluride alloys [77]. The improved efficiency mostly originates from a reduction in thermal conductivity according to the quantum confinement and surface phonon scattering.

Using molecular dynamics, Yu *et al.* [78] studied thermal conductivity of Bi_2Te_3 nanowires oriented along the [1 1 0] direction, which is typical for Bi_2Te_3 NWs. For each NW, the atomic structure is initially constructed from bulk Bi_2Te_3. The structure of the NWs is constructed by retaining all the atoms within a virtual cage in the bulk and removing all atoms outside the virtual cage. Figure 35.16 shows the schematics of Bi_2Te_3 NWs. Bi_2Te_3 has a rhombohedral crystal structure with one chemical formula per unit cell. The pseudohexagonal unit cell is often used with lattice parameters of $a = 4.4$ Å and $c = 30.5$ Å. The hexagonal cell is formed by stacking layers of atoms perpendicular to the c-axis following the sequence (quintet) of Te_1–Bi–Te_2-Bi–Te_1, with each quintet being bonded to the neighboring one by Te–Te interaction. Spheres in different colors as shown in Figure 35.16 represent different types of atoms in Bi_2Te_3. In the MD simulation, the empirical potential developed by Qiu and Ruan [79] was adopted, which is comprised of a Morse-type potential to

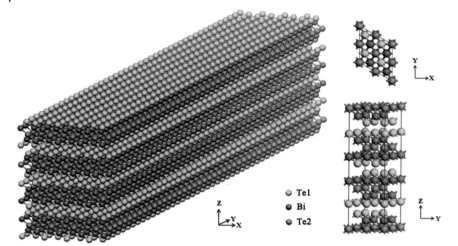

Figure 35.16 The schematics of Bi_2Te_3 NWs. Spheres in different colors represent different types of atoms in Bi_2Te_3 [78].

describe the short-range term and a Coulombic term for the long-range electrostatic interaction.

The length dependence of thermal conductivity is shown in Figure 35.17. It is known that for the thermal conductivity of semiconductor nanowires, according to the kinetic theory of phonon transport, the thermal conductivity can be approximated as follows:

$$\frac{1}{\kappa} = \frac{1}{\kappa_0}\left(1 + \frac{\lambda}{L}\right). \tag{35.5}$$

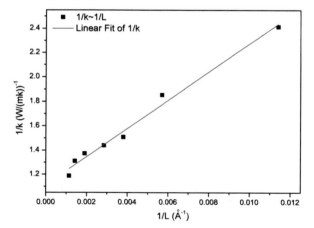

Figure 35.17 Thermal conductivity of Bi_2Te_3 NWs as a function of nanowire length. The data obtained from NEMD for a series of nanowires are used to plot $1/\kappa$ as a function of $1/L$ [78].

Here, κ is the length-dependent thermal conductivity, L is the length of nanowire, κ_0 is the converged thermal conductivity with infinite length, and λ is the phonon mean free path (MFP). This relationship implies that a plot of the inverse of the thermal conductivity κ versus the inverse of the nanowire length L should be linear and its intercept should be the inverse of the thermal conductivity of the infinitely large system. As shown in Figure 35.17, from the intercept and slope values of the plot, the value of κ_0 is found to be about 0.89 W/mk and the phonon mean free path λ of Bi_2Te_3 nanowire is about 10 nm.

In a recent experiment, electrical resistivity is found to be 1.1 mΩ − cm for Bi_2Te_3 nanowire [80]. We can estimate the electronic contribution to the thermal conductivity of the nanowire using the Wiedemann–Franz law $\kappa_e = \sigma L_W T$, with a Lorenz number $L_W = 2(k_B/e)^2 = 1.45 \cdot 10^{-8} V^2/K^2$, where T is 300 K. The electronic contribution to thermal conductivity κ_e is about 0.4 W/mK. Thus for Bi_2Te_3 nanowires, the lattice thermal conductivity is higher than its electron counterpart. Furthermore, the total thermal conductivity is about 1.29 W/mK, which is very close to the value measured experimentally [80].

It is obvious that lattice thermal conductivity of a Bi_2Te_3 nanowire is lower than that of the bulk single-crystal Bi_2Te_3 (1.73 W/mK), highlighting the advantage of using Bi_2Te_3 nanowires for thermoelectric applications. However, in contrast to the 100-fold reduction in the thermal conductivity of Si nanowires, the reduction in the thermal conductivity in Bi_2Te_3 is minor. In semiconductors, phonon transport is dominated by phonon–phonon Umklapp scattering and edge scattering. Phonon MFP in silicon nanowire is on the order of 60–240 nm, which leads to a significant reduction in the thermal conductivity of silicon nanowires. However, in a Bi_2Te_3 nanowire, the intrinsic phonon MFP according to phonon–phonon Umklapp scattering is only about 10 nm. As the MFP is comparable to the diameter of the Bi_2Te_3 NW, it leads to a minor reduction in thermal conductivity compared to the bulk Bi_2Te_3. Thus, there will be a significant reduction in thermal conduction only when the diameter of Bi_2Te_3 nanowire is reduced down to 10 nanometers. In this range, the edge boundary scattering becomes greatly enhanced.

References

1 International Energy Agency (2014) *Technology Roadmap Solar Photovoltaic Energy*, International Energy Agency.

2 Kyocera *Kyocera KU260-6MCA Solar Module Specifications*. URL www .kyocerasolar.com/ (2016). URL accessed 4/Nov/2016.

3 Yue, D., You, F., and Darling, S. (2014) Domestic and overseas manufacturing scenarios of silicon-based photovoltaics: life cycle energy and environmental comparative analysis. *Solar Energy*, **105**, 669–678.

4 Green, M. (2007) Thin-film solar cells: review of materials, technologies and commercial status. *J. Mater. Sci. Mater. Electron.*, **18**, 15–19.

5 Hänni, S., Bugnon, G., Parascandolo, G., Boccard, M., Escarré, J., Despeisse, M., Meillaud, F., and Ballif, C. (2013) High-

efficiency microcrystalline silicon single-junction solar cells. *Prog. Photovoltaics*, **21**, 821–826.

6 Henini, M. (ed.) (2008) *Handbook of Self Assembled Semiconductor Nanostructures for Novel Devices in Photonics and Electronics*, Elsevier.

7 Wang, Z. (ed.) (2008) *Self-Assembled Quantum Dots*, Springer.

8 Kolasinski, K. (2006) Catalytic growth of nanowires: vapor–liquid–solid, vapor–solid–solid, solution–liquid–solid and solid–liquid–solid growth. *Curr. Opin. Solid State Mater. Sci.*, **10**, 182–191.

9 Wen, C., Reuter, M., Tersoff, J., Stach, E., and Ross, F. (2010) Structure, growth kinetics, and ledge flow during vapor–solid–solid growth of copper-catalyzed silicon nanowires. *Nano Lett.*, **10**, 514–519.

10 Garnett, E., Liang, W., and Yang, P. (2007) Growth and electrical characteristics of platinum-nanoparticle-catalyzed silicon nanowires. *Adv. Mater.*, **19**, 2946–2950.

11 Wang, Y., Schmidt, V., Senz, S., and Gösele, U. (2006) Epitaxial growth of silicon nanowires using an aluminium catalyst. *Nat. Nanotechnol.*, **1**, 186–189.

12 Ross, F. (2010) Controlling nanowirer structures through real time growth studies. *Rep. Prog. Phys.*, **73**, 114501.

13 Wu, Y., Cui, Y., Huynh, L., Barrelet, C., Bell, D., and Lieber, C. (2004) Controlled growth and structures of molecular-scale silicon nanowires. *Nano Lett.*, **4**, 433.

14 Lu, W. and Lieber, C.M. (2006) Semiconductor nanowires. *J. Phys. D*, **39** (21), R387–R406.

15 Yan, R., Gargas, D., and Yang, P. (2009) Nanowire photonics. *Nat. Photonics*, **3**, 569–576.

16 Kodambaka, S., Tersoff, J., Reuter, M.C., and Ross, F.M. (2006) Diameter-independent kinetics in the vapor–liquid–solid growth of Si nanowires. *Phys. Rev. Lett.*, **96** (9), 096105.

17 Park, W.I., Zheng, G., Jiang, X., Tian, B., and Lieber, C.M. (2008) Controlled synthesis of millimeter-long silicon nanowires with uniform electronic properties. *Nano Lett.*, **8** (9), 3004–3009.

18 Hannon, J.B., Kodambaka, S., Ross, F.M., and Tromp, R.M. (2006) The influence of

the surface migration of gold on the growth of silicon nanowires. *Nature*, **440** (7080), 69–71.

19 Ross, F.M., Tersoff, J., and Reuter, M.C. (2005) Sawtooth faceting in silicon nanowires. *Phys. Rev. Lett.*, **95** (14), 146104.

20 Wen, C.Y., Tersoff, J., Hillerich, K., Reuter, M.C., Park, J.H., Kodambaka, S., Stach, E.A., and Ross, F.M. (2011) Periodically changing morphology of the growth interface in Si, Ge, and GaP nanowires. *Phys. Rev. Lett.*, **107** (2), 025503.

21 Schwarz, K.W., Tersoff, J., Kodambaka, S., Chou, Y.C., and Ross, F.M. (2011) Geometrical frustration in nanowire growth. *Phys. Rev. Lett.*, **107** (26), 1–5.

22 Allen, J., Hemesath, E., Perea, D., Lensch-Falk, J., Li, Z., Yin, F., Gass, M., Wang, P., Bleloch, A., Palmer, R., and Lauhon, L. (2008) High-resolution detection of au catalyst atoms in Si nanowires. *Nat. Nanotechnol.*, **3** (3), 168.

23 Kelzenberg, M.D., Turner-Evans, D.B., Kayes, B.M., Filier, M.a., Putnam, M.C., Lewis, N.S., and Atwater, H.a. (2008) Photovoltaic measurements in single-nanowire silicon solar cells. *Nano Lett.*, **8** (2), 710–714.

24 Kim, B.J., Tersoff, J., Kodambaka, S., Jang, J.S., Stach, E.A., and Ross, F.M. (2014) Au transport in catalyst coarsening and Si nanowire formation. *Nano Lett.*, **14** (8), 4554–4559.

25 Kawashima, T., Mizutani, T., Nakagawa, T., Torii, H., Saitoh, T., Komori, K., and Fujii, M. (2008) Control of surface migration of gold particles on Si nanowires. *Nano Lett.*, **8** (1), 362–368.

26 Hillerich, K., Dick, K.a., Wen, C.Y., Reuter, M.C., and Kodambaka, S., and Ross, F.M. (2013) Strategies to control morphology in hybrid group III–V/group IV heterostructure nanowires. *Nano Lett.*, **13** (2), 1–5.

27 Wu, Y., Fan, R., and Yang, P. (2002) *Nano Lett.*, **2**, 83–86.

28 Gudiksen, M., Lauhon, L., Wang, J., Smith, D., and Lieber, C. (2002) *Nature*, **415**, 617.

29 Lauhon, L., Gudiksen, M., Wang, C., and Lieber, C. (2002) *Nature*, **420**, 57–61.

30 Shi, L., Jiang, J., Zhang, G., and Li, B. (2012) *Appl. Phys. Lett.*, **101**, 233114.

31 Wen, C.Y., Reuter, M., Bruley, J., Tersoff, J., Kodambaka, S., Stach, E., and Ross, F. (2009) *Supramol. Sci.*, **326**, 1247.

32 Wen, C.Y., Reuter, M.C., Ross, F.M., Su, D., and Stach, E.A. (2015) Strain and stability of ultrathin Ge layers in Si/Ge/Si axial heterojunction nanowires. *Nano Lett.*, **15** (3), 1654–1659.

33 Vastola, G., Shenoy, V.B., and Zhang, Y.W. (2012) Controlling the interface composition of core–shell and axial heterojunction nanowires. *J. Appl. Phys.*, **112** (6), 064311.

34 Peng, K., Xu, Y., Wu, Y., Yan, Y., Lee, S.T., and Zhu, J. (2005) Aligned single-crystalline Si nanowire arrays for photovoltaic applications. *Small*, **1** (11), 1062–1067.

35 Han, H., Huang, Z., and Lee, W. (2014) Metal-assisted chemical etching of silicon and nanotechnology applications. *Nano Today*, **9** (3), 271–304.

36 Hochbaum, A. and Yang, P. (2010) Semiconductor nanowires for energy conversion. *Chem. Rev.*, **110** (1), 527–546.

37 Jung, J.Y., Guo, Z., Jee, S.W., Um, H.D., Park, K.T., and Lee, J.H. (2010) A strong antireflective solar cell prepared by tapering silicon nanowires. *Opt. Express*, **18**, A286–A292.

38 Kempa, T.J., Cahoon, J.F., Kim, S.K., Day, R.W., Bell, D.C., Park, H.G., and Lieber, C.M. (2012) Coaxial multishell nanowires with high-quality electronic interfaces and tunable optical cavities for ultrathin photovoltaics. *Proc. Natl. Acad. Sci. USA*, **109** (5), 1407–1412.

39 Garnett, E. and Yang, P. (2010) Light trapping in silicon nanowire solar cells. *Nano Lett.*, **10** (3), 1082–1087.

40 Garnett, E.C., Brongersma, M.L., Cui, Y., and McGehee, M.D. (2011) Nanowire solar cells. *Ann. Rev. Mater. Res.*, **41** (1), 269–295.

41 Kempa, T.J., Tian, B., Kim, D.R., Jinsong, H., Xiaolin, Z., and Lieber, C.M. (2008) Single and tandem axial p-i-n nanowire photovoltaic devices. *Nano Lett.*, **8** (10), 3456–3460.

42 Tian, B., Zheng, X., Kempa, T.J., Fang, Y., Yu, N., Yu, G., Huang, J., and Lieber, C.M. (2007) Coaxial silicon nanowires as solar cells and nanoelectronic power sources. *Nature*, **449** (7164), 885–889.

43 Xie, P., Hu, Y., Fang, Y., Huang, J., and Lieber, C.M. (2009) Diameter-dependent dopant location in silicon and germanium nanowires. *Proc. Natl. Acad. Sci. USA*, **106** (36), 15254–15258.

44 Krogstrup, P., Jørgensen, H.I., Heiss, M., Demichel, O., Holm, J.V., Aagesen, M., Nygard, J., and Fontcuberta i Morral, A. (2013) Single-nanowire solar cells beyond the Shockley–Queisser limit. *Nat. Photonics*, **7** (4), 306–310.

45 Tian, B., Kempa, T., and Lieber, C. (2009) Single nanowire photovoltaics. *Chem. Soc. Rev.*, **38**, 16–24.

46 Fuyuki, T., Kondo, H., Yamazaki, T., Takahashi, Y., and Uraoka, Y. (2005) Photographic surveying of minority carrier diffusion length in polycrystalline silicon solar cells by electroluminescence. *Appl. Phys. Lett.*, **86**, 262108.

47 Pickett, E., Gu, A., Huo, Y., Garnett, E., Hu, S., Sarmiento, T., Thombare, S., Liang, D., Li, S., Cui, Y., Mcgehee, M., Mcintyre, P., and Harris, J. (2010) Faceting and disorder in nanowire solar cell arrays. *IEEE*, 35th, 1848–1853.

48 Würfel, P. (2005) *Physics of Solar Cells, From Principles to New Concepts*, Wiley-VCH Verlag GmbH, Weinheim.

49 Kim, S.K., Day, R.W., Cahoon, J.F., Kempa, T.J., Song, K.D., and Park, H.G., and Lieber, C.M. (2012) Tuning light absorption in core/shell silicon nanowire photovoltaic devices through morphological design. *Nano Lett.*, **12**, 4971–4976.

50 Cao, L., Fan, P., Vasudev, A.P., White, J.S., Yu, Z., Cai, W., Schuller, J.A., Fan, S., and Brongersma, M.L. (2010) Semiconductor nanowire optical antenna solar absorbers. *Nano Lett.*, **10** (2), 439–445.

51 Sandhu, S., Yu, Z., and Fan, S. (2014) Detailed balance analysis and enhancement of open-circuit voltage in single-nanowire solar cells. *Nano Lett.*, **14** (2), 1011–1015.

52 Kim, S.K., Song, K.D., Kempa, T.J., Day, R.W., Lieber, C.M., and Park, H.G. (2014)

Design of nanowire optical cavities as efficient photon absorbers. *ACS Nano*, **8** (4), 3707–3714.

53 Fan, Z., Razavi, H., Do, J.w., Moriwaki, A., Ergen, O., Chueh, Y.L., Leu, P.W., Ho, J.C., Takahashi, T., Reichertz, L.A., Neale, S., Yu, K., Wu, M., and Ager, J.w., and Javey, A. (2009) Three-dimensional nanopillar-array photovoltaics on low-cost and flexible substrates. *Nat. Mater.*, **8** (8), 648–653.

54 Kelzenberg, M.D., Boettcher, S.W., Petykiewicz, J.A., Turner-Evans, D.B., Putnam, M.C., Warren, E.L., Spurgeon, J.M., Briggs, R.M., Lewis, N.S., and Atwater, H.a. (2010) Enhanced absorption and carrier collection in Si wire arrays for photovoltaic applications. *Nat. Mater.*, **9** (3), 239–244.

55 Peng, K.Q., Huang, Z.P., and Zhu, J. (2004) Fabrication of large-area silicon nanowire p–n junction diode arrays. *Adv. Mater.*, **16** (1), 73–76.

56 Zhu, J., Yu, Z., Burkhard, G.F., Hsu, C.M., Connor, S.T., Xu, Y., Wang, Q., McGehee, M., Fan, S., and Cui, Y. (2009) Optical absorption enhancement in amorphous silicon nanowire and nanocone arrays. *Nano Lett.*, **9**, 279–282.

57 Zhu, J., Xu, Y., Wang, Q., and Cui, Y. (2010) Amorphous silicon core–shell nanowire Schottky solar cells. 2010 35th IEEE Photovoltaic Specialists Conference, pp. 000453–000456.

58 Li, J., Yu, H., Wong, S.M., Li, X., Zhang, G., Lo, P.G.Q., and Kwong, D.L. (2009) Design guidelines of periodic Si nanowire arrays for solar cell application. *Appl. Phys. Lett.*, **95** (24), 2009–2011.

59 Yablonovitch, E. and Cody, G.D. (1982) Intensity enhancement in textured optical sheets for solar cells. *IEEE Trans. Electron. Dev.*, **29** (2), 300–305.

60 Andreani, L.C. and Gerace, D. (2006) Photonic-crystal slabs with a triangular lattice of triangular holes investigated using a guided-mode expansion method. *Phys. Rev. B*, **73**, 235114.

61 Tsakalakos, L., Balch, J., Fronheiser, J., Shih, M.Y., LeBoeuf, S.F., Pietrzykowski, M., Codella, P.J., Korevaar, B.A., Sulima, O., Rand, J., Davuluru, A., and Rapol, U. (2007) Strong broadband optical absorption in silicon nanowire films. *J. Nanophoton.*, **1** (1), 013552.

62 Shah, A.V., Schade, H., Vanecek, M., Meier, J., Vallat-Sauvain, E., Wyrsch, N., Kroll, U., Droz, C., and Bailat, J. (2004) Thin-film silicon solar cell technology. *Prog. Photovoltaics*, **12**, 113–142.

63 Sivakov, V., Andrä, G., Gawlik, A., Berger, A., Plentz, J., Falk, F., and Christiansen, S.H. (2009) Silicon nanowire-based solar cells on glass: synthesis, optical properties, and cell parameters. *Nano Lett.*, **9** (4), 1549–1554.

64 Yazawa, K. and Shakouri, A. (2011) Cost-efficiency trade-off and the design of thermoelectric power generators. *Environ. Sci. Tech.*, **45**, 7548–7553.

65 Zhang, G. and Zhang, Y.W. (2015) *Mech. Mater.*, **91**, 382–398.

66 Rowe, D.M. (ed.) (2006) *Thermoelectrics Handbook: Macro to Nano*, Taylor & Francis Group, New York.

67 Heremans, J., Dresselhaus, M., Bell, L., and Morelli, D. (2013) When thermoelectrics reached the nanoscale. *Nat. Nanotechnol.*, **8**, 471–473.

68 Zhang, G. and Manjooran, N. (ed.) (2014) *Nanofabrication and its Application in Renewable Energy*, Royal Society of Chemistry Publishing.

69 Hochbaum, A., Chen, R., Delgado, R., Liang, W., Garnett, E., Najarian, M., Majumdar, A., and Yang, P. (2008) *Nature*, **451**, 163.

70 Li, D., Wu, Y., Kim, P., Shi, L., Yang, P., and Majumdar, A. (2003) *Appl. Phys. Lett.*, **83**, 2934.

71 Zhou, J. and Balandin, A. (2001) *J. Appl. Phys.*, **89**, 2932.

72 Martin, P., Aksamija, Z., Pop, E., and Ravaioli, U. (2010) *Nano Lett.*, **10**, 1120.

73 Lim, J., Hippalgaonkar, K., Andrews, S., Majumdar, A., and Yang, P. (2012) *Nano Lett.*, **12**, 2475.

74 Shi, L., Yao, D., Zhang, G., and Li, B. (2009) *Appl. Phys. Lett.*, **95**, 063102.

75 Shi, L., Yao, D., Zhang, G., and Li, B. (2010) *Appl. Phys. Lett.*, **96**, 173108.

76 Peng, H., Dang, W., Cao, J., Chen, Y., Wu, D., Zheng, W., Li, H., Shen, Z.X., and Liu, Z. (2012) *Nat. Chem.*, **4**, 281.

77 Poudel, B., Hao, Q., Ma, Y., Lan, Y., Minnich, A., Yu, B., Yan, X., Wang, D., Muto, A., Vashaee, D., Chen, X., Liu, J., Dresselhaus, M., Chen, G., and Ren, Z. (2008) *Supramol. Sci.*, **320**, 634.

78 Yu, C., Zhang, G., Lian-Mao, P., Wenhui, D., and Zhang, Y.W. (2014) *Appl. Phys. Lett.*, **105**, 023903.

79 Qiu, B. and Ruan, X. (2009) *Phys. Rev. B*, **80**, 165203.

80 Hsin, C.L., Wingert, M., Huang, C.W., Guo, H., Teng, H.S., Suh, J., Wang, K., Wu, J., Wu, W.W., and Chen, R. (2013) *Nanoscale*, **5**, 4669.

36

Nanoliquid Metal Technology Toward High-Performance Energy Management, Conversion, and Storage

Jing Liu

Chinese Academy of Sciences, Technical Institute of Physics and Chemistry, Beijing Key Lab of CryoBiomedical Engineering & Key Lab of Cryogenics, Zhong Guan Cun Dong Lu, No.29, Beijing 100190, China

36.1
Introduction

Since the first industrial revolution, various technologies and inventions have come into being, which significantly drive the world civilization to an ever unprecedented height. Meanwhile, they also brought about tough challenges due to huge energy consumption. With the rapid depletion of the natural energy resources such as fossil fuel and oil, it now becomes increasingly urgent to find new ways for better utilization of energy, spanning from investigations on finding renewable energy sources, innovating energy storage and management strategy, to improvement of current energy utilization efficiency and so on [1]. Clearly, an idealistic energy material would provide a most possible way for tackling the remission issues of energy.

Aiming to improve the thermal conductivity of the heat transfer fluids widely used in industry, functionalized nanoparticles smaller than 100 nm in diameter were proposed to load into base fluid to form various dilute suspensions. This leads to the important technical concept of nanofluids [2], which becomes a basic strategy to make new functional fluids. Due to increasing values in a wide variety of energy areas, the number of researches on nanofluids has undergone an explosive growth, covering from characterizing the physical effects of species [3], concentrations [4], shapes [5,6], and sizes [7,8] of nanoparticles to the preparation of base fluid [9–11], working temperature [7], coating [8], and so on. A series of physical models have thus been established to interpret the mechanisms [12–14] of the enhanced heat conduction of nanofluids. Several typical relations used for characterizing the effective thermal conductivity of nanofluid are listed in Table 36.1. The base fluid of those conventional nanofluids mainly includes water, ethylene glycol [4,5,15–17], and oil [18,19]. It has been proven that even a small amount of addition of the nanoparticles would evidently

Nanotechnology for Energy Sustainability, First Edition. Edited by Baldev Raj, Marcel Van de Voorde, and Yashwant Mahajan.
© 2017 Wiley-VCH Verlag GmbH & Co. KGaA. Published 2017 by Wiley-VCH Verlag GmbH & Co. KGaA.

Table 36.1 Typical models for effective thermal conductivity of nanofluids.

Models	Expressions	Remarks
Maxwell [21]	$\dfrac{k_{\mathrm{eff}}}{k_{\mathrm{f}}} = 1 + \dfrac{3(\alpha - 1)\phi}{(\alpha + 2) - (\alpha + 1)\phi}$	Spherical particles
Hamilton-Crosser [22]	$\dfrac{k_{eff}}{k_{\mathrm{f}}} = \dfrac{\alpha + (n - 1) - (n - 1)(1 - \alpha)\phi}{\alpha + (n - 1) + (1 - \alpha)\phi}$	$\alpha = 3$ for spheres; $\alpha = 6$ for cylinders
Jeffrey [23]	$\dfrac{k_{\mathrm{eff}}}{k_{\mathrm{f}}} = 1 + 3\beta\phi + \left(\dfrac{15}{4}\beta^2 + \dfrac{9}{16}\beta^3 \dfrac{\alpha + 2}{2\alpha + 3} + \cdots \right)\phi^2$	High-order terms present pair interaction of randomly dispersed spheres
Bruggeman [24]	$\dfrac{k_{\mathrm{eff}}}{k_{\mathrm{f}}} = \dfrac{1}{4}\left[(3\phi - 1)\dfrac{k_{\mathrm{p}}}{k_{\mathrm{f}}} + (2 - 3\phi) + \sqrt{\Delta} \right]$ $\Delta = (3\phi - 1)\left(\dfrac{k_{\mathrm{p}}}{k_{\mathrm{f}}} \right)^2 + (2 - 3\phi)^2 + [2 + 9\phi(1 - \phi)]\dfrac{k_{\mathrm{p}}}{k_{\mathrm{f}}}$	Considering clustering of nanoparticles and surface adsorption

k_{eff}, effective thermal conductivity of solid/liquid suspensions; k_{f}, thermal conductivity of base fluid; k_{p}, thermal conductivity of nanoparticles; $\alpha = k_{\mathrm{p}}/k_{\mathrm{f}}$, thermal conductivity ratio; $\beta = (\alpha - 1)/(\alpha + 2)$; n, particle shape factor; and ϕ, particle volume fraction.

increase the poor thermal conductivity of the original fluid, say a growth rate of exceeding 150% at 1 vol% nanotube-in-oil [18] and 40% at 0.3 vol% Cu nanoparticles in ethylene glycol [5]. However, it should be noted that such improvement is in fact rather limited due to the inherent characteristics of the adopted base liquid. Therefore, there is an urgent need to find further ideal base liquid that should be highly conductive. The concept of nanoliquid metal was thus proposed in such a situation [20] whose major role at the initial stage was to serve as coolant. However, the technical concept as stimulated by the nanoliquid metal is rather diverse and has in fact generalized purpose to be extended to more application areas.

In principle, the nanoliquid metal is a suspension of liquid metal and its alloy containing nanometer-sized particles. The room-temperature liquid metal and its alloy were first introduced in the area of cooling high-heat flux devices in 2002 [25], which now has already become a commercial reality. Their increasing applications have been found in a variety of engineering areas such as chip cooling [26–28], waste heat recovery [29,30], kinetic energy harvesting [31], thermal interface material (TIM) [32], even printed electronics [33–35], owing to their multimode capabilities such as low melting point, high thermal and electrical conductivity as well as more other desirable physical or chemical properties. Particularly, with much larger surface tension and density, such liquid metal-based fluid could suspend with much higher addition ratio of nanoparticles. Therefore, to fulfill various specific needs, a wide variety of nanomaterials loading options can be developed to mix particles with the liquid metal or its alloy. This opens a highly efficient start-up approach for innovating existing energy materials. To promote the research and development of this new area, this chapter is dedicated to present an overview of the nanoliquid metal technology through illustrating its most basic features and applications in the field of energy. The related fundamental and technical issues will be discussed. It can be anticipated that nanoliquid metal as an emerging functional material would offer many new chances for developing future energy technologies.

36.2
Typical Properties of Nanoliquid Metal

In technological principle, nanoliquid metal consists of liquid metal and nanoparticles that can be functional materials [1]. Figure 36.1 presents the appearance of liquid metal alloy, water, and nanoparticles. Among the many material candidates, the well-known gallium-based liquid metal is a low-melting-point metal that stays in liquid state at room temperature up to over 2300 °C and exhibits a strong tendency to supercool below its melting point of 29.8 °C. The high thermal and electrical conductivity of the liquid metal promises its broad applications in energy management, conversion, storage, and printed electronics [33]. Compared to the conventional base fluid, liquid metal and its alloys have a much higher conductivity, which can be of higher orders. Figure 36.2 lists a comparison regarding the thermal properties of typical liquid metals and conventional base fluids. Clearly, the liquid metal owns much superior thermal conductivity while having a modest volumetric thermal capacity. Furthermore, it has been revealed that the large density difference between the nanoparticles and the conventional base fluid seriously restricts the maximum volume fraction for the addition of nanoparticles as they would easily deposit if overloaded [20]. However, unlike that, the liquid metal owns an extremely high surface tension that is approximately 7–10 times larger than that of water and ethylene glycol. This

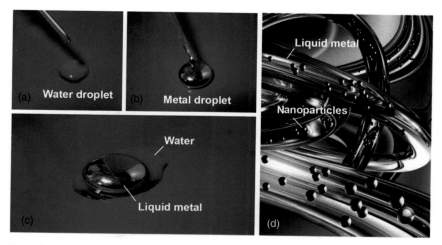

Figure 36.1 Comparison between droplets of water (a) and liquid metal alloy (b); (c) mixture of liquid metal alloy and water; (d) schematic of suspending nanoparticles inside liquid metal.

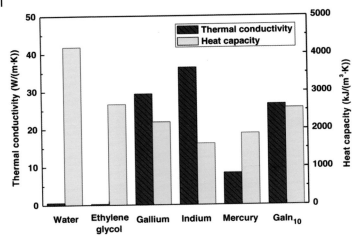

Figure 36.2 Comparison of thermal conductivity and heat capacity per unit volume between typical liquid metals and conventional base fluids. (Reprinted with permission from Ref. [1].)

offers the possibility of having a much larger volume fraction of loading nanoparticles, which further indicates a much larger space of enhancing and improving nanoliquid metal's performances.

Concerning the nanoparticles, they generally comprise metals that can be gold, silver, copper, aluminum, iron, and nickel, and oxides such as copper oxide, alumina, silicon oxide, titanium oxide, and so on, as well as carbon nanotubes [36], graphene, nitride, and carbide. Nanoparticles with different shapes, sizes, and concentrations play a key role in the modification of liquid metal and its alloy, forming composite materials with specific functions more than just thermal conductivity alone. When materials change to nanosize, they usually turn to have a highly unsaturated nature due to a large number of defects and dangling bonds on surface atoms, resulting in active chemical reactivity and high surface energy; when the particle size is close to the wavelength of light and De Broglie wave, projected depth, and other dimensions of the physical characteristics, periodic boundary conditions will be destroyed and new size effects will appear on the properties of electricity, magnetism, acoustics, optics, and thermodynamics [37]. Clearly, if loading such nanomaterials with desirable properties into the liquid metal, the fabricated nanoliquid metal would display multifunctions that can be electrically, magnetically, acoustically, optically, chemically, and thermally optimized. Table 36.2 presents a comparison on physical properties between conventional base liquid and typical metals with low melting points.

There remain lots of key issues to be solved before the applications of nanoliquid metal, one of which is to uniformly disperse nanoparticles in liquid metal. To date, only very limited trials have been made. Some particles have already been successfully loaded in liquid gallium [20,41–43] and mercury [44–46]. However, according to the experiments, it is in fact not easy for the "naked" metal to be directly dispersed as powders in a liquid metal. To resolve this

Table 36.2 Physical properties for conventional base fluid and typical metals with low melting points [38–40].

	Thermal conductivity	Heat capacity	Density	Surface tension	Melting point	Boiling point
	W/(m K)	J/(kg K)	kg/m³	mN/m	°C	°C
Water	0.6	4183	1000	72.8	0	100
Ethylene glycol	0.258	2349	1132	48.4	−12.6	197.2
Gallium	29.4[a]	370[a]	5907[a]	707[a]	29.8	2204.8
Indium	36.4[b]	230	7030[b]	550[m]	156.8	2023.8
Mercury	8.34[c]	139[c]	13 546[c]	455[c]	−38.87	356.65
Cesium	17.4[d]	236[d]	1796[d]	248[d]	28.65	2023.84
Sodium	86.9[d]	1380[d]	926.9[d]	194[d]	97.83	881.4
Potassium	54.0[e]	780[e]	664[e]	103[d]	63.2	756.5
GaIn$_{20}$	26.58[f]	403.5[f]	6335[f]		16	

a) 50 °C.
b) 160 °C.
c) 25 °C.
d) 100 °C.
e) at melting point.
f) 20 °C.

difficulty, prior to the preparation of the nanoliquid metal, silica-coated metal powders such as nickel, iron, and FeNbVB can be prepared via a chemical method. Silicon has a high affinity for liquid gallium and it can help improve the dispersion stability of liquid metal-based suspensions. Moreover, the silica layer can provide the core of the particles a good protection from oxidation and corrosion. The pH value may affect the coating process [47]. Figure 36.3 presents a typical modification procedure for nanoparticles in the preparation of the liquid metal-based magnetic fluid (MF). Conventional physical dispersion methods such as ultrasonic and mechanical vibration can be used in the synthesis of nanoliquid metal. It has been previously reported that any exposed region of the galinstan alloy in the air or even as little as 0.2% volume of oxygen will create a solid thin film of gallium oxide on the surface [48], which may prevent further oxidation [49]. Therefore, the synthesis process of nanoliquid metal is suggested to be in a protective or vacuum atmosphere. Based on the law that energized conductor can be driven by ampere force generated in a magnetic field, an alternating voltage of different directions and a perpendicular magnetic field is supposed to induce vibrations of liquid metal, consequently dispersing nanoparticles. If a complex system is allowed, the direction of magnetic field and electrodes can be designed to gain multidirectional vibrations. However, there is a limitation for the operation of such method in controlling nanoparticles with low electrical conductivity. Actually, nanoparticles with conductive,

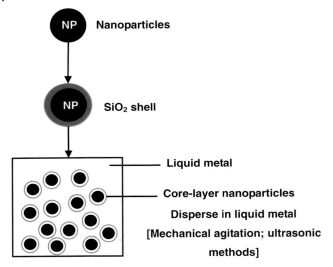

Figure 36.3 The modification procedure for preparation of the liquid metal-based MF. (Modified from Ref. [47].)

semiconducting, or isolating properties can also be loaded into liquid metal as an electrical property adjustable approach.

36.3
Emerging Applications of Nanoliquid Metal in Energy Areas

36.3.1
Energy Management

With the rapid development of science and technology, energy issues are becoming increasingly prominent and a large number of heat transfer problems have been raised in the engineering fields such as power, metallurgy, petroleum, chemical industry, and materials, as well as aviation, aerospace, and other high-tech fields [1]. The heat load and transfer intensity of heat exchangers are increasing in such a high speed that it is relatively difficult for the conventional pure working fluid to meet the requirements of heat transfer and cooling under some special conditions. In this aspect, small heat transfer coefficient of the working fluid has become a major barrier to develop new-generation high-profile heat transfer technology.

To improve heat transfer capability of the working fluids itself, one can naturally come up with two routes: one is to find innovative alternatives; the other is to modify the existing working fluids. Along the first route, liquid metal and its alloy have been proposed and successfully used in many occasions of high heat flux [26–29,50,51] owing to their superior thermal conductivity, electromagnetic

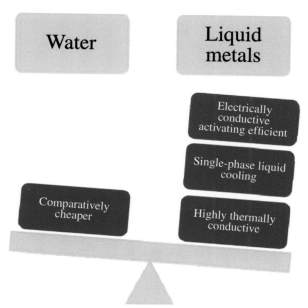

Figure 36.4 Comparison between water and liquid metal. (Reprinted with permission from Ref. [38].)

field drivability, and extremely low power consumption. Figure 36.4 displays several overwhelming advantages of liquid metal compared to water. In addition, Gao and Liu demonstrated that the thermal conductivity of a gallium-based thermal interface material could reach ~13.07 W/(m K) at room temperature, which is significantly higher than that of the best conventional thermal greases [32]. Figure 36.5 presents liquid metal as thermal interface materials that could display as metal pad, paste, or a component of different alloy and particles. Along the second route, adding particles is an important feasible way to enhance heat transfer as the suspended nanoparticles increase the thermal conductivity of the fluids and the chaotic movement of ultrafine particles would stimulate fluctuation and turbulence of the fluids, which accelerates the energy exchange process [52]. Clearly, many more supermetal fluidic materials with varied specific properties can be achieved via nanotechnology in future.

Thermal conductivity enhancement ratio as a function of volume fraction for different nanoparticles in liquid gallium suspensions by theoretical calculations are shown in Figure 36.6. Studies on nanofluids mainly focus on two aspects: experiments and numerical simulations. However, the thermal conductivity of nanofluids still depends on experimental measurements due to the large deviation between the mixed phase prediction model and the complicated phenomena that coexist [1]. According to the existing measurements, a conclusion can be drawn that the properties and behavior of nanofluids significantly depend on the properties of the base liquid and the dispersed phases, particle

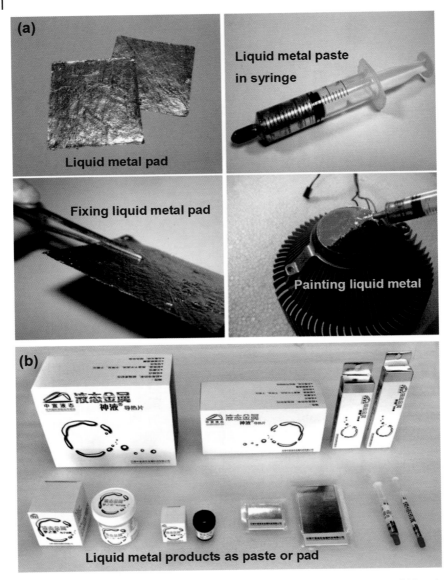

Figure 36.5 Liquid metals as thermal interface materials. (a) Prototyping sample [32]. (b) Practical products for thermal interface materials. (Reprinted with permission from Ref. [32].)

morphology, size, and concentration, as well as the presence of surfactants or dispersants [53]. In order to improve the heat transfer performance, a kind of material with higher coefficient of thermal conductivity is often taken into consideration. However, it has been confirmed that it is not always sensible to select materials in this way. For example, as is revealed, with the same volume fraction

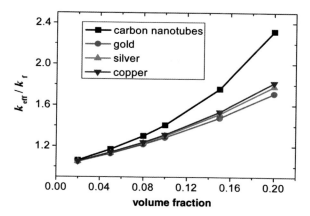

Figure 36.6 Thermal conductivity enhancement ratio as a function of volume fraction for different nanoparticles in liquid-gallium suspensions. (Reprinted with permission from Ref. [20].)

in the same base fluid, thermal conductivity of Fe nanofluids looks better than Cu nanofluids [53]. It is worth noting that the liquid metal itself can be driven by magnetic methods or thermosyphon effect [54]. Therefore, nanoliquid metal will exhibit an immense potential for making compacter cooling system with low noise and high efficiency.

Not only the thermal conductivity should be the core of the study, even other fluid dynamics characteristics such as viscosity and wettability as well as specific heat are critically significant. It is clear that the increase of thermal conductivity might be offset by the decrease of the effective heat capacity, the increase of viscosity, or the variation of wettability [55]. Some approaches are implemented to improve properties of nanofluids besides thermal conductivity. Indium-in-PAO (Poly Alpha Olefin) nanofluid was designed to use phase change nanoparticles instead of common solid nanoparticles and the effective volumetric heat capacity of the nanofluid containing 8 vol% indium nanoparticles was increased by up to 20% [56].

It should be pointed out that nanoliquid metal can have much more complex styles among different phases and species. In certain complex situations, special requests may be made to fabricate the thermal interface material with capabilities, for example, of maintaining thermal conductivity and electrical resistivity for the safe running of the target device. In a recent study [57], Mei *et al.* successfully made a thermally conductive and highly electrically resistive grease, the liquid metal poly (LMP) grease, through homogeneously dispersing liquid metal droplets in methyl silicone oil (Figure 36.7). There, the room temperature liquid metal was used as a new kind of liquid filler, instead of conventional solid particles. Through directly mixing and stirring in air, liquid metal micrometer-droplets were accidentally discovered to be homogeneously distributed and sealed in the matrix of methyl silicone oil (Figure 36.8). The experimental results indicate that the original highly electrically conductive liquid metal can be turned into a

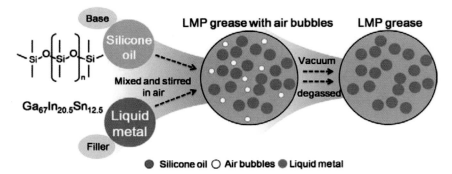

Figure 36.7 A schematic diagram for the fabrication process of the liquid metal poly grease. (Reprinted with permission from Ref. [57].)

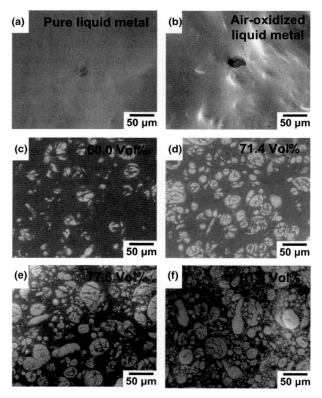

Figure 36.8 Plain-view ESEM images of (a) the pure liquid metal; (b) the air-oxidized liquid metal; (c)–(f) the four different volume ratio LMP greases. (Reprinted with permission from Ref. [57].)

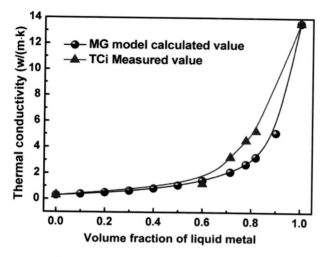

Figure 36.9 Experimentally measured and theoretically calculated thermal conductivities of the LMP thermal greases. (Reprinted with permission from Ref. [57].)

highly electrically resistive composite (Figures 36.9 and 36.10), by simply blending with methyl silicone oil. When the filler content comes up to 81.8 vol%, the thermal conductivity, viscosity, and volume resistivity read 5.27 W/(m·°C), 760 Pa·s, and 1.07×107 Ω·m, respectively. Compared to the pure liquid metal, the obtained composite materials own several evident advantages, including low cost and high electrical resistance. Furthermore, as the experiments revealed, the

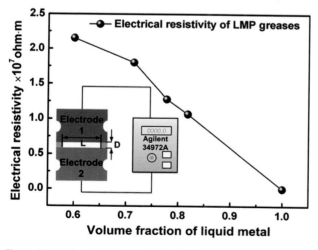

Figure 36.10 Electrical resistivity of the LMP greases and liquid metal (*inset*: Agilent 34972A meg-ohm test system). (Reprinted with permission from Ref. [57].)

greases presented no obvious corrosion effect, compared with pure liquid metal. In order to get thermal interface materials (TIMs) with higher thermal conductivity and electrical resistivity, additional procedures can be adopted for the optimization of ternary composites, consisting of nanoparticles, silicone oil, liquid metal, and so on. This method is expected to make it more practicable for liquid metals to be modified into a much safer TIMs in cooling electronics. It suggests a feasible approach to flexibly modify the material behaviors of the room-temperature liquid metals. The resulted thermally conductive, however highly electrically resistive, greases can be significant in a wide variety of electronic packaging applications.

36.3.2
Energy Conversion

While the original idea of nanofluids was to enhance the thermal conductivities of the heat transfer fluids, the influence of nanoparticles has been found even more profound than mere thermal property [1]. For example, nanoliquid metal is supposed to be an ideal working fluid competent for conversions between various energy forms.

One kind of magnetic fluid based on liquid metal has especially attracted much interest over the years. So far, most of the investigated base fluids for making magnetic nanofluids are focused on the organic oil or water. Figure 36.11 depicts the appearance of conventional ferrofluid on glass with a magnet underneath. Two typical organic solvents of ferrofluids are toluene and cyclohexane, whose applications are limited for relatively high toxicity and high volatility [43], whereas water can remain liquid only in a narrow range of temperature. As an alternative, liquid metal provides a superior thermal or electrical conductivity

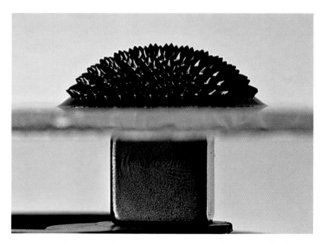

Figure 36.11 Ferrofluid on glass, with a magnet underneath. (*Source:* http://www.answers.com/topic/ferrofluid). (Reprinted with permission.)

over the conventional fluid, which is advantageous as carried liquid. Liquid metals such as gallium fluids usually have extremely high boiling point, above 2000 °C, and very low evaporation pressure. This enables such liquid metal-based magnetic nanofluids to maintain stable function in high temperature situations. One of the promising applications is the use of magnetic nanoliquid metal in different heat exchangers as well as in devices for magnetocaloric energy conversion. A loop of ferrofluid in a magnetic field can change the angular momentum and affect the rotation of the spacecraft, which is attributed to the conversion of magnetic energy with mechanical power. Furthermore, magnetic nanoliquid metal can be injected into the voice coil magnetic air gap to improve performance of the loudspeaker due to its cooling and damping effects [58]. Ferrofluid speakers have excellent characteristics such as high efficiency, big power, low distortion, and good performance at low frequency. Such magnetic nanoliquid metal improves the conversion efficiency of electromagnetic energy [59].

Fabrication of magnetic nanoliquid metal fluid appears rather intriguing. Researches on adding Ni or Fe nanoparticles into mercury have been carried out since years [45]. A pity lies in that due to serious toxicity, practical application of mercury-based magnetic fluids is rather limited. Recently, the room-temperature liquid metal gallium was identified as a perfect substitute of mercury. Its high boiling point and low melting point ensure the material an excellent working fluid within a very wide temperature range. In addition, the thermal conductivity of such a fluid is orders higher than that of water or organic base fluid. Hence, gallium-based magnetic nanofluids can be very promising in the magnetocaloric energy conversion and electromagnetic applications. However, due to the very large surface tension of the liquid metal, suspending nanoparticles uniformly into such base fluids remains a tough issue.

Until now, very few researches on gallium-based magnetic fluids are available. Iron alloys [43], micrometer or submicrometer Ni [44] or Fe particles have been successfully mixed with gallium in several works. Chemically synthesized FeNbVB nanoparticles were added into the liquid gallium after being coated by a layer of SiO_2. Over the temperature range of 293–353 K, it was found that the magnetization of the magnetorheological fluid varied with the temperature [43]. In addition, the suspension moves under magnetic gradient, and such movement was influenced by temperature [42].

Recently, based on a finding [32] that the wettability and compatibility of gallium with various materials can be significantly improved by a slight oxidization processing, Xiong et al. [59] revealed that loading Ni nanoparticles into the partially oxidized gallium and its alloys can help make desired magnetic nanofluid. Their experiments further showed that the Ni nanoparticles sharply increased the freezing temperature and latent heat of the obtained magnetic nanoliquid metal fluid, while the melting process was less affected (Figure 36.12). Clearly, the variation due to particles loading is rather evident. Because of the extremely small amount of the oxygen content, the difference between DSC curves of pure gallium and the magnetic fluid with 10 vol% Ni particles in gallium was mainly

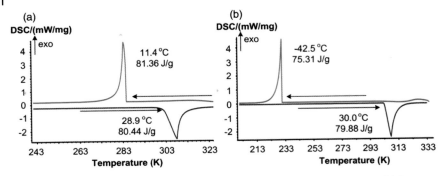

Figure 36.12 DSC curves of Ga+10% Ni and pure gallium [59]. (a)Gallium +10% Ni. (b) Pure gallium. (Reprinted with permission from Ref. [59].)

due to the Ni particles [59]. The surface interaction between base fluid and particle, as well as the size effects, and the Brown movements of the nanoparticles may contribute to such variation. For the gallium sample added with 10 vol% coated Ni particles, a hysteresis loop was observed and the magnetization intensity decreased with the increase of the temperature (Figure 36.13). The slope for the magnetization–temperature curve within 10–30 K was about 20 times of that from 40 to 400 K (Figure 36.13). Furthermore, the dynamic impact experiments of striking magnetic liquid metal droplets on the magnet revealed that the regurgitating of the leading edge of the liquid disk and the subsequent wave that often occurred in the gallium–indium droplets would disappear for the magnetic fluids case due to attraction force of the magnet (Figure 36.14).

When talking about energy conversion with nanofluids, solar energy conversion cannot be ignored. Recent studies also indicate that selected nanofluids may improve the efficiency of direct absorption solar thermal collectors [60–63]. Taking the advantage of nanofluids' special optical properties, they have been used in the solar energy absorption of full-spectrum and selective band absorption. In the 1970s, a technique of direct absorption collection (DAC) was

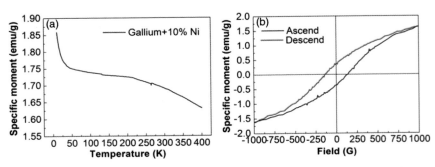

Figure 36.13 Properties of gallium-based magnetic nanofluid containing 10 vol% coated Ni nanoparticles [59]. (a) Temperature dependence of magnetization. (b) Hysteresis loop. (Reprinted with permission from Ref. [59].)

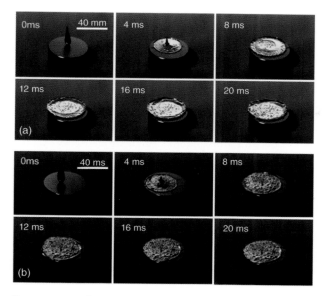

Figure 36.14 High-speed videos for dynamic impact of striking liquid metal droplets on a magnet plate [59]. (a) GaIn$_{24.5}$. (b) GaIn$_{24.5}$-based magnetic fluid. (Reprinted with permission from Ref. [59].)

proposed based on the application of mixed particles in working fluid in solar energy utilization [64]. The technique is now using working fluids, which can be molten salt and water containing ultrafine particles to absorb solar radiation directly in whole or in part for photothermal conversion. However, as liquid metal always has high reflection and is almost opaque, it is hard for nanoliquid metal to absorb sunlight but adding nanoparticles can still be a feasible way to modifying this property. Furthermore, in a DAC system, nanoliquid metal can collect heat in the flow channel at the bottom of solar panels, which is considered to improve the efficiency of the photothermal conversion.

36.3.3
Energy Storage

The application of phase change materials (PCMs) grew rapidly in the past few years, especially in areas such as solar energy, thermal comfort control, green building, environmental conservation, electronic cooling, and so on [1]. For example, phase change materials for cool storage (PCMCT) are the functional materials used for heat or cold storing by way of phase change process in the energy storage system. They can adjust the peaks and valleys of the energy usage through absorbing and releasing cold. Not all materials are suitable for cold storage and the degree of supercooling should not be too large. The process of PCMCT contains the occurrence of supercooling phenomenon, at which time cold is stored in the way of sensible heat, consequently reducing the cold storage. In order to reduce the degree of supercooling, homologous nucleating agents or impurities were added to the phase change materials [65]. Khodadadi and Hosseinizadeh elucidated through the method of numerical calculation and simulation that nanoparticle-enhanced phase change materials have great potential for improved thermal energy storage [66]. The process of phase changing is essentially crystallization of liquid molecules. The reduction of the degree of supercooling by nanoparticles can be explained by the nonhomogeneous nucleation theory. So far, tremendous efforts have been made for finding new powerful PCMs [67] or improving performance of the currently available PCMs (Figure 36.15), which are generally subject to inherent defects, such as low thermal conductivity, poor stability after millions of repeated solidifying and melting processes, easy phase separation during transition, and narrow temperature span between the melting point and the evaporation state.

It has been demonstrated that liquid gallium dispersing with 1.0 wt % silica nanoparticles of 10 nm in size can remain stable for more than 400 days at 276–277 K and mass fraction must be strictly limited [68]. There is a lot of solidification-released heat that is desired to be moved away in a short time and a relatively high thermal conductivity is therefore needed. The feasibility of the low-melting-point liquid metals has been investigated as phase change materials for its large heat transfer capacity, excellent reversibility of phase transition, and small phase expansion [67]. The present authors' laboratory has already introduced the low-melting-point metals or their alloys, including nanoparticle components as PCM for the thermal management of a group of mobile electronics such as USB flash memory [69] and smartphone [70]. The specific heat per unit volume was 473 509.2 kJ/m^3 for gallium, larger than conventional phase change materials such as $Na_2SO_4 \cdot 10\,H_2O$, N-eicosane, and paraffin [69]. Such low-melting-point metals should work well as a very promising PCM candidate for next-generation industrial heat exchangers because of their many favorable properties, such as high thermal conductivity, good electrical conductivity, low vapor pressure, and small volume expansion during the phase transition. Figure 36.16 illustrates gallium PCM-based thermal management of a group of mobile devices. According to experiments, the instantaneous heat generated from the electronics system such as USB, mobile phone, and so on during working could be well absorbed by the gallium PCM filled inside. A specific volume of gallium PCM will be enough to maintain the target temperature below safe temperature for an acceptably long time. This is sufficient to guarantee the safe running of the device. Furthermore, selective loading of phase change nanoparticles can improve heat capacity and finally create a much superior liquid material for heat storage via phase changing. The large latent heat and heat capacity per unit volume of nanoliquid metal may bring about a volume reduction of phase change capsules. With such merits, it is evident that nanoliquid metal can be promising for its super thermophysical properties and large latent heat per unit volume.

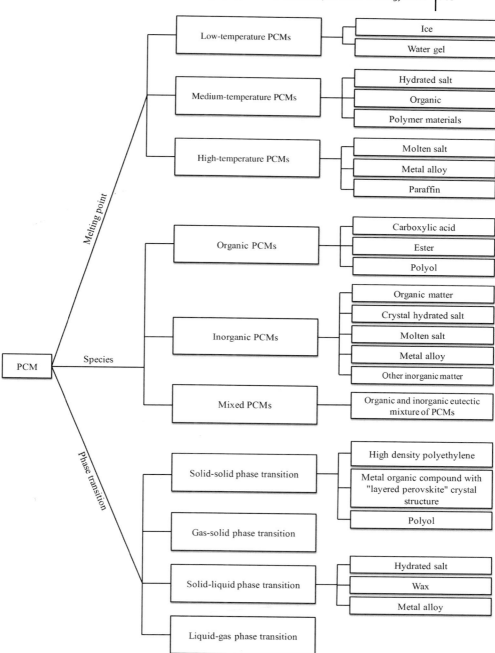

Figure 36.15 Category of PCM based on melting point, species, and phase transition mechanism. (Reprinted with permission from Ref. [67].)

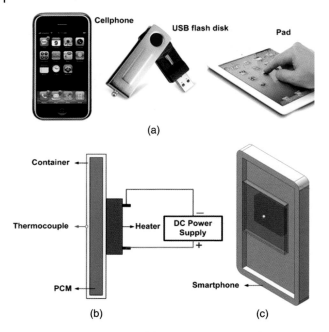

Figure 36.16 (a) Common mobile devices demanding for thermal management. (b) Schematic for the experimental setup for gallium PCM-based thermal management of smartphone. (c) Isometric view of the container adhered to the smartphone [69,70]. (Reprinted with permission from Refs [69,70].)

Several typical thermophysical properties of additional low-melting point metal or alloy used as PCM have been listed in Table 36.3. It can be observed that the temperature gap between the melting temperature and evaporation point of metal was extremely large. This guarantees the rather wide suitability of such kind of PCM in many fields.

Different from the routine approaches, using metals as PCM can efficiently solve the problems emerged in traditional PCMs, owing to their super thermophysical property. Classified by melting point, metal PCM can generally be divided into three categories: high temperature, middle temperature, and low temperature as listed in Table 36.4. Their application fields are also provided in the table.

36.4
Challenging Scientific and Technological Issues

The high performance of nanoliquid metal warrants its significant applications in a wide variety of energy areas. However, before the applications are completely approved in industry, lots of fundamental issues and technological

Table 36.3 Thermophysical properties of several typical metals or alloys with low melting point as PCM [38,67].

Liquid metals	Melting point (°C)	Evaporation point (°C)	Specific heat (kJ/(kg °C))	Density (kg/m^3)	Thermal conductivity (W/(m °C))	Enthalpy of fusion (kJ/kg)
Rubidium	38.85	685.73	0.363[a]	1470[a]	29.3[a]	25.74
$Bi_{44.7}Pb_{22.6}In_{19.1}Sn_{8.3}Cd_{5.3}$	47	—	0.197	9160	15	36.8
$Bi_{49}In_{21}Pb_{18}Sn_{12}$	58	—	0.201	9010	10	28.9
$Bi_{50}Pb_{26.7}Sn_{13.3}Cd_{10}$	70	—	0.184	9580	18	39.8
$Bi_{52}Pb_{30}Sn_{18}$	96	—	0.167	9600	24	34.7
$Bi_{58}Sn_{42}$	138	—	0.201	8560	19	44.8
$Sn_{91}Zn_9$	199	—	0.272	7270	61	32.5
Tin	232	2622.8	0.221	730[b]	15.08[c]	60.5[c]
Bismuth	271.4	1560	0.122	979	8.1	53.3

a) At melting point.
b) At 100 °C.
c) At 200 °C.

challenges must be overcome. The following are a group of important issues that should be carefully addressed [1].

Suspended particles bring additional troubles such as particle deposition, conglomeration, susceptibility to fouling, degeneration of solution quality, and possible flow jamming over the channels [1]. The stability issues about nanofluid suspension include those on thermodynamic, fluidic, and aggregation stability, respectively [71]. Interaction between the suspended nanoparticles causes the tendency of agglomeration of the particles. Once the particles aggregate, it is difficult to separate them, causing further aggregation. Subsequently, aggregation gradually becomes larger, forming nanoparticle clusters and thus the uniform dispersity reduces. With reference to nanoliquid metal, there still exist many problems related to synthesis of nanofluids as it is difficult to blend liquid metal with nanoparticles in the process of moistening and the stratification may affect the uniformity of dispersion [1]. Based on these considerations, a few of the directions to solve the problems can be the following: treatment and

Table 36.4 Metal PCM category [67].

Category	High temperature	Middle temperature	Low temperature
Temperature scale	Above 200 °C	40~200 °C	0~30 °C
Typical metal PCM	Other metal alloy	Bi-based metal alloy	Ga-based metal alloy
Application	Solar energy	Wasted heat recovery and solar energy	Thermal management, building energy conservation, and thermal comfort

modification of nanoparticles' surface, development of the dispersing agents and stabilizers with new superior performance, exploration of the dispersion conditions, and the optimization of the preparation process. Furthermore, metal particles should be loaded carefully into the base liquid and coated with isolation layer due to the causticity of liquid metal on some solid metal [72,73].

In order to guarantee the applications of nanoliquid metal, fundamental studies are greatly needed to understand the physical mechanisms. A better interpretation of heat transfer enhancement of the nanofluids will accelerate its practical application. However, the current understanding on nanofluids is still somewhat limited. Nanoliquid metal is seriously neglected but innovative energy material calls for establishments of more theoretical models. The existing equations for the conventional nanofluids may be modified to adapt to nanoliquid metal. As thermal conductivity should not be a mere evaluation standard, an integrated perspective to understand nanoliquid metal from the aspects of heat transfer, rheology, and phase change is seriously needed.

Except for the thermofluidic issues, additional energy conversion mechanisms and possible applications in nanoliquid metal such as those among mechanical, electrical, magnetic, acoustic, optical and thermal energies, and so on are worth investigation. This may lead to new findings for future better energy utilization. And some latest works on developing nanoenergy such as piezoelectricity generation [74] or multimode micro/nanoscale heat transfer enhancement [75] can shed light on such endeavor. Overall, despite the extensive research interest that nanofluids have been attracting, there is still currently a strong lack of consistent criteria to assess repeatability and reliability of the results of the experiments. For example, with respect to the researches on thermal conductivity of the conventional nanofluids, the inconsistencies between theory and the experiment data still remain due to different measuring approaches such as transient hot wire, steady-state heated plates, oscillating temperature and thermal lensing, the often-incomplete characterization of the nanofluid samples, and the differences in the synthesis processes used to prepare those samples [76]. Besides, the highly conductive nanoliquid metal should display a series of very different behaviors with that of conventional fluids, which unfortunately reserved unclear right now. Clearly, in order to quantify the effects of the different factors in affecting the nanoliquid metal materials, more comparative experiments as well as theoretical justifications have to be made to differentiate the contribution of stabilizer, nanoparticles, and liquid metal ingredients and a database is supposed to be built in the near future.

36.5
Summary

It can now be well understood that nanoliquid metal owns immense potential as emerging energy material. The higher thermal and electrical conductivity of liquid metal than that of conventional base liquid promises its significant values in cooling devices with much better performance [1]. The applications of nanoliquid metal controlled by magnetic methods can bring about compacter cooling system with low noise and high efficiency. This offers opportunities for engineers to develop highly effective heat transfer equipment. In the aspect of energy conversion, the possible usage of nanoliquid metal covers magnetic, electrical, acoustic, and solar energy. In the process of energy conversion, nanoliquid metal acts as working fluid carrying heat or a component to improve the conversion efficiency. And in terms of energy storage, nanoliquid metal is supposed to be an excellent phase change material for compact cold or heat storage benefiting from its strong heat transfer capacity, excellent reversibility of phase transition, and small phase expansion. However, lots of scientific and technological challenges still remain to be solved, which requires interdisciplinary collaboration among nanomaterials, colloid science, physics, and engineering. The synthesis approaches, suspension stability, special properties of nanoliquid metal calls for more investigations. In order to significantly extend the applications of nanoliquid metal, tremendous efforts are needed to better understand the involved physical mechanisms. Unlike the conventional nanofluids, the fundamental properties and applications of nanoliquid metal have been seriously ignored in the past few years. It is also for such reason that a fruitful future for developing the nanoliquid metal technologies will be coming.

Acknowledgment

This work is partially supported by the Dean's Research Funding of the Chinese Academy of Sciences.

References

1 Zhang, Q. and Liu, J. (2013) Nanoliquid metal as an emerging functional material in energy management, conversion and storage. *Nano Energy*, **2**, 863–872.

2 Choi, S.U.S. and Eastman, J.A. (1995) Enhancing thermal conductivity of fluids with nanoparticles, developments and applications of non-Newtonian flows. International Mechanical Engineering Congress and Exhibition, San Francisco, CA, pp. 99–105.

3 Razi, P., Akhavan-Behabadi, M.A., and Saeedinia, M. (2011) Pressure drop and thermal characteristics of CuO-base oil nanofluid laminar flow in flattened tubes under constant heat flux. *Int. Commun. Heat Mass Transf.*, **38**, 964–971.

4 Kwak, K. and Kim, C. (2005) Viscosity and thermal conductivity of copper oxide nanofluid dispersed in ethylene glycol. *Korea-Australia Rheol. J.*, **17**, 35–40.

5 Eastmana, J.A., Choi, S.U.S., Li, S., Yu, W., and Thompson, L.J. (2001) Anomalously increased effective thermal conductivities of ethylene glycol-based nanofluids containing copper nanoparticles. *Appl. Phys. Lett.*, **78**, 718–720.

6 Gao, L., Zhou, X.F., and Ding, Y. (2007) Effective thermal and electrical conductivity of carbon nanotube composites. *Chem. Phys. Lett.*, **434**, 297–300.

7 Das, S.K., Putra, N., Thiesen, P., and Roetzel, W. (2003) Temperature

dependence of thermal conductivity enhancement for nanofluids. *J. Heat Transf.*, **125**, 567–574.

8 Patel, H.E., Das, S.K., Sundararajan, T., Nair, A.S., George, B., and Pradeep, T. (2003) Thermal conductivities of naked and monolayer protected metal nanoparticle based nanofluids: manifestation of anomalous enhancement and chemical effects. *Appl. Phys. Lett.*, **83**, 2931–2933.

9 Wang, X., Xu, X., and Choi, S.U.S. (1999) Thermal conductivity of nanoparticle–fluid mixture. *J. Thermophys. Heat Transf.*, **13**, 474–480.

10 Xie, H., Lee, H., Youn, W., and Choi, M. (2003) Nanofluids containing multiwalled carbon nanotubes and their enhanced thermal conductivities. *J. Appl. Phys.*, **94**, 4967–4971.

11 Hwang, Y., Lee, J.K., Lee, C.H., Jung, Y.M., Cheong, S.I., Lee, C.G. *et al.* (2007) Stability and thermal conductivity characteristics of nanofluids. *Thermochim. Acta*, **455**, 70–74.

12 Keblinski, P., Phillpot, S.R.E., Choi, S.U.S., and Eastman, J.A. (2002) Mechanisms of heat flow in suspensions of nano-sized particles (nanofluids). *Int. J. Heat Mass Transf.*, **45**, 855–863.

13 Evans, W., Fish, J., and Keblinski, P. (2006) Role of Brownian motion hydrodynamics on nanofluid thermal conductivity. *Appl. Phys. Lett.*, **88**, 093116.

14 Chandrasekara, M. and Suresha, S. (2009) A review on the mechanisms of heat transport in nanofluids. *Heat Transf. Eng.*, **30**, 1136–1150.

15 Harish;, S., Ishikawa, K., Einarsson, E., Aikawa, S., Chiashi, S., Shiomi, J., and Maruyama, S. (2012) Enhanced thermal conductivity of ethylene glycol with single-walled carbon nanotube inclusions. *Int. J. Heat Mass Transf.*, **55**, 3885–3890.

16 Sharma, P., Baek, I.-H., Cho, T., Park, S., and Lee, K.-B. (2011) Enhancement of thermal conductivity of ethylene glycol based silver nanofluids. *Powder Technol.*, **209**, 7–19.

17 Yu, W., Xie, H.Q., and Chen, L.F. (2010) Investigation on the thermal transport properties of ethylene glycol-based nanofluids containing copper nanoparticles. *Powder Technol.*, **197**, 218–221.

18 Choi, S.U.S., Zhang, Z.G., Yu, W., Lockwood, F.E., and Grulke, E.A. (2001) Anomalous thermal conductivity enhancement in nanotube suspensions. *Appl. Phys. Lett.*, **79**, 2252.

19 Choi, C., Yoo, H.S., and Oh, J.M. (2008) Preparation and heat transfer properties of nanoparticle-in-transformer oil dispersions as advanced energy-efficient coolants. *Curr. Appl. Phys.*, **8**, 710–712.

20 Ma, K.Q. and Liu, J. (2007) Nanoliquid-metal fluid as ultimate coolant. *Phys. Lett. A*, **361**, 252–256.

21 Maxwell, J.C. (1881) *A Treatise on Electricity and Magnetism*, 2nd edn, Clarendon Press, Oxford, UK.

22 Hamilton, R.L. and Crosser, O.K. (1962) Thermal conductivity of heterogeneous two component systems. *Ind. Eng. Chem. Fundam.*, **1**, 187–191.

23 Jeffrey, D.J. (1973) Conduction through a random suspension of spheres. *Proc. R. Soc. Lond. A*, **335**, 355.

24 Wang, B.-X., Zhou, L.-P., and Peng, X.-F. (2003) A fractal model for predicting the effective thermal conductivity of liquid with suspension of nanoparticles. *Int. J. Heat Mass Transf.*, **46**, 2665–2672.

25 Liu, J. and Zhou, Y.X. (2002) Chip cooling equipment with low melting point metal and its alloy acting as working fluid. China Patent 021314195.

26 Ma, K.Q. and Liu, J. (2007) Heat-driven liquid metal cooling device for the thermal management of a computer chip. *J. Phys. D Appl. Phys.*, **40**, 4722–4729.

27 Deng, Y.G. and Liu, J. (2010) A liquid metal cooling system for the thermal management of high power LEDs. *Int. Commun. Heat Mass Transf.*, **37**, 788–791.

28 Deng, Y.G. and Liu, J. (2010) Hybrid liquid metal–water cooling system for heat dissipation of high power density microdevices. *J. Heat Mass Transf.*, **46**, 1327–1334.

29 Dai, D., Zhou, Y.X., and Liu, J. (2011) Liquid metal based thermoelectric generation system for waste heat recovery. *Renew. Energy*, **36**, 3530–3536.

30 Li, H.Y., Zhou, Y., and Liu, J. (2014) Liquid metal based printable thermoelectronic

generator and its performance evaluation. *Sci. China Technol. Sci.*, **44**, 407–416.

31 Liu, J. (2012) Piezoelectric thin film electricity generator and its fabrication method. China Patent 2012103225845.

32 Gao, Y.X. and Liu, J. (2012) Gallium-based thermal interface material with high compliance and wettability. *Appl. Phys. A*, **107**, 701–708.

33 Zhang, Q., Zheng, Y., and Liu, J. (2012) Direct writing of electronics based on alloy and metal (DREAM) ink: a newly emerging area and its impact on energy, environment and health sciences. *Front. Energy*, **6**, 311–340.

34 Gao, Y.X., Li, H.Y., and Liu, J. (2012) Direct writing of flexible electronics through room temperature liquid metal ink. *PLoS One*, **7**, e45485.

35 Zheng, Y., He, Z.Z., Yang, J., and Liu, J. (2014) Personal electronics printing via tapping mode composite liquid metal ink delivery and adhesion mechanism. *Sci. Rep.*, **4**, 4588.

36 Ding, Y., Alias, H., Wen, D., and Williams, R.A. (2006) Heat transfer of aqueous suspensions of carbon nanotubes (CNT nanofluids). *Int. J. Heat Mass Transf.*, **49**, 240–250.

37 Liu, J.P. and Hao, X.Y. (2009) *Modification of Polymer-Based Materials (in Chinese)*, 1st edn, Science Press, Beijing.

38 Li, H.Y. and Liu, J. (2011) Revolutionizing heat transport enhancement with liquid metals: proposal of a new industry of water-free heat exchangers. *Front. Energy*, **5**, 20–42.

39 Iida, T. and Guthrie, R.I.L. (1993) *The Physical Properties of Liquid Metals*, Clarendon Press, Oxford.

40 Shimoji, M. (1977) *Liquid Metals: An Introduction to the Physics and Chemistry of Metals in the Liquid State*, Academic Press, New York.

41 Ito, R., Dodbiba, G., and Fujita, T. (2005) MR fluid of liquid gallium dispersing magnetic particles. *Int. J. Mod. Phys. B*, **19**, 1430–1436.

42 Fujita, T., Park, H.-S., Ono, K., Matsuo, S., Okaya, K., and Dodbiba, G. (2011) Movement of liquid gallium dispersing low concentration of temperature sensitive magnetic particles under magnetic field. *J. Magn. Magn. Mater.*, **323**, 1207–1210.

43 Dodbiba, G., Ono, K., Park, H.S., Matsuo, S., and Fujita, T. (2011) FeNbVB alloy particles suspended in liquid gallium: investigating the magnetic properties of the MR suspension. *Int. J. Mod. Phys. B*, **25**, 947–955.

44 Popplewell, J., Charles, S.W., and Hoon, S.R. (1980) Aggregate formation in metallic ferromagnetic liquids. *IEEE Trans. Magn.*, **16**, 191–196.

45 Linderoth, S., Rasmussen, L.H., and Mørup, S. (1991) New methods for preparing mercury based ferrofluids. *J. Appl. Phys.*, **69**, 5124.

46 AbuAljarayesh, I., Bayrakdar, A., Yusuf, N.A., and AbuSafia, H. (1993) AC susceptibility of cobalt in mercury magnetic fluids. *J. Appl. Phys.*, **73**, 6970.

47 Park, H.S., Cao, L.-F., Dodbiba, G., and Fujita, T. (2014) *Preparation and Properties of Silica-Coated Ferromagnetic Nano Particles Dispersed in a Liquid Gallium Based Magnetic Fluid*, The University of Tokyo, Japan.

48 Liu, T.Y., Sen, P., and Kim, C.J. (2012) Characterization of nontoxic liquid-metal alloy galinstan for applications in microdevices. *J. Microelectromech. Syst.*, **21**, 443–450.

49 Tostmann, H., Shpyrko, O., Pershan, P.S., Dimasi, E., Ocko, B., and Deutsch, M. (1999) Surface structure of liquid metals and the effect of capillary waves: X-ray studies on liquid indium. *Phys. Rev. B*, **59**, 783–791.

50 Deng, Y.G. and Liu, J. (2010) Design of practical liquid metal cooling device for heat dissipation of high performance CPUs. *J. Electron. Packaging*, **132**, 031009.

51 Deng, Z.S. and Liu, J. (2006) Capacity evaluation of a MEMS based micro cooling device using liquid metal as coolant. The 1st IEEE International Conference on Nano/Micro Engineered and Molecular Systems, Zhuhai, China, pp. 1311–1315.

52 Daungthongsuk, W. and Wongwises, S. (2007) A critical review of convective heat transfer of nanofluids. *Renew. Sustain. Energy Rev.*, **11**, 797–817.

53 Hong, T.-K. and Yang, H.-S. (2005) Study of the enhanced thermal conductivity of Fe nanofluids. *J. Appl. Phys.*, **97**, 064311.

54 Li, P.P. and Liu, J. (2011) Harvesting low grade heat to generate electricity with thermosyphon effect of room temperature liquid metal. *Appl. Phys. Lett.*, **99**, 094106.

55 Sarkar, J. (2011) A critical review on convective heat transfer correlations of nanofluids. *Renew. Sustain. Energy Rev.*, **15**, 3271–3277.

56 Han, Z.H. (2008) Nanofluids with enhanced thermal transport properties, PhD thesis, University of Maryland.

57 Mei, S.F., Gao, Y.X., Deng, Z.S., and Liu, J. (2014) Thermally conductive and highly electrically resistive grease through homogeneously dispersing liquid metal droplets inside methyl silicone oil. *J. Electron. Packag.*, **136**, 011009.

58 Athanas, L.S. (1994) Loudspeaker utilizing magnetic liquid suspension of the voice coil. U.S. Patent 5335287.

59 Xiong, M.F., Gao, Y.X., and Liu, J. (2013) Fabrication of magnetic nano liquid metal fluid through loading of Ni nanoparticles into gallium or its alloy. *J. Magn. Magn. Mater.*, **354**, 279–283.

60 Taylor, R.A., Phelan, P.E., Otanicar, T.P., Adrian, R., and Prasher, R. (2011) Nanofluid optical property characterization: towards efficient direct absorption solar collectors. *Nanoscale Res. Lett.*, **6**, 1–11.

61 Lenert, A. and Wang, E.N. (2012) Optimization of nanofluid volumetric receivers for solar thermal energy conversion. *Sol. Energy*, **86**, 253–265.

62 Otanicar, T., Taylor, R.A., and Phelan, P.E. (2009) Impact of size and scattering mode on the optimal solar absorbing nanofluid. ASME 2009 3rd International Conference on Energy Sustainability, 1, pp. 791–796.

63 Taylor, R.A., Phelan, P.E., Otanicar, T., Walker, C.A., and Nguyen, M. (2011) Applicability of nanofluids in high flux solar collectors. *J. Renew. Sustain. Energy*, **3**, 023104.

64 Hunt, A.J. (1978) Small particle heat exchangers. Lawrence Berkeley Laboratory Paper LBL-7841.

65 Ona, E.P., Ozawa, S., Kojima, Y., Matsuda, H., Hidaka, H., Kakiuchi, H. *et al.* (2003) Effect of ultrasonic irradiation parameters on the supercooling relaxation behavior of PCM. *J. Chem. Eng. Jpn.*, **36**, 799–805.

66 Khodadadi, J.M. and Hosseinizadeh, S.F. (2007) Nanoparticle-enhanced phase change materials (NEPCM) with great potential for improved thermal energy storage. *Int. Commun. Heat Mass Transf.*, **34**, 534–543.

67 Ge, H.S., Li, H.Y., Mei, S.F., and Liu, J. (2013) Low melting point liquid metal as a new class of phase change material: an emerging frontier in energy area. *J. Renew. Sustain. Energy Rev.*, **21**, 331–346.

68 Cao, L.F., Park, H., Dodbiba, G., Ono, K., Tokoro, C., and Fujita, T. (2011) Keeping gallium metal to liquid state under the freezing point by using silica nanoparticles. *Appl. Phys. Lett.*, **99**, 143120.

69 Ge, H.S. and Liu, J. (2012) Phase change effect of low melting point metal for an automatic cooling of USB flash memory. *Front. Energy*, **6**, 207–209.

70 Ge, H.S. and Liu, J. (2013) Keeping smartphones cool with gallium phase change material. *J. Heat Transf.*, **135**, 054503–054508.

71 Xuan, Y.M. and Li, Q. (2010) *Energy Transfer Theory and Application of Nanofluids* (in Chinese), Science Press, Beijing.

72 Deng, Y.G. and Liu, J. (2009) Corrosion development between liquid gallium and four typical metal substrates used in chip cooling device. *Appl. Phys. A*, **95**, 907–915.

73 Surmann, P. and Zeyat, H. (2005) Voltammetric analysis using a self-renewable non-mercury electrode. *Anal. Bioanal. Chem.*, **383**, 1009–1013.

74 Wang, Z.L. (2011) *Nanogenerators for Self-Powered Devices and Systems*, Georgia Institute of Technology, Atlanta, GA.

75 Liu, J. (2006) *Micro/Nano Scale Heat Transfer* (in Chinese), 4th edn, Science Press, Beijing.

76 Buongiomo, J., Venerus, D.C., Prabhat;, N., McKrell, T., Townsend, J., and Christianson, R. (2009) A benchmark study on the thermal conductivity of nanofluids. *J. Appl. Phys.*, **106**, 094312.

37

IoNanofluids: Innovative Agents for Sustainable Development

Carlos Nieto de Castro, Xavier Paredes, Salomé Vieira, Sohel Murshed, Maria José Lourenço, and Fernando Santos

Universidade de Lisboa, Faculdade de Ciências, Centro de Química Estrutural, Campo Grande, 1749-016 Lisboa, Portugal

37.1
Introduction

Our first paper on the thermal properties of IoNanofluids (INFs) was published in 2009 [1]. Since then significant attention has been paid to it by research groups and companies. This can be explained not only by the interesting properties and behavior of these nanosystems but also by the number of potential industrial applications, triggered by the nanofluids research, and as a result of the tremendous opportunities given by the use of ionic liquids.

As in all high-tech fields, the impact in the first years is very impressive, even though the attitude toward research is not of high quality, and the results obtained are not very well discussed, regarding the physical nature of the systems and the current understanding of the relation between property and structure.

As explained in our recent book [2], there are still many areas in nanofluids research that need further attention. It is necessary to understand completely the structure of the nanosystems, as they dominate all the potential applications, making ideas commercially valuable. This is the case with long-term stability, reliability, and accuracy of experimental systems used to acquire experimental data, consistency, and resolving uncertainties in measured data and findings, lack of fundamental understanding of heat transfer mechanisms, and underlying mechanisms from macro- to nanoscale.

These problems are aggravated with IoNanofluids, because the structure of the base liquids, the ionic liquids (ILs), is by far more complex than the common molecular solvents such as water and glycols, and also because their interactions with nanomaterials are more complex, due to specific interactions between the ions and the nanoparticles.

Nanotechnology for Energy Sustainability, First Edition. Edited by Baldev Raj, Marcel Van de Voorde, and Yashwant Mahajan.
© 2017 Wiley-VCH Verlag GmbH & Co. KGaA. Published 2017 by Wiley-VCH Verlag GmbH & Co. KGaA.

However, there are several reasons to study these complex systems:

- Enhanced thermal properties for heat transfer and heat storage
- Nanoregions created by complex interactions enhance reactivity and selectivity of chemical reactions (nanocatalysis)
- IoNanofluids are designable and fine-tuneable through base ILs to meet any specific application or task requirement
- Nonflammability and nonvolatility at ambient conditions, environmentally friendly solvents, and reaction fluids.

The aim of this chapter is to provide the reader an understanding of the problems that need research, with special emphasis on preparation, properties, and applications of IoNanofluids. More detailed information about IoNanofluids can be found in recent reviews by our group [3–6].

37.2
IoNanofluids: Nature, Definitions, Preparation, and Structure Characterization

An IoNanofluid is defined as a stable dispersion of nanomaterial particles (tubes, rods, spheres, etc.) in an ionic liquid. Some authors also called them nanoparticles-enhanced ionic liquids (NEILs). It involves joining three realities – nanomaterials, ionic liquids, and nanofluids. If the nanomaterial is from a natural source (biota), we can obtain IoBiofluids. This rationale can be schematized in Figure 37.1, using the three-component ternary diagram. Therefore, to understand the behavior of IoNanofluids, we need to understand not only the interactions between the nanoparticles, the ionic liquid ions, but also the interaction between the ions and the nanoparticles.

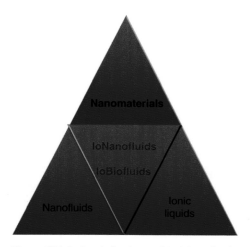

Figure 37.1 Rationale for the work in IoNanofluids.

Nanofluids are not simply mixtures of solids in liquids, and therefore the thermodynamics of mixing (and the transport properties of mixtures) cannot be used. *A fortiori*, the nanofluid success depends on the method of preparation and on the purity of the starting materials. The simpler nanofluids are in fact biphasic systems with a solid phase dispersed in a liquid phase, forming systems that go from simple dispersions to gels and micellar solutions (different nanostructure organization), which require a fundamental property for any engineering or medical application – to be stable over a long period of time. However, today it is still not possible to interpret completely the stability mechanisms of nanofluids necessary for the wide range of applications foreseen (energy, biomedical, and mechanical), a deemed requisite for its success. Other special and essential requirements are the durability of the suspension (time frame), the prevention of agglomeration, and the constancy of the fluid chemistry.

This was not the first conception of nanofluids, sometimes considered by "mechanistic" approaches of solid particles mixed with a liquid and, because of that, there were many publications showing tremendous differences between thermal conductivity data supposedly obtained for the "same" nanofluid [7], which of course are meaningless. In fact the fluids used were different, not counting also that the measuring methods used could not be the most appropriate.

In order to discuss the preparation of nanofluids, we have to analyze the starting nanomaterials, how they were produced, and what is known about its characterization (shape of the particles, geometric dimensions, state of aggregation).

37.2.1
The Nanomaterial Problem

Nanomaterials are nowadays a common expression for the scientific community that spread to the general public. Their usefulness for industry is unmistakable and accepted by scientists and industrialists. The field of application is vast and the continuous development in science and technology estimates their global market at 11 million tons at a market value of €20 billion and the direct employment in this sector is estimated at 300 000–400 000 in Europe [8]. Although these values combine materials such as metals, metal oxides, ceramics, polymers, and so on, the market is still dominated by nanocarbons and synthetic amorphous silica. Its size is still growing fast every year, especially because many commercial products incorporating nanomaterials are being developed, mainly due to new applications of existing nanomaterials than to new ones. The development of new high-tech products typically takes high-cost, long-timeframe, and high-risk activity, and sometimes it is difficult to achieve commercial success with new materials. The manufacturers tend to sell nanomaterials without mentioning their origin or method of preparation, apart from not giving material physical and chemical data, as some are still unknown.

The properties and characteristics of the nanomaterials are an extremely important issue. Most of the nanomaterials are spherical, rod or oblong in shape.

Current manufacturing procedures can result in batch to batch variability as the degree of outer shell nanomaterial functionalization caused during manufacture and geometries of the nanomaterials can change. In a majority, the producers of nanoparticles do not disclose the real structure of the particles, the presence of polymer coatings, or oxidative type reactions, as it happens with nanosilver [6], where 80% of the manufacturers use polymer coatings of nondisclosed thickness and properties, which modifies the expected heat transfer properties of the nanomaterials. Many publications on nanofluid systems previously reported were probably not aware of these problems and therefore those results have to be confirmed.

It is difficult to produce nanomaterials on a large scale, because the mass of material produced is very small. One gram of nanoparticles is approximately equivalent to one billion particles. The main methods of preparation of nano-materials are the top-down and bottom-up approaches. As examples of top-down approaches we can have attrition, milling, and lithography – etching. As bottom-up approaches colloidal dispersion and lithography – growth of thin films, nanolithography, and nanomanipulation are examples.

Once produced, the nanomaterial must be characterized, not only morpholog-ically (shape and size of particles, cluster dimensions, and particle size distribu-tion), but also property-wise. Fundamental research has been done on the synthesis of nanomaterials but the difficulty to characterize them, particularly in thermophysical, mechanical, and toxicological properties, still leaves most Mate-rial Safety Data Sheets only with information regarding the appearance, specific gravity and bulk density, melting point, solubility in common solvents, pH, and chemical composition. Some of the their most remarkable features are related to their durability, thermal and mechanical properties, and other functions such as low density, permeability, hydrophobicity, and so on, so their accurate character-ization is fundamental in order to enhance or optimize a product performance.

For the nanofluids field the solid entities (particle, tube, and rod) or its aggre-gates are fundamental in the intermolecular–ionic interaction with the base fluid, especially for IoNanofluids. Therefore, all the available information about the nanomaterial is necessary to understand the nanofluids behavior. While they remain poorly understood or completely unknown, they can never be properly valuable, unless we trust only the final results for each application. Technology must rely on science and not on chance or true empiricism.

37.2.2
The Nanofluid Preparation

Nanofluid preparation is a fundamental step for stability, characterization, and use. Generally, the existence of two methods of preparing/obtaining nanofluids, the one-step and two-step methods, are accepted. A comprehensive discussion about these methods can be found in the work by Lourenço and Vieira [9]. The first consists in the combination of the process of synthesis and dispersion of nanoparticles in the base fluid in a single step, usually using evaporation of nanoparticles in vacuum conditions and subsequent condensation in the base fluid (physical path), or by chemical reaction in the base fluid. These techniques reduce the effects of the agglomeration of nanoparticles, but have a very high operational cost as a disadvantage. In addition, they need to be developed in batches, that is, normally in small quantities, making the commercial production of nanofluids nonprofitable.

In the two-step method, nanoparticles are synthesized in the form of ultrathin dry powder and further dispersed in the base fluid through physical/mechanical processes, such as intense agitation, magnetic force, sonication, ultrasonic agitation, or high-pressure homogenization. The main advantage of these methods is the variety of nanofluids that can be obtained, using various materials with different sizes and geometry. However, the quality of the nanomaterial obtained is fundamental, namely, the geometry of the nanoparticles, its purity, and its aggregation behavior, both in the solid state and in contact with the base fluid. The cost is substantially less than that in the one-step method, more economical to produce large-scale nanofluids, because the synthesis techniques of nanopowders have already been scaled-up to industrial production levels. Table 37.1 displays the main advantages and disadvantages of each method. Examples of the preparation of IoNanofluids by the two methods can be found elsewhere [10–12].

For more detailed information on methods of preparing/obtaining nanofluids bibliographical analysis, see Refs [13–15].

The ultrasonic energy methods, such as sonication, applied mostly to two-step method, deserve a further discussion, as they are, in our opinion, the best way to produce stable suspensions, especially in ionic liquids. In fact, acoustic methods are very efficient, precise and fast, and less prone to contamination compared to traditional techniques used for mixing/dispersing particles in a liquid medium, since there is no such risk of bringing more residues into a fresh sample.

Sonication or the use of ultrasonic energy can be applied by two methods: a probe or an ultrasonic bath. A sketch about the energy dissipation in the fluid has been provided in Figure 37.2. The same effect can be achieved with totally different applied power.

Table 37.1 Comparison of nanofluids preparation methods.

Method	Advantages	Disadvantages
One step	Nanoparticle agglomeration minimized (transportation, storage, drying, and dispersion of nanoparticles is avoided)	Use of low vapor pressure fluids Difficult to identify all the species present in solution
Two step	Practical and economic process	Stability (powders aggregate due to strong van der Walls forces). Internal stress, surface defects, contaminations May require surfactants

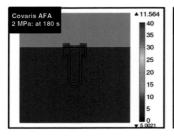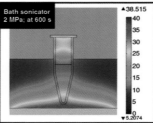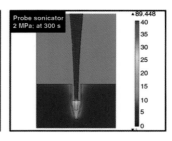

Figure 37.2 Sketch of the energy field (or temperature) in an ultrasonic bath and in a probe sonicator, using finite-element analysis software modeling. (Courtesy of covarisinc.com [16].)

Direct sonication is recommended for dispersing powders since the energy output received by the materials is higher than the one sensed when the sample is in an ultrasonic bath, which typically operates at lower energy levels than those attained with probes. The main disadvantage of probe sonicator is the potential for sample contamination by erosion of the probe tip. Sonication with baths is suggested for resuspensions of samples that were previously sonicated directly or when higher energy levels can cause unintentional alterations or damages to the particles, or to small volume samples [9]. However, if the different variables that influence the sonication success are studied, nanofluids and IoNanofluids where the integrity of the starting nanomaterial is maintained can be achieved. The mechanisms of sound propagation in the nanofluid generate a large amount of energy, increasing, sometimes drastically, the temperature of the samples, which can easily destroy or at least modify the original characteristics of the nanomaterials, especially if we use carbon structures such as carbon nanotubes (CNTs), graphene, or fullerenes.

Consequently, some parameters must be well defined for a given system and by certain equipment, by evaluating the particle size and suspension situation. They are temperature, sonication time, frequency and operation mode, sample volume and concentration, probe and vessel, probe to tip distance, aerosol and foaming, and solvent properties (surface tension, viscosity, density, velocity of the sound wave, chemical composition, and solubility). Most of the published works in the field of nanofluids were not aware of the ill-effects that sonication can create, and therefore, many property measurements done in an expected nanofluid refer in fact to another system, not characterized.

It is not the purpose of this chapter to discuss all these effects. They are just an alert to future users of the sonication technique. Excellent works on ultrasonic methods have been published and the readers are referred to them [17–19]. A complete discussion on the effects described, along with tips for optimizing sonication (time, volume, concentration, and power output), can also be found in reference [9]. Also, difficulties associated with equipment suppliers are analyzed [9].

37.2.3
Nanofluid Stability

As already mentioned, a key factor for reliable property measurement is the nanofluid stability. On the other hand, in terms of industrial/laboratory production the lack of nanofluid stability could compromise the storage for long periods or impose very special and expensive methods. Research on nanofluid stability is critical because their properties are influenced by it, making it urgent to study and analyze the factors that influence dispersion stability in these nanosystems.

A digression in the literature of nanofluids shows that the use of terms such as mixture, solution, dispersion, and colloid is made very imprecisely and a brief explanation is useful for the reader. Let us first consider solutions and dispersions. For the case of solutions, the mixing process is spontaneous, reversible, and thermodynamically stable ($\Delta G_{mix} < 0$), and the inhomogeneities exist only at the molecular/ionic levels. The properties of solutions are independent of the way these solutions are prepared. However, for dispersions, the mixing is nonspontaneous, irreversible, thermodynamically unstable ($\Delta G_{mix} > 0$) (require a stabilization agent) and unmix spontaneously. Inhomogeneities appear on scales much larger than the molecular dimensions. In addition, the dispersion properties are strongly dependent on the way they are prepared, which implies the use of empirical preparation procedures to achieve repeatability. However, if the unmixing procedure is slow (timescale to be defined), *kinetically stable dispersions* can be obtained.

Several phenomena can occur, depending on the nature of the particles and their interparticle forces (and also on the forces between the particles and the base liquid). This has been shown in Figure 37.3. The unstable system can separate by sedimentation (gravity driven), or its particles can form aggregates of increasing sizes that may settle out also under the influence of gravity. For

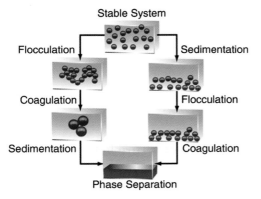

Figure 37.3 Example of phenomena that can occur in an unstable dispersion [20].

example, aggregation of nanoparticles in ionic liquids is caused by a competition between van der Walls (VDW) or polar forces and double-layer forces screening caused by ions (DL), at the particle–IL interface. VDW forces are attractive and DL forces are repulsive. Balance between these forces controls aggregation of NPs in ILs, and therefore the stability of a dispersion in the IoNanofluid. The process of this formation is called flocculation, can be reversed and the "floc" can separate or not from the mother solution. If the aggregate changes to a much denser form – coagulation, an irreversible process. These forms are not stable and separate in two-phase systems, either by sedimentation or by flocculation of the much denser phase (creaming). In the end, aggregation can originate phase separation.

The nanofluids/IoNanofluids systems research must be driven to search for kinetically stable dispersions, long-lived, because the success of its use in scientific or industrial applications depends on how we analyze and characterize the emulsion prepared and its stability (as particle aggregation can destroy our purpose). In fact, the result of the nanoparticles aggregation is manifested not only by precipitation and blockage of microchannels but also in the reduction of its thermal conductivity.

There are several methods of stability assessment and some processes to improve stability mechanisms in dispersions [21]. We shall mention the main ones.

First, the *spectral absorbency analysis*. The temporal decrease of the absorbency is a strong sign of phase separation by flocculation or coagulation. The sedimentation kinetics can also be followed by these measurements, using UV-vis or NIR spectroscopies. Second, the *electrokinetic potential analysis*, as many of the important properties of these colloidal systems are determined by the electrical charge (or potential) of the nanoparticle. Zeta potential can be defined as the potential difference between the dispersion medium and the stationary layer of fluid attached to the dispersed particle. If the forces are repulsive, particles tend not to aggregate, while if they are attractive, they will easily aggregate. The zeta potential indicates the degree of repulsion between adjacent, similarly charged particles in dispersion. The significance of zeta potential is that its value can be related to the stability of colloidal dispersions. For small molecules and particles, a high zeta potential will confer stability, that is, the solution or dispersion will resist aggregation. When the potential is low, attraction exceeds repulsion and the dispersion will break and flocculate. So, colloids with high zeta potential (negative or positive) are electrically stabilized, while colloids with low zeta potentials tend to coagulate or flocculate. Colloids with zeta potential from 40 to 60 mV are believed to be very stable, and those with more than 60 mV have excellent stability. At <25 mV, flocculation or coagulation takes place, even within short timescales. Therefore, the zeta potential is an important parameter characterizing colloidal dispersion and nanofluids [9].

All these comments apply to IoNanofluids and the use of the spectral absorbency methods and zeta potential determination are recommended for quality control of these complex systems.

37.2.4
Methods to Improve Nanofluids Stability

The most employed technique to improve the stability of nanofluids is by far the addition of surfactants/dispersants, in order to modify the surface characteristics of the nanoparticles. Adding dispersants in the two-phase system is an easy and economic method to improve wettability, enhancing the interaction between the liquid and the particles and therefore decreasing the particle–particle interaction. When the base fluid of nanofluids is a polar solvent, like most of the ionic liquids, we should select water-soluble surfactants; otherwise, we will select oil soluble ones. However, surfactants might cause several problems, changing the structure of the liquid phase and contaminating the heat transfer media or producing foams when heated. In addition, surfactant molecules attaching on the surfaces of nanoparticles can enlarge the thermal resistance between the nanoparticles and the base fluid, which may limit the enhancement of the effective thermal conductivity, for some of the applications. Ionic liquids can produce stable IoNanofluids without the use of surfactants, and therefore are normally not used.

The surfactant-free technique is a promising approach to achieve long-term stability of nanofluid with functionalized nanoparticles. The first record of functionalization of CNTs in ionic liquids was presented by Aida and coworkers [22,23], who discovered that carbon nanotubes and ILs could be blended to form gels termed as "Bucky gels," now considered in the category of IoNanofluids. The chemical modification to functionalize the surface of CNTs is a common method to enhance the stability of CNTs in solvents. Here, we recommend a review about the surface modification of CNTs [24], using infrared spectrum and zeta potential measurements. Other methods, such as plasma treatment, ball milling, and centrifugal bead mill processes have also been used, but not on IoNanofluids.

The understanding of the mechanisms that avoid nanoparticle aggregation is an active area of research. Stability means that the particles do not aggregate at a significant rate. The rate of aggregation is in general determined by the frequency of collisions and the probability of cohesion during collision. DVLO (after Derjaguin, Landau, Verwey, and Overbeek) theory deals with colloidal stability [25,26], suggesting that the stability of a particle in solution is determined by the sum of VDW attractive and electrical double layer repulsive forces between particles, as they approach each other due to the Brownian motion they are undergoing. If the attractive force is larger than the repulsive force, the two particles will collide, aggregate, and the suspension is not stable. If the particles have a sufficiently high repulsive force, the suspensions will be stable. For stable nanofluids or colloids, the repulsive forces between particles must be dominant. According to the types of repulsion, the fundamental mechanisms that affect colloidal stability are divided into two kinds, one is steric repulsion, and another is electrostatic (charge) repulsion [27]. In addition, a better understanding of particle growth (aggregation) in liquids, namely, the role of solvation forces in the preferential attachment of nanoparticles, using *in situ* tunneling emission microscopy (TEM) imaging and steered molecular dynamics (SMD) simulations [28] looks a very promising way to study nanofluid structure, and, *a fortiori*, IoNanofluids. Work done by Szilagyi *et al.* [29], studying the article aggregation kinetics of nanosized sulfate latex spheres (50 nm radius) in several ionic liquids (containing the anions tetrafluoroborate, BF_4^-, dicyanamide, $N(CN)_2^-$ and thiocyanate, SCN^-) with 1-buthyl-3-methylimidazolium, C_4mim^+ as the cation, and dicyanamide with 1-buthyl-3-methylpyridinium, C_4mpyr^+ and 1-buthyl-3-methylpyrrolidinium, C_4mpyr^+ cations with their water mixtures, arrived at equations that can help to distinguish between stable and unstable colloidal suspensions. When the interaction potential acting between the particles is known, aggregation rates can be estimated more accurately. Aggregation rates were studied by dynamic light scattering (DLS) and time-resolved light scattering, showing that the smaller the particles, the slower the rate of aggregation. Although their conclusions have to be extended to carbon and other nanomaterials, they open several possibilities to understand the aggregation phenomena in ionic liquids, giving better clues to make very stable suspensions, with lifetimes much greater than 10^8 s (~3 years), for low particle concentration.

37.2.5
Structural Characterization of Nanofluids

From the above discussion it is clear that a good work in nanofluids area needs the structural (or spatial organization) characterization of the nanofluid, namely, obtaining the particle size and distribution in the nanofluid. Ideally, we should have a very narrow or even single dimension, but this is actually impossible, with the available techniques for nanomaterial production. This fact conditions the particle distribution in the base liquid, especially if we have a structured fluid as an ionic liquid, as the particles might aggregate, originating clusters of different size, with specific interactions between the particles/clusters and ionic liquid. In addition, the interface between a nanomaterial such as MWCNT and $[C_4mim][BF_4]$ is bound to be much larger than that expected by many theories [30]. The structural characterization can be achieved by dynamic light scattering (DLS) techniques, giving the particle distribution directly, or by TEM imaging and subsequent counting of number of particles with average sizes. Both these methods, important for the nanomaterials (powders, films), are well understood, but have many limitations for nanofluids, namely, for those that are nonlight transparent such as the carbon nanofluids in DLS, and because TEM imaging can characterize solids and even solid–liquid interfaces, but never the liquid itself. All the efforts to resolve these problems will be welcome.

37.3
IoNanofluids Properties

If clustering of nanoparticles can promote micro-phase separation and therefore, decrease drastically the lifetime of a nanofluid, it can also create preferential paths for heat transfer, enhancing the thermal properties of the nanofluid, as reviewed recently [31]. Therefore, the properties of the IoNanofluids (as for nanofluids in general) are strongly influenced by the amount of nanoparticles dispersed (concentration, mass, or volume), by energetic interaction factors at a molecular level, macroscopically affected by temperature. Although no pressure dependency studies are available, it is expected that this quantity will affect the properties of the liquid phases, namely, at high pressures. As the addition of surface active agents to promote nanofluid stability changes completely its organization, all the discussion will be based in surfactant-free IoNanofluids.

The main properties that have been measured of IoNanofluids are density, heat capacity, viscosity, and thermal conductivity. There are some controversies to be solved in the values of the properties. This is caused in our opinion by ill definition of the IoNanofluids and their organization and morphological characterization, as discussed in the previous sections, and by the quality of the measurements performed. As already discussed in a previous paper, dedicated to ionic liquids [32], the measurement of thermophysical properties of these systems is not straightforward. Until recently, this was justified by two main factors: one directly influenced by the structure and properties of ionic liquids, like the sample and its chemical reactivity/solubility power, and the other, directly dependent on the quality of the mathematical modeling of the instrumentation used. Both these points apply to IoNanofluids, namely, the presence of impurities in the samples, from which water is significant, and the bad use of available instrumentation in a given laboratory. This fact is highly enhanced for viscosity, whereby the particle concentration in flow processes can become inhomogeneous due to the flow pattern, and therefore different in the measuring sensor and in the bulk IoNanofluid. In thermal conductivity, many measurements using transient hot wire probes ignore the electrical conductivity of the samples and the possible accumulation of particles near the hot probes, as well as the possibility of additional convective heat transfer. As already discussed in previous sections, the IoNanofluid preparation, a fact not normally discussed in the published data, is crucial for the stability and influences therefore the properties measured.

In a recent review by Arthur and Karim [33], written for high-temperature applications in solar energy, a comprehensive analysis of the enhancements in thermophysical and rheological properties of 10 ionic liquids ([C$_4$mpyrr] [(CF$_3$SO$_2$)$_2$N], [C$_4$mim][(CF$_3$SO$_2$)$_2$N], [C$_4$mim] [N(SO$_2$C$_2$F$_5$)$_2$], [C$_4$mim] [CF$_3$SO$_3$], [C$_4$mim][BF$_4$], [C$_6$mim][BF$_4$], [C$_6$mim][(CF$_3$SO$_2$)$_2$N], [C$_8$mim] [(CF$_3$SO$_2$)$_2$N], [C$_4$mmim][(CF$_3$SO$_2$)$_2$N], and [(C$_1$)$_3$NC$_4$][(CF$_3$SO$_2$)$_2$N]), with several nanomaterials such as carbon black, MWCNTs, graphene, and Al$_2$O$_3$ has been presented. In addition, work by Ferreira *et al.* [34], with phosphonium ILs [(C$_6$)$_3$PC$_{14}$)][PH$_2$O$_2$], [(C$_4$)$_3$PC$_1$)][C$_1$SO$_4$], [(C$_6$)$_3$PC$_{14}$] [(CF$_3$SO$_2$)$_2$N], and [(C$_6$)$_3$-PC$_{14}$][(C$_2$F$_5$)$_3$PF$_3$], and work by our group [1,10,11,30] complement most of the published data on IoNanofluids.

From Arthur and Karim analysis, it can be concluded that the results for heat capacity are extremely confusing, as several systems show a decrease in heat capacity instead of an enhancement, especially those with a very low mass fraction of the nanomaterial ([C_6mim][BF_4] [35], (0.03–0.06 wt%, graphene, and MWCNTs), but also for those using carbon black (up to −40%). The remaining data depicts 10–30% increase in specific heat, but was obtained only with Al_2O_3 nanoparticle (<50 nm, 1–3% wt%). Data obtained in our laboratory for [C_4mim)] [PF_6] with MWCNTs [1] and TiO_2 (not yet published) show an increase in the heat capacity, less than 8% (1% wt%). Ferreira *et al.* data show that the heat capacity of [(C_6)$_3$PC$_{14}$][PH_2O_2] and [(C_6)$_3$PC$_{14}$] [(CF_3SO_2)$_2$N] + MWCNT show enhancements of the order of 7–8%, for low weight fractions (0.01–0.02%). Sheveylyova *et al.* [36] measured the heat capacity of [C_4mim][BF_4] and [C_4mim] [PF_6] IoNanofluids with MWCNTs produced by gas-phase catalytic precipitation at hydrocarbons pyrolysis over the temperature range of 80–370 K, covering the same liquid range than our previous work. However, the INFs obtained had around 12 wt% of nanomaterial concentration, but surprisingly obtaining the same type of enhancement (8%) that we obtained with 1 wt%. This again can be derived from the different characteristics of the MWCNTs used in both studies and from the different method of preparation. Unfortunately, there has not been a systematic study of the heat capacity of selected ionic liquids and nanomaterials so that we can conclude on the real effect of nanoparticles and their concentration dependence.

Thermal conductivity and viscosity are the properties that have been studied extensively in several ionic liquids, mostly with MWCNTs, but also with graphene and Al_2O_3. For the latter, it is first necessary to demonstrate if the IoNanofluid is a Newtonian fluid, as some ILs are.

We start by analyzing the thermal conductivity enhancement of IoNanofluids. From all the data published, we can conclude that the enhancement is significant, even for low nanoparticles loads. In the review of Arthur and Karim, thermal conductivity was found to increase with mass fraction, ranging from 3 to 35%. As the thermal conductivity of the ILs is a weak increasing function of temperature, data for the IoNanofluids also show weak dependence on temperature and strong dependence on concentration of NPs. However, some IoNanofluid present a strong dependence on temperature and a weaker dependence on concentration, as shown in Figure 37.4 for [C_2mim][$C(CN)_3$] IoNanofluids [37], but with a sudden decrease at temperatures around 340 K.

In Table 37.2 we present a summary of the enhancements obtained in our laboratory with several ionic liquids, for a mass fraction of 1% of MWCNT, at 303 K [1,10,11,30,37].

As can be seen, the enhancements are significant, but less than 15%. The only exceptions are the results obtained for [C_4mim][(CF_3SO_2)$_2$N], for which two values were obtained, for IoNanofluids prepared in different occasions with different batches of MWCNTs [1,30], a strong evidence that the IoNanofluids were not the same, because the method of preparation was different, showing the importance of nanofluid preparation and its influence in the properties. Another significant result is that the attempts to prepare stable IoNanofluids with the phosphonium salts [(C_6)$_3$PC$_{14}$][Br] and [(C_6)$_3$PC$_{14}$][(CN)$_2$N] were not successful, which did not happen with the [(C_6)$_3$PC$_{14}$][PH_2O_2], [(C_6)$_3$PC$_{14}$][C_1SO_4]and

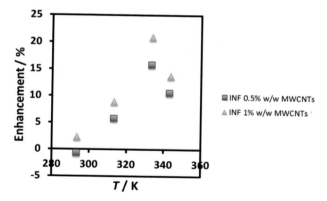

Figure 37.4 Thermal conductivity enhancement for [C₂mim][C(CN)₃] with 0.5 and 1% of MWCNTs. INF stands for IoNanofluid.

[(C₆)₃PC₁₄][(CF₃SO₂)₂N] of Ferreira *et al.* work [38]. Our IoNanofluids were stable for more than 24 h, but after this, phase separation and deposition was observed. The influence of the different anion is therefore important for the IoNanofluid stability, whereas the cation seems to influence more the thermal conductivity enhancement.

The compounds studied can be separated into headgroup (imidazolium, pyrrolidinium, or phosphonium), an aliphatic side chain and an anion, as shown in Figure 37.5.

The change of the head group does not seem to affect the enhancement very much, although the influence of the anion is bigger, with the [(CF₃SO₂)₂N] salt having the smallest and the [C₂H₅OSO₃] the highest. The same result is obtained when we compare the liquids based on the [C₄mim] cation, with the [(CF₃SO₂)₂N] salt having smaller enhancement than the [N(CN)₂] and [C(CN)₃] ones. The data obtained also justify that by changing the side chain length, the enhancement in the thermal conductivity increases, possibly due to a better interaction between this chain and the carbon nanotube surface (nonpolar entities) that facilitates heat transfer. The influence of the C₄mim cation, giving

Table 37.2 Summary of thermal conductivity enhancements for some ionic liquids at 303 K and 1 wt% of MWCNT.

	BF_4	PF_6	Br	CF_3SO_3	$(CF_3SO_2)_2N$	$C_2H_5\ OSO_3$ ECOENG	$(CN)_2N$	$(CN)_3$
C_2mim	—	—	—	—	2.4	8.6	7.4	4.9
C_4mim	5.8	3.4	—	8.5	35.6 [1]; 13.7 [35]	—	8.8	
C_6mim	3.8	1.4	—	—	6.4	—	—	
C_8mim	—	—	—	—	6.2	—	—	
C_2mpyr	—	—	—	—	5.1	—	8.8	

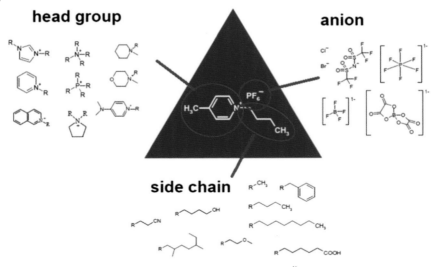

Figure 37.5 Approach to the molecular structure of ionic liquids[1].

always rise to a bigger enhancement, was the subject of a discussion in Ref. [30]. More data on these types of systems are necessary to support well all these conclusions and infer the best IoNanofluids for a given heat transfer application.

In relation with density, the effect of the nanomaterial is purely mechanic (mass effect). Results obtained in our laboratory for the $[N(CN)_2]$ IoNanofluids [10] show that the enhancement is very small, reaching its peak at 0.65% for the case of the IoNanofluid $[C_4mim]$ $[N(CN)_2]$ at 1 wt% at 313 K. The change of the density enhancement seems to decrease with increasing temperature and increase, at the same temperature, with the nanomaterial concentration.

Finally, we have viscosity. This property is very important in many applications, not just in heat transfer fluids, as described in the following section. From the point of view of property measurement, it is strongly affected by impurities of the ionic liquid, and also by the quality of the nanomaterial and its ability to make 3D networks, which increases the viscosity of the fluid. It can be seen that there is a general trend of increasing viscosity with increasing concentration of nanoparticles. This is as expected as the changes in viscosity have been attributed to agglomeration of nanoparticles, where a larger amount of aggregates will be present in higher concentrations [34,38]. However, an interesting phenomenon has been observed by Liu *et al.* [39] and Wang *et al.* [35] in which the addition of MWCNT and graphene nanoparticles resulted in a reduction in the viscosity. Visser *et al.* [40] found that $[C_4mim][N(SO_2C_2F_5)_2]$ had the greatest increase in viscosity and attributed this phenomenon to the nanoparticles interaction with the cations and anions of the ionic liquids studied. Temperature and

---

1) Adapted from Stefan Stolte (University of Bremen) talk, in (Eco)toxicity and biodegradation of ionic liquids, REGALIS Summer School, Grove, Pontevedra, June 9–12, 2013.

Brownian motion of nanoparticles that can cause better particle contact and agglomeration at high temperatures seem to compete, but viscosity can increase with temperature as a net result. Enhancements of 10–400% were found for Al_2O_3 nanoparticles, but MWCNTs and graphene have shown enhancements in the lower rim, even negative in the work of Wang [35]. The larger (bulkier) the anions, the bigger the enhancement found.

Ionic liquids are known to have at least some Newtonian liquid behavior – linear proportionality – in wide ranges of shear stress, between shear rate and shear stress. However, many of the IoNanofluids are non-Newtonian fluids. Usually they can have a Bingham plastic behavior, whereby viscosity appears to be infinite until a certain shear stress is achieved (or until a yield stress is overcome), or shear thinning, when the viscosity decreases as shear rate/stress is increased (pseudo plastic). The system $[C_4mpyrr][N(CN)_2]$ was studied in our laboratory, with MWCNTs with 0.5 and 1 wt%, for temperatures between 293 and 343 K. The results obtained are displayed in Figure 37.6, along with the demonstration that pure ionic liquid is Newtonian, while the INFs demonstrate

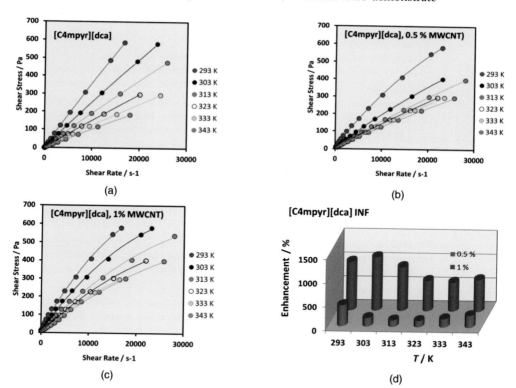

Figure 37.6 Variation of shear stress with shear rate with temperature for $[C_4mpyrr]$ $[N(CN)_2]$ (Newtonian) and its IoNanofluids (INFs) with 0.5 and 1 wt% of MWCNTs. (a) Pure ionic liquid. (b) 0.5 wt% INF. (c) 1 wt% INF. (d) Viscosity enhancement as a function of temperature for the two INFs studied (dca means dicyanamide).

a shear thinning behavior. Also displayed is the enhancement found as a function of temperature.

From the results presented, we can conclude that the pure ionic liquid is Newtonian in the temperature range. The enhancement is very big (up to 1200% at 303 K) and its variation with temperature seems to decrease (Arrhenius type) first and then increase. The nearly Newtonian zone of the IoNanofluid extends from two to three hundreds in shear rate to a few thousands, depending on the base fluid and the nanoparticle size and type. Further research on the rheology of the IoNanofluids is necessary in order to understand the effect of the nanoparticle size distribution in the flowing behavior of these fluids.

37.4
Applications of IoNanofluids

Applications of IoNanofluids to several areas of science and technology rely on the unique properties of the ionic liquids merged with those of nanosystems (particles, rods, tubes, etc.). The properties of the nanomaterials can be modified through chemical/physical modification of their surface with ionic liquids, and consequently ionic liquid structures can be functionalized with the nanosystems, enhancing the molecular–ionic interactions in the IoNanofluid, generating structures that are more efficient in catalysis, heat transfer, and electrochemistry [41]. Here, good proton-transfer ability of ionic liquids and high electron conductivity afforded by metal nanoparticles, carbon nanotubes, and graphene are essential for many applications as electrodes or electrolytes in cell batteries, as soon as the solid interphases and the suspensions are thermally, electrically, and kinetically stable. Electrochemical sensors and actuators with ionic liquid–nanoparticle composites can obtain high selectivity, high sensitivity, and fast responses.

Three fields deserve our attention and will be tackled in this chapter. They all enhance the liquid phase, or multiphasic behavior of the IoNanofluids, and not on pastes or solid depositions. Other CNT hybrid applications can be found in the review by Tunckol *et al.* [42]. Also, due to the price of ionic liquids and of nanomaterials, we restrict to applications whereby low concentrations of nanomaterial can be used. These are new engineering fluids (heat transfer fluids, fluids for boiling and condensation processes, and new absorbing refrigeration fluids, including high temperature ones for solar energy applications such as solar collectors), nanocatalysts with high efficiency and selectivity, namely, those based on ionic liquid-supported polymeric membranes (SILP), with ease of cheap recycling, and new paints for solar absorbing panels based on natural nanomaterials.

37.4.1
IoNanofluids as New Engineering Fluids

Engineering fluids occupy a significant amount of chemical compounds and mixtures in our world. Their application in the chemical, manufacturing, and mining industries requires thermal, chemical, rheological, and flow properties. However, environmental and health (toxicological) requirements of our society, allied to new technological developments and to the world economic situation, justifies the search for new fluids, with new properties, to replace/improve current chemicals. Recent studies have demonstrated that ionic liquids are promising alternatives to classical heat transfer fluids: aliphatic, aromatic, and silicone oils [1,43]. IoNanofluids can increase the thermal conductivity of an ionic liquid by more than 30% and heat capacity by 10%. The heat capacity storage, associated with better heat transfer coefficients, will improve the heat transfer efficiency, in addition to contributing to the replacement of nondegradable chemical compounds. It is known that if the heat transfer variables (thermal conductivity, viscosity, and heat capacity) are well chosen, the heat transfer efficiency can increase up to 30%. Needless to say that viscosity will be a fundamental parameter of the ionic liquid (and therefore of the IoNanofluids), because its value is critical for pumping energy requirements and heat transfer coefficients are very sensitive to it.

Another area of impact of IoNanofluids is the use as absorbents for vapor absorption refrigeration. Energy-efficient cooling concepts play an important role in numerous applications in the area of comfort and industrial cooling, covering the cooling demand in a sustainable manner with a drastically reduced carbon footprint. An enormous amount of heat energy is daily dissipated to the atmosphere as low-temperature waste heat from industrial and electric power plants. This low-temperature energy cannot be utilized in any useful heating operation but can be used as driving heat of absorption for heat pumps in refrigeration. Recently, the research on new working pairs for absorption chillers considering IL as absorbents appears promising in overcoming the drawbacks of the state-of-the-art absorbent LiBr, such as crystallization, corrosion, and instability [44,45]. Systems with ILs + water and nanomaterials can enhance the properties such as thermal conductivity, while water, helping also in the thermal conductivity side, decreases the viscosity, thus decreasing the necessary pumping power.

Nanofluids can enhance boiling and condensation processes [46], as the conventional method to increase the cooling rate by extending heat transfer surfaces or area is not sufficient to meet the ever increasing cooling demand by many advanced devices and industries and implies increases in capital investment. Although boiling of fluids is a complex and elusive process, it is a very efficient mode of heat transfer in various energy conversion and heat exchange systems as well as cooling of high-energy density electronic components. There are two common types of boiling: pool boiling and flow or forced convective boiling. Pool boiling refers to boiling on a heater surface submerged in a pool of initially quiescent liquid [47,48], while flow boiling is boiling in a flowing stream of fluid, where the heating surface may be the channel wall confining the flow [49,50]. In addition, studies on nanofluid droplets impingement and spreading on heated substrate surfaces are of great importance for their application as an advanced coolant particularly in spray cooling systems [51,52]. The impact of the nanofluids in this area also envisages excellent effects if IoNanofluids are used, although there are not many applications of ionic liquids in this area [53], and none of IoNanofluids.

The use of nanofluids/IoNanofluids for solar energy applications, namely, for the efficient absorption of the solar radiation and for its transmission to heat/cooling systems, has been studied by several authors [54–56]. An example was the use of $[C_6mim][BF_4]$/graphene nanofluids in direct absorption solar collectors (DACs) [57], where the receiver efficiency increased with the solar concentration and receiver height, but decreased with the graphene concentration. The authors concluded that the potential for the utilization of these novel fluids is very high, leading to a significant increase in the performance of the solar energy conversion processes (thermal or photovoltaic). In addition to the thermal properties, the excellent optical properties of these fluids and their thermal stability have an added value in their use in solar collectors [58,59].

37.4.2
IoNanofluids as Nano-Driven Catalysts

Nanocatalysis is a fast growing field that involves the use of nanomaterials as catalysts for a wide variety of catalytic reactions. Heterogeneous catalysis represents one of the oldest commercial applications of nanoscience; metal nanoparticles, semiconductors, oxides, and other compounds have been widely used for important chemical reactions (of high industrial interest). The key objective of the research in nanocatalysis is to produce catalysts with 100% selectivity, extremely high activity, low power consumption, and long life, by controlling precisely the size, shape, spatial distribution, composition and surface electronic structure, and thermal and chemical stability of individual nanoingredients.

The metal nanocatalysts prepared by adsorption of nanoparticles in supports involve functionalization of these supports to adsorb nanoparticles. Membranes (catalytic) may also be used. Nanocarbons, being very different among them in structure and properties, and on their chemistry of surfaces, are excellent catalyst supports, as they can anchor active phases, in addition to possible functionalization, namely, by ionic liquids.

One example is the immobilization without covalent bonding using a thin film of ionic liquid containing the homogenous catalyst immobilized on a large-surface-area support material (SILP (supported ionic liquid phase)) with CNTs, graphite, and activated carbon [60]. Better selectivity and enantioselectivity in Diels–Alder and Mukaiyama aldol reactions was found. The system used is a combination of an ionic liquid and a solid support (CNTs or other), and therefore the increase in thermal conductivity of the reaction media plays a significant role, namely, in exothermic reactions. Another example is using (PIILP (polymer immobilized ionic liquid phase)) catalysts [61], generated from a range of linear and cross-linked ionic polymers for asymmetric carbon–carbon bond formation. Thin $[C_2mim]$ $[(CF_3SO_2)_2N]$ film coated onto CNTs and SiO_2 supports were used. A different framework used step-one IoNanofluids [62]. A composite of graphene oxide (GO) with manganese oxide (MnCo) was prepared by using a solvothermal method and during the synthesis, both the reduction of GO and growth of metal oxides took place simultaneously. The prepared composite material was highly selective for the liquid-phase oxidation of *p*-cresol to form *p*-hydroxybenzaldehyde, increasing the redox potential of the mixed oxide after being supported on GO, leading to higher activity of the catalyst for the oxidation reaction. Finally, Kessler presents a very complete work in catalysis with IoNanofluids [63]. Cu_2O nanoparticles were synthesized in tetra-*n*-butylphosphonium acetate, $[(C_4)_4P][CH_3COO]$ an ionic liquid with high stabilizing potential and low melting point. These nanoparticles could be used as a recyclable decarboxylation catalyst for 2-nitrobenzoic acids and as a catalyst for Buchwald–Hartwig reactions. Depending on the reaction parameters or functional groups of the substrates, the catalyst showed moderate to excellent activity and recyclability [64,65].

37.4.3
Development of New Paints for Solar-Absorbing Panels Based on Natural Nanomaterials

As discussed on the use of nanofluids and IoNanofluids for solar energy applications, the absorbing plate of solar collectors is generally covered by a black surface in order to capture elevated quantities of radiation for greater heat conversion. Ideally, this surface should present high solar absorbance in the UV-vis regions and low emissivity in the thermal infrared (spectrally selective). Paint coatings are a cheap and practical way of achieving these absorbers and pigments, one of the main components of paint coatings is essential for the optical and thermophysical characteristics of the finished product. The idea of using nanocarbon structures dispersed in ionic liquids originate the development of three new pigments for paints in solar applications, employing MWCNTs, $[C_4mim][PF_6]$, and crystal violet IoNanofluids with high solar absorbance and thermal emissivity compared to the base paint [66]. The pigments were synthesized using two solvent-free methods, microwave and ultrasound techniques, functionalizing the carbon nanotubes with 1-methylimidazole/crystal violet mixtures. An increase up to 11% in the absorbance and a net gain in efficiency of 0.45–0.57 were obtained. The paint coatings produced with these pigments were found to be spectrally selective, a necessary requisite for future solar panel collector applications.

Natural nanomaterials open a completely new field in heat transfer as they have unusual thermal properties derived from its structure. It is in principle possible to use natural products from the biosphere, marine or plant origin, to be used as heat transfer enhancers. There is a plethora of materials, all renewable and biodegradable, that can be tested and their thermal properties are to be determined. From the present knowledge, some of them are already applied in domestic applications such as cherry stones and grape seeds, whereby several minutes of heating in a microwave oven can be sufficient to heat a fleece bag of these seeds to 70 °C and let it last for several hours at 50 °C, a self-kept warming pillow. This result is quite surprising and can only be explained by the morphology of the cherry stones [3]. In the case of marine wastes, an application has been developed with melanin, a natural biopolymer that is widely distributed in living organisms. They act mainly as photoprotectors. Due to the regular shape of these nanoparticles [67], the heat capacity of the melanins is greater than common silicon-based heat transfer liquids and ionic liquids, and their possible replacements can be used in developing new pigments for solar paints, based on IoNanofluids, biodegradable, and environment-friendly.

37.5
Challenges in IoNanofluids Research

As quoted in reference [2] the major challenges of nanofluids research include achieving long-term stability of nanofluids, reliability and accuracy of experimental systems used to measure various properties of nanofluids, consistency and resolving uncertainties in measured data and findings, lack of fundamental understanding of heat transfer mechanisms, model development considering underlying mechanisms in macro- to nanoscale, and practical application-based research. Most of these challenges apply to IoNanofluids research, but there are some problems related to the use of ionic liquids, especially concerning applications.

One of the main problems is related to the large volume/tonnage applications, as the large-scale production of ionic liquids has not yet been attained, and only some ILs are being produced in big quantities. However, when it should, we would be able to replace actual nonbiodegradable and environmental harmful heat transfer and reactive fluids in industrial scale heat exchangers, heat and mass transfer equipment and reactors, as in other small capacity heat exchangers such as microchannels and research laboratories. Second, we have the sample preparation and reliable measurements, keeping in mind that the systems are electrically conducting, once the long-time stability (preferably without surfactants) is achieved. From the point of view of thermodynamics (Gibbs formulation), we are dealing with a suspension or emulsion with different degrees of aggregation of the nanoparticles and therefore the phase is not homogeneous especially when surfactants are used to stabilize the systems and to avoid microphase separation and stratification. Therefore, any theory to be developed to interpret the behavior of these fluids (i.e., nanoscale systems) is forced to apply thermodynamics of multiphase (at least biphasic) heterogeneous systems. This approach has not been followed by the current research and its solution is one of the most significant challenges that nanofluid research faces within the next years, trying to derive the heat transfer and thermodynamic properties of these nanosystems from constitutive equations and by involving statistical thermodynamics of "solvent" and nanomaterials shape (particles, spheres, rods, films, etc.). Hypotheses as property additivity, found applicable in some cases to properties such as density and heat capacity, have to be tested at molecular level. However, transport properties pose a much bigger challenge, as in liquid state theory of molecular fluids the actual knowledge of the special organization and interactions for liquid densities is far from being solved.

Again, ionic liquids and their interaction with the nanomaterials, especially with the contribution of the interphase cation/anion/nanomaterial, and its extension in space, have to be obtained either by molecular simulation with density functional theory (DFT) and molecular dynamics or by developing approximate models for the thermal conductivity and viscosity of IoNanofluids, based on real parameters. However, the characterization of the nanomaterials (knowing from the manufacturers the syntheses process), including the particle/clusters distribution analysis, and the study of the interface nanofluid/nanomaterial by imaging techniques, such as TEM, is a very important issue, necessary for the minimal understanding of their behavior and properties. The routing for using natural nanomaterials can have a strong development if this characterization is well done and if the biological synthesis mechanisms are well understood.

Understanding heat transfer mechanisms in static or flowing IoNanofluids, through chaotic Brownian motion or through preferred aggregates/clusters/paths, is also an enormous challenge.

Finally, one very important point yet not dealt in this work, the environmental and human toxicity of ionic liquids and nanomaterials, requires the attention of all those who want to work in the IoNanofluid area. This problem was analyzed some years ago in Ref. [6], where we wrote, ". . . this problem must be tackled as a logical sequence from laboratory synthesis to industrial production. The priority aspects will be the ecological, human health, and waste elimination besides the costs of the full operation, as the functionalities of IoNanofluids are many and extremely varied. The most important parameter to be considered is the interference parameter for the mixtures (interfacial behaviour), a logic consequence of the primordial role of the nanomaterial – IL interaction [68]. A wide range of physicochemical properties are relevant to toxicology, like particle size distribution, morphology, chemical composition, solubility and surface chemistry and reactivity" [69].

Having posed the fundamental challenges in this field, the success of many technological applications of IoNanofluids requires a pragmatic approach, whereby the experiment, theoretical development, and correlation (theoretical based) can help to construct a methodology for property estimation for process design, namely, in industrial scale heat exchangers, heat and mass transfer (and storage) equipment, and reactors.

37.6
Challenges to Industrial Applications

We would like to finish by challenging industrial companies, namely, nanomaterial producers and chemical and metallurgical manufacturers. The application of

IoNanofluids in such areas as heat transfer engineering, catalytic reaction, and high temperature applications can contribute to decrease our carbon footprint, increase the energy efficiency of several chemical and metallurgical processes, and contribute to a clean and "green" chemical processing. However, much that it is being done at the industrial level is under secrecy agreements and under development studies. One of the main drawbacks is the cost of the nanoparticles and its effect on the nanofluid cost. Michaelides [70] quoted that the cost of a 4l nanofluid with a volume fraction of 0.02 of CNTs would be between $1000 and $14 000 (maximum of $125/g of CNTs). This is a very high price to be paid by a commercial coolant.

The problem can increase if ionic liquids are used. The correct choice of the ionic liquid used, due to their thermal stability and properties, paves the way to strong developments, in economic and environmental terms. The cost/kg of IoNanofluids production at high volume still needs to decrease for these fluids to be competitive with fluids now used in industry that degrade above 200 °C and possess environmental problems. However, in applications and devices with a high profit margin, or that use small quantities of IoNanofluids, the way is completely open. IoNanofluids can therefore contribute not only for energy sustainability, but also for a better society. Industrial companies have a significant role in this uptake of nanofluids, and recent patent filing is a strong sign of this wealthy evolution.

Acknowledgments

The authors would like to thank Fundação para a Ciência e Tecnologia (FCT), Portugal for funding to Centro de Ciências Moleculares e Materiais through PEST OE/QUI/UI0536/2014 and Centro de Química Estrutural (CQE) PEST OE/QUI/UI100/2013. Xavier Paredes would like to thank for his research grant SFRH/BPD/103352/2014 and Salomé Vieira for her research grant SFRH/BD/64974/2009.

References

1 Nieto de Castro, C.A., Lourenço, M.J.V., Ribeiro, A.P.C., Langa, E., Vieira, S.I.C., Goodrich, P., and Hardacre, C. (2010) Thermal properties of ionic liquids and IoNanofluids of imidazolium and pyrrolidinium liquids. *J. Chem. Eng. Data*, **55** (2), 653–661.

2 Sohel Murshed, S.M. and Nieto de Castro, C.A. (eds) (2014) *Nanofluids: Synthesis, Properties and Applications*, NOVA Science Publishers, Inc., New York.

3 Ribeiro, A.P.C., Vieira, S.I.C., França, J.M.P., Queirós, C.S., Langa, E., Lourenço, M.J.V., Murshed, S.M.S., and Nieto de Castro, C.A. (2011) Thermal properties of ionic liquids and IoNanofluids, in *Ionic Liquids: Theory, Properties, New Approaches* (ed. A. Kokorin), InTech, Rijeka, Croatia, pp. 33–60.

4 Murshed, S.M.S. and Nieto de Castro, C.A. (2012) Nanofluids as advanced coolants, in *Green Solvents I: Properties and*

Applications in Chemistry (eds A. Mohammad and Inamuddin), Springer, Dordrecht, The Netherlands, pp. 397–415.

5 Nieto de Castro, C.A., Murshed, S.M.S., Lourenço, M.J.V., Santos, F.J.V., Lopes, M.L.M., and França, J.M.P. (2012) IoNanofluids: new heat transfer fluids for green process development, in *Green Solvents I: Properties and Applications in Chemistry* (eds A. Mohammad and Inamuddin), Springer, Dordrecht, The Netherlands, pp. 233–249.

6 Nieto de Castro, C.A., Ribeiro, A.P.C., Vieira, S.I.C., França, J.P.M., Lourenço, M.J.V., Santos, F.V., Murshed, S.S.M., Goodrich, P., and Hardacre, C. (2013) Synthesis, properties and physical applications of IoNanofluids, in *Ionic Liquids: New Aspects for the Future* (ed. J.-i. Kadokawa), Intech, pp. 165–193.

7 Tertsinidou, G., Assael, M.J., and Wakeham, W.A. (2015) The apparent thermal conductivity of liquids containing solid particles of nanometer dimensions: a critique. *Int. J. Thermophys.*, **36** (7), 1367–1395.

8 Nanotechnology (2014) European Commission Website http://ec.europa.eu/nanotechnology/index_en.html (retrieved April 14, 2014).

9 Lourenço, M.J. and Vieira, S.I. (2014) Nanofluids preparation methodology, in *Nanofluids: Synthesis, Properties and Applications* (eds S.M. Sohel Murshed and C.A. Nieto de Castro), NOVA Science Publishers, Inc., New York, pp. 1–28.

10 França, J.M.P., Reis, F., Vieira, S.I.C., Lourenço, M.J.V., Santos, F.J.V., Nieto de Castro, C.A., and Pádua, A.H. (2014) Thermophysical properties of ionic liquid dicyanamide (DCA) nanosystems. *J. Chem. Thermodyn.*, **79**, 248–257.

11 França, J.M.P., Vieira, S.I.C., Lourenço, M.J.V., Murshed, S.M.S., and Nieto de Castro, C.A. (2013) Thermal conductivity of $[C_4mim][(CF_3SO_2)_2N]$ and $[C_2mim]$ $[EtSO_4]$ and their IoNanofluids with carbon nanotubes: experiment and theory. *J. Chem. Eng. Data*, **58** (2), 467–476.

12 Patil, V., Cera-Manjarre, A., Salavera, D., Rode, C., Patil, K., Nieto de Castro, C.A., and Coronas, A. (2016) Ru-imidazolium halide IoNanofluids: synthesis, structural, morphological and thermophysical properties. *J. Nanofluids*, **5** (2), 191–208.

13 Haddad, Z., Abid, C., Oztop, H.F., and Mataoui, A. (2014) A review on how the researchers prepare their nanofluids. *Int. J. Therm. Sci.*, **76**, 168–189.

14 Kharissova, O.V., Kharisov, B.I., and Ortiz, E.G. de C. (2013) Dispersing carbon nanotubes with ionic surfactants under controlled conditions: comparisons and insight. *RSC Adv.*, **3**, 24812–24852.

15 Richter, K., Birkner, A., and Mudring, A.V. (2010) Stabilizer-free metal nanoparticles and metal–metal oxide nanocomposites with long-term stability prepared by physical vapor deposition into ionic liquids. *Angew. Chem., Int. Ed.*, **49** (13), 2431–2435.

16 Covaris Inc. Adaptive Focused Acoustics™ (AFA) Technology, Woburn, MA. Available at http://covarisinc.com/pre-analytical/afa-vs-sonicators/ (accessed May 31, 2016).

17 Berlan, J. and Mason, T.J. (1992) Sonochemistry: from research laboratories to industrial plants. *Ultrasonics*, **30**, 203–212.

18 Suslick, K.S. (1988) *Ultrasound: Its Chemical, Physical and Biological Effects*, Wiley-VCH Verlag GmbH, Weinheim, Germany.

19 Mason, T.J. and Peters, D. (2003) *Practical Sonochemistry: Power Ultrasound Uses and Applications*, 2nd edn, Woodhead Publishing Ltd., Cambridge.

20 Huber, A. (2014) Physical chemistry I: structure and matter. University of Graz, Austria (open access slides).

21 Tadros, T.F. (ed.) (2010) Colloid stability, in *The Role of Surface Forces: Part I*, Colloids and Interface Science Series, vol. 1, Wiley-VCH Verlag GmbH, Weinheim, Germany.

22 Fukushima, T., Kosaka, A., Ishimura, Y., Yamamoto, T., Takigawa, T., Ishii, N., and Aida, T. (2003) Molecular ordering of organic molten salts triggered by single-walled carbon nanotubes. *Science*, **300** (5628), 2072–2074.

23 Fukushima, T. and Aida, T. (2007) Ionic liquids for soft functional materials with carbon nanotubes. *Chem. Eur. J.*, **13** (18), 5048–5058.

24 Wepasnick, K.A., Smith, B.A., Bitter, J.L., and Fairbrother, D.H. (2006) Chemical and structural characterization of carbon nanotube surfaces. *Anal. Bioanal. Chem.*, **396**, 1003–1014.

25 Israelachvili, J. (1992) *Intermolecular and Surface Forces*, 2nd edn, Academic Press, London.

26 Popa, I., Gillies, G., Papastavrou, G., and Borkovec, M. (2010) Attractive and repulsive electrostatic forces between positively charged latex particles in the presence of anionic linear polyelectrolytes. *J. Phys. Chem. B*, **114**, 3170–3177.

27 Yu, W. and Xie, H. (2012) A review on nanofluids: preparation, stability mechanisms, and applications. *J. Nanomater.*, **2012**, 435873.

28 Welch, D.A., Wohel, T.J., Park, C., Faller, R., and Evans, J.E. (2016) Understanding the role of solvation forces on the preferential attachment of nanoparticles in liquid. *ACS Nano*, **10** (1), 181–187.

29 Szilagyi, I., Szabo, T., Desert, A., Trefalt, G., Oncsika, T., and Borkovec, M. (2014) Particle aggregation mechanisms in ionic liquids. *Phys. Chem. Chem. Phys.*, **16**, 9515–9524.

30 Ribeiro, A.P.C., Vieira, S.I.C., Goodrich, P., Hardacre, C., Lourenço, M.J.V., and Nieto de Castro, C.A. (2013) Thermal conductivity of $[C_n mim][(CF_3SO_2)_2N]$ and $[C_4 mim][BF_4]$ IoNanofluids with carbon nanotubes: measurement, theory and structural characterization. *J. Nanofluids*, **2**, 55–62.

31 Angayarkanni, S.A. and Philip, J. (2015) Review on thermal properties of nanofluids: recent developments. *Adv. Colloid. Interface Sci.*, **225**, 146–176.

32 Nieto de Castro, C.A. (2010) Thermophysical properties of ionic liquids: do we know how to measure them accurately? *J. Mol. Liq.*, **156**, 10–17.

33 Arthur, O. and Karim, M.A. (2016) An investigation into the thermophysical and rheological properties of nanofluids for solar thermal applications. *Renew. Sustain. Energy Rev.*, **55** 739–755.

34 Ferreira, A.G.M., Simões, P.N., Ferreira, A.F., Fonseca, M.A., Oliveira, M.S.A., and Trino, A.S.M. (2013) Transport and thermal properties of quaternary phosphonium ionic liquids and IoNanofluids. *J. Chem. Thermodyn.*, **64**, 80–92.

35 Wang, F., Han, L., Zhang, Z., Fang, X., Shi, J., and Ma, W. (2012) Surfactant-free ionic liquid-based nanofluids with remarkable thermal conductivity enhancement at very low loading of graphene. *Nanoscale Res. Lett.*, **7**, 314–320.

36 Shevelyova, M.P., Paulechka, Y.U., Kabo, G.J., Blokhin, A.V., and Kabo, G. (2013) Physicochemical properties of imidazolium-based ionic nanofluids: density, heat capacity, and enthalpy of formation. *J. Phys. Chem. C*, **117**, 4782–4790.

37 França, J.M.P., Murshed, S.M.S., Lourenço, M.J.V., Nieto de Castro, C.A., Quinteros-Lama, H.E., and Segura, H. (2015) Thermal conductivity of ionic liquids and IoNanofluids: experiment and prediction. Presented at the 19th Symposium on Thermophysical Properties, June 21–25, Boulder, CO.

38 Jo, B. and Banerjee, D. (2014) Viscosity measurements of multi-walled carbon nanotubes-based high temperature nanofluids. *Mater. Lett.*, **122**, 212–215.

39 Liu, J., Wang, F., Zhang, L., Fang, X., and Zhang, Z. (2014) Thermodynamic properties and thermal stability of ionic liquid-based nanofluids containing graphene as advanced heat transfer fluids for medium-to-high-temperature applications. *Renew. Energy*, **63**, 519–523.

40 Visser, A.E., Bridges, N.J., Garcia-Diaz, B.L., Gray, J.R., and Fox, E.B. (2012) Al_2O_3-based nanoparticle-enhanced ionic liquids (NEILs) for advanced heat transfer fluids, in *Ionic Liquids: Science and Applications*, ACS Symposium Series, vol. **1117**, ACS, New York, pp. 259–270.

41 He, Z. and Alexandridis, P. (2015) Nanoparticles in ionic liquids: interactions and organization. *Phys. Chem. Chem. Phys.*, **17**, 18238–18261.

42 Tunckol, M., Durand, J., and Serp, P. (2012) Carbon nanomaterial–ionic liquid hybrids. *Carbon*, **50** (4), 4303–4334.

43 França, J.M.P., Nieto de Castro, C.A., Nunes, V.M.B., and Matos Lopes, M.L.S. (2009) The influence of thermophysical properties of ionic liquids in chemical

process design. *J. Chem. Eng. Data*, **54**, 2569–2575.

44 Seiler, M., Kühn, A., Ziegler, F., and Wang, X. (2013) Sustainable cooling strategies using new chemical system solutions. *Ind. Eng. Chem. Res.*, **52**, 16519–16546.

45 Zheng, D., Dong, L., Huang, W., Wu, X., and Nie, N. (2014) A review of imidazolium ionic liquids research and development towards working pair of absorption cycle. *Renew. Sustain. Energy Rev.*, **37**, 47–68.

46 Murshed, S.M.S. and Nieto de Castro, C.A. (2013) Boiling heat transfer and droplet spreading of nanofluids. *Recent Pat. Nanotechnol.*, **7** (3), 216–223.

47 Prakash, N.G., Anoop, K.B., and Das, S.K. (2007) Mechanism of enhancement/ deterioration of boiling heat transfer using stable nanoparticles suspensions over vertical tubes. *J. Appl. Phys.*, **102**, 074317.

48 Park, K.J., Jung, D., and Shim, S.E. (2009) Nucleate boiling heat transfer in aqueous solutions with carbon nanotubes up to critical heat fluxes. *Int. J. Multiphase Flow*, **35**, 525–532.

49 Henderson, K., Park, Y.G., Liu, L., and Jacobi, A.M. (2010) Flow-boiling heat transfer of R-134a-based nanofluids in a horizontal tube. *Int. J. Heat Mass Transf.*, **53**, 944–951.

50 Kim, T.I., Chang, W.J., and Chang, S.H. (2011) Flow boiling CHF enhancement using Al_2O_3 nanofluid and an Al_2O_3 nanoparticle deposited tube. *Int. J. Heat Mass Transf.*, **54**, 2021–2025.

51 Murshed, S.M.S. and Nieto de Castro, C.A. (2011) Spreading characteristics of nanofluid droplets impacting onto a solid surface. *J. Nanosci. Nanotechnol.*, **11**, 3427–3433.

52 Mitra, S., Saha, S.K., Chakraborty, S., and Das, S. (2012) Study on boiling heat transfer of water–TiO_2 and water–MWCNT nanofluids based laminar jet impingement on heated steel surface. *Appl. Therm. Eng.*, **37**, 353–359.

53 He, G.-H., Fang, X.-M., Xu, T., Zhang, Z.-G., and Gao, X.-N. (2015) Forced convective heat transfer and flow characteristics of ionic liquid as a new heat transfer fluid inside smooth and microfin tubes. *Int. J. Heat Mass Transf.*, **91**, 170–177.

54 Bridges, N.J., Visser, A.E., and Fox, E.B. (2011) Potential of nanoparticle-enhanced ionic liquids (NEILs) as advanced heat-transfer fluids. *Energy Fuels*, **22**, 4862–4864.

55 Mahian, O., Kianifar, A., Kalogirou, S.A., Pop, I., and Wongwises, S. (2013) A review of the applications of nanofluids in solar energy. *Int. J. Heat Mass Transf.*, **57**, 582–594.

56 Verma, S.K. and Tiwari, A.K. (2015) Progress of nanofluid application in solar collectors: a review. *Energy Convers. Manage.*, **100**, 324–346.

57 Liu, J., Ye, Z., Zhang, L., Fang, X., and Zhang, Z. (2015) A combined numerical and experimental study on graphene/ionic liquid nanofluid based direct absorption solar collector. *Sol. Energy Mater. Sol. C*, **136**, 177–186.

58 Karami, M., Bahabadi, M.A.A., Delfani, S., and Ghozatloo, A. (2014) A new application of carbon nanotubes nanofluid as working fluid of low-temperature direct absorption solar collector. *Sol. Energy Mater. Sol. C*, **121**, 114–118.

59 Zhang, L., Liu, J., He, G., Ye, Z., Fang, X., and Zhang, Z. (2014) Radiative properties of ionic liquid-based nanofluids for medium-to-high-temperature direct absorption solar collectors. *Sol. Energy Mater. Sol. C*, **130**, 521–528.

60 Goodrich, P., Hardacre, C., Paun, C., Ribeiro, A., Kennedy, S., Lourenço, M.J.V., Manyar, H., Nieto de Castro, C.A., Besnea, M., and Pârvulescu, V.I. (2011) Asymmetric carbon–carbon bond forming reactions catalysed by metal(II) bis (oxazoline) complexes immobilized using supported ionic liquids. *Adv. Synth. Catal.*, **353**, 995–1004.

61 Doherty, S., Knight, J.G., Ellison, J.R., Goodrich, P., Hall, L., Hardacre, C., Muldoon, M., Ribeiro, A., Lourenço, M.J.V., Nieto de Castro, C.A., Davey, P., and Park, S. (2014) An efficient Cu(II)-bis (oxazoline)-based polymer supported ionic liquid phase catalyst for asymmetric carbon–carbon bond formation. *Green Chem.*, **16**, 1470–1479.

62 Jha, A., Patil, S.H., Solanki, B.P., Ribeiro, A.P.C., Nieto de Castro, C.A., Patil, K.R., Coronas, A., and Rode, C.V. (2015) Reduced graphene oxide composite with redoxible MnCo-oxide for *p*-cresol oxidation using molecular oxygen. *ChemPlusChem*, **80** (7), 1164–1169.

63 Kessler, M.T. (2014) Nanocatalysis in ionic liquids – syntheses, characterization and application to nanoscale catalysts. Dissertation, University of Koln, Germany.

64 Kessler, M.T., Gedig, C., Sahler, S., Wand, P., Robke, S., and Prechtl, M.H.G. (2013) Recyclable nanoscale copper(I) catalyst in ionic liquid media for selective decarboxylative C–C bond cleavage. *Catal. Sci. Technol.*, **3**, 992–1001.

65 Kessler, M.T., Robke, S., Sahler, S., and Prechtl, M.H.G. (2016) Ligand-free copper(I) oxide nanoparticle-catalysed amination of aryl halides in ionic liquids. *Catal. Sci. Technol.*, **4**, 102–108.

66 Vieira, S.I., Lourenço, M.J.V., Maia Alves, J., and Nieto de Castro, C.A. (2013) Using ionic liquids and MWCNT's

(IoNanofluids) in pigment development. *J. Nanofluids*, **1** (2), 148–154.

67 Vieira, S.I.C., Araújo, M., André, R., Madeira, P., Humanes, M., Lourenço, M.J.V., and Nieto de Castro, C.A. (2013) Sepia melanin: a new class of nanomaterial with anomalously high heat storage capacity obtained from a natural nanofluid. *J. Nanofluids*, **2**, 104–111.

68 Commission of European Communities (2008) Regulatory Aspects of Nanomaterials: Summary of Legislation in Relation to Health, Safety and Environment Aspects of Nanomaterials, Regulatory Research Needs and Related Measures. COM (2008) 366 final, 2008, pp. 1–45.

69 Sayes, C.M. and Warheit, D.B. (2009) Characterization of nanomaterials for toxicity assessment. *WIREs Nanomed. Nanobiotechnol.*, **1** (6), 660–670.

70 Michaelides, E.E. (Statis) (2014) *Nanofluidics: Thermodynamic and Transport Properties*, Springer, Heidelberg, Germany.

Part Five
Energy Conservation and Management

38

Silica Aerogels for Energy Conservation and Saving

Yamini Ananthan, Keerthi Sanghamitra K., and Neha Hebalkar

Advanced Research Centre for Powder Metallurgy and New Materials (ARCI), Centre for Nanomaterials, Balapur, Hyderabad 500005, Telangana, India

38.1
Introduction

Energy is a basic need for the existence and survival of life on Earth. In the modern world, energy is not only a crucial factor for living but it also plays a vital role in faster technological, industrial, economic, and social growth, thereby improving our standard of living. The need of energy is increasing exponentially in every field, such as agriculture, construction, automobiles, communication, all kinds of industries, and of course, for domestic use. In the present situation, with limited availability of natural resources of fuel, the world is working on war footing to attain high "energy efficiency," which is nothing but a reduction in the energy consumption. The increased energy production needs innovations in generating renewable energy sources, safer production of atomic energy, and efficient energy-generating equipment. The reduction in energy consumption depends on effective energy management strategies, which balance the demand and supply of the energy. The policies for minimizing the energy consumption have a core goal of energy conservation. Heat management involves energy conservation as a key parameter. Improving the efficiency of all the thermal processes that are used in energy generation, transfer, transmission, and recovery can contribute significantly to save energy. Typically, these involve boilers, hot fluid pipe lines, furnaces, ovens, heaters, storage tanks, air conditioners, and so on. Designing of proper thermal insulation is a means of huge energy conservation and cost saving.

Thermal insulation materials are those that allow minimum heat transfer through them and possess low thermal conductivity. These materials show lower heat transfer for all of the heat transport mechanisms, namely, conduction, convection, and radiation. However, all the insulation materials lose heat by some means. Although complete heat loss prevention is not achievable, every possible reduction in it counts for the energy saving. Development and use of new,

Nanotechnology for Energy Sustainability, First Edition. Edited by Baldev Raj, Marcel Van de Voorde, and Yashwant Mahajan.
© 2017 Wiley-VCH Verlag GmbH & Co. KGaA. Published 2017 by Wiley-VCH Verlag GmbH & Co. KGaA.

efficient thermal insulation materials compared to those conventionally available will be highly effective to serve the purpose.

38.2
Thermal Insulation Materials

Thermal insulation materials are used for performing various functions such as preventing heat loss or gain, controlling surface temperature, facilitating temperature control of process, increasing the operating efficiency of any system at hot or cold conditions, protection from surrounding temperature, and so on. Insulations are used in applications for a wide range of temperatures from cryo to refractory. Conventionally used thermal insulation materials are in the form of blocks, sheets, bricks, blankets, cements, or foams. These forms are made up of low thermal conductivity materials, fibers, granules, and so on. Ideally, the best thermal insulation material should have properties such as low thermal conductivity, low emissivity, corrosion resistance, chemical resistance, good compressive strength, lightweight, moisture resistance, fire resistance, antifungal, low shrinkage, soundproof, nontoxic, and eco-friendly. Choice of the material depends upon the application temperature and design conditions. There are different insulation challenges for different temperature conditions. At cryogenic temperature, freezing of absorbed moisture and damage of insulation material is a usual problem. The insulation material to be used at these temperatures should possess the best vapor barrier property. For high-temperature applications, very few options for materials are available that have thermal stability and all other properties mentioned earlier.

Table 38.1 lists the thermal conductivity and an application temperature range of various commercially used insulation materials; the information is collected from various sources. One observes the presence of a new material in the table known as "silica aerogel," which has low thermal conductivity and wider temperature range among all. This silica aerogel is a new-generation thermal insulation material having unique properties and can be used at cryogenic as well as high temperature applications. This chapter gives an overview of the properties, preparation, and thermal insulation applications of silica aerogel, and explains how this material is the best choice for insulation applications in energy conservation and storage.

38.3
Aerogels

38.3.1
What Are Aerogels?

As the name implies, it is a gel filled with air. Aerogels were first synthesized by Kistler [1,2] in the early 1930s. These are the lightest man-made solid materials. The lowest-density silica aerogel ever made has 99.8% porosity [3], which means

Table 38.1 Commercially used thermal insulation materials, their thermal conductivity, usage temperature range, and cost (compiled from various sources).

Insulation material	Thermal conductivity at 25 °C (W/(m K))	Temperature range (°C)	Thermal insulation applications	Cost (US$)
Glass-fiber mat	0.035	−200 to 500	Industrial, building, automotive, heat/cold storages	1.5–2.5/m^2
Cellular glass	0.048	−200 to 400	Industrial, pipeline, cryo, building	2.6–5.7/m^2
Polyurethane	0.036	−40 to 105	Building, cold storage, aviation, automotive, pipeline	2.5/kg
Calcium silicate	0.065	Up to 650	Industrial, furnaces	2–20/m^2
Mineral fiber	0.037	Up to 650	Building and high-temperature pipe	0.5–4.2/m^2
Expanded perlite	0.076	Up to 650	Building, heat/cold storage	0.32/kg
Polystyrene	0.037	−40 to 80	Building, cold storage	3.5–5.2/m^2
Ceramic fiber	0.08	Up to 1260	Industrial, furnaces	2.5–4.6/m^2
Silica aerogel	0.01	−200 to 800	Building, heat/cold storage, clothing	20–60/m^2
Cellulose	0.035		Building	7.18–7.45/kg
Vacuum insulation panels	0.003		Building, cryo	5–28/m^2
Gas-filled panels			Cold storage, transportation, automotive and buildings	—
1. Air	0.0361			
2. Argon	0.0492			
3. Krypton	0.0867			
Phase change materials		50 °C to 150 °C	Building	75.24/m^2
1. Water	1.6			
2. Sodium Silicate	0.103			
3. Paraffin	0.25			
Aerogel sheets	0.013	Up to 600 °C	Industrial, aerospace/defense, textile, buildings, heat/cold storage, automotive	20–60/m^2

99.8% by volume is nothing but air and just 0.2% is solid. They appear translucent or transparent yellow when seen against bright light and blue if viewed against a dark background. Figure 38.1a shows the scanning electron microscope (SEM) image of the nanoporous aerogel and Figure 38.1b and c shows blue aerogel and yellow aerogel due to Rayleigh scattering and Mie scattering,

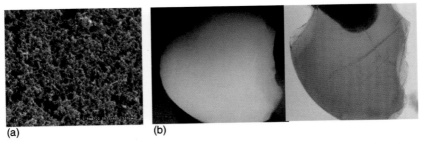

Figure 38.1 (a) SEM of nanoporous aerogel. (b) Rayleigh scattering and Mie scattering when silica aerogel is observed against dark and light background.

respectively. Hence, aerogel gets synonyms such as frozen smoke, solid smoke, solid air, blue smoke, and so on. Other than the most popular silica aerogels, these can be made from various chemical compounds such as organic polymers [4], biological polymers [5], most of the transition metal oxides [6–10], semiconductor nanostructures [11,12], carbon [13], carbon nanotubes [14], graphene [15,16], and so on. The world's lightest aerogel was reported recently with density $0.16 \, \text{mg/cm}^3$ and was made up of graphene [15]. Notably, the world's first 20-karat nugget of gold aerogel consisting of 98% air has been produced using milk protein by Gustav et al., which can float on cappuccino foam [17]. While it retains the metallic sheen of its regular form, it is thousand times lighter than the conventional gold alloys.

38.3.2
General Properties

Aerogels exhibit fascinating combination of properties that no other materials possess. General properties for silica aerogel having density $0.1 \, \text{g/cm}^3$ compiled by Hrubesh [18] are given in Figure 38.2. Importantly, the ultralow dense silica aerogel provides insulation for heat, sound, and electricity and can work for wide temperature range from cryo to high temperature. Huge surface area, low dielectric constant, and low refractive index are some other important properties that make them a potential material in several applications.

38.3.3
General Applications

Although thermal insulation is the most popular application of aerogels, its use in many other fields is possible. Figure 38.3 gives the general applications of aerogels. There are some excellent review articles on preparation, properties, and applications of aerogels [14,18–21].

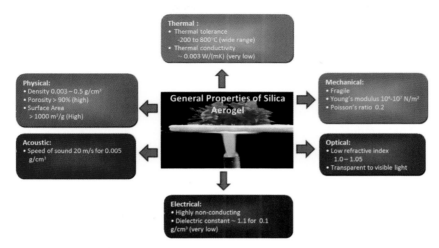

Figure 38.2 Properties of silica aerogel.

Figure 38.3 General applications of aerogel.

38.3.4
Thermal Insulation in Aerogels

The high thermal insulation property of aerogel is possible due to the nature and extent of nanoporosity present in it, which suppresses the conduction and convection heat flow through it. The thermal conduction in aerogels is explained by Lu *et al.* [22]. In porous materials such as aerogels, the total thermal conductivity (λ_t) is a summation of all three modes of thermal conduction, namely, gaseous (λ_g), radiative (λ_r), and solid (λ_s) $(\lambda_t = \lambda_g + \lambda_r + \lambda_s)$.

Thermal conduction in aerogel through solid is drastically reduced due to the ultralow density and hence very less solid content in it. Large amount of nanopores restrict the heat flow in aerogels. The pores in the aerogel are filled with air. Therefore, porosity and pore diameter are the important factors that decide the gaseous thermal convection. The mean free path of air at 1 bar is around 70 nm, which is larger than the typical value of average pore size in aerogels. Therefore, the thermal conductivity due to air is suppressed even at atmospheric pressure. One option to further reduce gaseous thermal transport is replacing the air in the pores by a light gas such as helium having a higher mean free path. In another way, the mean free path can be further elongated by reducing the air pressure inside the pores by slight evacuation. Thermal conductivity of air-filled silica aerogel is reduced to half when it is evacuated [23]. In case of silica aerogels, there is a small infrared absorption in the range of 3–5 µm [19]. This gives radiation leakage even at room temperature. Thus, the radiative mode of thermal conduction is high compared to other modes in silica aerogels and hence cannot be neglected. If the silica aerogels are opacified with a strong absorber of IR radiations such as carbon black, specific extinction coefficient, e(T), can be increased, which will reduce this mode of thermal conduction. A number of research groups have worked to study thermal conductivity in silica aerogel [24–27].

38.4
Preparation

38.4.1
Gel Formation

Mostly, aerogels are derived by a well-known sol–gel method, followed by super critical drying. This method involves the hydrolysis of the precursor in the presence of a catalyst to prepare a suspension of monomer particles called as sol and concurrently allowing these to condense into cross-linked polymers in the form of gel having an average pore size in the nanometer regime.

Gel is a three-dimensional solid structure with liquid trapped in its pores. The gel is further aged in its mother solution where further condensation takes place to increase the connectivity in the network to get strengthened and stiffened gel. The gel is "washed" several times to exchange the pore liquid with a pure solvent

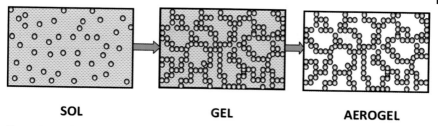

SOL **GEL** **AEROGEL**

Figure 38.4 Schematic representation of sol, gel, and aerogel.

and then dried to get a nanoporous solid where liquid in the gel is replaced by air, as schematically shown in Figure 38.4.

38.4.2
Drying of Gel

If the gel is dried by evaporation of solvent, the capillary pressure in the pores rises due to the surface tension of liquid in gel pores owing to ultrasmall pore size. This leads to breakage of thin pore walls and shrinks to give a dried product called "xerogel." To retain the pore walls intact, the capillary pressure buildup needs to be avoided. Mainly three types of drying processes are followed: supercritical drying [1,28], freeze-drying [29,30], and subcritical drying [29,31–34]. Table 38.2 gives details of these gel-drying processes. Of these processes, the supercritical drying is the most efficient method and it is commonly used by many aerogel manufacturers too. Figure 38.5 shows the schematic of an autoclave and a typical temperature–pressure curve followed during ethanol-based supercritical drying.

38.5
Aerogels in Various Forms: Monoliths, Granules, and Sheets

In spite of all the attractive properties in aerogels, the commercial use of it was restricted because of its high production cost and poor mechanical stability. Aerogels in pure form are highly fragile and hence suffer wastage of material due to breaking. This lead to the development of new and usable form of aerogels. Outcome of this was the development of silica aerogel in the form of flexible sheets and granules. The silica aerogel flexible sheets are generally made by infiltrating aerogel into the pores of fiber mat where the fibers can be of any type such as organic, inorganic, woven, nonwoven, and so on. This form of aerogel makes it easy to wrap around the object to be insulated and can be cut in any shape for easy installation on curved and complex shapes. Aerogel granules are another form of silica aerogel that is used as an insulation material by filling it appropriately around the object to be insulated. Granules and sheets can be

Table 38.2 Common gel drying processes for silica aerogel, their operational conditions, advantages, and disadvantages.

Name of the drying process	Description/principle	Process conditions	Equipment required	Advantages	Disadvantages	References
Supercritical drying	Removal of pore liquid above its critical point	S = Ethanol, CO_2. *Ethanol* $T_c = 243\,°C$ $P_c = 6.3\,MPa$ *CO_2* $T_c = 31.1\,°C$ $P_c = 7.39\,MPa$	Autoclave with temperature and pressure controlling facility	Highly porous aerogel, less time-consuming	High critical constants, risky due to flammability of organic solvents; high initial capital cost and operational costs; expensive	[1,18]
Freeze-drying	Removal of pore liquid by deep freezing and then sublimation under vacuum	S = t-butanol *Freezing* $T = 25.6\,°C$ Rate = $16.9\,kg/m^2\,h$	Deep freezer and vacuum dryer	Quick process, t-butanol is less toxic	Decrease of porosity due to crystallization of solvent during freezing	[19,20]
Subcritical drying	Surface modification, network strengthening, and then ambient drying	SMA = hexamethyl disilazane, tri-methyl-chlorosilane	No special equipment required	No energy required for drying, risk-free	Low porosity aerogel, large amount of solvent and significant time required for solvent exchange and drying	[19,21–24]

S: solvent; T_c = critical temperature; P_c = critical pressure; SMA: surface modification agents.

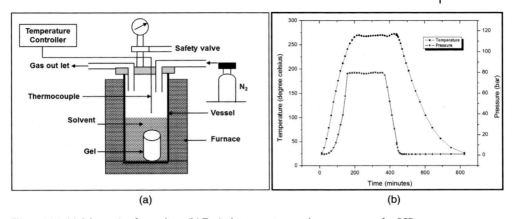

Figure 38.5 (a) Schematic of autoclave. (b) Typical temperature and pressure curve for SCD.

sandwiched between or encapsulated in substrates such as glass plates, wooden panels, fabric sheets, and so on for their use. Table 38.3 lists the companies that are the major players in the manufacturing of aerogel products, names of the products, and their properties and applications as mentioned on their respective Web sites.

All these forms of aerogel show thermal insulation property similar to monolithic aerogels with minor increase in the thermal conductivity. Comparison of the heat conduction mechanism in the above mentioned popular forms of silica aerogel, namely, monolithic, granular, and sheet, is shown in the Figure 38.6. Typical thermal conductivity of monolithic silica aerogel is as low as 0.01 W/(m K) [28] due to reduced solid content and pore size smaller than 100 nm, on the order similar to mean free path of air molecules. In case of aerogel granules, the thermal conductivity slightly increases from that of monoliths due to the intergranular gaps that contain air. Thermal conductivity of air is 0.025 W/(m K), which is greater than that of pure monolithic aerogel, and it contributes in the apparent thermal conduction in aerogel granules. It was reported that the silica aerogel granules filled without any compaction showed thermal conductivity of 0.024 W/(m K), which was reduced to 0.013 W/(m K) after compaction [35]. The effect of compaction pressure on their thermal conductivity showed decrease in the thermal conductivity value with increase in pressure, until most of the air gaps were removed, however beyond optimum pressure, the thermal conductivity increased through solid conduction. The third form of silica aerogel – sheet or blanket – is made by infiltrating aerogel in the pores of fiber mat where the aerogel content is about 50–60%. Hence, the thermal transport through fiber has an additional contribution in the overall thermal conductivity of the composite sheet. Thermal conductivity of various commercially available aerogel blankets is seen in the range of 0.014–0.031 (see Table 38.3).

Various silica aerogel-based products are developed in the author's laboratory for several thermal insulation applications. The products include monoliths,

Table 38.3 List of major aerogel manufacturing companies, their products, properties, and applications.

Name of the company and its Web site	Name of the product	Types of product	Property	Application
A Proctor Group Ltd, UK www.proctorgroup.com	Spacetherm®	Aerogel and fiber composite	$t = 5$ and 10 mm $\rho = 0.15$ g/cm^3 $K = 0.015$ W/(m K)	Construction, thermal insulation for walls, windows, and so on
Active Aerogels Coimbra, Portugal www.activeaerogels.com	Aeroflex	Silica aerogel and fiber composite	$t = 8$–30 mm $K = 0.031$ W/(m K) TR $= -180$ to 350 °C	Thermal insulation on oil ducts, pipelines, buildings, atmospheric balloons, re-entry vehicles, launchers
	Sil	Powder, beads	$K = 0.025$ W/(m K), hydrophobic	Dielectric filling, hydrocarbons absorption
	Sprayflex	Particles	$K = 0.040$ W/(m K), hydrophobic	Used on complex geometry
Aerogel Technologies LLC, USA www.aerogeltechnologies.com	Airloy™	Machinable mechanically strong aerogel	$\rho = 0.11$ g/cm^3 hydrophobic CS $= 260$–900 MPa.	Lightweight structures, multifunctional superinsulation, oil spill reclamation, ultracapacitors, sensors, catalyst supports
	SilicaTM	Monolithic silica aerogel	Hydrophobic, shape control	Chemisorption, chromatography, diffusion control media, and so on
AeroSafe Global (*formerly* American Aerogel Corporation, New York) www.americanaerogel.com	AeroSafe™	Insulation boxes using Vacuum insulation panel and Phase Change Materials	Controlled Room Temperature 2–8 °C ≤ -65 °C and ≤ -20 °C	Insulated shipping, temperature-sensitive payloads

Company	Product	Form	Properties	Applications
Aspen Aerogels Inc, Northborough, MA www.aerogel.com	Pyrogel® XT-E, Pyrogel® XTF	Silica aerogel and fiber composite	$t=5$ and 10 mm $K=0.021$ W/(m K)	High temperature steam pipes, vessels, equipments, and aerospace and defense systems.
	Cryogel® Z, Cryogel® x201	Silica aerogel and fiber composite	$t=5$ and 10 mm $K=0.017$ W/(m K)	Subambient and cryogenic pipelines, cold storage, and aerospace
	Spaceloft® Subsea	Silica aerogel and fiber composite	$t=5$ and 10 mm $K=0.0145$ W/(m K)	Medium- to high-temperature offshore oil pipelines
	Spaceloft®	Silica aerogel and fiber composite	$t=5$ and 10 mm $K=0.0165$ W/(m K)	Ambient temperature walls, floors, and roofs
Cabot Corporation Boston, MA www.cabotcorp.com	ENOVA®	Aerogel granules	$K=0.012$ W/(m K) hydrophobic	Insulative coatings, ultralow gloss, light diffusion, and skin and beauty care
	LUMIRA®	Aerogel particles	U Value $= 0.75$ W/(m²K) per 25 mm hydrophobic	Architectural daylighting applications
	P100, 200, 300 particles	Aerogel particles	Porosity >90%, hydrophobic, $SA=600$–800 m²/g	High-performance insulating composites such as boards, blankets, and plasters
	Thermal Wrap™ (TW350, 600, 800)	Silica aerogel granules within nonwoven fibers	$t=3.5/6/8$ mm $TR=-200\,°C$ to $125\,°C$	Applications from building and construction to oil and gas pipe lines
Fixit AG, www.fixit.ch	Fixit 222, Fixit 244	Aerogel granules	$\rho=220$ kg/m³, $K=0.028$ W/(m K), $t=3$–150 mm Application time: 30 min	Spray coating for facades
JIOS Aerogel Corporation Gyeonggi-do, Korea www.jiosaerogel.com	JIOS AeroVa® aerogel	Aerogel powder	$SA=600$–1000 m²/g $TR=-200$ to $1600\,°C$	Blankets, coatings, cement and perlite composites, textiles, cosmetics
	JIOS AeroVa® fire resistant coating	Aerogel coating	Viscosity 500–2000 cPs 24 h curing time at 20 °C TR: 20 °C–1,000 °C	Fireproof coating (compatible on steel, iron, wood plastic, and concrete)

(continued)

Table 38.3 (*Continued*)

Name of the company and its Web site	Name of the product	Types of product	Property	Application
	JIOS AeroVa® – blanket H	Aerogel blanket insulation	$t = 5/10$ mm, $K = 0.019$ W/(m K), TR $= -40$ to 650 °C	Oil and gas pipe insulation, sub-sea pipe insulation, building and construction, automotive equipment, home appliances
Nano science and Technology Co., Ltd., Shaoxing, China (*formerly* Shaoxing City Nano High Tech Co., Ltd.,) www.nanuo.cn	FM series	Aerogel Felts	TR $= -50$ to 650 °C, $K = 0.015$–0.020 W/(m K)	Insulation for pipe lines, vessels, and equipment
	IP650	Aerogel Panels	$K = 0.020$ W/(m K), TR $= -50$ to 650 °C	Industrial furnace insulation
	SS650	Aerogel special shaped parts	$K = 0.020$ W/(m K), TR $= 100$ to 650 °C	Metallurgy and chemical pipe insulation
	VP100	Vacuum-insulated panels	$K = 0.004$ W/(m K), TR $= -200$ to 100 °C	LNG equipment, fridge, cooler, and cold insulation
	Powders and particles	Silica aerogel particles	$K = 0.020$ W/(m K), TR $= 100$ to 650 °C	High-temperature pipes, valves, and flanges insulation
Nanopore, Incorporated Albuquerque, New Mexico www.nanopore.com	Nanoglass©	Thin films	dielectric constant $1.1 < K < 2.5$	Semiconductors
Oros www.orosapparel.com	Aerogel apparel	Aerogel clothing	TR: -50 to 148 °C	Apparel – jackets, track pants
Shivershield LLC, USA www.shivershield.com	Shivershield apparel	Aerogel flexible panels	—	Apparel – jackets, track pants, and insoles

Company	Product	Form	Properties	Applications
Surnano Aerogel Co. Ltd. Zhejiang, China www.china-aerogel.com	SNP	Silica aerogel particles and powder	$\rho \leq 100\,kg/m^3$ $K = 0.016\,W/(m\,K)$ $TR = -80$ to $200\,°C$	Functional structure interlayer, filling layer, and composite layer for insulation, air clean, water treatment
	SNF 200, 400, 400 C, 650, 1000	Aerogel and fiber composites	$\rho = 220\,kg/m^3$ $K \leq 0.020\,W/(m\,K)$ $T_{max} = 550\,°C$	Thermal insulation for central air-conditioning pipe, petroleum pipe, and so on
	SAP 400, SAP 650	Aerogel panel	$\rho = 200\,kg/m^3$ $K = 0.016\,W/(m\,K)$ $TR = -270$ to $650\,°C$	Metallurgy, chemical engineering, defense, aerospace and building
TAASI Aerogel Technologies, USA www.taasi.com	Pristina™	Powder, granule, beads, extruded forms	$\rho = 0.2$–$0.6\,g/cm^3$	Removal of pollutants from water and air; adsorbents and catalyst; catalytic conversion of carbon monoxide
Thermablok Aerogels Ltd, UK www.thermablok.co.uk	Thermablok Aerogel Blanket	Aerogel and fiber composite	$t = 2.5, 5$ and $10\,mm$ $\rho = 150\,kg/m^3$ $TR = -200$ to $200\,°C$ $K = 0.014\,W/(m\,K)$	Building Constructions and Energy efficiency retrofits

K: thermal conductivity at $25\,°C$; TR: temperature range; t: thickness; T_{max}: maximum temperature; SA: surface area; CS: compressive strength; P: porosity; ρ: density.

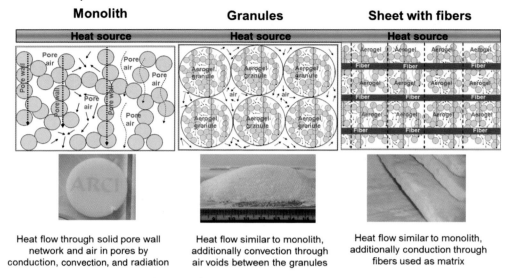

Figure 38.6 Schematic of heat transport mechanism in different forms of aerogel.

flexible sheets, granules, composite building materials, aerogel formulation as an additive, and so on. The processes used to produce these products are novel and innovative. The silica aerogel-based flexible sheets developed have shown the highest commercial demand and hence features of this product are presented as below. Wide ranges of silica aerogel flexible sheets are prepared by infiltrating silica aerogel within the matrix of fibers. Fibers used to make flexible sheets are of inorganic, organic, and/or blending of both. As a representative example, Figure 38.7 shows the scanning electron microscopy images of ceramic refractory fiber mat before and after aerogel infiltration in it. It is clearly seen that the aerogel is completely filled in the empty space of the fiber matrix. The sheets can be produced in single or multilayered structure, with the pure silica aerogel granules being sandwiched between the aerogel sheets, where many combinations of thickness and number of layers are possible. The process developed to produce such sandwich structured aerogel sheet is novel. Various lamination layers are available to encapsulate the aerogel sheet depending upon the requirement of the application. Silica aerogel used for infiltration in all the above types of mats is stable up to 800 °C as seen in terms of stability of pores, which is directly related to its thermal insulation performance. This is evident from the nitrogen adsorption isotherm for silica aerogel heated at 800 °C, which is very similar to that of as-prepared shown in Figure 38.8a. Due to the huge porosity, the aerogel-infiltered sheets show large surface area in the range of 365–600 m^2/g (Figure 38.8b) depending upon the type of the sheet. These values indicate high aerogel content in the sheet and its highly porous nature. Such porosity leads to achieve thermal conductivity as low as 0.016 W/(m K) at 37 °C (as per ASTM C 177). The silica aerogel incorporated in the fiber matrix has a special property of suppressing thermal

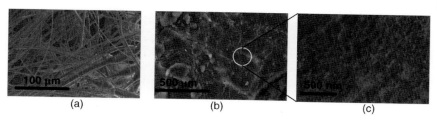

(a)   (b)   (c)

Figure 38.7 Scanning electron microscope image of (a) bare refractory fiber mat with density 120 kg/m³; (b) silica aerogel-infiltered refractory mat; (c) magnified image of silica aerogel infiltered in fiber mat.

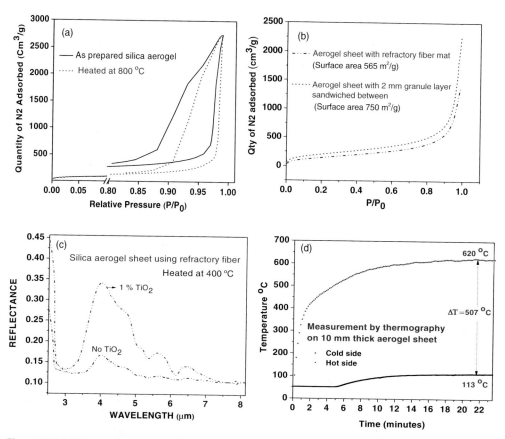

Figure 38.8 (a) N₂ gas adsorption isotherm for as-prepared silica aerogel and after heating at 800 °C in air. (b) N₂ gas adsorption isotherm for plane aerogel sheet and sandwich aerogel sheet. (c) Infrared radiation reflectance of aerogel sheet heated at 400 °C. (d) Thermal insulation performance of 10 mm aerogel sheet measured by thermography technique.

conductivity at higher temperatures due to infrared reflectance as shown in Figure 38.8c. This property was achieved by uniformly dispersing extremely fine metal oxide particles in a small concentration, which is produced by a novel, *in situ* process during the production of silica aerogel. The aerogel sheet, with the above-mentioned features, shows best thermal insulation performance. Figure 38.8d is a graph depicting insulation property of 10 mm aerogel sheet when subjected to 600 °C in a steady-state condition. The difference in the temperature seen between hot and cold side of 10 mm thick sheet was about 500 °C, which is substantial.

Other product features include low density ($0.2 \, g/cm^3$), hydrophobic yet breathable in nature, possibility to produce in variable thickness, easy to cut, fire and chemical resistant, and noncorrosive. All these properties make the product a potential material for a wide range of thermal insulation applications. The sheets are further customized for their use at high as well as cryo temperature applications.

Granular form of silica aerogel developed at author's laboratory is another useful product that can be sandwiched between plates, fabrics, plastic sheets, and so on to make an insulating panel or sheet with superior thermal insulation properties. This also acts as an additive for paints, cements, and so on to make other thermal insulation products. These technologies are in progress for its commercialization. Figure 38.9 shows the photographs of the aerogel products developed at author's laboratory and some of its features.

38.6
Thermal Insulation Applications

38.6.1
Industrial

Industrial applications of aerogels get benefited by the fact that aerogel installations and operational costs show significant economic benefits and energy conservation. Lower thermal conductivity of aerogels pays back in terms of thin

(a) (b)

Figure 38.9 (a) Silica aerogel flexible sheet roll having excellent thermal insulation, hydrophobicity, and fire-resistant property. (b) *Top:* A heap of aerogel granules prepared in the author's laboratory. *Bottom:* Silica aerogel granules seen under optical microscope.

insulation layers required to achieve the expected performance. As seen in Table 38.3, most of the products are suitable for industrial applications.

Oil and gas refineries and other industries that use high-temperature processes or fluids need insulation to safeguard pipelines, boilers, heaters, pumps, valves, and so on from atmospheric moisture, chemical corrosion, prevention against leakage, and other kinds of derogatory effects. These insulation systems directly contribute to the industry in terms of energy savings by reducing heat loss and also by reducing the total amount of fuel burned to maintain such high temperature, thereby directly reducing the total CO_2 emitted into the atmosphere.

Aerogels have been the most sought-after insulation systems for oil and gas companies. This is mainly because of the lightweight and yet superior thermal performance of the composites wherein a thinner sheet can be employed in place of a thicker insulating foam. By this, the total volume occupied by the insulation system reduces greatly. Indirect cost reduction is seen due to saving in transportation, storage of aerogel insulation, and labor charges. Aspen Aerogels Inc. and Cabot Corporation have been providing aerogel-based insulation blankets that can be seen from the case studies mentioned in their Web sites. The superior thermal performance, chemical and pressure resistance of the products combined with the assembly cost savings make them most suitable for this application. Also the operational cost of the pipelines decreases when compared to conventional insulation systems as aerogel-based products have more longevity and improved degradation resistance.

A number of industries, including power generation plants, use high-temperature and high-pressure boilers, heaters, and so on and employ pipelines to carry very hot fluids. Aerogel blankets and other composite products reduce the heat loss and increase the lifetime of these process machineries and pipelines and hence are now the most preferred materials.

Industries and plants that were designed and built with minimal insulation in the yesteryears (when energy crisis was unheard of) need insulation to reduce the total energy spent. Compared to other commercially available insulation systems, aerogels prove advantageous for such situations as thin aerogel blankets that are hydrophobic and still breathable can easily cover up the entire system without cramming up space.

38.6.2
Architectural

Buildings face a pressing need to keep the energy loss to a bare minimum as around 20–40% of global energy consumption is contributed by the residential and commercial buildings [36]. Insulating the HVAC (heating, ventilation, and air-conditioning) systems with highly efficient insulation reduces the use of energy to maintain thermal comfort.

Being the most susceptible part of buildings, windows require insulation that minimizes heat transmission and has maximum transparency for daylight. Conventional insulation for windows used either vacuum insulation panels or triple-glazed systems, whose overall heat transfer coefficient (U-value) is usually less than $1\,W/(m^2\,K)$. However, to further reduce the U-value, the total number of glass panes has to be increased (for triple-glazed systems) and this will have a negative impact on solar light and day light transmission into the building [37]. Research has been carried out to coat a layer of silica aerogel on the window glazing units as an alternative [38]. Window glazing, which is filled with aerogel, has been studied extensively by many groups [39–41]. Studies reveal that evacuated aerogel glazing of dimensions ($0.58\,m \times 0.58\,m \times 0.58\,m$) with 20 mm glass distance can offer a U-value of $0.42\,W/(m^2\,K)$ and solar energy transmittance above 0.75 [42] The optical transparency of the aerogel can be tuned between 25 and 80% during the synthesis procedure and a corresponding U-value of $0.014–0.021\,W/(m^2\,K)$ can be obtained [26].

In recent past, historic buildings have been retrofitted in order to preserve them. However, many a times, only the internal surface can be insulated due to valuable façade designs and to preserve other heritage requirements. Conventional insulators tend to be thicker and occupy space. Alternatively, aerogel-incorporated insulation sheets with minimal thickness can be used even externally. Also, hydrophobicity of the aerogel helps in safeguarding the brick wall construction from moisture and retains a lot of internal space too. A villa from 1921 was thermally insulated using aerogel blanket, which also improves fire protection (fire resistance Class A – ASTM E84) and enhances breathability of the building [43].

Recent developments in plastering have given rise to different types of renders based on silica aerogel. Stahl *et al.* developed a breathable external plaster based on aerogel in a cement-free binder with a very low thermal conductivity as well as low water vapor transmission resistance. They compared it with commercially available plasters and observed superior properties of their product [44]. Research is underway to incorporate aerogel directly into the cement matrix so as to obtain thermally and mechanically suitable composition for slimmer building blocks [45].

Reduction of U-value by 19 and 15% was observed when aerogel-based plaster was applied as a process of refurbishment to a stone wall and brick wall (internally and externally lime plastered), respectively. A thermographic analysis (Figure 38.10) of a 5 mm aerogel-based plaster applied on the inner surface of a

Figure 38.10 View of the investigated building and the correspondent infrared imagine: The aerogel-based plaster was applied in the internal wall of the third floor.

multistoried family house in Pordenne, Italy, revealed that the floor with aerogel-based plaster showed a reduction of 2 °C when compared to conventional plastering. The markings M1–M8 (denoted by X_{M1}–X_{M8}) in Figure 38.10 represent the data points at which the temperature reading was observed [46].

38.6.3
Automotive

When it comes to automobiles, insulation is aimed mainly for (i) restricting engine heat from reaching the driver and the passengers, (ii) thermal insulation of vehicle body from outside heat or cold, and (iii) attenuating the sound from the engine, road, and wind to provide comfort for the travelers. Also, any noise produced inside the passenger compartment should be dissipated.

The two most common solutions are either a two-layer trimming (a poroelastic layer covered by a waterproof heavy layer) or one or two layers of felt (good absorption despite poor insulation performance). Silica aerogels are proven to have high thermal insulation and low emissivity as well as acoustic absorption due to its high porosity and specific surface area [47–49]. The nonorganic nature of aerogel makes it noncombustible and is able to withstand high temperature. And being the lightest insulation material, the total load on the system reduces, thereby increasing fuel efficiency.

The aforementioned properties make aerogels an excellent multifunctional alternative for trimming parts in cars, buses, tractors, railway coaches, and airplanes. Specially designed aerogel composites can also be used for insulating the fuel tanks and also to provide water-resistant, corrosion-resistant, transparent coating to the electronic parts of the automotives as well as for the exterior finishing of the same.

38.6.4
Cold Storages and Cryogenic Insulation

Chillers, refrigerators, food storage containers and transport containers that are used for biodegradable materials and certain kinds of chemicals, all can be considered under one roof as cold storage devices.

Unlike hot systems, cold storage systems have a different need for insulation. Refrigeration systems face condensations that are detrimental. Preventing condensation and moisture forms the basis for determining the insulation system to be used.

A cold storage system usually operates between −10 and −45 °C. They can use a variety of refrigerants such as freon, ammonia, glycol, brine, and other specialty liquids in piping materials made up of copper, iron, stainless steel, and so on. Insulation in such systems is mainly required to minimize heat gain in the internal fluids and to control surface condensation. Also, the insulation should be effective enough to prevent ice accumulation. Important features for consideration of insulating refrigeration systems are thermal conductivity, water vapor transmission property, water absorption properties, fire and smoke performance, and so on. At present, closed cell foam materials are used [50].

Aerogel composites can be a good add-on to such existing insulation as it can greatly reduce the thickness of the foam employed. Also, refrigeration piping that is used outdoor can be cladded with specifically designed aerogel blankets to reduce power consumption and control moisture. Breathable hydrophobic aerogel composites ensure that any moisture from outside does not reach the surface of the piping and at the same time releases moisture from the surface of the pipe to the atmosphere (if any).

Cryotechnology that comprises transport and storage of liquefied gases or frozen biomedical specimens, aerogel-based protection, proves advantageous. Besides its superinsulation properties, aerogel tends to embrittle less with decreasing temperature than, for example, polymer foam insulation. Fesmire *et al.* has worked extensively on aerogel for cryoinsulation [51–60].

38.6.5
Aerospace

Being developed mainly for aerospace applications at NASA, aerogel found its way to space way back in 1997. The Mars Rover, "Soujourner" of NASA, which encounters large temperature fluctuations during its course, used aerogel to keep its battery pack insulated so that the battery remained functional at 21 °C, which is well within its operation temperature of −40 °C to +40 °C. Aerogels reinforced with flexible sheets are being studied for usage in extra vehicular activity suits. This is because aerogel can withstand extreme vacuum conditions and undergo high mechanical flex cycling and still retain its thermal behavior [61]. NASAs Kennedy Space Center has studied the thermal and physical properties of aerogel under various cryogenic vacuum conditions for use in space-launch applications [51]. Comparative study between conventional materials such as polyurethane sprays foam, perlite powder, and multilayer insulation (MLI) revealed that the aerogel-based insulation had better advantage owing to its low density and high insulation capacity even with reduced thickness. Also, tunable hydrophobicity of aerogel can effectively stop the formation of frost or ice in the pores.

The author's laboratory has also worked on inorganic fiber-reinforced aerogel sheets for an Indian aerospace program and demonstrated that a 10 mm sheet remained intact and helped in temperature reduction of 450 °C even when the sheets were rapidly heated up to 900 °C under severe vibration of up to 2 kHz.

38.6.6
Defense

Aerogel insulation has significant performance advantages in defense applications due to its lighter weight, enhanced insulation, and smaller thickness. To list a few, thermal insulation in tank engines, generators, aircraft engine components, exhausts, fuel tanks, a covering layer to hide from the IR cameras, in shipboards, personal shelters, clothing, sleeping bags and footwear for extreme low-temperature conditions, food and water containers, and so on can be considered as possible application areas. This can minimize fuel consumption largely, reduce the weight burden on soldiers and make their life more comfortable, and increase their safety.

38.6.7
Textiles

Aerogels finds its use in fabrics to cater the needs of different professionals. Soldiers and tourists at subzero temperature zones are currently equipped with bulky clothing to save themselves from harsh climates that are uncomfortable and are an extra load to their body weight causing muscle strain and work load. Nuckols *et al.* worked on aerogel-infiltrated fabrics and had a comparative study with the then available commercial fabrics for cold water diving garments. They found a 76% thermal improvement with the former for the same thickness and lesser weight [62].

Aspen Aerogels tested their product Pyrogel with the help of Pierre Mielcarek and Philippe Remeniéras who wore the aspen aerogel jackets and took a three Mont Blac route. They found that when they reached the top, they were surprisingly still comfortable with no cold sensation other than their hands.[1] Pierre Mielcarek reported that insoles made from aerogels kept his feet warm even when there was extremely violent wind at −20 to −25 °C blowing at 70–80 km/h outside.

OROS' Lukla Jackets have incorporated aerogel in jackets and is commercializing them as a high-altitude sports gear for skiing and mountain climbing.[2]

In another study, fire fighters' garments were altered by incorporating superhydrophobic aerogel in blended fabric. A 68.64% increase in thermal resistance was found with 2% coating of aerogel and reduced air permeability of up to 45.46%. Also, the aerogel-coated fabric next to skin acted as a moisture management fabric, thereby enabling better comfort for the fire fighter [63].

Zrim *et al.* have found that with a suitable laminate, the silica–aerogel composite of thickness 2.7 mm and mass 500 g/m^2 retained its original thermal conductivity and water permeability of 0.016 W/(m K) and 1.31 mg/cm^{2}h, respectively [64].

Venkataraman *et al.* fabricated a new equipment to study the thermal properties of nonwoven fabrics that have aerogels incorporated in them. The system is capable of operating at subzero temperature also [65].

1) //www.aerogel.com/_resources/common/userfiles/file/Case_Study_Mt_Blanc_web.pdf (2007)
2) //www.orosapparel.com/pages/oros-technology

38.7
Energy Saving and Conservation Using Aerogel Products

As seen in various reports, by 2040, about 50% increase in the energy demand is predicted for the industrial sector accompanying the economic growth and rise in the middle class of the society. In addition to this, industry will have a greater share in the use of electricity due to advancement in the manufacturing processes, level of automation, and control technologies. Major growth is also visualized for the building and construction industry as living standards are rising very fast. Considering the fact of major share of energy demand in the industrial sector, here are some examples stated to give a flavor on the impact of aerogel insulation on energy saving.

The conventional thermal insulation systems used in the heavy industry include mineral wool, calcium silicate, fiber glass, and so on. The discussions with the experts in industrial insulation revealed a major problem of heat losses due to the old thermal insulation materials used in industrial equipment, and replacing them with new ones is the most expensive task. The insulation performance of these materials deteriorates with time due to the absorption of water in the form of moisture. The absorbed water vaporizes on the hot side of the insulation, travels toward the outer layers of insulation, and condenses due to colder temperature. This affects the insulation property as the convective heat transfer increases. As noted in the discussions, the heat loss can increase up to two to three times than the performance in initial days and similar impact on the energy cost is observed. The option of replacing the whole insulation system is highly expensive as it includes material cost, its transportation cost, labor cost for removing existing bad insulation material, and installing a new one. As per the new concept, regeneration of insulation performance in the existing insulation system can be done just by wrapping a single layer of aerogel insulation sheet. This sheet would possess a property of repelling water but still is breathable to the water vapor. This not only provides further insulation but also allows the water vapor in the old insulation to evaporate and restricts water to enter from outside. After this process, the deteriorated insulation performance can be recovered up to 50%. From the Internet search, it was known that Aspen Aerogel, USA, under trade name "ReGenaWrap" has introduced this concept to some heavy industries.

As per the information shared by the competent sources, huge energy and cost saving is observed when aerogel insulation is used in industrial application in place of mineral wool. As an example in a boiler, where estimated requirement of conventional mineral wool insulation material is 2500 m³, if this is replaced by aerogel insulation, the quantity of aerogel insulation can be reduced by at least 2.7 times. Although the cost of aerogel insulation can be high, up to eight times than the mineral wool, a 300% saving due to heat loss is expected and hence the payback period reduces drastically. Lesser space and weight requirement is the other added advantage. Enhancement in the life of the insulation is expected due to the hydrophobic nature, which provides protection from water.

Nanotechnology Research Center, Research Institute of Petroleum Industry, Tehran, Iran, had published a case study where a simulation study comparing performance of various insulation materials was used for insulating a pipeline carrying steam [66]. The length of the pipeline used was 1500 ft having 2 in. diameter. The temperature of the vapor was 200 °C with 180 psi pressure and 2500 lb/h flow rate. If no insulation is used, the vapor fraction at the outlet reduces to about 54% of that at the inlet. Out of polystyrene, glasswool, rock-wool, glass fiber, polyurethane foam, and aerogel sheet tested, the best perform-ance was shown by aerogel. A 4 mm thick aerogel sheet retained 95% of the vapors at the outlet. Other materials required much higher thicknesses of 160, 22, 18, 15, and 10 mm for polystyrene, glasswool, rockwool, glass fiber, and poly-urethane foam, respectively. The vapor fraction at the outlet could not cross beyond 88% when other materials were used [67].

Oak Ridge National Laboratory and Plasma Physics Laboratory in the United States jointly tested the aerogel insulation performance for the vacuum chamber in the "Stellarator," which is a device used in nuclear reactor to confine the hot plasma in a vacuum chamber using magnetic field in order to control the nuclear fusion reaction. The entire system is surrounded by the cryostat to per-mit operation of the magnetic coils at liquid nitrogen temperature. The role of insulation to the vacuum chamber becomes very important to reduce the heat loss and decrease the heavy load on cryostat. Various candidate insulation mate-rials were tested, which include perlite, glass beads, mineral/insulation pellets, and aerogel. Two types of commercially available aerogel products were tested, namely, Pyrogel ®, a product of Aspen Aerogel Inc., and the pourable insulation, Nanogel®, a granular product produced by Cabot Corp., under license to Aspen Aerogel Inc. All other materials were found to be inferior to aerogel. During operation, with the aerogel insulation, the load on heating of vacuum chamber reduced from 16.5 to 8.2 kW and the load on the cryostat was reduced from 11.3 to 3.1 kW. Similar observation was found during idle conditions also.

38.8
Challenges and Future Perspectives

As mentioned in Table 38.3, there are a number of products available on the market for various thermal insulation applications. However, they are limitedly exploited in commercial use due to its high cost and this is the biggest challenge in maximizing the use of aerogel products to achieve larger energy conservation. The higher cost is due to the production step – supercritical drying. This step needs a specialized high-pressure vessel to produce aerogels. There are no alter-nate drying processes found till now that can give the equivalent quality of aero-gel as formed by supercritical drying. This issue stands as the most important barrier to exploit its maximum advantage in energy conservation. An alternative cost-effective drying method to produce the best quality aerogels is a need of time! However, it is also true that the higher cost is compensated by the advantages of aerogel insulation in terms of extended life, energy saving, requirement of less material, and hence lowering transportation, storage, labor cost, and return on investment can be achieved much faster. There is an urge to popularize the aerogel insulation among all types of industries globally. Efforts by the existing market players are remarkable in this direction. In future, many more aerogel products should enter into the market with higher performance and functionalities, fulfilling the needs of various applications with customized solutions and higher volume of production, matching the increased demand for supply.

38.9
Safety and Hazard Measures

Silica aerogel is basically sand if compared by its chemical composition. Silica is an abundantly occurring natural substance and hence is eco-friendly and non-hazardous. However, in aerogel form, it may become dusty while handling, cutting, or crushing. Besides, the basic building unit of aerogel is a nanoparticle of a size about 5–10 nm. These are the reasons to lay down safety measures while using and handling the aerogel-based products. Excessive inhalation of aerogel dust may cause damage to respiratory track. The skin and eye contact may produce irritation and create dryness. The US OSHA standard for permeable exposure limit for synthetic amorphous silica is not available. Hence, the standards similar for silica (CAS Number 7631-86-9) are considered for aerogel safety guidelines. As per this, the permeable exposure limit for amorphous silica is 15 mg/m^3 (total dust) and 5 mg/m^3 (respirable fraction). Exposure to the aerogel dust may be controlled by following standard industrial hygiene practices.

38.10
Summary

Silica aerogel is the world's most efficient thermal insulation material that can sustain at cryo as well as at high temperatures. This is made by the conventional sol–gel method followed by supercritical drying. Along with thermal insulation, it possesses other properties, namely, sound and electrical insulation, huge surface area, ultralow density, low dielectric constant and refractive index, and higher transparency. Due to these extraordinary set of properties in single material, silica aerogel has found place in a variety of applications. The challenge of fragile nature was tackled by making them in the granular form or infiltering them in fiber mat to give a form of flexible sheet. This revolutionized its commercial use and now the manufacturing of aerogel products is carried out by some companies mainly in the United States, Europe, and Asia; the products are gaining popularity due to their superior thermal insulation performance. The use of silica aerogel flexible sheets has been established in industries such as oil and gas, power plants, refineries, and aerospace, and has shown potential in building insulation and textiles. The increasingly gaining attention toward this highly efficient thermal insulation material is a hope to significantly contribute to achieving targets of energy conservation and saving.

Acknowledgments

The authors would like to thank the host organization, ARCI, Hyderabad, India, for all the support provided for the research and development of aerogel products. The cost and energy saving data were received from various credible sources who have vast experience in the industrial insulation. Authors sincerely thank them for their valuable help in sharing data and their experiences about aerogel product performance. The authors sincerely thank Dr. Y.R. Mahajan for his encouragement and technical support throughout the research and development work of aerogel products. His help is gratefully acknowledged for critically reviewing the chapter and providing valuable suggestions.

References

1 Kistler, S.S. (1931) Coherent expanded aerogels and jellies. *Nature*, **127**, 741.
2 Kistler, S.S. (1932) Coherent expanded aerogels. *J. Phys. Chem.*, **36** (1), 52–64.
3 Tillotson, T.M. and Hrubesh, L.W. (1982) Transparent ultralow-density silica aerogels prepared by a two-step sol–gel process. *J. Non Cryst. Solids*, **145**, 44–50.
4 Lu, X., Caps, R., Fricke, J., Alviso, C.T., and Pekala, R.W. (1995) Correlation between structure and thermal conductivity aerogels of organic. *J. Non Cryst. Solids*, **188**, 226–234.
5 Gavillon, R. and Budtova, R. (2008) Aerocellulose: new highly porous cellulose prepared from cellulose–NaOH aqueous solutions. *Biomacromolecules*, **9**, 269–277.
6 Abecassis-Wolfovich, M., Rotter, H., Landau, M.V., Korin, E., Erenburg, A.I., Mogilyansky, D., and Gartstein, E. (2003) Texture and nanostructure of chromia aerogels prepared by urea-assisted homogeneous precipitation and low-temperature supercritical drying. *J. Non-Cryst. Sol.*, **318**, 95–111.
7 Long, J.W., Logan, M.S., Carpenter, E.E., and Rolison, D.R. (2004) Synthesis and characterization of Mn–FeO$_x$ aerogels

with magnetic properties. *J. Non Cryst. Solids*, **350**, 182–188.
8 Krompiec, S., Mrowiec-Biało, J., Skutil, K., Dukowicz, A., Pajazk, L., and Jarzezbski, A.B. (2003) Nickel–alumina composite aerogel catalysts with a high nickel load: a novel fast sol–gel synthesis procedure and screening of catalytic properties. *J. Non Cryst. Solids*, **315**, 297–303.
9 Huang, R., Hou, L., Zhou, B., Zhao, Q., and Ren, S. (2005) Formation and characterization of tin oxide aerogel derived from sol–gel process based on tetra(*n*-butoxy) tin(IV). *J. Non Cryst. Solids*, **351**, 23–28.
10 Tang, P.E., Sakamoto, J.S., Baudrin, E., and Dunn, B. (2004) V$_2$O$_5$ aerogel as a versatile host for metal ions. *J. Non Cryst. Solids*, **350**, 67–72.
11 Kalebaila, K.K., Georgiev, D.G., and Brock, S.L. (2006) Synthesis and characterization of germanium sulfide aerogels. *J. Non Cryst. Solids*, **352**, 232–240.
12 Mohanan, J.L. and Brock, S.L. (2004) A new addition to the aerogel community: unsupported CdS aerogels with tunable optical properties. *J. Non Cryst. Solids*, **350**, 1–8.

13 Saliger, R., Fischer, U., Herta, C., and Fricke, J. (1998) High surface area carbon aerogels for supercapacitors. *J. Non Cryst. Solids*, **225** (1), 81–85.

14 Gesser, H.D. and Goswami, P.C. (1989) Aerogels and related porous materials. *Chem. Rev*, **89**, 765–788.

15 Sun, H., Xu, Z., and Gao, C. (2013) Multifunctional, ultra-flyweight, synergistically assembled carbon aerogels. *Adv. Mater.* doi: 10.1002/adma/201204576

16 Hüsing, N. and Schubert, U. (1998) Ultra light and highly compressible graphene aerogels. *Angew. Chem., Int. Ed.*, **37**, 22–45.

17 Gustav, N., Maria, P.F.R., Sreenath, B., Marco, M., and Raffaele, M. (2015) Amyloid template gold aerogels. *Adv. Mater.* doi: 10.1002/adma.201503465

18 Hrubesh, L.W. (1998) Aerogel applications. *J. Non Cryst. Solids*, **225**, 335–342.

19 Fricke, J. and Tillotson, T. (1997) Aerogels: production, characterization and applications. *Thin Solid Films*, **297**, 212–223.

20 Pajonk, G.M. (1991) Silica aerogel. *Appl. Catal.*, **72** (2), 217–266.

21 Soleimani, D.A. and Abbasi, M.H. (2008) Silica aerogels: synthesis, properties and characterization. *J. Mater. Process. Technol.*, **199**, 10–26.

22 Lu, X., Caps, R., Fricke, J., Alviso, C.T., and Pekala, R.W. (1995) Correlation between structure and thermal conductivity of organic aerogels. *J. Non Cryst. Solids*, **188**, 226–234.

23 Zeng, S.Q., Hunt, A., and Greif, R. (1995) Transport properties of gas in silica aerogel. *J. Non Cryst. Solids*, **186**, 264–270.

24 Rettelbach, Th., Säuberlich, J., Korder, S., and Fricke, J.E. (1995) Thermal conductivity of silica aerogel powders at temperatures from 10 to 275K. *J. Non Cryst. Solids*, **186**, 278–284.

25 Gonzalez, F.J., Ashley, C.S., Clem, P.G., and Boreman, G.D. (2004) Antenna-coupled microbolometer arrays with aerogel thermal isolation. *Infrared Phys. Technol.*, **45**, 47–51.

26 Ackerman, W.C., Vlachos, M., Rouanet, S., and Fruendt, J. (2001) Use of surface treated aerogels derived from various silica precursors in translucent insulation panels. *J. Non Cryst. Solids*, **285**, 264–271.

27 Haranath, D., Pajonk, G.M., Wagh, P.M., and Rao, A.V. (1997) Effect of sol–gel processing parameters on thermal properties of silica aerogels. *Mater. Chem. Phys.*, **49**, 129–134.

28 Aegerter, M.A., Leventis, M., and Koebel, M.M. (eds) (1994) *Aerogels Handbook*, Springer, New York.

29 Egeberg, E.D. and Engell, J. (1989) Freeze-drying of silica gels from siliciumethoxide. *Rev. Phys. Appl.*, **24** (C4), 23–28.

30 Rabinovich, E.M., Johnson, D.W., MacChesney, J.B., and Vogel, V.M. (1983) Properties of high-silica glasses from colloidal gels: I. Preparation for sintering and properties of sintered glasses. *J. Am. Ceram. Soc.*, **66**, 683–688.

31 Arount, I. and David, J.P. (1998) Uniform macroporous ceramics and plastics by emulsion templating. *Adv. Mater.*, **10** (9), 697–700.

32 Pamela, J.D., Brinker, C.J., Smith, D.M., and Assink, R.A. (1992) Pore structure evolution in silica gel during aging/drying. *J. Non Cryst. Solids*, **142**, 197–207.

33 Hæreid, S., Ejnarsrud, M.A., and Sherer, G.W. (1994) Mechanical strengthening of TMOS-based aerogels by aging in silane solution. *J. Sol-Gel Sci. Technol.*, **3**, 199–204.

34 Deshpande, R., Smith, D.M., and Brinker, C.J. (1992) Preparation of low-density aerogels at ambient pressure. *Mater. Res. Soc. Symp. Proc.*, **271**, 567–572.

35 Neugebauer, A., Chen, K., Tang, A., Allgeier, A., Glicksmana, L.R., and Gibson, L.J. (2014) Thermal conductivity and characterization of compacted, granular silica aerogel. *Energy Build.*, **79**, 47–57.

36 Lombard, L.P., Ortiz, J., and Pout, C. (2008) A review on buildings energy consumption information. *Energy Build.*, **40**, 394–398.

37 Duer, K. and Svendsen, S. (1998) Monolithic silica aerogel in superinsulating glazings. *Sol. Energy*, **63**, 259–267.

38 Kim, G.S. and Hyun, S.F H. (2003) Synthesis of window glazing coated with silica aerogel films via ambient drying. *J. Non Cryst. Solids*, **320**, 125–132.

39 Buratti, C. and Moretti, E. (2012) Experimental performance evaluation of aerogel glazing systems. *Appl. Energy*, **97**, 430–437.

40 Ihara, T., Grynning, S., Gao, T., Gustavsen, A., and Jelle, B.P. (2015) Impact of convection on thermal performance of aerogel granulate glazing systems. *Energy Build.*, **88**, 165–173.

41 Berardi, U. (2015) The development of a monolithic aerogel glazed window for an energy retrofitting project. *Appl. Energy*, **154**, 603–615.

42 Schultz, J.M. and Jensen, K.I. (2008) Evacuated aerogel glazings. *Vacuum*, **82**, 723–729.

43 Zagorskas, J., Zavadskas, E.K., Turskis, Z., Burinskienė, M., Andra, B., and Dagnija, B. (2014) Thermal insulation alternatives of historic brick buildings in Baltic Sea Region. *Energy Build.*, **78**, 35–42.

44 Stahl, T., Brunner, S., Zimmermann, M., and Ghazi, W.K. (2012) Thermohygric properties of a newly developed aerogel based insulation plastering for both exterior and interior applications. *Energy Build.*, **44**, 114–117.

45 Ng, S., Jelle, B.P., Sandberg, L.I.C., Gao, T., and Wallevik, O.H. (2015) Experimental investigations of aerogel-incorporated ultra-high performance concrete. *Constr. Build. Mater.*, **77**, 307–316.

46 Buratti, C., Moretti, E., Belloni, E., and Agosti, F. (2014) Development of innovative aerogel based plasters: preliminary thermal and acoustic performance evaluation. *Sustainability*, **6**, 5839–5852.

47 Gibiat, V., Lefeuvre, O., Woignier, T., Pelous, J., and Phalippou, J. (1995) Acoustic properties and potential applications of silica aerogels. *J. Non Cryst. Solids*, **186**, 244–255.

48 Forest, L., Gibiat, V., and Hooley, A. (2001) Impedance matching and acoustic absorption in granular layers of silica aerogels. *J. Non Cryst. Solids*, **5**, 230–235.

49 Caponi, S., Fontana, A., Montagna, M., Pilla, O., Rossi, F., Terki, F., and Woignier, T. (2003) Acoustic attenuation in silica porous systems. *J. Non Cryst. Solids*, **322**, 29–34.

50 Schmidt, R. (2006) Types of insulation for refrigeration applications. Available at http://www.insulation.org/articles/article.cfm (December 13, 2015).

51 Fesmire, J. (2006) Aerogel insulation systems for space launch applications. *Cryogenics*, **46**, 111–117.

52 Johnson, W.L., Fesmire, J.E., and Demko, J.A. (2010) Analysis and testing of multilayer and aerogel insulation configurations. *Adv. Cryog. Eng.*, **1218**, 780–787.

53 Coffman, B.E., Fesmire, J.E., Augustynowicz, S.D., Gould, G., and White, S. (2010) Aerogel blanket insulation materials for cryogenic applications. *Adv. Cryog. Eng.*, **1218**, 913–920.

54 Koravos, J.J., Miller, T.M., Fesmire, J.E., and Coffman, B.E. (2010) Nanogel aerogel as a load bearing insulation material for cryogenic systems. *Adv. Cryog. Eng.*, **1218**, 921–927.

55 Smith, T.M., Williams, M.K., Fesmire, J.E., Sass, J.P., and Weiser, E.S. (2009) Polyimide-aerogel hybrid foam composites for advanced applications, in *Polyimides and Other High Temperature Polymers: Synthesis, Characterization and Applications*, vol. **5** (ed. K.L. Mittal), CRC Press, pp. 295–304.

56 Begag, R. and Fesmire, J.E. (2008) Nonflammable, hydrophobic aerogel composites for cryogenic applications. Thermal Conductivity 29, DEStech Publications, Lancaster, PN, pp. 323–333.

57 Fesmire, J.E. and Sass, J.P. (2008) Aerogel insulation applications for liquid hydrogen launch vehicle tanks. *Cryogenics*. doi: 10.1016/j.cryogenics.2008.03.014

58 Scholtens, B.E., Fesmire, J.E., Sass, J.P., and Augustynowicz, S.D. (2008) Cryogenic thermal performance testing of bulk-fill and aerogel insulation materials. *Adv. Cryog. Eng.*, **53**, 152–159.

59 Fesmire, J.E., Augustynowicz, S.D., and Rouanet, S. (2002) Aerogel beads as cryogenic thermal insulation system. *Adv. Cryog. Eng.*, **47**, 1541–1548.

60 Fesmire, J.E., Rouanet, S., and Ryu, J. (1998) Aerogel-based cryogenic superinsulation. *Adv. Cryog. Eng.*, **44**, 219–226.

61 Bheekhun, N., Talib, R.A., and Hassan, M. F R. (2013) Aerogels in aerospace: an overview. *Adv. Mater. Sci. Eng.*, **(2013)**, 406065.

62 Nuckols, M.L., Hyde, D.E., Wood-Putnam, J.L., Giblo, J., Caggiano, G.J., Henkener, J.A., and Stinton, B. (eds) (2009) Design and evaluation of cold water diving garments using super-insulating aerogel fabrics. Proceedings of the American Academy of Underwater Sciences 28th Symposium, March 13–14, 2009, Atlanta, GA, pp. 237–244.

63 Shaid, A., Wang, L., and Padhye, R. (2014) The thermal protection and comfort properties of aerogel and PCM-coated fabric for firefighter garment. *Chem. Mater. Eng.*, **2**, 37–43.

64 Zrim, K., Mekjavic, I.B., and Rijavec, T. (2015) Properties of laminated silica aerogel fibrous matting composites for footwear applications. *Text. Res. J.* doi: 10.1177/0040517515591781.

65 Venkataraman, M., Rajesh, M., Wienera, J., Militkya, J., Kotreshb, T.M., and Vaclavikc, M. (2015) Novel techniques to analyse thermal performance of aerogel-treated blankets under extreme temperatures. *J. Text. Inst.*, **106**, 736–747.

66 Sourena, S., Roghayeh, L., and Vahideh, B. (2008) Energy conservation opportunity by the use of aerogel nanomaterials in steam pipes. 3rd IASME/WSEAS International Conference on Energy & Environment, University of Cambridge, UK, February 23–25.

67 Goranson, P., Freudenberg, K., McGinnis, G., Dudek, L., and Zarnstorff, M. (2009) Application of high-performance aerogel insulating materials (analysis & test results). *Fusion Eng.* doi: 10.1109/FUSION.2009.5226454

39

Nanotechnology in Architecture

George Elvin

Ball State University, College of Architecture and Planning, Architecture Bldg, Muncie, IN 47306, USA

39.1
Nanotechnology and Green Building

39.1.1
Green Building

Buildings account for nearly 40% of global energy consumption. They are therefore a prime focus of our efforts to reduce energy use. Often referred to as green building, these efforts are beginning to have an impact. Green buildings, as defined by the Leadership in Energy and Environmental Design Council, use 32% less electricity. The American Solar Energy Society suggests that 40% of the energy savings required to achieve necessary CO_2 reductions could come from the building sector [1]. Better building insulation, lighting, and on-site energy storage are just a few of the opportunities to make the building industry the leader in CO_2 reduction and energy conservation.

But for the building industry to achieve its potential as the leader in sustainable development, new materials are urgently needed. Many current and emerging products incorporating nanoengineered materials can improve the environmental performance of buildings. With environmental standards for buildings growing more stringent every day, global demand for green buildings that use less energy and result in less waste and CO_2 emissions will only rise (Table 39.1).

39.1.2
The Role of Nanotechnology

By working at the molecular level, nanotechnology opens up new possibilities in material design. Because nanoparticles have unique mechanical, electrical, optical, and reactive properties distinct from larger particles, some nanomaterials can change color, shape, and phase much more easily than at the macroscale.

Nanotechnology for Energy Sustainability, First Edition. Edited by Baldev Raj, Marcel Van de Voorde, and Yashwant Mahajan.
© 2017 Wiley-VCH Verlag GmbH & Co. KGaA. Published 2017 by Wiley-VCH Verlag GmbH & Co. KGaA.

Table 39.1 Potential sources of EU CO_2 emission reductions.

Improvement	CO_2 reduction (tons/year)
Thermal insulation	174–196
Glazing standards	50
Lighting efficiency	50
Controls	26

Buildings have the potential to become leading sources of CO_2 reductions. (*Source:* CALEB Management Services, "Assessment of the potential savings of CO_2 emissions in European building stock.")

Fundamental properties such as strength, surface-to-mass ratio, conductivity, and elasticity can be designed in to create distinct materials with new performance capabilities.

But taking attributes such as strength and lightness to new extremes or combining them in new ways is only part of nanotechnology's impact. Another is the emergence of dynamic new "smart" materials. Some electrochromic windows, for example, exploit tungsten oxide molecules' ability to change reflectivity when stimulated with an electrical charge. With the flip of a switch, these windows can change from transparent to opaque. Self-healing concrete uses embedded nanoepoxy bubbles that rupture as cracks develop, enabling structural members to repair themselves from the inside out. Quantum dot lighting and other nano-enhanced lighting technologies could bring significant efficiency improvements and energy savings.

Demand for nanomaterials in the US construction industry in 2006 totaled less than $20 million [2]. Today, demand exceeds $1 billion per year. Hundreds of companies offer nanomaterials for green building – in paints, sealants, wallboard, solar panels, and more. Self-cleaning windows, smog-eating concrete, and toxin-sniffing nanosensors are all on the market. Some nanoengineered building products can improve the strength, durability, and versatility of building materials, reduce material toxicity and waste, and improve building insulation. Nanosensors allow us to monitor the performance of building materials and systems as well as environmental conditions and user activity, and the information they generate can help us to reduce building energy use. Recent advances in nanotechnology are also contributing to building energy conservation and on-site renewable energy, reducing dependence on nonrenewable resources.

Other products still in development offer even more promise for improving the environmental and energy performance of buildings. Photovoltaic windows, roofing, and façade panels now entering the market, for example, could turn future buildings into large-scale solar collectors. The development of photocatalytic coatings points to a future of self-cleaning buildings, and advances in nanoscale adhesives may make future building materials self-healing (Table 39.2).

In green building, the potential for reduced energy consumption, waste, toxicity, nonrenewable resource consumption, and carbon emissions through the architectural applications of nanotechnology is significant. These environmental

Table 39.2 Ranking of "environmentally beneficial" nanotechnologies.

Rank	Technology
1	Electricity storage
1	Engine efficiency
2	Hydrogen economy
3	Photovoltaics
3	Insulation
4	Thermovoltaics
4	Fuel cells
4	Lighting
6	Lightweighting
6	Agriculture pollution reduction
7	Drinking water purification
8	Environmental sensors
8	Remediation

Most "environmentally beneficial" nanotechnologies are well-suited to use in buildings. (*Source:* From Ref. [3].)

benefits will be led by current improvements in photovoltaics, insulation, coatings, air and water purification, followed by advances in lighting, and, eventually, breakthroughs in structural materials and adhesives. Thin-film organic solar cells are cutting the cost of solar energy and making solar cells more adaptable to a wide range of architectural applications. And nanocomposites, hybrid materials merging nanomaterials with conventional ones such as steel and concrete, will improve strength-to-weight ratios in building materials, meaning less material will be required for the same application.

The ability to design materials to have specific properties is transforming the very nature of material engineering. As BASF polymer researcher Franz Brandstetter said, "Now, instead of asking, 'What will this material do?' we can ask, What properties do we want?' " His comment suggests that tomorrow's architects will be able to define the performance criteria they are seeking and then have materials designed to meet them.

39.2
Energy

39.2.1
Energy Conservation

39.2.1.1 Insulation
Across all industries, the single most cost-effective strategy for reducing CO_2 emissions is better building insulation. Therefore, even a slight improvement in

insulation could greatly reduce CO_2 emissions and energy consumption world-wide. Home heating in the European Union, for instance, is responsible for nearly one quarter of EU carbon emissions. Better insulation in Germany, The Netherlands, Italy, the United Kingdom, Spain, and Ireland could reduce EU carbon emissions by 100 million metric tons per year [4]. In the United States, improved insulation could save 2.2 quadrillion Btu of energy (3% of total energy use) and reduce carbon emissions by 294 billion pounds annually [5].

Current building insulation is estimated to save roughly 12 quadrillion Btu annually or 42% of the energy that would be consumed without it [5]. Nanotechnology has already been employed to make some of the most efficient insulating materials ever known. Manufacturers estimate that current insulating materials derived from nanotechnology are approximately 30% more efficient than those made from conventional materials [6]. Nanoscale materials hold great promise as insulators because of their extremely high surface-to-volume ratio. This gives them the ability to trap still air within a material layer of minimal thickness. Insulating nanomaterials may be sandwiched between rigid panels, applied as thin films, or painted as coatings.

39.2.2
Aerogel

Aerogel is an ultralow density solid, a gel in which the liquid component has been replaced with gas. Nicknamed "frozen smoke," aerogel has a content of just 5% solid and 95% air, and is said to be the lightest solid in the world. The aerogels used in building insulation are silica aerogels with a cross-linked internal pattern of silicon dioxide chains creating a network of air-filled pores just 5–70 nm in diameter. This myriad of nanopores traps still air, giving aerogel its insulating ability. Silica aerogels are some of the most effective insulators known, roughly five times more effective than the polystyrene panels prevalent in building insulation today. Because nanoporous aerogels can be sensitive to moisture, however, they are often marketed sandwiched between wall panels that repel moisture. Aerogel panels are available with up to 75% translucency, and their low thermal conductivity means that a 9 cm thick aerogel panel can offer an R-value of R-28, a value previously unheard of in a translucent panel [7].

Architectural applications of aerogel include windows, skylights, translucent wall panels, and blanket insulating sheets. Spaceloft by Aspen Aerogels is a 10 mm thick opaque aerogel blanket with a thermal conductivity of 13.1 mW/(m K) at 273 K, resulting in an insulating value more than double that of traditional insulation materials. ZAE Bayern offers windows consisting of granular aerogel sandwiched between polymethylmethacrylate panels. These panels are separated from the two outermost glass panels by a gap filled with krypton or argon gas.

Because aerogel insulation is elaborated elsewhere in this book, it will not be covered in detail in this section. However, its contribution to building energy conservation cannot be understated. It has been hailed as "one of the most if not the most promising thermal insulation material of the last decades," and its contribution to building energy reduction is only expected to rise if its cost can be reduced [8].

39.2.3
Insulating Coatings

Insulating nanocoatings can also be applied as thin films to glass and fabrics. Insulating nanoparticles can be applied to substrates using chemical vapor deposition, dip, meniscus, spray, or plasma coating to create a layer bound to the base material. Insulation can also be painted or sprayed on in the form of a coating. This is a tremendous advantage over more conventional bulk insulators such as fiberglass, cellulose, and polystyrene boards, which often require the removal of building envelope components for installation. Because they trap air at the molecular level, insulating nanocoatings even a few thousands of an inch thick can have a significant effect. The manufacturer of one nanocoating used in the world's largest airport suggests that the average surface temperature difference when applied correctly is approximately 30 degrees Fahrenheit for three coats [9].

Some nanoinsulations use silica, titania, and carbon in a 3D, highly branched network of particles 2–20 nm in diameter to create a unique pore structure. In the form of vacuum insulation panels, these can have thermal resistance values as high as R-40/in., seven to eight times greater than conventional foam insulation materials, according to manufacturer claims. This performance improvement is due to the unique shape and small size of its large number of pores. Solid-phase conduction is low due to the material's low density and high surface area [10]. Nanoparticles with extreme insulating value can also be incorporated into conventional paints. Some of these use hollow ceramic spheres, with each containing a vacuum inside. These spheres create a thermal barrier by refracting, reflecting, and dissipating heat [11].

Insulating nanocoatings have great potential for reducing energy use in existing buildings. Currently, retrofitting existing buildings with wall insulation can be cost-prohibitive because it requires tearing open the walls, installing insulation boards, batts, or fill, and replacing the walls. Coatings, in contrast, can be simply and economically applied to wall surfaces. Of course, surface coatings are typically quite thin, which limits their insulating ability. However, as the efficiency of insulating nanocoatings continues to improve, so will their contribution to energy savings in existing buildings.

39.2.4
Thin-Film Insulation

Heat absorbing films can be applied to windows as well. Some of these incorporate a nanofilm "interlayer," which, according to manufacturers, offers more cost-effective control of heat than previously available laminating systems. Nanotech-based window films can also reduce heat and ultraviolet light penetration. Some can reject up to 97% of the sun's infrared light and up to 99.9% of UV rays. Infrared (IR) rays can also be blocked using transparent IR-absorbing coatings for heat-absorbing films for windows. Used in transparent film coatings, these coatings can improve solar absorption properties while maintaining optical transparency, according to manufacturers. The result is improved heat management and greatly reduced energy consumption by air conditioners.

39.2.5
Emerging Insulation Technologies

Aerogel insulation continues to advance, and new technologies promise to challenge their place as the world's best insulators. Research is underway to create, for example, nanotech-based spray-on aerogels and inkjet printer-based aerogels. NASA is currently developing X-Aerogels for use as cryogenic rocket fuel storage tanks and space suits. According to NASA engineer Nicholas Leventis, "The most striking feature of X-Aerogels is that for a nominal 3-fold increase in density (still an ultra-lightweight material), the mechanical strength can be up to 300 times higher than the strength of the underlying native aerogel." This means that X-Aerogels could, according to Leventis, "play the dual role of the thermal insulator and structural material." In buildings, where current aerogels must be protected by rigid enclosing materials, this raises the possibility of flexible, ultra-lightweight, superinsulating structural materials for integration into tomorrow's buildings [12].

Graphene is nature's thinnest elastic material, comprised of a one-atom thick sheet of carbon atoms. It has been found to have exceptional strength, thermal conductivity and electric conductivity. But graphene aerogel can be an effective insulator. In fact, the lightest material known is graphene aerogel that is seven times lighter than air. When graphene is organized in specific layers, its insulating properties can be turned on and off to create a dynamic insulation material (DIM). Although still in the concept stage of their development for architecture, nanoscientists are already developing DIMs in the electronics industry. When they are developed for architecture, DIMs could block heat loss from buildings in winter and block its entry in summer, reducing the amount of energy required to heat and cool buildings [13,14].

39.2.6
Future Market for Nanoinsulation

New nanotechnologies for insulation could lead to future buildings where insulation could double as structural members or, conversely, it could be made more flexible. In either case it could also be turned on and off. Nanotechnology could make insulation more efficient, less reliant on nonrenewable resources, and less toxic. If the performance of nanoinsulation products lives up to manufacturer claims, these products could foster significant improvements in energy savings and carbon reduction. However, independent testing of insulating nanomaterials and products in use will be necessary to verify manufacturer claims and convince potential buyers of their effectiveness. Some manufacturers are already making the results of such tests public, with encouraging results.

One of the greatest potential energy-saving characteristics of nanocoatings and thin films is their applicability to existing surfaces for improved insulation. They can be applied directly to the surfaces of existing buildings, whereas the addition of conventional insulating materials such as cellulose fiber, fiberglass batts, and rigid polystyrene boards to existing buildings typically requires expensive and invasive access to wall cavities and remodeling. Nanocoatings could also make it much easier to insulate solid-walled buildings, which make up approximately one-third of the UK's housing stock. And unlike cellulose fiber, fiberglass batts, and rigid polystyrene boards, nanocoatings can be made transparent. Their application to existing structures could lead to considerable energy savings, and they do not appear to raise some of the environmental and health concerns attributed to fiberglass and polystyrene.

39.2.7
On-Site Renewable Energy

39.2.7.1 Solar
Because building-integrated photovoltaics (BIPV) are covered elsewhere in this book, this section will focus on the architectural implications of their use. BIPV utilizing nanolayers or nanorods include thin-film materials, conducting polymeric films, new nanocrystalline materials, and silicon solar enhancement. Organic thin films, or plastic solar cells, use low-cost materials primarily based on nanoparticles and polymers. They are formed on inexpensive polymer substrates that can take advantage of the relatively inexpensive "roll-to-roll" production methods used in newspaper presses. The other dramatic advantage of organic thin films is their flexibility, which enables their integration into far more building applications than conventional flat glass panels. Architects often struggle to integrate rigid, flat glass solar panels into building facades and roofs in an esthetically pleasing way. Flexible thin-film panels integrated into building envelopes, however, could overcome these esthetic objections while also generating electricity from a larger surface area.

Many thin-film solar technologies rely on nonsilicon semiconductor materials such as copper, indium, gallium, and selenium (CIGS). BIPV systems using thin-film nanotechnologies were marketed by companies such as Nanosolar and Konarka (both no longer extant) over a decade ago. Nanosolar employed semiconductor quantum dots and other nanoparticles in their SolarPly BIPV panels to create large-area, solar–electric "carpet" for integration with commercial roofing membranes. SolarPly was utilized in a variety of building products because the cells were both nonfragile and bendable. Konarka manufactured light-activated "power plastic" that could be coated or printed onto a surface. Their photovoltaic fibers and durable plastics brought power-generating capabilities to structures including tents, awnings, roofs, windows, and window coverings [15].

Quantum dot technology could also play a role in solar's future. In silicon, one photon of light frees one electron from its atomic orbit. But researchers at the National Renewable Energy Laboratory have now demonstrated that quantum dots of lead selenide can produce up to seven electrons per photon when exposed to high-energy ultraviolet light. These dots would be far less costly to incorporate into solar cells than the large crystalline sheets of silicon used today. A photovoltaic device based on quantum dots could have an efficiency of 42%, far better than silicon's typical efficiency of 12% [16].

Currently, these technologies are less efficient than traditional crystalline silicon panels. However, they are less expensive, thanks to their low-temperature processing, use of plastic substrates, and low-cost "roll on" installation method. This combination of lower cost and lower efficiency makes them an attractive alternative when large surface areas are available for their deployment. This may be the case with the roofs of large buildings, but not when compared to large-scale installation on open land.

Nanotechnology is not only an alternative to silicon-based solar. It is also contributing significantly to today's silicon-based solar market. Innovalight, for example, developed a technology they said has the potential to greatly reduce the cost of silicon-based solar cells. The company, now owned by DuPont, has developed a silicon nanocrystalline ink that can make flexible solar panels as much as 10 times less expensive than current solutions. Their silicon process lends itself to low-cost, high-throughput manufacturing [17].

Solaicx designed and built a proprietary single-crystal silicon wafer production system for the silicon-based photovoltaic manufacturing industry. The company, now owned by MEMC, reported that their technology allows the manufacture of low-cost, high-quality single-crystal silicon ingots at high volume for conversion into solar wafers. Solaicx expects their process to be up to five times more productive than traditional methods. They also anticipate that silicon utilization will be greatly improved because they will be able to slice thinner silicon wafers of between 300 and 150 µm, allowing excess silicon to be recycled back into the manufacturing process [18].

SolarWindow is developing what they call a first-of-its-kind transparent glass window capable of generating electricity using silicon nanoparticles. While conventional photovoltaic solar cells lose about 50% of incident energy as heat, silicon nanocrystals can produce more than one electron from a single photon of sunlight, providing a way to convert some of the energy lost as heat into additional electricity [19].

One potential effect of improved solar cell efficiency will be to make buildings one of our primary sources of energy. Given their relatively large surface area and solar exposure, buildings are excellent platforms for solar arrays. These arrays will not be limited to the roofs of buildings. Already solar collecting windows and façade panels are available or in development. SolarWindow, for example, uses a liquid solar collecting coating over vision glass to combine the benefits of solar energy collection with see-through glass. When applied to windows on towers, the company estimates, these windows can generate up to 50 times the power of conventional rooftop solar systems. The Liberta solar façade system uses opaque black glass façade panels and copper indium gallium selenide collecting technology to gather energy from the sun. These commercially available panels can deliver 125 Watt/m^2. Solar windows and façade panels can be expected to grow in efficiency and availability, making future buildings an increasing source of clean power generation [20].

Organic thin-film solar cells, many of which incorporate nanotechnology, could open new architectural possibilities and overcome the esthetic concerns some architects hold against rigid flat panels, which can hardly be integrated into building facades. Thanks to their flexibility and thinness, thin films could be integrated into windows, roofs, and façades, potentially making almost the entire building envelope a solar collector. These advances will allow for greater integration of solar collecting hardware with architectural features. Solar panels will no longer be the ugly appendages that mar the rooflines of many buildings. Solar windows and façade panels, along with already more commercially available solar roof tiles and panels, in effect, hide their solar collecting technologies within more pleasing architectural elements. Other nanotechnology advances such as black silicon, nanoscale antenna arrays, solar collecting paint, and quantum dots could potentially improve solar cell efficiency by up to 40% over the next 10 years (Figure 39.1) [21].

39.2.8
Energy Storage

Improved energy storage can reduce our dependence on fossil fuels, lowering carbon dioxide emissions from energy production. Currently, energy for homes and offices is not stored on-site. Instead, it is delivered on an as-needed basis from power lines. However, the separation of energy source from its point of use, as when the energy in subterranean coal deposits must be converted and transported to coal-burning power plants, then transmitted along power lines to homes and offices, wastes most of the energy latent in the original fuel source. This inefficiency can be overcome by producing energy at the point of use, as in the case of building-integrated photovoltaics.

Altairnano is one of the most established companies using nanotechnology to develop new batteries. The company entered into the battery industry roughly 10 years ago when their material scientists identified novel ways to use nanoscale technologies to process lithium titanate oxide materials. They have since commercialized large format, nanolithium–titanate battery cells, which they claim offer key advantages over other lithium-ion battery technologies. The Smart Nanobattery by mPhase Technologies uses tubes to provide a "superhydrophobic nanostructured surface" topped by an electrolyte. When the electrolyte is made to fall between the tubes, a current is generated. According to the company, "Conventional batteries utilize nickel-cadmium, Ni-zinc, and lithium-ion batteries. While these are effective for normal day-to-day use, they are quite

Figure 39.1 Smart textile curtains in the Soft House by Kennedy & Violich Architecture utilize "organic photovoltaic nanotechnology." These energy-harvesting curtains, which respond to sun angle to block or collect the sun's rays, are capable of generating up to 16 000 Wh of electricity daily. (Courtesy of Kennedy & Violich Architecture.)

inefficient when used as high performance power sources. In addition, the electrolytes used in these batteries are toxic and produce significant amounts of battery waste that are hazardous to the environment."

Increasing energy input from building façades and roofs requires greater on-site energy storage capability in buildings. Environmentally harmful lead-acid batteries are still prevalent in on-site energy storage, but more efficient lithium-ion batteries are becoming more common. The promise of nanotechnology in improving on-site energy storage is well covered elsewhere in this book (see Part Two, Energy Storage and Distribution). In fact, one ranking of environment-friendly nanotechnologies put energy storage at the top of the list. Most environment-friendly nanotechnologies are well-suited to use in buildings. The architectural significance of their development will be to reduce the area required within buildings for energy storage, and potentially the quantities of toxic battery chemicals required [3].

39.2.9
Lighting

Lighting and appliances consume approximately one-third of the energy used in building operation. Not only do lighting fixtures consume electricity, but most produce heat, which can add to building cooling costs. Incandescent lights, for example, waste as much as 95% of their energy as heat. Fluorescent lights use less energy and produce less heat, but can contain trace amounts of mercury. Because of the heat generated by lighting, most office buildings run air conditioning when the outside air temperature is above 12 °C (55 °F). In fact, the cores of most buildings over $20\,000\,\text{ft}^2$ require cooling even during the winter heating season. Because of this effect, every 3 W of lighting energy conserved saves approximately additional 1 W of air cooling energy. The energy-saving potential in more efficient lighting is therefore significant [22].

One of the most promising technologies for energy conservation in lighting is light-emitting diodes (LEDs). In a global lighting market of $21 billion, the current market for high-brightness LEDs exceeds $4 billion. Current uses of LEDs include civil works such as traffic lights and signs, as well as some building applications such as the façade of the Galleria Shopping Mall in Seoul by the architecture firm, UN studio.

Some LEDs are projected to have a service life of roughly 100 000 h and offer the lowest long-term cost of operation available. Potential energy savings from LEDs are estimated at 82–93% over conventional incandescent and fluorescent lighting. LEDs could save 3.5 quadrillion BTUs of electricity and reduce global carbon emissions by 300 million tons per year, potentially cutting global lighting energy demand in half by 2025. The principal obstacle to greater adoption of LEDs, however, is cost; they currently cost at least 10 times more than fluorescent ceiling lights [23].

Among the most promising nanotechnologies for energy conservation in lighting are organic light-emitting diodes (OLEDs). When electricity is run through the strata of organic materials making up an OLED, atoms within them become excited and emit photons. OLEDs are highly efficient, long-lived natural light sources that can be integrated into extremely thin, flexible panels. Their introduction in the marketplace has so far been limited to small electronic components such as cell phone displays, but their applications continue to grow in scale. OLEDs offer unique features, including extreme flexibility, transparency when turned off, and tunability to produce variable colored light.

OLEDs could be used to create windows and skylights that mimic the look and feel of natural light after dark. And they could be applied to any surface, flat or curved, to make it a source of light. With this technology, walls, floors, ceilings, curtains, cabinets, and tables could become light sources. Carbon nanotube–organic composites could even lead to structural panels capable of integrating lighting. This multifunctional ability of surfaces integrating OLEDs could lead to energy savings not only because OLEDs are more efficient than today's lighting technologies but also because of efficient integrating of lighting into other building components. Scientists in Germany, for example, recently developed OLEDs that are transparent. Transparent OLEDs could be embedded into laminated glass, enabling windows to switch between transparent glazing and informational display panels, or act as both simultaneously.

Quantum dots are nanoscale semiconductor particles that can be tuned to brightly fluoresce at virtually any wavelength in the visible and infrared portions of the spectrum. They can be used to convert the wavelength, and therefore the color, of light emitted by LEDs. Quantum dot lighting results in brighter displays with less power than any other known light-emitting technology, according to one manufacturer. And while mercury and cadmium pose an environmental concern, the same manufacturer argues that the light fixture they offer "contains such a small amount of cadmium and is so power efficient that the technology can actually decrease the amount of cadmium in the environment through reduced emissions at power plants, as the burning of coal releases elemental cadmium into the air" [24].

As costs decline, experts anticipate that LEDs will take an increasing share of the task lighting market (for reading and other activities requiring bright, focused light), while OLEDs will be increasingly popular for ambient lighting (low-light conditions such as hotel lobbies and high-end restaurants). As the transition from conventional lighting to solid-state LEDs and OLEDs evolves, solid-state lamps will be made to fit existing incandescent and fluorescent fixtures. These advanced light fixtures will offer users the ability to change room color with the turn of a conventional dimmer switch, as is already possible with LED lighting in some hotels and night clubs. Mass commercialization of LEDs and OLEDs, however, will depend on improvements in their efficiency. Most current LEDs and OLEDs provide efficiency of roughly 30–160 lumens, whereas 1000 lumens will be required for their widespread adoption [25].

39.3
Air and Water

39.3.1
Air Purification

Americans spend up to 90% of their time indoors, and in 90% of US offices the number one complaint is lack of outdoor air. The US Environmental Protection Agency (EPA) estimates that poor indoor air quality results in $60 billion per year in medical expenses. But indoor air quality can be improved by using materials that emit few or no toxins and volatile organic compounds (VOCs) and resist moisture, thereby inhibiting the growth of mold. It can also be improved by adding systems, equipment, and products that identify indoor air pollutants or enhance air quality.

Nanotechnology is contributing to indoor air quality on all of these fronts. Nano-enhanced e-HEPA (for electric high-efficiency particulate arrest) filtration systems can sift the air to filter particles, eliminate undesirable odors, and kill airborne health threats. Some of these systems use a metal dust filter that has been coated with 8 nm silver particles and can kill 99.7% of influenza viruses. Up to 98% of odors may be eliminated, and other nanofilters can eliminate all noxious VOC fumes from paint, varnishes, and adhesives [26].

39.3.2
Depolluting Surfaces

Self-cleaning surfaces enabled by nanotechnology offer energy savings by reducing the energy consumed in cleaning building facades. They also reduce the runoff of environmentally hazardous cleansers. As surfaces self-clean, they are "depolluting," removing organic and inorganic air pollutants, such as nitrogen oxide from the air, and breaking them down into relatively benign elements. Depolluting nanocoatings show promise in cleansing indoor air and reducing instances of sick building syndrome (SBS). The World Health Organization estimates that up to 30% of new or renovated energy-efficient buildings may suffer from SBS [27]. The EPA estimates that SBS costs the US economy $60 billion per year in medical expenses, absenteeism, lost revenue, reduced productivity, and property damage [28].

One current drawback to photocatalytic coatings utilizing titanium dioxide, however, is that they require ultraviolet light for activation, reducing their effectiveness indoors. New research and development in "visible light photocatalysis" using nanoparticles other than titanium is bringing the depolluting advantages to building interiors. Coatings using layered double metal hydroxides, for example, can be applied to indoor surfaces to improve the indoor climate and reduce ventilation requirements, thereby improving the building's energy efficiency [29]. Nanoscientists at the Institute for Nanoscale Technology, Toshiba, and other organizations are developing visible light photocatalysts that can be activated by a standard lightbulb [30].

Outdoors, photocatalytic coatings such as the ones used in the Jubilee Church in Rome suggest the possibility of smog-eating facades, roads, and bridges for reducing outdoor air pollution. Italcementi, Skanska, and other global materials and construction companies have developed catalytic cement and concrete products coated with depolluting titanium dioxide [31]. These products have proven effective in reducing nitric oxide (NO_x), a major pollutant generated by auto exhaust. Tests of this technology in The Netherlands found that the NO_x levels were reduced by 25–45% [32].

39.3.3
Water Conservation and Purification

Water is the source of all life on Earth, and yet 1.3 billion people do not have access to safe drinking water. Furthermore, water is implicated in 80% of all sickness and diseases according to the World Health Organization. And less than 1% of the world's drinking water is actually fit for drinking. Water contaminants can include metals such as cadmium, copper, lead, mercury, nickel, zinc, chromium, and aluminum; nutrients including phosphate, ammonium, nitrate, nitrite, phosphorus, and nitrogen; and biological elements such as bacteria, viruses, parasites, and biological agents from weapons. UV light is an effective purifier, but is energy-intensive, and application in large-scale systems is sometimes considered cost-prohibitive. Chlorine, also commonly used in water purification, is undesirable because it is one of the world's most energy-intensive industrial processes, consuming about 1% of the world's total electricity output in its production [33].

Nanotechnology is opening new doors to water decontamination, purification, and desalinization, and providing improved detection of waterborne harmful substances. "We envision that nanomaterials will become critical components of industrial and public water purification systems," said Dr. Mamadou Diallo, Director of Molecular Environmental Technology at the California Institute of Technology, recipient of an EPA grant for nanotechnology research [34]. For example, iron nanoparticles have a high surface area and reactivity, and can be used to detoxify carcinogenic chlorinated hydrocarbons in groundwater. They can also render heavy metals such as lead and mercury insoluble, reducing their contamination. Dendrimers, with their sponge-like molecular structure, can clean up heavy metals by trapping metal ions in their pores. Nanoscale filters have a charged membrane, enabling them to treat both metallic and organic contaminant ions via both steric filtration based on the size of openings and Donnan filtration based on electrical charge. They can also be self-cleaning [35].

Gold nanoparticles coated with palladium have proven to be 2200 times better than palladium alone for removing trichloroethylene from groundwater. In addition, photocatalytic nanomaterials enable ultraviolet light to destroy pesticides, industrial solvents, and germs. Titanium dioxide, for example, can be used to decontaminate bacteria-ridden water. When exposed to light, it breaks down bacterial cell membranes, killing bacteria such as *Escherichia coli* [36]. Purification and filtration of water can also be achieved through nanoscale membranes or using nanoscale polymer "brushes" coated with molecules that can capture and remove poisonous metals, proteins, and germs.

39.4
Materials

Material strength is critical in a building, defining its structure, longevity, and resistance to gravity, wind, earthquake, and other loads that act to tear it down. Strength is equally important in nonstructural components, such as windows and doors, for security and durability. A load-bearing structural material's strength/weight ratio is particularly important because stronger, lighter materials can carry greater loads per unit of material. A higher strength/weight ratio means fewer

mode	3/4/2016	HV	spot	WD	HFW		40 μm
SE	2:02:51 AM	2.00 kV	2.0	19.4 mm	99.5 μm		Caltech Greer Group SEMentor

Figure 39.2 This 3D Kagome nanolattice developed by the Greer Group at the California Institute of Technology has demonstrated novel fracture toughness and crack-blunting mechanisms. With structural members just 300 nm wide, such materials could provide the building blocks for a new breed of high-performance building materials. (Scanning electron microscope image by Lucas Meza. Courtesy of Julia Greer, California Institute of Technology.)

materials, which in turn means less resources and energy consumed in production. Nanotechnology promises significant improvements in structural materials in two ways. First, nano-reinforcement of existing materials such as concrete and steel will lead to nanocomposites, materials produced by adding nanoparticles to a bulk material in order to improve the bulk material's properties. Eventually, when cost and technical know-how permit, we will see structures made from altogether new materials such as carbon nanotubes (Figure 39.2).

39.4.1
Concrete

Concrete is the world's most widely used manufactured material; about 1 ton of concrete is produced each year for every human being in the world. Global annual trade in concrete is estimated at \$13–14 trillion. Energy consumption, carbon emissions, and waste are all major environmental concerns connected with concrete production and use. Portland cement, the dry powder "glue" that holds aggregate, water, and lime together to make concrete, accounts for roughly 12% of concrete's volume, but 92% of its energy demand. For every ton of cement produced, 1.3 tons of CO_2 is released into the atmosphere. Worldwide, cement production generates over 1.6 billion tons of carbon, more than 8% of total carbon emissions. Waste is also considerable, as concrete accounts for more than two-thirds of construction and demolition waste with only 5% currently recycled [37].

Nanotechnology shows promise in addressing several of the environmental concerns associated with concrete. The addition of carbon nanotubes, silver dioxide, and ferrous oxide (Fe_2O_3) can be added to concrete during its mixing to improve compressive strength and other mechanical properties. Titanium dioxide nanoparticles can give concrete the ability to absorb carbon dioxide and air pollutants. Nanosensors can also be added to monitor acidity and chloride ions, the principal causes of deterioration and failure through corrosion. Similarly, nano-engineering of conventional concrete ingredients can improve performance [38].

However, in part because the bulk synthesis of nanoparticles such as carbon nanotubes is still too expensive for widespread use and concrete is such a high-volume material, few commercial products incorporate them. One is a concrete repair mortar with improved bond strength, tensile strength, density, and impermeability, as well as reduced shrinkage and cracking, according to its manufacturer. Others, described below, are in development [39].

Nanotechnology is leading to new low-energy cements and stronger ones through improved particle packing. The addition of nanoparticles can also improve concrete's durability through physical and chemical interactions such as pore filling. Conventional concrete must be reinforced with steel to resist tension loads, and placing steel "rebar" in forms prior to the introduction of wet concrete is a time-consuming and expensive process. Nanofiber reinforcement, including the introduction of carbon nanotubes, has been shown to improve the strength of concrete significantly. Even simply grinding Portland cement into nanoscale particles has been shown to increase compressive strength four-fold [40]. Carbon nanofibers, according to researchers at Vanderbilt University, could one day be added to concrete bridges, heating them during winter or allowing them to self-monitor for cracks because of the fibers' ability to conduct electricity (Table 39.3) [41].

Table 39.3 Nanobinders can double concrete's compressive strength.

Compressive strength (mpa)	
w/nanobinders	91.7
w/o nanobinders	45.2

The compressive strength of concrete with (top) and without (bottom) nanobinders after curing for 28 days. (*Source:* From Ref. [42].)

Carbon nanotubes also have the potential to effectively hinder crack propagation in cement composites. Reinforcing concrete with nanofibers will produce tougher concretes by interrupting crack formation as soon as it is initiated. Development of low-energy cements will also contribute to increased use of supplementary cementing materials such as fly ash and slag while making concrete production more environmentally sustainable [43]. Self-healing concrete is another possibility, made possible by the addition of a microencapsulated healing agent and a catalytic chemical trigger. When cracks form in the concrete, the microcapsules rupture, releasing the healing agent, which solidifies on contact with the chemical trigger also contained in the concrete [44].

Nanobinders are another area of promise for more sustainable concrete. "Development of nano-binders can lead to more than 50% reduction of the cement consumption," report Konstantin Sobolev and Miguel Ferrada-Gutiérrez, "capable to offset the demands for future development and, at the same time, combat global warming." The results of their experiments studying the mechanical properties of cement-based materials with nano-SiO_2, TiO_2, and Fe_2O_2 demonstrated an increase in compressive and flexural strength of mortars containing nanoparticles [42]. Adding nano-SiO_2, or nanosilica, to concrete also promises benefits. It can, for example, improve concrete's mechanical properties by creating denser particle packing of the micro and nanostructure. Nanosilica can also improve durability by reducing calcium leaching in water and blocking water penetration. It can even allow for more fly ash to be added to the concrete without sacrificing strength and curing speed, which can improve concrete durability and strength while reducing the overall volume of cement required.

Nanoparticulate titanium dioxide can also improve the environmental performance of concrete. It can, for example, be added to cement to enhance sterilization since it breaks down organic pollutants, volatile organic compounds, and bacterial membranes through powerful catalytic reactions. It can even reduce airborne pollutants when applied to outdoor surfaces. Additionally, it is hydrophilic, giving self-cleaning properties to surfaces to which it is applied. Carbon nanotubes are also likely to play an important role in the future of concrete. Adding small amounts of carbon nanotubes can improve compressive and flexural strength compared to unreinforced concrete. The high defect concentration on the surface of the oxidized multiwalled carbon nanotubes could also create better linkage between nanostructures and binders, thereby improving the mechanical properties of the composite. Improved linkage is important because of carbon nanotubes' propensity for clumping together and the lack of cohesion between them and the surrounding bulk material. Cost is the other great obstacle, as they can cost as much as $200 000/lb. But considerable industry, government, and academic resources are being devoted to reducing their cost, which will continue to drop, making carbon nanotube composites more cost-effective.

Experimentation is also underway on self-healing concrete. When self-healing concrete cracks, embedded microcapsules rupture and release a healing agent into the damaged region through capillary action. The released healing agent contacts an embedded catalyst, polymerizing to bond the crack face closed. In fracture tests, self-healed composites recovered as much as 75% of their original strength. They could increase the life of structural components by as much as two or three times [45].

39.4.2
Steel

Steel is a major component in reinforced concrete construction as well as a primary construction material in its own right. Light gauge steel framing for residential buildings is the fastest-growing use of steel. The United States consumes approximately 130 million tons of steel per year, and more than half of annual spending for steel is on residential framing. Adding nanoparticles of copper, magnesium, or calcium to steel during processing can reduce cracking and increase strength and corrosion resistance.

Several forms of steel using nanoscale processes are available today. A brand of steel-reinforcing bar for concrete construction, for example, is now marketed as MMFX steel. MMFX steel is, according to its manufacturer, five times more corrosion-resistant and up to three times stronger than conventional steel. MMFX steel products are used in structures across North America: bridges, highways, parking structures, and residential and commercial buildings. The added strength of MMFX steel results in a decrease in the amount of conventional steel required [46]. Steel produced using MMFXs technology has a unique laminated lath structure resembling plywood, which limits the formation of microgalvanic cells, the primary corrosion initiator driving the corrosion reaction. MMFXs "plywood" effect reportedly makes the steel very strong and increases corrosion resistance, ductility, and toughness.

Another steel product employing nanotechnology, although not yet available in structural dimensions, is Sandvik Nanoflex. This material offers a high modulus of elasticity combined with extreme strength, resulting in thinner and lighter components than those made from aluminum and titanium. Sandvik Nanoflex was first used in medical equipment such as surgical needles and dental tools. It has since been used in larger scale applications including ice axes. The strength and surface properties of Sandvik Nanoflex are also used in the automotive industry, replacing hard-chromed low-alloy steels, thereby eliminating the need for hard-chromizing processes [47].

39.4.3
Wood

While concrete is the most consumed construction material by weight, on a volume basis, wood is the most used construction material in the United States. Nearly 2 million housing units are constructed of wood in the United States every year. Wood can be attractive from an environmental standpoint because it is renewable and can be readily recycled and reused. Wood performance can be improved by the introduction of nanoparticles or nanosensors, or by manipulating cell structure at the nanoscale. And because wood is made up of nanotubes, or lignocellulosic fibrils, which are twice as strong as steel, the possibility of a naturally grown, easily recycled material stronger than steel is not out of the question. Nanosensors could relay information about material performance and environmental conditions such as structural loads, temperatures, moisture content, and decay [38].

Nanotechnology promises to improve the structural performance and serviceability of wood by giving scientists control over fiber-to-fiber bonding at a microscopic level and nanofibrillar bonding at the nanoscale. It could also reduce or eliminate the formation of the random defects that limit the performance of wood today [38]. Some experts foresee nanotechnology as "a cornerstone for advancing the biomass-based renewable, sustainable economy." Nanocatalysts that induce chemical reactions and make wood more multifunctional than it is today, nanosensors to identify mold, decay, and termites, quantum dot fiber tagging, natural nanoparticle pesticides and repellents, self-cleaning wood surfaces, and photocatalytic degradation of pollutants are all envisioned by today's wood engineers [48].

One of the great problems facing wood construction is rot. Pressure-treating wood can delay the problem, but the metallic salts sometimes employed can pose a health and environmental hazard. Safer organic insecticides and fungicides, however, are often insoluble, making it difficult for them to permeate the lumber. Scientists at Michigan Technological University's School of Forest Resources and Environmental Science have discovered a way to embed organic compounds in nanoscale plastic beads. The beads can permeate wood fibers because of their tiny size. This technology could allow the industry to use more environment-friendly biocides [49].

39.4.4
Glass

Reducing heat loss and heat gain through windows is critical to reducing energy consumption in buildings. Energy lost through residential and commercial windows costs US consumers roughly $25 billion per year. Nanotechnology is reducing heat loss and heat gain through glazing, thanks to thin-film coatings and thermochromic, photochromic, and electrochromic technologies. Electrochromic windows, which change opacity when an electrical current is passed through them, are composed of several layers of nanomaterials. Some combine thin metallic coatings of nickel or tungsten oxide bound between two transparent electrical conductors. Passing a voltage between these conductors generates a distributed electrical field, which causes coloration ions to move from the ion storage film and into the electrochromic film. The result is a significant change in window opacity, which can be used to control indoor heat loss and gain. Using a dimmer switch, a user can select any level of opacity within the range possible for any given window. This allows for clear vision glass at night or when the window is not in direct sun, and a much darker surface when in direct sun, which could add unwanted heat to the building [38].

Thin-film coatings are spectrally sensitive surface applications for window glass. They filter out unwanted infrared light to reduce heat gain in buildings. Thermochromic technologies are being studied, which react to changes in temperature and provide thermal insulation to give protection from heating while maintaining adequate lighting. Photochromic technologies react to changes in light intensity by increasing their light absorption. All of these applications are intended to reduce energy use in cooling buildings and can help bring down energy consumption in buildings.

39.4.5
Plastics

Vinyl (polyvinyl chloride (PVC)), which is used in a wide range of building materials, has come under fire recently as detrimental to human health. Phthalates, used to make PVC flexible, have been cited as bronchial irritants and potential asthma triggers. In addition, PVC production is the world's largest consumer of chlorine gas, using about 16 million tons of chlorine per year worldwide [50]. The polymer matrices that make up plastics can be modified through the introduction of nanoparticles or nanoscale modification of current components. For example, using a spray pyrolysis process, glass microspheres or microballoons can be cast in a polymer matrix, increasing compressive strength while also reducing density. Nanocomposite plastics can also be formed, such as nano-reinforced polyester. This material has demonstrated excellent thermal and electrical insulation qualities while remaining strong and lightweight. It is corrosion resistant, has a high fatigue limit, good impact strength, and fine surface finish. Tests show it could be used as a load-bearing structural material in bridges, doors, windows, facades, and structural systems [44].

The plastics common in buildings are typically so flammable that they require the addition of flame-retardant chemicals, many of which come with their own health and environmental concerns. The state of Washington has even banned one class of flame-retardants from use in household items. Scientists from the University of Massachusetts Amherst created a synthetic polymer that requires no flame-retardants because it simply will not burn. Their polymer used bishy-droxydeoxybenzoin as a building block, which releases water vapor instead of hazardous gases when it burns. The synthetic polymer is clear, flexible, durable, and much cheaper to make than high-temperature, heat-resistant plastics in current use, which tend to be brittle and dark in color [51].

Elsewhere, a team of University of Virginia researchers used carbon nanotubes to unite the virtues of plastics and metals in a new ultra-lightweight, conductive material. This nanocomposite material is a mixture of plastic, carbon nanotubes, and a foaming agent, making it extremely lightweight, corrosion-proof, and cheaper to produce than metal. Their experiments revealed that while the nanotubes make up only 1–2% of the nanocomposite, they increase its electrical conductivity by 10 orders of magnitude. The addition of carbon nanotubes also increased the material's thermal conductivity, improving its capacity to dissipate heat (Figure 39.3) [52].

Figure 39.3 Kimsooja's *A Needle Woman: Galaxy Was a Memory, Earth Is a Souvenir* shimmers with an ever-changing iridescence. The effect is created by acrylic panels treated with an iridescent nanopolymer developed at Cornell University, where the 46 ft. tall sculpture is located. (Photo by Aaron Wax. Courtesy of Cornell Council for the Arts, Ithaca, New York, and Kimsooja Studio. Kimsooja, *A Needle Woman: Galaxy Was a Memory, Earth Is a Souvenir*, 2014, steel, custom acrylic panel, laminated polymer film, circular mirror, 46 ft. × 4.5 ft. (diameter) with collaboration from Jaeho Chong (architect), Professor Ulrich Wiesner (materials science and engineering), Commissioned by Cornell Council for the Arts, Ithaca, New York.)

39.4.6
Gypsum

The average new American home contains more than 7 metric tons of gypsum, making gypsum one of the most prevalent materials in construction today. North America alone produces 40 billion ft^2 of gypsum board (drywall) per year. But drywall raises many environmental issues. Panels must be dried at 260 °C (500 °F), making their processing energy consumption a concern. Drywall also consumes 100 million metric tons of calcium sulfate, a nonrenewable resource, per year. Synthetic gypsum avoids this problem, but its processing by flue gas desulfurization releases mercury [53]. Waste is yet another concern, since as much as 17% of all drywall is lost during manufacturing and installation. Finally, drywall can be a breeding ground for Stachybotrys and other harmful molds [54].

Nanotechnology shows promise in the manufacture of lighter yet stronger drywall. UltraLight Panels by Sheetrock, for instance, are reportedly 30% lighter than conventional gypsum boards. The company reportedly employed nanotechnology to "change the way gypsum crystals bond to one another, creating stronger bonds that made it possible to reduce the amount of gypsum in each board without compromising strength" [55]. Laboratory experiments elsewhere on nanosized gypsum confirm up to three times higher hardness of nanogypsum as compared to conventional micron-sized gypsum [56]. Nanotechnology is also the key to so-called air-cleaning gypsum board systems. AirRenew gypsum board, according to manufacturer, CertainTeed, "actively removes formaldehyde from the air, converting it into a safe, inert compound." [57]

39.4.7
Roofing

Asphalt shingles make up more than 80% of the $30 billion US roofing market. Heating their asphalt binders, however, can pose health hazards and the release of hazardous air pollutants [58]. In addition, many asphalt shingles are reinforced with fiberglass, which has its own environmental and health hazards. Adding nanoparticles such as nanoclay, nanosilica, and nanotubes can improve asphalt durability, which would be a major benefit because of the harsh environmental conditions endured by asphalt roofs [59]. Greater durability could lead to reduced asphalt production and, in turn, fewer negative environmental side effects. Outside of asphalt, nanotechnology is beginning to make an impact on roofing. Manufacturers offer what they refer to as self-cleaning clay roof tiles. The tile's burned-in surface finish destroys dirt particles, grease deposits, soot, moss, and algae with the aid of sunlight [60]. The integration of photovoltaics into roof tiles and shingles is another possibility. According to researchers at the University of Toronto, nanoengineered colloidal quantum dots "could be mixed into inks and painted or printed onto thin, flexible surfaces, such as roofing shingles, dramatically lowering the cost and accessibility of solar power for millions of people" [61].

Nanotechnology could also assist in bringing sunlight into buildings through the roof. Cabot Corporation, for example, has a supply and marketing agreement with Centerpoint Translucent Systems for the use of Nanogel translucent aerogel in energy-efficient daylighting roofing systems. The Nanogel daylighting material combines high light transmission with energy efficiency and sound insulation. It can be incorporated into polycarbonate panels made specifically for translucent roofing applications. The combined panel provides more than five times the energy efficiency of glass panels typically used in residential sloped glazing. Centerpoint's roofing structure is engineered to allow penetration of natural, filtered daylight into home living areas without the energy loss and increased heating and cooling costs associated with traditional glass roof inserts [62].

39.4.8
New Structural Materials

While the introduction of nanomaterials into building structural components has begun with the reinforcement of conventional materials such as wood, concrete, and steel, breakthrough materials made primarily from nanomaterials are changing smaller scale products, including sporting equipment, and will eventually scale up to impact the building industry. Nanotubes, nanofibers, and nanosheets of carbon and similar materials may eventually form the structural skeletons of new buildings. For example, researchers at the University of Texas at Dallas together with an Australian colleague have produced transparent carbon nanotube sheets that are stronger than the same-weight steel sheets. These can be made so thin that a square kilometer nanotube sheet would weigh only 30 kg [63]. The prospect of transparent sheet materials stronger than steel not only holds tremendous energy-saving potential, it also promises to dramatically transform conventional assumptions about the relationship between building structure and skin. Could, for example, a superthin nanotube sheet serve as both skin and structure, eliminating the need for conventional structural systems altogether?

The prospect of transparent sheet materials stronger than steel and highly conductive of heat and electricity vividly illustrates one of the key energy-saving attributes of emerging nanomaterials – their versatility. For example, nanotubes can be used as electrodes for bright organic light-emitting diodes (OLEDs). They can be lighter, more energy-efficient, and allow for a more uniform level of brightness than current cathode ray tube and liquid crystal display technology. They could be used to illuminate surfaces in buildings that also serve to support the structure.

Nanomaterials and nanoreinforcement of existing materials could greatly extend the durability and lifespan of building materials, resulting in reduced maintenance and replacement costs as well as energy conservation. Researchers at the University of Bayreuth, for instance, recently developed aggregated diamond nanorods that have replaced natural diamonds as the world's hardest substance. While they may never be used structurally, these materials together with similar ones such as carbon nanotubes suggest that their use as reinforcing and

Figure 39.4 Developments in nanotechnology could eventually make possible futuristic buildings like those envisioned by designer Mihai Potra. (Concept Structure, designed and rendered by Mihai Potra.)

eventually stand-alone materials may dramatically extend the lifespan, durability, strength, and sustainability of many building materials (Figure 39.4) [64].

39.5
Nanosensors

While nanomaterials are bringing performance improvements to building materials, nanotechnology's most dramatic impact may come in the area of nanosensors. Nanosensors embedded in building materials will gather data on the environment, building users, and material performance, and even interact with users and other sensors until buildings become networks of intelligent, interacting components. Through the use of nanosensors, building components will initially become smarter, gathering data on temperature, humidity, vibration, stress, decay, and a host of other factors. This information will be invaluable in monitoring and improving building maintenance and safety. Dramatic improvements in energy conservation can be expected as well, as when environmental control systems recognize patterns of building occupancy and adjust heating and cooling accordingly. Similarly, windows will self-adjust to reflect or let pass solar radiation. Eventually, networks of embedded sensors will interact with those worn or implanted in building users, resulting in "smart environments" that self-adjust to individual needs and preferences. Everything from room temperature to wall color could be determined based on invisible, passive communication between sensors.

Work on smart environments is already underway. Leeds NanoManufacturing Institute (NMI), for example, was part of a €9.5 million EU-funded project to develop a house designed with special walls to contain wireless, batteryless sensors and radio frequency identity tags to collect data on stresses, vibrations, temperature, humidity, and gas levels. "If there are any problems, the intelligent sensor network will alert residents straightaway so they have time to escape," said NMI chief executive Professor Terry Wilkins. The self-healing house walls were built from novel load bearing steel frames and high-strength gypsum board, and contained nanopolymer particles that turn into a liquid when squeezed under pressure, flow into the cracks to harden, and form a solid material [65].

Nanosensors could also contribute to improvements in building materials. Concrete, for example, is attacked by carbon dioxide and chloride ions, resulting in corrosion and separation of reinforcing steel. Chinese researchers have created sensors that monitor reinforced concrete for acidity and chloride ions. These sensors can be embedded directly into the concrete mix to enable monitoring in place throughout the life of a structure [66]. Nanosensors can also be integrated directly into concrete to collect performance data on concrete density and viscosity, curing and shrinkage, temperature, moisture, chlorine concentration, pH, carbon dioxide, stresses, reinforcement, corrosion, and vibration. They could even monitor external conditions such as seismic activity, building loads, and, in roadways, traffic volume and road conditions. The latter are examples of "smart aggregates," in which micro-electromechanical devices are cast directly into concrete roadways. Valuable information from these sensors can be gathered by monitoring vehicles or monitored wirelessly [67].

39.6
Environmental and Health Concerns

Buildings will be one of the primary points of contact between people and nanomaterials. People know they will be in constant contact with materials and products allowed into their homes and offices. Concerns surrounding the widespread use of nanomaterials in buildings include the disruption of cell wall integrity associated with some single-walled nanotubes, nucleic acid damage relating to multiple-walled nanotubes, oxidative stress induced by some titanium dioxide, release of toxic heavy metals or other components, and direct oxidation upon contact with cell constituents. Inhalation of TiO_2 particles has resulted in lung inflammation, and SiO_2 nanoparticles have been reported as human carcinogens [38].

The pervasiveness, uncertainty, complexity, and rapid development of nanotechnologies for building combine to create a potentially volatile environment. Some environmental groups, for example, have warned that nanotechnology could prove to be "the next asbestos," a reminder of the grave health consequences wrought by a once-promising building technology. Because nanotechnology is a new and powerful technology full of uncertainty, care should be taken to listen to the concerns expressed by consumers, workers, and building users.

The uncertainty surrounding the effects of nanoparticles on the environment and the human body is sure to continue as a concern in the development from experimental nanoscience to marketplace products. Reports find, for example, that ultrafine particles behave differently and can be more toxic than equivalent larger sized particles of a given material at similar doses per gram of body weight [68,69]. Regulation of nano-based products based solely on particle size, however, is proving extremely difficult. Consumers of nanotechnology's architectural applications will undoubtedly be concerned about potential environmental and human health hazards, and the fear of them, whether justified or not, could impede the spread of nanotechnology in the marketplace.

Like any new technology, nanotechnology raises concerns. By virtue of their size, for example, nanoparticles are more readily absorbed into the body than larger particles. In addition, little is known about how they accumulate in the body or the environment. Silver nanoparticles, for instance, are proven antibacterial agents incorporated into many nano-enhanced paints and coatings. Samsung even coats some of its appliances with silver nanoparticles to kill germs [70]. But concerns that nanosilver could accumulate in the environment, killing beneficial bacteria and aquatic organisms, as well as human health concerns, have led the EPA to make products containing silver nanoparticles the subject of the first EPA regulations applying to nanotechnology. Now, any company wishing to sell products advertised as germ-killing and containing nanosilver or similar nanoparticles will first have to provide scientific evidence that the product does not pose an environmental risk. But the EPA has long regulated silver because it is a heavy metal known to cause health and environmental problems in sufficient quantities [71].

Because of the large number of people employed in the construction industry, workplace regulation of nanoengineered materials and processes could also become a concern. The harmful side effects of carbon nanotube manufacturing, for example, have been described in one study. Researchers found cancer-causing compounds, air pollutants, toxic hydrocarbons, and other substances of concern. They are now working with four major US nanotube producers to help develop strategies for more environment-friendly production [72]. At present, however, the National Institute for Occupational Safety and Health only offers guidelines for workplace safety for workers in contact with nanomaterials [73]. Since buildings are the primary source of contact between people and materials through both dermal and respiratory absorption, architects and engineers along with manufacturers will need to stay attuned to regulations affecting nanotechnology. Responsible research, development, testing, and regulation will be key, as will consumer and architect education.

References

1 Kutscher, C.F. (ed.) (2007) *Tackling Climate Change in the U.S*, American Solar Energy Society, Boulder, CO.

2 The Freedonia Group (2007) Nanotechnology in construction forecasts to 2011, 2016 & 2025, Study #218, Cleveland, OH, May.

3 Oakdene Hollins (2007) Environmentally beneficial nanotechnologies. Available at http://www.oakdenehollins.co.uk/pdf/ 098_Environmentally_beneficial_ nanotechnologies.pdf (accessed April 11, 2016).

4 European Insulation Manufacturers Association (2011) Tackling climate

change. Available at http://www.eurima. org/energy-efficiency-in-buildings/ tackling-climate-change.html (accessed March 30, 2016).

5 Energy Conservation Management, Inc. *et al.* (2016) Green and competitive: the energy, environmental, and economic benefits of fiber glass and mineral wool insulation products. Available at http:// insulationinstitute.org/wp-content/ uploads/2016/01/GREEN.pdf (accessed March 30, 2016).

6 Cientifica (2007) Nanotech: Cleantech – Quantifying the Effect of Nanotechnologies on CO_2 Emissions, White Paper, London.

7 Cabot Corp. (2016) Aerogel. Available at http://www.cabotcorp.com/solutions/ products-plus/aerogel (accessed March 30, 2016).

8 Jelle, B.P. (2011) Traditional, state-of-the-art and future thermal building insulation materials and solutions: properties, requirements and possibilities. *Energy Build.*, **43**, 2549–2563.

9 Industrial Nanotech, Inc. (2016) Nansulate Homeprotect. Available at http://www. industrial-nanotech.com/product/ nansulate-home-protect-clear-coat-2/ (accessed March 30, 2016).

10 Nanopore Thermal Insulation (2016) Available at www.nanopore.com/thermal. html (accessed March 30, 2016).

11 Insuladd Environmental Products (2016) Insuladd insulation values. Available at http://www.insuladd.com/product-info/ insulation-values/ (accessed March 30, 2016).

12 Leventis, N. (2005) Mechanically Strong Lightweight Materials for Aerospace Applications (X-Aerogels). NASA Technical Report IAC-05-C2.7.09. Available at http://ntrs.nasa.gov/archive/ nasa/casi.ntrs.nasa.gov/20060013346.pdf (accessed April 9, 2016).

13 Cartwright, J. (2015) Graphene finally gets an electronic on/off switch. Chemistry World, September 29, 2015. Available at http://www.rsc.org/chemistryworld/2015/ 09/graphene-band-gap-electronics-transistors-semiconductor (accessed April 11, 2016).

14 Jelle, B.P. (2011) Traditional, state-of-the-art and future thermal building insulation materials and solutions: properties, requirements and possibilities. *Energy Build.*, **43** (10), 2549–2563

15 Konarka Technologies, Inc. (2007) About Konarka. Available at http://www.konarka. com/about/.

16 Talbot, D. (2007) TR10: nanocharging solar. Technology Review. Available at http://www.technologyreview.com/ Energy/18285/ (accessed March 12).

17 Innovalight Inc. (2007) Available at www. innovalight.com/.

18 Solaicx Creating a revolution (2010). Available at http://www.solaicx.com/ pages/pv.htm.

19 Octillion Corp. (2007) U.S. government researchers validate high energy capability of nanoparticles: key to Octillion's NanoPower Windows. Available at www. octillioncorp.com/OCTL_20070821.html (accessed August 21, 2007).

20 Solarwindow Technologies (2015) Solarwindow accelerates product-durability testing following promising early results. Available at http:// solarwindow.com/2015/03/solarwindow-accelerates-product-durability-testing-following-promising-early-results/ (accessed March 30, 2016).

21 Kiger, P.J. (2013) Sun Plus Nanotechnology: can solar energy get bigger by thinking small? Available at http://news.nationalgeographic.com/news/ energy/2013/04/130429-nanotechnology-solar-energy-efficiency/ (accessed 30 March 2016).

22 Lighting Design Lab (2006) Available at www.lightingdesignlab.com/, Seattle (accessed March 30, 2016).

23 United Nations Environmental Programme (2007) Buildings can play a key role in combating climate change, Oslo, Norway. Available at http://www. unep.org/Documents.Multilingual/Default. asp?DocumentID=502&ArticleID=5545 &l=en (accessed March 30, 2016).

24 Nanosys Inc. (2016) ROHS compliance for quantum dot display components expanded and extended by European Union. Available at http://www .nanosysinc.com/environment/ Milpitas, CA (accessed March 30, 2016).

25 Lebby, M. (2007) Greentech lighting: seeking efficiencies through solid state

technologies. Available at http://www
.signallake.com/innovation/LEDLighting
081507.pdf (accessed March 30, 2016).

26 Azonano.com (2004) Samsung launches
nano e-HEPA air purifier system.
Available at www.azonano.com/details.
asp?ArticleID=560 (accessed April 9,
2016).

27 U.S. Environmental Protection Agency
(1995) The Inside Story: A Guide to
Indoor Air Quality. EPA 402-K-93-007,
EPA, Washington.

28 Axlerad, R. (1989) Economic implications
of indoor air quality and its regulation and
control. NATO/CCMS Pilot Study on
Indoor Air Quality: The Implications of
Indoor Air Quality for Modern Society.
Report on meeting in Erice, Italy, February
1989, pp. 89–116.

29 Andersen, M.M. (2007) NanoByg: a survey
of nanoinnovation in Danish construction,
Risø-R-1602, Risø National Laboratory,
Roskilde, Denmark.

30 Todras-Whitehill, E. (2006) Nanotech
toilets could clean themselves. Popular
Science, June 14.

31 Wired.com (2005) Scrubbing bubbles hit
the streets . Available at http://archive.
wired.com/science/planetearth/news/
2005/07/68282 (accessed April 9, 2016).

32 Portland Cement Association (2016)
Building a better (cleaner) world in the
21st century. Available at http://www.
cement.org/cement-concrete-basics/
products/self-cleaning-concrete (accessed
April 9, 2016).

33 Thornton, J. (2002) Environmental
impacts of polyvinyl chloride building
materials. Healthy Building Network,
Washington.

34 Cosier, S. (2006) Big problems, little
solutions. Scienceline. Available at http://
scienceline.org/2006/09/22/env-cosier-
nanotech/ (accessed April 9, 2016).

35 Mann, S. (2006) Nanotechnology and
construction. Nanoforum.

36 Fulekar, M.H., Pathak, B., and Kale, R.K.
(2014) *Environment and Sustainable
Development*, Springer, New York.

37 International Organization for
Standardization (2005) Concrete,
reinforced concrete and pre-stressed
concrete, Geneva.

38 Sev, A. and Ezel, M. (2014)
Nanotechnology innovations for the
sustainable buildings of the future. *Int. J.
Civil Environ. Struct. Constr. Archit. Eng.*,
8 (8), 886–896.

39 Arcon Construction Supplies (2015)
EMACO nanocrete. Available at www.
arconsupplies.co.uk/emaco-nanocrete
(accessed April 9, 2016).

40 Garcia-Luna, A. and Bernal, D.R. (2005)
High strength micro/nano fine cement.
2nd International Symposium on
Nanotechnology in Construction,
Bilbao, Spain, November 13–16,
pp. 285–292.

41 Vanderbilt UniversityVanderbilt engineer
receives National Science Foundation
'CAREER' award for nano-fiber concrete
research. Press release, July 12, 2005.
Available at http://engineering.vanderbilt.
edu/news/2005/vanderbilt-engineer-
receives-national-science-foundation-
career-award-for-nano-fiber-concrete-
research/ (accessed April 9, 2016).

42 Sobolev, K. and Ferrada-Gutiérrez, M.
(2005) How nanotechnology can change the
concrete world: Part 2. American Ceramic
Society Bulletin, No. 11, pp. 16–19.

43 Institute for Research in Construction
(2005) Nanotechnology and concrete:
small science for big changes National
Research Council, Ottawa, Canada.
Available at http://www.nrc-cnrc.gc.ca/
eng/achievements/highlights/2005/
nanotechnology_concrete.html (accessed
April 9, 2016).

44 Mohamed, A.S.Y. (2015) Nano-innovation
in construction, a new era of sustainability.
2015 International Conference on
Environment and Civil Engineering
(ICEACE 2015), Pattaya, Thailand
(accessed April 24–25, 2015).

45 Kloeppel, J.E. (2001) Mimicking biological
systems, composite material heals itself.
Press release, University of Illinois at
Urbana-Champaign, February 14.
Available at https://www.sciencedaily.
com/releases/2001/02/010215075006.html
(accessed April 2016).

46 MMFX Technologies Corporation (2016)
Patented microstructure. Available at
http://www.mmfx.com/technology/
(accessed April 9, 2016).

47 Azonano.com (2003) Ultra High Strength Stainless Steel Using Nanotechnology," December 19. Available at www.azonano.com/details.asp?ArticleID=338 (accessed April 9, 2016).

48 Wegner, T. (2007) *Nanotechnology for the Forest Products Industry*, US Forest Service Forest Products Laboratory, Madison, WI.

49 Michigan Technological University (2006) Treating it right: using nanotechnology to preserve wood. Michigan Technological University Faculty/Staff Newsletter, May 10.

50 Thornton, J. (2002) Environmental impacts of polyvinyl chloride (PVC) building materials.

51 UMass (2007) UMass Amherst scientists create fire-safe plastic. Press release, May 30. Available at https://www.umass.edu/newsoffice/article/umass-amherst-scientists-create-fire-safe-plastic (accessed April 9, 2016).

52 University of Virginia (2007) U.Va. Engineering School-developed nanocomposite material wins award. University of Virginia News, June 28. Available at https://news.virginia.edu/content/uva-engineering-school-developed-nanocomposite-material-wins-award (accessed April 9, 2016).

53 Heebink, L.V. and Hassett, D.J. (2003) Mercury release from FGD. 2003 International Ash Utilization Symposium, Center for Applied Research, University of Kentucky.

54 Centers for Disease Control and Prevention (2012) Facts about *Stachybotrys chartarum* and other molds. Available at http://www.cdc.gov/mold/stachy.htm (accessed October 13, 2016).

55 Drywall gets high-tech and greener, September 17, 2012. Available at http://www.greenbuildermedia.com/blog/drywall-gets-high-tech-and-greener#sthash.9ASo071E.dpuf (accessed April 9, 2016).

56 Osterwalder, N. *et al.* (2007) Preparation of nano-gypsum from anhydrite nanoparticles: strongly increased vickers hardness and formation of calcium sulfate nano-needles. *J. Nanopart. Res.*, **9** (2), 275–281.

57 Now there's an air-renew gypsum board for every room. Available at www.certainteed.com/airrenew (accessed April 9, 2016).

58 Calkins, M. (2006) Greening the blacktop. Landscape Architecture Magazine, October. Available at http://www.asla.org/lamag/lam06/october/ecology.html (accessed April 9, 2016).

59 Yang, J. and Susan, T. (2013) A review of advances of nanotechnology in asphalt mixtures. *Procedia Soc. Behav. Sci.*, **96**, 1269–1276.

60 Erlus AG (2006) Erlus Lotus. Available at http://www.erlus.de/ModelleSelbstreinigend/lotus/ (accessed April 9, 2016).

61 (2014) New class of nanoparticle brings cheaper, lighter solar cells outdoors. Press release, June 9. Available at http://media.utoronto.ca/media-releases/new-class-of-nanoparticle-brings-cheaper-lighter-solar-cells-outdoors/ (accessed April 9, 2016).

62 (2005) Cabot Corp and Centerpoint LLC agree to produce translucent nanogel-filled roofing systems. Press release, May 10. Available at http://www.prnewswire.com/news-releases/cabot-corp-and-centerpoint-llc-agree-to-produce-translucent-nanogelr-filled-roofing-systems-to-bring-comfortable--efficient-daylighting-to-new-home-interiors-54342307.html (accessed April 9, 2016).

63 McGregor, S. (2005) U.T. Dallas-led research team produces strong, transparent carbon nanotube sheets. UT Dallas News Release, Aug. 18. Available at http://www.utdallas.edu/news/archive/2005/carbon-nanotube-sheets.html (accessed April 9, 2016).

64 Zyga, L. (2009) Scientists discover material harder than diamond. Available at http://phys.org/news/2009-02-scientists-material-harder-diamond.html (accessed April 9, 2016).

65 Azonano.com (2007) Special house walls containing nano polymer particles Available at www.azonano.com/news.asp?newsID=3930 (accessed April 9, 2016).

66 Du, R.-G. (2006) *In situ* measurement of Cl-concentrations and pH at the reinforcing steel/concrete interface by combination sensors. *Anal. Chem.*, **78** (9), 3179–3185.

67 Mann, Surinder (2006) Nanotechnology and Construction.

68 Oberdörster, G., Eva, O., and Jan, O. (2005) Nanotoxicology: an emerging discipline evolving from studies of ultrafine particles. *Environ. Health Perspect.*, **113** (7), 823–839.

69 Wiesner, M.R. (2006) Responsible development of nanotechnologies for water and wastewater treatment. *Water Sci. Technol.*, **53** (3), 45–51.

70 Samsung (2016) Silver nano health system. Available at http://www.samsung.com/ph/ consumer/learningresources/ washingmachine/silver_nano/site.html (accessed April 9, 2016).

71 Weiss, R. (2006) EPA to regulate nanoproducts sold as germ-killing. Washington Post, November 23. Available at http://www.washingtonpost.com/ wp-dyn/content/article/2006/11/22/ AR2006112201979.html (accessed April 9, 2016).

72 Petkewich, R. (2007) Nanotube synthesis emits toxic by-products. Chemical & Engineering News, Aug. 27. Available at http://pubs.acs.org/iapps/wld/cen/results. html?line3=Nanotube%2BSynthesis% 20Emits%20Toxic%20By-Products (accessed April 9, 2016).

73 National Institute for Occupational Safety and Health (2005) Approaches to safe nanotechnology: an information exchange with NIOSH. Available at http://www.cdc.gov/niosh/topics/ nanotech/pdfs/Approaches_to_Safe_ Nanotechnology.pdf (accessed April 9, 2016).

40
Nanofluids for Efficient Heat Transfer Applications

Baldev Raj,[1] S.A. Angayarkanni,[2] and John Philip[2]

[1]*National Institute of Advanced Studies, Indian Institute of Science Campus, Bangalore 560 012, Karnataka, India*
[2]*Indira Gandhi Centre for Atomic Research, SMARTS, Metallurgy and Materials Group, HBNI, Kalpakkam 603102, Tamil Nadu, India*

40.1
Introduction

Nanofluids, suspensions of nanomaterials, have some unique features that are quite different from dispersions of mm- or µm-sized particles. Compared to the conventional cooling liquids such as water, kerosene, ethylene glycol, and microfluids, nanofluids do not block flow channels, induce only a very small pressure drop during flow, and exhibit higher thermal conductivities, which are beneficial aspects for heat transfer applications. Three properties that make nanofluids promising coolants are the increased thermal conductivity, the increased heat transfer, and the increased critical heat flux. Studies have shown that relatively small amounts of nanoparticles can enhance thermal conductivity of base fluids to a large extent. Different types of nanofluids are used for a broad range of engineering applications such as in automobiles, coolants [1], brake fluids [2], domestic refrigerators [3], solar devices [4,5], cosmetics, drug delivery [6], defect sensors [7], optical filters [8], hyperthermia [9], sealant [10], and so on. A schematic representation of applications of nanofluids in diverse fields is shown in Figure 40.1.

Nanofluids have been a topic of great interest during the last one decade primarily due to the initial reports of extremely large thermal conductivity (k) enhancement in nanofluids with a small percentage of nanoparticles [11–13]. However, systematic studies on thermal properties of nanofluids have shown only a modest enhancement in thermal conductivity in conventional nanofluids. Therefore, the present research effort is on tailoring of novel nanofluids with large thermal conductivities. Dispersions of nanomaterials with larger aspect ratio and morphologies (e.g., nanorods, CNT, graphene), magnetic nanofluids, and phase change materials (PCMs) are some of the new materials that have

Nanotechnology for Energy Sustainability, First Edition. Edited by Baldev Raj, Marcel Van de Voorde, and Yashwant Mahajan.

Figure 40.1 Schematic representation of nanofluids and their applications of different fields.

shown extraordinary thermal conductivities. Magnetic nanofluids offer interesting applications in electronics cooling, especially in miniature devices, such as nano- and microelectromechanical systems, due to tenability of their thermal properties. Studies show that the direct absorption solar collector efficiency can be increased by using suitable nanofluids as absorbing medium. Similarly, phase change materials with nanoinclusions have found interesting applications in thermal energy storage. This chapter focuses on the recent progress in the nanofluids.

Studies show that the thermal conductivity of nanofluids depends on many factors such as particle volume fraction, particle material, particle size, particle shape, and so on. Several studies have shown that the thermal conductivity increase was beyond the EMT predictions [14–22]. One of the important physical parameters that play a major role in nanofluid thermal properties and stability is the nanoparticles size. Some reports have shown an increase in k with a decrease in nanoparticle size [23–26], whereas others have shown a decrease in k with increase in particle size [14,27,28]. Several studies have shown an increase in k enhancement with increase in particle aspect ratio [29–35]. Recent studies reveal that the effective heat conduction through nanoparticle agglomerates can result in an enhanced k in nanofluids [36–42], and the sedimentation of aggregates in base fluids results in a lower k [43,44]. In general, the thermal conductivity of the nanofluid will be higher when the dispersed nanoparticles have a higher k [45–49]. Most of the reports show that an optimum concentration of additives can enhance the k of nanofluids and their stability [13,50,51]. In addition to the above mentioned factors, pH of the nanofluids [50,52–54], sonication time [31,45,55–60], and base fluid properties [18,40,45,48,60–63] also influence the k of nanofluids.

40.2
Traditional Nanofluids

An initial study by Eastman *et al.* [13] showed a 40% enhancement in thermal conductivity in 0.3 vol% of CuO nanofluid. An enhancement of 32.4% in the effective thermal conductivity was observed in water-based Al_2O_3 nanofluids for a volume fraction of ~4.3% [64]. Increase in k with decrease in particle size was reported in water-based Al_2O_3 nanofluids of three different particle diameters of 20, 50, and 100 nm [65]. Minsta *et al.* [25] reported zero k enhancement in water-based Al_2O_3 nanofluids of particle size 36 and 47 nm at room temperature but at elevated temperature an increase in thermal conductivity was observed in nanofluids with smaller particle size than larger one. Cho *et al.* [66] reported an anomalous increase in thermal conductivity of 18% with a very low particle loading of Ag nanoparticle of 10 000 ppm. Buongiorno *at al.* [61] observed that the k enhancement of nanorods of Al_2O_3/PAO was greater than that of PAO-based alumina nanofluids with spherical particles. The k studies in Al_2O_3 nanofluids with different particle morphologies showed that the k enhancement coefficients follow the sequence cylinders > bricks > platelets ~ blades [35]. Timofeeva *et al.* [14] showed a larger k with time in water-based Al_2O_3 nanofluids due to the formation of larger agglomerates. Methanol-based Al_2O_3, silicon dioxide (SiO_2) and TiO_2 nanofluids showed an increase in k with increase in φ [67]. The k enhancement was found to decrease with decreasing particle size in water and ethylene glycol-based Al_2O_3 nanofluids of particle size varying from 8 to 282 nm [68]. Several studies have shown that the thermal conductivity increases with particle loading beyond the EMT predictions [14–22]. Lee *et al.* [69] observed 20% k enhancement for the 4 vol% for ethylene glycol-based CuO nanofluid. An increase in k with increase in concentration was observed in EG-based Cu nanofluid of particle size 12 and 3 nm [70]. Xuan and Li [71] reported that the thermal conductivity ratio increases from 1.2 to 1.8 as the φ of the particle increases from 2.5 to 8% in transformer oil (TO)-based Cu nanofluid. For water-based Cu nanofluid, k/k_f was varied from 1.1 to 1.6 as φ of the particle increased from 1 to 5% [71]. Shin and Lee [72] studied the thermal conductivity of polyethylene and polypropylene particles dispersed in a mixture of silicon oil and kerosene. They have observed a thermal conductivity enhancement of 13% for 10 vol% particle loading. A maximum thermal conductivity enhancement of 74% was observed at a particle loading of 0.3 vol% in water-based Cu nanofluids [43]. An enhancement of 16.5% was observed in ethylene glycol-based iron (Fe) nanofluid for a particle loading of 0.3 vol% [47]. A 30% k enhancement is observed in water-based iron oxide (Fe_3O_4) nanofluid as φ increases from 0 to 0.04 [73]. At a particle loading of 4 vol%, water-based Fe_3O_4 nanofluids exhibited a 38% enhancement in thermal conductivity, which was attributed to particle clustering [74]. Our studies on water-based aggregating (TiO_2) and nonaggregating (γ-Al_2O_3) nanofluid suggest that nonaggregating nanofluids exhibit k within Maxwell limit, while aggregating nanofluids exhibit k enhancement beyond Maxwell limit. This was in good agreement with some of the earlier experimental findings [75]. Figure 40.2 shows the variation of thermal conductivity as a function of volume fraction (φ) for γ- Al_2O_3 and TiO_2 along with the EMT theoretical fit. The maximum k enhancement for γ-Al_2O_3 and TiO_2 nanofluids showed a thermal conductivity enhancement of 16 and 9% at a particle loading of 4 vol% [76].

An increase in k with decrease in particle size was reported in water-based Al_2Cu and ethylene glycol-based Ag_2Al [77]. A decrease in k with increase in particle size was reported in water-based TiO_2 nanofluids [78]. Similarly, the thermal conductivity study in ethylene glycol-based nanofluid showed an increase in k with a decrease in particle size [46]. A larger k was observed for larger particle size in water-based silicon carbide (SiC) nanofluids with four different particle size 20, 30, 35, and 90 nm [79]. Studies on water-based CuO, MWCNT, and SiO_2 nanofluids showed that MWCNT has the highest k enhancement and SiO_2 has the lowest k enhancement [48]. A decrease in k with

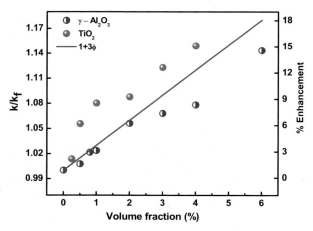

Figure 40.2 Variation of thermal conductivity as a function of volume fraction (ϕ) for Al_2O_3 and TiO_2 along with the EMT theoretical fit.

decrease in particle size (in the size range of 2–40 nm) was reported in water-based gold nanofluids [80]. Similarly, a decrease in k with increase in particle size from 2.8 to 9.5 nm was observed in kerosene-based Fe_3O_4 nanofluid [81]. For the same particle loading of 5 vol% rod-shaped TiO_2 nanoparticle in water with an aspect ratio of 4 showed an enhancement of 32%, whereas nanofluids containing spherical nanoparticles of diameter 15 nm showed an enhancement of 29% [29]. Our time-dependent thermal conductivity studies on stable (non-aggregating) and unstable (aggregating) nanofluids show that the nonaggregating nanofluids exhibit time-independent k/k_f, whereas aggregating nanofluids exhibit a time-dependant k/k_f [76]. Figure 40.3a shows the variation of k/k_f and the percentage of enhancement in k with time, after 2 min of sonication at two different particle concentrations of 0.5 and 4 vol% for γ-Al_2O_3 nanofluid. Here, k/k_f increases with the particle loading and is found to be time independent. Figure 40.3b shows the variation of k/k_f and the percentage of enhancement in k with time, after 2 min of sonication at four different particle concentrations (0.7 and 2 vol%) for α-Al_2O_3 nanofluid.

The k/k_f increases with time and reaches a peak value after a time interval of about 3 h and then starts decreasing with time. Both the increase and decrease in k/k_f can be explained on the basis of cluster formation. Immediately after sonication, particles are well dispersed. With time, the particles start forming clusters

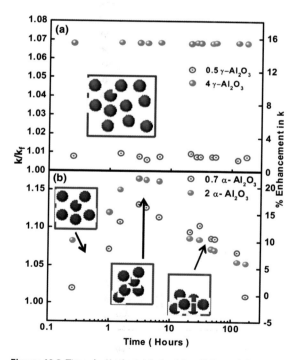

Figure 40.3 Time dependent k/k_f for (a) γ-Al_2O_3 and (b) α-Al_2O_3 nanofluids for different volume fractions.

that lead to an increase in thermal conductivity because the conduction is effective through percolating conducting path [42]. With increasing time, due to strong interaction between the clusters, the clusters tend to come closer and form thicker agglomerates, which eventually sediment due to gravitational force, resulting in a decrease in k.

Some researchers reported that the thermal conductivity of the nanoparticle is not a primary factor for the observed enhancement in nanofluid k [74,82,83,84]. For the same particle loading of 4 vol%, water-based Fe_3O_4 nanofluids exhibited 38% k enhancement, while water-based nanofluids containing TiO_2 or Al_2O_3 nanoparticles showed 30% enhancement even though bulk Fe_3O_4 crystal has a lower k than Al_2O_3, CuO, and TiO_2 crystals [74]. For the same concentration of nanoparticle (1 vol%), the k enhancement for water-based TiO_2 was higher than that of Al_2O_3 nanofluids, though the bulk k is higher for Al_2O_3 nanoparticle [47]. Chopkar *et al.* [85] reported that for the same particle concentration, the k enhancement for water-based Ag_2Al nanofluid is slightly more than Ag_2Cu nanoparticle suspensions. Sinha *et al.* [86] reported a k enhancement of 48–70% for the same concentration of 1 vol% in Cu nanofluids, whereas Fe nanofluids showed a k enhancement of around 21–33%. Wang *et al.* [50] reported that water-based Cu nanofluid have a higher k enhancement compared to water-based Al_2O_3 nanofluids, and the reason for the larger k enhancement in Cu nanofluid was attributed to the higher k of copper nanoparticles. Patel *et al.* [46] observed that metallic nanofluids exhibit higher k enhancement than the oxide nanofluids, which was proved to be wrong recently. Eastman *et al.* [13] observed an enhancement in k in ethylene glycol-based Cu nanofluid stabilized with thioglycolic acid, compared to nonacid-containing nanofluids. Wang *et al.* [50] also reported that an optimal concentration of sodium dodecylbenzenesulfonate (SDBS) can result in an enhanced k in water-based Cu and Al_2O_3 nanofluids. Figure 40.4 shows the variation of k/k_f and the percentage (%) of decrease in thermal conductivity (TC) as a function of surfactant concentration for SDS, CTAB, and NP10 in water. In all the three systems, the k/k_f is found to decrease with increase in surfactant concentration. Micellar system is a disordered system. As the surfactant concentration increases, the number of micelles in the system increases, leading to increase in the degree of disorder, which has resulted in a decrease in k/k_f with increase in surfactant concentration. Muller [87] has suggested that the strength of the interatomic binding force is related to thermal conductivity. Greater the strength of this binding force, greater is the thermal conductivity. The binding force is weak in the case of micellar systems due to its disordered structures leading to a lower thermal conductivity. The EMT fit is shown in Fig. 4 by a solid line. It can be seen that the experimental data show large deviation from effective medium theory at low-volume fractions, especially for CTAB and NP9. At low-volume fraction, the number micelles in the system will be less and therefore the interfacial thermal resistance is also less. The lower interfacial thermal resistance at lower concentration might be a probable reason for the observed large deviation from theoretical fit at lower volume fractions. For volume fraction above 0.02, the

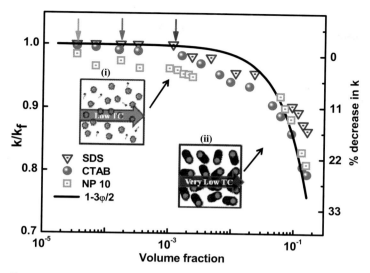

Figure 40.4 Variation of thermal conductivity as a function of volume fraction for SDS, CTAB, and NP10 along with the theoretical fit. (The CMC values of surfactants are shown by arrows of respective symbol colors). The insets show the schematic representation of nanofluids: (i) Soft system with spherical randomly arranged ($k_p < k_f$). (ii) Cylindrical micelles randomly arranged ($k_p \ll k_f$) leading to lower thermal conductivity.

experimental data fit well with the effective medium theory for all three surfactants. The long alkyl chain group of nonionic surfactant micelles is found to be a very poor thermal conductor at very low concentrations, with large interfacial tension, compared to their anionic counterparts [88]. The inset in Figure 40.4 shows the schematic representation of nanofluids based on the experimental findings. Figure 40.4i depicts a soft micellar system with spherical micelles ($k_p < k_f$). This scenario arises when the surfactant concentration is above the critical micellar concentration (cmc). Because $k_p < k_f$, the system shows a $k/k_f < 1$. Figure 40.4ii renders cylindrical micelles randomly arranged at extremely large concentrations of surfactants ($c \gg$ cmc).

The effect of surfactant at nanoparticle/base fluid interface on the thermal conductivity of nanofluids was also studied. Figure 40.5 shows the variation of k/k_f as a function of silica nanoparticle volume fraction with and without SDS and NP9 along with its best fits with and without interfacial resistance. The k/k_f value for 10 CMC surfactants is also shown. With and without surfactant (10 CMC fixed), the k/k_f enhancement was almost the same. The effective medium theory fit on the experimental data is shown by the solid line, which shows a very good agreement with the experimental data. It can be seen that the value of k/k_f with pure surfactant was negative, while it was positive at all other concentration of nanoparticles. This shows that the thermal property of a nanofluid in the presence of surfactant and nanoparticles simply follows the thermal conductivity of

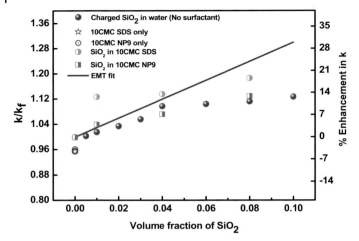

Figure 40.5 Variation of k/k_f as a function of silica nanoparticle volume fraction with and without SDS/NP 9 and its best fit with and without interfacial resistance. The k/k_f value for 10 CMC surfactant is also shown.

nanoparticles in the fluid and the addition of surfactant, even beyond optimal concentration, would not hamper the thermal conductivity of nanofluids.

Water-based Al_2O_3 nanofluids (1 vol%) showed a k enhancement from 2 to 10% as the temperature was increased from 21 to 50 °C [89]. The increase in k with increase in temperature from 10 to 60 °C was observed in water and ethylene glycol-based Al_2O_3 nanofluid (5 vol%) [14]. The temperature-independent k enhancement was also reported in ethylene glycol and water-based nanofluids [32,62]. A temperature-independent k enhancement was observed in water-based SiC nanofluids [90]. Shima *et al.* [91] studied the thermal conductivities of base fluid and nanofluid with temperature (Figure 40.6a) for both aqueous and nonaqueous ferrofluids with average particle diameter of 8 nm over the temperature range of 25–50 °C. They observed that the thermal conductivity ratio was independent of temperature as shown in Figure 40.6b, which suggested a less dominant role of microconvection on thermal conductivity enhancement of nanofluids.

A constant k enhancement was observed in hexadecane (HD)-based Al_2O_3 in the temperature range 25–50 °C [92]. Similarly, a temperature-independent k enhancement was observed for kerosene-based Fe_3O_4 nanofluids [38,93]. There are reports showing a decrease in k with increase in temperature in water-based TiO_2 [94] and hexane-based Bi_2Te_3 nanorods [95]. Aqueous nanofluids over the temperature range 25–50 °C showed an increase in k with temperature, whereas a decrease in k was observed in nonaqueous (kerosene and hexadecane based) nanofluids [91]. Thermal conductivity of dibenzyl toluene-based MoS_2 nanofluids was found to increase from 1.175 to 1.375 as the temperature increased from 40 to 180 °C [96]. Recent studies reveal that the effective heat conduction through nanoparticle agglomerates can result in an enhanced k in

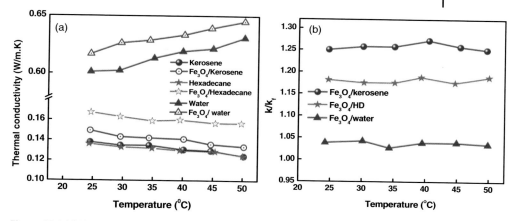

Figure 40.6 (a) Variation of thermal conductivity with temperature for kerosene, hexadecane, and water-based nanofluids. (b) The variation of k/k_f with temperature for kerosene, hexadecane, and water-based Fe_3O_4 nanofluids with $\varphi \sim 0.01$ and $d = 6$ nm.

nanofluids [36–42], and the sedimentation of aggregates in base fluids results in a lower k [43,44]. Water-based untreated diamond nanofluids showed a decrease in k with time, whereas plasma-treated stable diamond nanofluids showed a time-independent k [97]. Nasiri et al. [98] observed a decrease in k with time due to the formation of agglomerates. A time independent k was observed in ethylene glycol-based ZnO [99], graphene oxide nanofluids [17], and kerosene-based Fe_3O_4 nanofluids [93]. Table 40.1 gives a summary of the reported values of percent enhancement at different concentration for different nanofluids.

Table 40.1 Summary of the reported values of % enhancement at different concentration for different nanofluids.

Nanoparticle	Base fluid	Concentration φ%	% enhancement	Reference
Silver	Epoxy	25	790	[100]
Al_2O_3	Water	4.3	32	[64]
CuO	EG	4	20	[69]
Polyethylene	Silicon oil	10	13	[72]
Copper	TO	1.5	12	[71]
		8	45	
Copper	Water	2.5	23	
		7.5	78	
Silver	EG	10 000 ppm	18	[66]
Copper	Water	0.3	74	[43]
SiC	Water	1	4	[90]
		4	22	
Al_2O_3	PAO	0.4	5	[61]
		2.8	15	

(continued)

Table 40.1 (Continued)

Nanoparticle	Base fluid	Concentration $\varphi\%$	% enhancement	Reference
Fe_3O_4	Kerosene	1.71	0.6	[75]
		5.03	12.5	
CuO	EG	0.18	7	
		0.92	12	
		1.14	14	
Nanodiamond	MO	0.5	2.5	[101]
		1.9	12	
	EG	0.2	1	
		0.85	11	
Fe_3O_4	Kerosene	0.25	4	[93]
		1	32	
Fe_3O_4	Kerosene	2	6	[38]
		5	18	
		0.05	21	
ND-Ni	Water	0.62	3	[21]
		3.03	20	
ND-Ni	EG	0.62	0.1	[21]
		3.03	8	
TiO_2	Water	1	8	[76]
		2	12	
		4	15	
CuO	Water		27	[102]
	0.01		53	
ZnO	0.05		9	
	0.02			
	0.08		12	
	0.06		33	
	0.2		11.9	
Fe	Water	0.003	0	[103]
		0.0025	1	
		0.01	5	
Cu	Tetradecane	0.001	2	[104]
		0.01	6	
		0.015	9	
		0.020	14	
Al_2O_3	Water	1	8	[26]
TiO_2	Water	0.6	3.5	[78]
Al_2Cu	Water	0.5	30	[77]
Al_2Cu	EG	0.5	45	[77]
Al_2O_3	Water	2	3.6	[68]
Al_2O_3	Water	14	13	[25]

Al	EG	1	11	[46]
Al	Water/EG	1	5.5	
Cu	Water	1	11.3	
Al_2O_3	Water	2	4.5	
CuO	Transformer oil		11	
Cu	EG		17	
Al_2O_3	Water	0.51	6	[65]
α- SiC	Water	4.1	12.4	[79]
SiO_2	Water	8	11	[105]
Al_2O_3	Water	0.5	7.5	[106]
Cu	EG	0.03	15	[70]
Al_2O_3	Water	0.51	2	[107]
Al_2O_3	Water	1	2	[89]
Bi_2Te_3	HD	1	13	[95]
Al_2O_3	Water	5	8	[14]
Al_2O_3	HD	2	11	[92]
TiO_2	Water	1	1	[94]
MoS_2	Dibenzyl toluene	—	17.5	[96]
SiO_2	Water	1	3	[48]
CuO			5	
Fe_3O_4	Water	4	38	[74]
TiO_2			30	
Al_2O_3			30	
Al_2O_3	Water	1	14	[47]
TiO_2			3	
Ag_2Al	Water	1.8	140	[85]
Al_2Cu			120	
Cu	EG	1	60	[86]
Fe			27	
Cu	Water	0.09	15	[50]
Fe_3O_4	Kerosene	0.01	2	[49]
Ag			4	
Fe_3O_4	Hexadecane	0.01	5	
Ag			5	
CuO	EG	4	40	[13]
Al_2O_3	Water	4	18	[88]
SiO_2	Water		10	

40.3
CNT-Based Nanofluids

An anomalous thermal conductivity enhancement of 150% was observed in synthetic polyalphaolefin oil-based MWCNTs nanofluid for a very low particle loading of 1 vol% [108]. Ethylene glycol-based carbon nanotube (CNT) nanofluids showed a thermal conductivity enhancement of 12% at a particle loading of 1 vol% and synthetic engine oil-based CNT nanofluids showed a 30% k enhancement at a particle loading of 2 vol% [60]. For water-based MWCNT nanofluid, the thermal conductivity enhancement was found to be 11% for a volume fraction of 1 vol% [48]. For ethylene glycol-based MWCNT nanofluid, a considerable thermal conductivity enhancement of 17% was observed at 1 vol% [62]. A 20% increase in k at 2.5 vol% of carbon nanotubes was observed in ethylene glycol-based CNT nanofluids [62]. A study on graphene oxide nanosheets [18] showed a thermal conductivity enhancement of 30, 62, and 76% at a particle loading of 5 vol% for three different base fluids, distilled water, propyl glycol, and liquid paraffin, respectively. The observed k enhancement was above the EMT prediction. Reports showed that the thermal conductivity of MWCNT nanofluids increases with increase in volume fraction and the k enhancement observed were well above the EMT predictions [29–35,44,59,109–111]. An increase in k with increase in nanotube length was reported in water-based MWCNT [32]. The k enhancement was about 14, 18, 38, and 48% for the nanotube length of 0.5, 1, 1.7, and 5 μm, respectively. For the same particle loading of 1 vol%, the CNT nanofluids with an aspect ratio of 666 showed a k enhancement of 2.08, whereas CNT with an aspect ratio of 18.8 showed a k/k_f of 1.43 [112]. These experimental results suggest that particle having higher aspect ratio can give a higher k enhancement than the spherical particle due to the effective heat transfer along the length of the rod-shaped particle. With time, a decrease in k is observed in water-based CNT nanofluids due to sedimentation and agglomeration [43]. Horton *et al.* [113] showed that the thermal conductivity of magnetic-metal-coated carbon nanotubes can significantly enhance under applied magnetic field due to the controlled aggregation of magnetic nanoparticle along the field direction. Similar results were reported in Ni and Fe_2O_3 incorporated in water-based SWCNT [39,41,114]. Recent studies show that the stable nanofluids exhibit thermal conductivity within EMT prediction, whereas aggregating nanofluids show enhancement beyond EMT predictions [75]. It was reported that CTAB has no effect on k enhancement in water-based CNT nanofluids [34]. Another report showed that an optimum concentration of polyisobutene succinimide (3 wt%) provides a better k enhancement in PAO-based MWCNT nanofluid [31]. Ghozatloo *et al.* [115] observed that the k of water-based surface modified CNT increases with increase in temperature. Table 40.2 gives a summary of the reported values of percent enhancement at different concentration for different CNT-based nanofluids.

Table 40.2 Summary of the reported values of % enhancement at different concentration for different CNT-based nanofluids.

Nanoparticle	Base fluid	Concentration (φ%)	% enhancement	Reference
MWCNT	Water	0.24	14	[32]
CNT	Water	1	108	[112]
MWCNT	Water	1	35	[34]
DWCNT	Water		3	
CNT			32	[31]
CNF	Water	2.5	8	[33]
CNT	Water	0.3	2	[62]
CNT	Ethylene glycol	1	12	[60]
CNT/ID:5-10 OD: 20-50	Engine oil	2	30	
MWCNT	Water	1	11	[48]

40.4
Magnetic Nanofluids

Magnetic nanofluids, suspensions of superparamagnetic nanomaterials in base fluids have been studied for their tunable thermal conductivity. Recent studies showed that the conduction through linear agglomerates of nanoparticle in base fluids can lead to a dramatic enhancement of k [1,36,81,113,116]. A 130% k enhancement was observed in kerosene-based Fe_3O_4 (0.0171φ%) nanofluids at an applied magnetic field of 350 G [117]. Shima and Philip [49] showed the effect of suspended particles thermal conductivity on enhancement of k in nano-fluids. They observed that in the dilute limit, the thermal conductivity of nano-particle does not influence the k enhancement and the k of nanofluids is solely dependent on the volume fraction of the nanoparticle. Li et al. [116] observed an enhancement of k/k_f in water-based magnetic nanofluids of size 26 nm in the presence of an external magnetic field along the field direction. Shima and Philip [81] investigated the effect of particle size ranging from 2.8 to 9.5 nm at a fixed concentration of $\varphi = 0.04$ on field-induced thermal conductivity enhance-ments in kerosene-based Fe_3O_4 nanofluids. Figure 40.7 shows k/k_f and the per-centage of enhancement in k as a function of magnetic field strength for kerosene-based Fe_3O_4 nanofluids having different particle sizes at $\phi = 0.04$. No change in k with external magnetic field strength is observed in nanofluids with particle size 2.8 nm. With the increase in nanoparticle size, a field-induced k enhancement is also increased. The nanofluid with largest particle size (9.5 nm) is found to exhibits largest k enhancement. For a magnetic field strength of 330G, the k enhancement was 16 and 240% for a nanofluid with particle size 2.8 and 9.5 nm, respectively, at a particle loading of $\varphi = 0.04$.

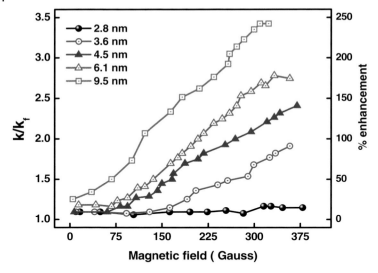

Figure 40.7 k/k_f and the % of enhancement in k as a function of magnetic field strength for kerosene-based Fe_3O_4 nanofluids with different particle sizes (2.8, 3.6, 4.5, 6.1, and 9.5 nm) at $\phi = 0.04$.

Philip *et al.* [36] observed a 300% k enhancement in kerosene-based Fe_3O_4 nanofluid at a particle loading of 6.3 vol.%, at an applied field of 80 G. The observed enhancement in k was shown to be due to the effective heat transport through chainlike aggregates of nanoparticle. Figure 40.8 shows the variation of k/k_f as a function of applied magnetic field, for kerosene-based Fe_3O_4 nanofluids with $\varphi = 0.063$. Here, an increase in thermal conductivity with magnetic field is observed up to a certain magnetic field (regions i–iii), due to the increase in the aspect ratio of the chainlike structure in the fluid, which carries the heat effectively. A decrease in thermal conductivity observed above a critical magnetic field strength is expected to be due to "zippering" of chains [118]. The inset in Figure 40.8 (i) shows the micrograph of kerosene-based magnetite nanofluids with $\varphi = 0.063$ in the absence of external magnetic field, where no aggregates are visible. Figure 40.8 (ii–v) shows the micrograph of the nanofluid in the presence of increasing magnetic strengths. At low magnetic field strength (Figure 40.8 (ii)), there is a formation of small dipolar chains. With increasing magnetic field strength, the chain length increases (Figure 40.8 (iii and iv)), resulting in evenly spaced single nanoparticle chains throughout the nanofluid volume. At very high magnetic field strength (Figure 40.8 (v)), zippering of the dipolar chains is evident in the microscopic images. Horton *et al.* [113] showed that the thermal conductivity of magnetic-metal-coated carbon nanotubes can significantly enhance under applied magnetic field due to the controlled aggregation of magnetic nanoparticle along the field direction. Similar results were reported in Ni and Fe_2O_3 incorporated in water-based SWCNT [39,41,114].

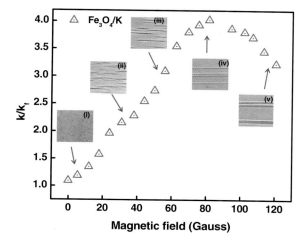

Figure 40.8 Variation of k/k_f as a function of applied magnetic field, for kerosene-based Fe_3O_4 nanofluids with $\varphi = 0.063$. Inset shows phase contrast optical microscopic images for (i) zero external magnetic field and (ii–v) with increasing magnetic field strengths.

Reversibly tunable thermal conductivity was reported in magnetic nanofluids [81]. Figure 40.9 shows the k/k_f of kerosene-based Fe_3O_4 nanofluid with $\varphi = 0.026$ during rise and decay of applied magnetic field. The enhancement starts above 20 G and with further increase in magnetic field a drastic enhancement in k is observed. A maximum enhancement of k (~128%) is observed at a magnetic field of ~95 G, above which the k value starts to decrease slightly.

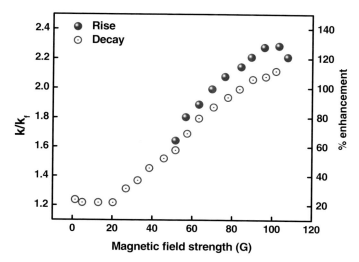

Figure 40.9 k/k_f and the % of enhancement in k as a function of applied magnetic field strengths (during rise and decay) for kerosene-based Fe_3O_4 nanofluids with $\phi = 0.026$.

While lowering the magnetic field, the k value shows a small hysteresis but comes back to the original value when the magnetic field is turned off, which shows the perfect reversibility of field-induced structures and thermal properties.

40.5
Graphene Nanofluids

Recently grapheme-based nanofluids have been studied for their thermal properties. Many reports suggest extremely large thermal properties in graphene-based nanofluids. Gallego *et al.* [119], who observed an increase in k with the addition of functionalized grapheme sheet (FGS) in epoxy resin, suggest that phonon coupling of the vibrational modes of the graphene and of the polymeric matrix plays a dominant role on the thermal conductivities of the liquid and solid states. FES in liquid media are not able to transfer the heat because the vibrational modes are not compatible and hence has lower k, whereas when the FGS are surrounded by the more rigid matrix, the differences between the frequencies of vibrational modes are smaller and enable phonon coupling, resulting in increase in k with the addition of FGS. Nika *et al.* [120] using molecular dynamics reported that the long mean free path of the long-wavelength acoustic phonons in graphene results in abnormal nonmonotonic dependence of the thermal conductivity on the length and width of a ribbon. Yu *et al.* [18] observed 30, 62, and 78% enhancement in k with the addition of 5φ% of graphene oxide nanosheets in water, propyl glycol, and paraffin, respectively. A k enhancement of 15% was observed in functionalized graphene–ethylene glycol (EG) + distilled water nanofluids with the 0.395φ% particle loading and the k of both base fluid and the nanofluid increases linearly with temperature [121]. Hadadian *et al.* [122] observed a 30% k enhancement in ethylene glycol-based graphene oxide (GO) nanofluid with mass fraction of 0.07 for all temperature. A molecular dynamic study shows that the fierce phonon scattering caused by edge roughness and junctions lead to the nonmonotonic dependence of thermal conductivity on the length of SGNRs [123]. Guo *et al.* [124] investigated the thermal conductivity of graphene nanoribbons (GNRs) with different edge shapes as a function of length, width, and strain using nonequilibrium molecular dynamics method. Strong length dependence of thermal conductivity was reported, indicating high thermal conductivities of GNRs. The effect of edge-passivation by hydrogen H-passivation and isotope mixture with random or superlattice distributions on the thermal conductivity of rectangular graphene nanoribbons GNRs of several nanometers in size was studied using molecular dynamics [125]. The thermal conductivity is considerably reduced by the edge H-passivation and isotope mixing. An enhancement of 11.3 and 13.7% with the addition of 0.04φ% of graphene wrapped MWCNT-based water and ethylene glycol nanofluids [126]. A 14% enhancement in k was observed in water-based exfoliated graphene nanofluid (0.056φ%) at 25 °C, which increases to about 64% at 50 °C [127]. Yu *et al.* [16] observed 82% enhancement in k with the addition of 5φ% graphene nanosheets

(GNS) in ethylene glycol. Lee *et al.* [128] reported 33% k enhancement with the addition of 4φ% in ethylene glycol-based graphene nanoplatelets (GNP) nanofluid. The k of water-based graphene nanoplatelets showed 27.64% with the addition of 0.1 wt% of GNPs with a specific surface area of 750 m^2/g [22]. The enhancement in k with 0.04φ% of aqueous graphene–f-MWNT nanofluids is 10.5% [129]. Gupta *et al.* [130] observed 9% k enhancement in water-based graphene nanosheets with 0.2φ particle loading. Ethylene glycol-based graphene nanosheets showed 61% enhancement for 5φ% particle loading [17]. An analytical model of effective thermal conductivity of graphene-based composites showed that the GNP composites show various nonlinear behaviors at different GNP concentrations, and a large interface thermal resistance across the GNP–matrix interface causes a significant degradation in the thermal conductivity enhancement [131]. Analytical model showed that the thermal conductivity enhancement mainly relies on a GNP percolation effect triggered by the presence of CNTs, rather than the isolated CNTs and GNPs, even at very low GNP percolation threshold [132]. Ethylene glycol-based graphene nanofluids showed a k enhancement quite independent of temperature [133]. EG-based nanodiamond nanofluid also showed a temperature-independent k [101].

40.6
Hybrid Nanofluid

Recently, hybrid nanofluids have been prepared by suspending different types of nanoparticles or composite nanoparticles in base fluids [43,109,134]. The hybrid nanofluids with optimal thermophysical characteristics (e.g., high k, lower η etc.) may find effective applications in heat transfer because of their optimal heat transfer characteristics. Several methods are used to prepare hybrid nanofluids [43,109,134–139]. Oil-based hybrid oxide sphere/CNT nanofluids were prepared by first preparing spherical oxide nanoparticles, followed by catalytic growth of CNTs and then the prepared hybrid oxide sphere/CNT were dispersed in oil [109]. Water-based CNT–AuNP hybrid nanofluids were prepared by mixing the water-based AuNP nanofluid with the CNT nanofluid of different concentration [43]. Hybrid γ-Al_2O_3/MWNT nanoparticles were prepared using solvothermal process in ethanol and the particles were redispersed in water using ultrasonicator [134]. Chen *et al.* [135] prepared water-based Ag–MWNT hybrid nanofluid. The functionalized MWNT and Ag nanoparticles were used to prepare Ag/MWNT nanofluid using silver mirror reaction. Botha *et al.* [136] prepared oil-based silver–silica hybrid nanofluids by one-step method. The alumina–copper hybrid particles were synthesized by a thermochemical method and dispersed in water with addition of sodium lauryl sulfate (SLS) using an ultrasonic vibrator [137]. Paul *et al.* [138] prepared ethylene glycol-based hybrid nanofluid. Al–Zn nanoparticles were prepared by mechanical alloying and the particles were dispersed in ethylene glycol using ultrasonicator. Wet chemical method was used to prepare hybrid silica and hybrid MWNT nanostructures and the prepared particles are dispersed in water using ultrasonicator [139]. The hybrid water-based suspensions were formulated by mixing the Al_2O_3 nanofluid with microencapsulated phase change material particles (MEPCM) suspension in an ultrasonic vibration bath [140]. A solution-free green method using focused solar electromagnetic radiation was used to synthesize graphene and graphene – multiwalled carbon nanotube composite and a stable nanofluid were prepared by dispersing the nanomaterials in DI water and EG base fluids by ultrasonication for 30 min [129]. Nanocrystalline alumina–copper hybrid ($Al_{12}O_3$–Cu) powder was prepared by a thermochemical synthesis method and were dispersed in deionized water with sodium lauryl sulfate (SLS) as dispersant by using an ultrasonic vibrator [141]. Han *et al.* [109] observed 21% enhancement in k in PAO-based hybrid sphere/CNT nanofluids with 0.2φ% of particle, which was much higher compared to k of nanofluid containing spherical nanoparticles of the same particle loading. Jana *et al.* [43] observed no k enhancement in the hybrid nanofluid with CNT–AuNP and CNT–CuNP. For the same particle loading, the water-based Ag/MWNT showed a 23% k enhancement, whereas nanofluid having MWNT showed 10% k enhancement [135]. Botha *et al.* [136] observed a 15% enhancement in k when 0.6 wt% of silver was added to 0.07 wt% of silica dispersed in transformer oil. A maximum k enhancement of 12.11% was observed in water-based Al_2O_3–Cu hybrid nanofluid at a particle loading of 2φ% [137]. Ethylene glycol-based Al–Zn hybrid nanofluid showed a k enhancement of 16% with a 0.1φ% particle loading [138]. Baghbanzadeh *et al.* [139] reported that the k enhancement of water-based silica–MWCNT hybrid nanofluid is intermediate between the k enhancement of MWCNT (23%) and silica (8%) nanofluids. Ho *et al.* [140] reported that compared to the pure PCM suspension, the hybrid suspension has higher thermal conductivity with the Al_2O_3 nanoparticle mass fraction. The graphene–MWCNT hybrid nanofluids showed 10.5% (0.04φ%) k enhancement, which was higher than pure graphene nanofluids [129]. Hybrid Cu–Cu_2O nanofluid showed a maximum thermal conductivity enhancement compared to nanofluids containing Cu nanoparticles or Cu_2O nanoparticles alone [142]. A maximum k enhancement of 28% was observed in water-based CuO–HEG (HEG – hydrogen exfoliated graphene) nanofluid for 0.05φ% [143]. The hybrid nanostructure of functionalized MWNT (f-MWNT) and functionalized HEG (f-HEG) dispersed in water showed a thermal conductivity enhancement of 20% for 0.05φ% particle loading [144]. Nine *et al.* [145] observed that water-based Al_2O_3–MWCNT hybrid nanofluid showed a higher k than Al_2O_3 nanofluid. They reported that the hybrid nanofluid having cylindrical particles showed higher enhancement than hybrid spherical nanofluid. Water-based Ag–HEG hybrid nanofluid showed an enhancement of 25% for 0.05φ% [146]. An 8% of k enhancement was observed in silver-decorated hybrid MWNT–HEG nanofluid for a volume fraction of 0.04% at 25 °C [147]. An enhancement in thermal conductivity of 11.3 and 13.7% were attained with 0.04φ% in graphene-wrapped MWCNT hybrid nanofluids with DI water- and EG-based, respectively [126]. The k enhancement observed in water-based nanofluid containing 0.05 wt% MWNTs and 0.02 wt% Fe_2O_3 nanoparticles was

27.75%, which was higher than that of nanofluid containing 0.2 wt% single MWNTs or Fe_2O_3 nanoparticles [148]. In summary, hybrid nanofluids are shown to exhibit higher thermal conductivity than normal nanofluids. However, more systematic studies are required to understand the mechanisms and the reproducibility.

40.7
Thermal Conductivity of Phase Change Material

To overcome the limitations of traditional heat transfer media, phase change materials (PCMs) are being considered as effective media for thermal energy storage [149]. Latent heat storage has been on the frontiers of research with an aim of obtaining a heat storage system with a high heat capacity and good heat transfer property. Use of appropriate material to store the heat is the key to the development of heat storage systems. The last decade witnessed several new methods that were employed to enhance the thermal conductivity of nanofluids using carbon nanotubes [150–152], metal oxides [29,76,78,153–155], graphene [120,156,157], magnetic materials [81], and composites [158–162]. The use of organic phase change materials as continuous phases (base fluids) has attracted much attention in the recent years due to their self nucleating properties, chemical stability, high heat of fusion, safe and nonreactive nature, and ability to freeze without much supercooling [163]. The energy charge/discharge rate is one of the most crucial factors to meet the cooling and heating capacity demand and the storage size. The PCMs used for ambient temperature-related applications are hydrocarbons with different chain lengths. Most of the phase change materials (organic) have low thermal conductivity, which limits their utility in practical applications. To enhance the thermal conductivity of pure PCM several methods have been attempted, such as placing a metal structure in PCM [164], impregnating porous material [165], and dispersing high thermal conductivity particles in PCM [166]. The increase of the thermal conductivity of organic phase change materials using different nanoinclusions has been a topic of research in recent years. The nanoinclusions can be ceramic particles [167], metallic particles [168], and carbon materials [169] with higher thermal conductivity. Reversible tuning of electrical and thermal conductivity using phase change material in percolated composite materials [170–172] has drawn considerable attention owing to their applications in heat management systems in various industrial sectors such as construction, textile, food packaging industry, medical packaging industry, automobile, transportation, and so on. There have been some studies on thermal conductivity of PCM with nanoinclusion in the recent years [171–173]. Molecular dynamic studies on PCM show that, upon crystallization, a nanocrystalline structure develops that results in the doubling of thermal conductivity [174]. A suitable phase change temperature and a large melting enthalpy are two major requirements of a phase change material. Sutherland *et al.* [175] and Powell *et al.* [176] observed that a sudden increase in the thermal conductivity of *n*-octadecance shows when it transformed from liquid to solid state. Lyeo *et al.* [177] showed that the thermal conductivity of thin film of the phase change material $Ge_2Sb_2Te_5$ increases irreversibly with increasing temperature and undergoes large changes with phase transformations. Zheng *et al.* [173] reported a large contrast in electrical and thermal conductivities using first-order phase transitions in percolated composite materials. Harish *et al.* [171] reported a large contrast in the thermal conductivity enhancement of phase change alkane in liquid and solid state with single-walled carbon nanotube inclusions. Tunable electrical and thermal conductivity through freezing rate control in hexadecane-based nanocomposite were reported by Schiffres *et al.* [178]. Thermal conductivity studies on CNT–Hexadecane composites showed a three times increment in thermal conductivity at the phase change point of hexadecane [172]. Thermal conductivity studies on surfactant–hexadecane composite showed that thermal conductivity enhancement between the solid and liquid phase in presence of inverse micelles of size ranging from 1.5 to 5 nm can vary between 111 and 185% (Figure 40.10a). The *k* increases with

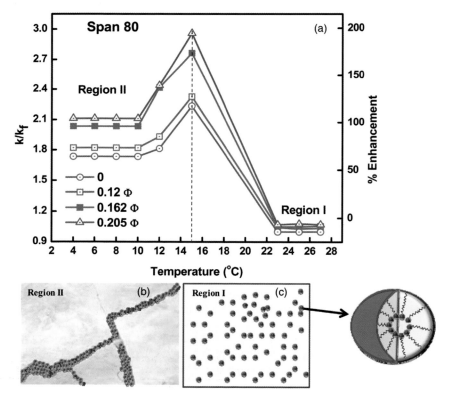

Figure 40.10 (a) Variation of k/k_f and the percentage of increase in thermal conductivity as a function of temperature for different volume fraction of Span 80 in hexadecane. (b and c) Configuration of the crystals with inverse micelles in solid state and inverse micelles in HD in liquid state. The magnified views of the inverse micelles are shown by arrows.

increase in Span 80 concentration in both regions I and II. In the liquid state, the k enhancement was nominal, that is, from 2 to 6%, as the Span 80φ increased from 0.12 to 0.205. In the solid state, immediately after freezing, the k enhancement increases from 88 to 143% as the Span 80φ increases from 0.12 to 0.205. With further decrease in temperature beyond the freezing point, the k is found to decrease a little and then becomes constant for all the concentrations of Span 80.

The molecular dynamics simulation studies show that the interfaces of organic liquids with surfactant tail groups are highly conductive [179]. The large conductance of the organic liquid–surfactant interface was attributed partly to the penetration of alkane molecule into the surfactant tail regions, which is more prominent for linear surfactant molecules. The very similar molecular and thus vibrational structures of the surfactant tails and the alkane molecules leads to the optimization of the van der Waals interactions between the chains in the two species. In the composite system, we have a matrix of alkane base fluid with surfactant micelles dispersed. As the net thermal resistance of inverse micelle is much lower than that of the alkanes because of the highly ordered packing of the linear chain surfactant within the sphere, the effective interfacial resistance along the conduction path should be much lower than that of pure alkane. When the alkane undergoes first-order phase transition during cooling, it form crystals [173] of different morphologies, whose aspect ratio can vary from micro- to millimeter scale with the rate of cooling [178]. It is postulated that during the freezing transition, the nanosized frozen inverse micelles with highly packed linear chain surfactant within the sphere are pushed to the grain boundaries filling the conduction path, thereby reducing the effective thermal resistance [171]. The results suggest that during the freezing transition, the inverse micelles are pushed to the grain boundaries of crystals, resulting in a reduced interfacial thermal resistance. Figure 40.10b and c show configuration of the crystals with inverse micelles in solid state and inverse micelles in HD in liquid state. The magnified views of the inverse micelles are shown by arrows. On freezing transition, the inverse micelles are pushed to the microcracks and the grain boundaries leading to a better heat conduction path. The alkane molecules may also undergo some ordering around the inverse micelles, which can restrict translational diffusions of alkane molecules [178].

The effect of aggregation of nanoinclusions on k enhancement during freezing shows that the functionized nanomaterials provide a higher k enhancement than that of bare nanofluids [170] as shown in Figure 40.11. It was known that the dispersibility of nanomaterials improves with functionalization, which was considered as the main reason for the observed enhancement in stabilized nanofluids. It is believed that the distribution of MWCNTs at the grain boundaries is uniform during the crystallization of alkanes in the case of F-MWCNT compared to the bundled one in the case of bare MWCNT. This finding showed that the well-dispersed MWCNTs are stable and provide better percolated nanofiber network pathways for the effective heat conduction [178]. The inset in Figure 40.11 shows the phase contrast optical microscopic images of (i) bare MWCNT and (ii) F-MWCNT in hexadecane. In both the cases, the volume

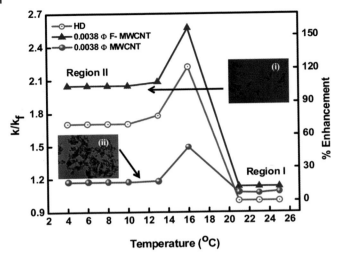

Figure 40.11 The variation of thermal conductivity as a function of temperature for bare MWCNTs and F-MWCNTs. The insets (i) and (ii) show the phase contrast optical microscopic images of bare MWCNT and F-MWCNTs, respectively, in hexadecane In all the cases, $\varphi = 0.0038$.

fraction of dispersed nanomaterial was the same ($\varphi = 0.0038$). The images confirm the presence of very large aggregates in the case of nanofluids with bare MWCNT compared to F- MWCNT, confirming better dispersibility under surface functionalization.

40.8
Conclusions

This chapter presents the recent developments in the field of nanofluids. Details of traditional nanofluids and the recently developed nanofluids are discussed. The details of various factors influencing thermal properties of nanofluids such as volume fraction, particle size, particle morphology, interfacial tension, temperature, base fluid, and agglomeration are discussed in detail. Most of the studies confirm that the thermal conductivity of stable nanofluids follow the effective mean field theoretical predictions and exhibit a temperature-independent thermal conductivity. Aggregation has a major influence on thermal conductivity where the effective heat conduction through percolating nanomaterials often yields a very high thermal conductivity. The parallel mode of heat transport through aligned nanochains in magnetic nanofluids offers extraordinary k enhancement. Nanomaterials with larger aspect ratio and phase change materials with nanoinclusions also show superior heat transfer properties. In the case of PCMs with nanoinclusions, during the freezing transition, the reduction in the effective thermal resistance due to the presence of nanomaterials at grain boundaries lead to an improved heat transport.

Acknowledgment

The authors thank the Board of Research Nuclear Sciences (BRNS) for support through a research grant for the advanced nanofluid development program.

References

1 Philip, J., Shima, P.D., and Raj, B. (2008) Nanofluid with tunable thermal properties. *Appl. Phys. Lett.*, **92**, 043108.

2 Kao, M.J. *et al.* (2007) Copper-oxide brake nanofluid manufactured using arc-submerged nanoparticle synthesis system. *J. Alloys Compd.*, **134–435**, 672–674.

3 Yu, W. and Xie, H. (2012) A review on nanofluids: preparation, stability mechanisms, and applications. *J. Nanomater.*, **2012**, 1–17.

4 Dudda, B. and Shin, D. (2013) Effect of nanoparticle dispersion on specific heat capacity of a binary nitrate salt eutectic for concentrated solar power applications. *Int. J. Therm. Sci.*, **69**, 37–42.

5 Mahian, O., Kianifar, A., Kalogirou, S.A., Pop, I., and Wongwises, S. (2013) A review of the applications of nanofluids in solar energy. *Int. J. Heat Mass Transf.*, **57**, 582–594.

6 Yang, D. *et al.* (2009) Hydrophilic multi-walled carbon nanotubes decorated with magnetite nanoparticles as lymphatic targeted drug delivery vehicles. *Chem. Commun.*, 4447–4449. doi: 10.1039/B908012K

7 Mahendran, V. and Philip, J. (2013) Naked eye visualization of defects in ferromagnetic materials and components. *NDT E Int.*, **60**, 100–109.

8 Taylor, R.A., Otanicar, T., and Rosengarten, G. (2012) Nanofluid-based optical filter optimization for PV/T systems. *Light Sci. Appl.*, **1**, 1–7.

9 Hernández, R. (2012) POLYSOLVAT-9, 9th International IUPAC Conference on Polymer–Solvent Complexes & Intercalates, Ukraine.

10 Raj, K. and Moskowitz, R. (1990) Commercial applications of ferroflus. *J. Magn. Magn. Mater.*, **85**, 233–245.

11 Xie, H., Lee, H., Youn, W., and Choi, M. (2003) Nanofluids containing multiwalled carbon nanotubes and their enhanced thermal conductivities. *J. Appl. Phys.*, **94**, 4967–4971.

12 Das, S.K., Putra, N., Thiesen, P., and Roetzel, W. (2003) Temperature dependence of thermal conductivity enhancement of nanofluids. *ASME J. Heat Transf.*, **125**, 567–574.

13 Eastman, J.A., Choi, S.U.S., Li, S., Yu, W., and Thompson, L.J. (2001) Anomalously increased effective thermal conductivities of ethylene glycol-based nanofluids containing copper nanoparticles. *Appl. Phys. Lett.*, **78**, 718–720.

14 Timofeeva, E.V. *et al.* (2007) Thermal conductivity and particle agglomeration in alumina nanofluids: experiment and theory. *Phys. Rev. E*, **76**, 061203.

15 Murshed, S.M.S., Leong, K.C., and Yang, C. (2008) Investigations of thermal conductivity and viscosity of nanofluids. *Int. J. Therm. Sci.*, **47**, 560–568.

16 Yu, W., Xie, H., Wang, X., and Wang, X. (2011) Significant thermal conductivity enhancement for nanofluids containing graphene nanosheets. *Phys. Lett. A*, **375**, 1323–1328.

17 Yu, W., Xie, H., and Bao, D. (2010) Enhanced thermal conductivities of nanofluids containing graphene oxide nanosheets. *Nanotechnology*, **21**, 055705.

18 Yu, W., Xie, H., and Chen, W. (2010) Experimental investigation on thermal conductivity of nanofluids containing graphene oxide nanosheets. *J. Appl. Phys.*, **107**, 094317.

19 Reinecke, B.N., Shan, J.W., Suabedissen, K.K., and Cherkasova, A.S. (2008) On the anisotropic thermal conductivity of magnetorheological suspensions. *J. Appl. Phys.*, **104**, 023507.

20 Cha, G., Ju, Y.S., Ahure, L.A., and Wereley, N.M. (2010) Experimental characterization of thermal conductance switching in magnetorheological fluids. *J. Appl. Phys.*, **107**, 09B505.

21 Sundar, L.S. *et al.* (2014) Enhanced thermal conductivity and viscosity of nanodiamond–nickel nanocomposite nanofluids. *Sci. Rep.*, **4**, 14.

22 Mehrali, M. *et al.* (2014) Investigation of thermal conductivity and rheological properties of nanofluids containing graphene nanoplatelets. *Nanoscale Res. Lett.*, **9**, 1–12.

23 Vajjha, R.S. and Das, D.K. (2009) Experimental determination of thermal conductivity of three nanofluids and development of new correlations. *Int. J. Heat Mass Transf.*, **52**, 4675–4682.

24 Kim, S.H., Choi, S.R., and Kim, D. (2007) Thermal conductivity of metal-oxide nanofluids: particle size dependence and effect of laser irradiation. *ASME J. Heat Transf.*, **129**, 298–307.

25 Mintsa, H.A., Roy, G., Nguyen, C.T., and Doucet, D. (2009) New temperature dependent thermal conductivity data for water-based nanofluids. *Int. J. Therm. Sci.*, **48**, 363–371.

26 Chon, C.H., Kihm, K.D., Lee, S.P., and Choi, S.U.S. (2005) Empirical correlation finding the role of temperature and particle size for nanofluid (Al$_2$O$_3$) thermal conductivity enhancement. *Appl. Phys. Lett.*, **87**, 153107.

27 Beck, M.P., Yuan, Y., Warrier, P., and Teja, A.S. (2010) The thermal conductivity of alumina nanofluids in water, ethylene glycol, and ethylene glycol+water mixtures. *J. Nanopart. Res.*, **12**, 1469–1477.

28 Beck, M.P., Yuan, Y., Warrier, P., and Teja, A.S. (2010) The thermal

conductivity of aqueous nanofluids containing ceria nanoparticles. *J. Appl. Phys.*, **107**, 066101.

29 Murshed, S.M.S., Leong, K.C., and Yang, C. (2005) Enhanced thermal conductivity of TiO$_2$–water-based nanofluids. *Int. J. Therm. Sci.*, **44**, 367–373.

30 Xie, H., Wang, J., Xi, T., and Liu, Y. (2002) Thermal conductivity of suspensions containing nanosized SiC particles. *Int. J. Thermophys.*, **23**, 571–580.

31 Yang, Y., Grulke, E.A., Zhang, Z.G., and Wu, G. (2006) Thermal and rheological properties of carbon nanotube-in-oil dispersions. *J. Appl. Phys.*, **99**, 114307.

32 Glory, J., Bonetti, M., Helezen, M., Hermite, M.M.L., and Reynaud, C. (2008) Thermal and electrical conductivities of water-based nanofluids prepared with long multiwalled carbon nanotubes. *J. Appl. Phys.*, **103**, 094309.

33 Lee, K.J., Yoon, S.H., and Jang, Y. (2007) Carbon nanofibers: a novel nanofiller for nanofluid applications. *Small*, **3**, 1209–1213.

34 Assael, M.J., Metaxa, I.N., Arvanitidis, J., Christofilos, D., and Lioutas, C. (2005) Thermal conductivity enhancement in aqueous suspensions of carbon multi-walled and double-walled nanotubes in the presence of two different dispersants. *Int. J. Thermophys.*, **26**, 647–664.

35 Timofeeva, E.V., Routbort, J.L., and Singh, D. (2009) Particle shape effects on thermophysical properties of alumina nanofluids. *J. Appl. Phys.*, **106**, 014304.

36 Philip, J., Shima, P.D., and Raj, B. (2007) Enhancement of thermal conductivity in magnetite based nanofluid due to chainlike structures. *Appl. Phys. Lett.*, **91**, 203108.

37 Li, Q., Xuan, Y., and Wang, J. (2005) Experimental investigations on transport properties of magnetic fluids. *Exp. Therm. Fluid Sci.*, **30**, 109–116.

38 Parekh, K. and Lee, H.S. (2010) Magnetic field induced enhancement in thermal conductivity of magnetite nanofluid. *J. Appl. Phys.*, **107**, 09A310.

39 Wensel, J. *et al.* (2008) Enhanced thermal conductivity by aggregation in heat transfer nanofluids containing metal oxide nanoparticles and carbon nanotubes. *Appl. Phys. Lett.*, **92**, 023110.

40 Xuan, Y., Huang, Y., and Li, Q. (2009) Experimental investigation on thermal conductivity and specific heat capacity of magnetic microencapsulated phase change material suspension. *Chem. Phys. Lett.*, **479**, 264–269.

41 Hong, H. *et al.* (2007) Enhanced thermal conductivity by the magnetic field in heat transfer nanofluids containing carbon nanotube. *Synth. Met.*, **157**, 437–440.

42 Prasher, R., Phelan, P.E., and Bhattacharya, P. (2006) Effect of aggregation kinetics on the thermal conductivity of nanoscale colloidal solutions (nanofluid) *Nano Lett.*, **6**, 1529–1534.

43 Jana, S., Khojin, A.S., and Zhong, W.H. (2007) Enhancement of fluid thermal conductivity by the addition of single and hybrid nano-additives. *Thermochim. Acta*, **462**, 45.

44 Kim, Y.J., Ma, H., and Yu, Q. (2010) Plasma nanocoated carbon nanotubes for heat transfer nanofluids. *Nanotechnology*, **21**, 295703.

45 Gowda, R. *et al.* (2010) Effects of particle surface charge, species, concentration, and dispersion method on the thermal conductivity of nanofluids. *Adv. Mech. Eng.*, **2010**, 1–10.

46 Patel, H.E., Sundararajan, T., and Das, S.K. (2010) An experimental investigation into the thermal conductivity enhancement in oxide and metallic nanofluids. *J. Nanopart. Res.*, **12**, 1015–1031.

47 Yoo, D.H., Hong, K.S., and Yang, H.S. (2007) Study of thermal conductivity of nanofluids for the application of heat transfer fluids. *Thermochim. Acta*, **455**, 66–69.

48 Hwang, Y.J. *et al.* (2006) Investigation on characteristics of thermal conductivity enhancement of nanofluids. *Curr. Appl. Phys.*, **6**, 1068.

49 Shima, P.D. and Philip, J. (2014) Role of thermal conductivity of dispersed nanoparticles on heat transfer properties of nanofluid. *Ind. Eng. Chem. Res.*, **53**, 980–988.

50 Wang, X.J., Zhu, D.S., and Yang, S. (2009) Investigation of pH and SDBS on

enhancement of thermal conductivity in nanofluids. *Chem. Phys. Lett.*, **470**, 107–111.

51 Zhu, D. *et al.* (2009) Dispersion behavior and thermal conductivity characteristics of Al_2O_3–H_2O nanofluids. *Curr. Appl. Phys.*, **9**, 131–139.

52 Wamkam, C.T., Opoku, M.K., Hong, H., and Smith, P. (2011) Effects of pH on heat transfer nanofluids containing ZrO_2 and TiO_2 nanoparticles. *J. Appl. Phys.*, **109**, 024305.

53 Younes, H., Christensen, G., Luan, X., Hong, H., and Smith, P. (2012) Effects of alignment, pH, surfactant, and solvent on heat transfer nanofluids containing Fe_2O_3 and CuO nanoparticles. *J. Appl. Phys.*, **111**, 064308.

54 Li, X.F. *et al.* (2008) Thermal conductivity enhancement dependent pH and chemical surfactant for Cu–H_2O nanofluids. *Thermochim. Acta*, **469**, 98–103.

55 Karthikeyan, N.R., Philip, J., and Raj, B. (2008) Effect of clustering on the thermal conductivity of nanofluids. *Mater. Chem. Phys.*, **109**, 50–55.

56 Meibodi, M.E. *et al.* (2010) The role of different parameters on the stability and thermal conductivity of carbon nanotube/water nanofluids. *Int. Commun. Heat Mass Transf.*, **37**, 319–323.

57 Amrollahi, A., Hamidi, A.A., and Rashidi, A.M. (2008) The effects of temperature, volume fraction and vibration time on the thermo-physical properties of a carbon nanotube suspension (carbon nanofluid). *Nanotechnology*, **19**, 315701.

58 Garg, P. *et al.* (2009) An experimental study on the effect of ultrasonication on viscosity and heat transfer performance of multi-wall carbon nanotube-based aqueous nanofluids. *Int. J. Heat Mass Transf.*, **52**, 5090–5101.

59 Assael, M.J., Metaxa, I.N., Kakosimos, K., and Constantinou, D. (2006) Thermal conductivity of nanofluids: experimental and theoretical. *Int. J. Thermophys.*, **27**, 999–1017.

60 Liu, M.S., Lin, M.C.C., Huang, I.T., and Wang, C.C. (2005) Enhancement of thermal conductivity with carbon

nanotube for nanofluids. *Int. Commun. Heat Mass Transf.*, **32**, 1202–1210.

61 Buongiorno, J. *et al.* (2009) A benchmark study on the thermal conductivity of nanofluids. *J. Appl. Phys.*, **106**, 094312.

62 Chen, L., Xie, H., Li, Y., and Yu, W. (2008) Nanofluids containing carbon nanotubes treated by mechanochemical reaction. *Thermochim. Acta*, **477**, 21–24.

63 Xie, H. *et al.* (2002) Thermal conductivity enhancement of suspensions containing nanosized alumina particles. *J. Appl. Phys.*, **91**, 4568–4570.

64 Masuda, H., Ebata, A., Teramae, K., and Hishinuma, N. (1993) Alteration of thermal conductivity and viscosity of liquid by dispersing ultra-fine particles (dispersion of γ-Al_2O_3, SiO_2, and TiO_2 ultra-fine particles). *Netsu Bussei*, 7, 227–233.

65 Teng, T.P., Hung, Y.H., Teng, T.C., Moa, H.E., and Hsu, H.G. (2010) The effect of alumina/water nanofluid particle size on thermal conductivity. *Appl. Therm. Eng.*, **30**, 2213–2218.

66 Cho, T., Baek, I., Lee, J., and Park, S. (2005) Preparation of nanofluids containing suspended silver particles for enhancing fluid thermal conductivity of fluids. *J. Ind. Eng. Chem.*, **11**, 400.

67 Mostafizur, R.M., Bhuiyan, M.H.U., Saidur, R., and Aziz, A.R.A. (2014) Thermal conductivity variation for methanol based nanofluids. *Int. J. Heat Mass Transf.*, **76**, 350–356.

68 Beck, M.P., Yuan, Y., Warrier, P., and Teja, A.S. (2008) The effect of particle size on the thermal conductivity of alumina nanofluids. *J. Nanopart. Res.*, **11**, 1129–1136.

69 Lee, S., Choi, S.U.S., Li, S., and Eastman, J.A. (1999) Measuring thermal conductivity of fluids containing oxide nanoparticles. *ASME J. Heat Transf.*, **121**, 280–289.

70 Jiang, H., Li, H., Xu, Q., and Shi, L. (2014) Effective thermal conductivity of nanofluids considering interfacial nano-shells. *Mater. Chem. Phys.*, **148**, 195–200.

71 Xuan, Y. and Li, Q. (2000) Heat transfer enhancement of nanofluids. *Int. J. Heat Fluid Flow*, **21**, 58–64.

72 Shin, S. and Lee, S.-H. (2000) Thermal conductivity of suspensions in shear flow fields. *Int. J. Heat Mass Transf.*, **43**, 4275–4284.

73 Khedkar, R.S., Kiran, A.S., Sonawane, S.S., Wasewar, K.L., and Umare, S.S. (2013) Thermo-physical properties measurement of water based Fe_3O_4 nanofluids. *Carbon Sci. Technol.*, **5**, 187–191.

74 Zhu, H., Zhang, C., Liu, S., Tang, Y., and Yin, Y. (2006) Effects of nanoparticle clustering and alignment on thermal conductivities of Fe_3O_4 aqueous nanofluids. *Appl. Phys. Lett.*, **89**, 023123.

75 Shima, P.D., Philip, J., and Raj, B. (2010) Influence of aggregation on thermal conductivity in stable and unstable nanofluids. *Appl. Phys. Lett.*, **97**, 153113.

76 Angayarkanni, S.A. and Philip, J. (2014) Effect of nanoparticles aggregation on thermal and electrical conductivities of nanofluids. *J. Nanofluids*, **3**, 17–25.

77 Chopkar, M., Kumar, S., Bhandari, D.R., Das, P.K., and Mannaa, I. (2007) Development and characterization of Al_2Cu and Ag_2Al nanoparticle dispersed water and ethylene glycol based nanofluid. *Mater. Sci. Eng. B*, **139**, 141–148.

78 He, Y. *et al.* (2007) Heat transfer and flow behaviour of aqueous suspensions of TiO_2 nanoparticles (nanofluids) flowing upward through a vertical pipe. *Int. J. Heat Mass Transf.*, **50**, 2272–2281.

79 Timofeeva, E.V. *et al.* (2010) Particle size and interfacial effects on thermo-physical and heat transfer characteristics of water-based α-SiC nanofluids. *Nanotechnology*, **21**, 215703.

80 Shalkevich, N. *et al.* (2010) On the thermal conductivity of gold nanoparticle colloids. *Langmuir*, **26**, 663–670.

81 Shima, P.D. and Philip, J. (2011) Tuning of thermal conductivity and rheology of nanofluids using an external stimulus. *J. Phys. Chem. C*, **115**, 20097–20104.

82 Hong, T.K., Yang, H.S., and Choi, C.J. (2005) Study of the enhanced thermal conductivity of Fe nanofluids. *J. Appl. Phys.*, **97**, 064311.

83 Li, C.H. and Peterson, G.P. (2006) Experimental investigation of

temperature and volume fraction variations on the effective thermal conductivity of nanoparticle suspensions (nanofluids) *J. Appl. Phys.*, **99**, 084314.

84 Hong, K.S., Hong, T.K., and Yang, H.S. (2006) Thermal conductivity of Fe nanofluids depending on the cluster size of nanoparticles. *Appl. Phys. Lett.*, **88**, 031901.

85 Chopkar, M., Sudarshan, S., Das, P.K., and Manna, I. (2008) Effect of particle size on thermal conductivity of nanofluid. *Metal. Mater Trans. A*, **39**, 1535–1542.

86 Sinha, K., Kavlicoglu, B., Liu, Y., Gordaninejad, F., and Graeve, O.A. (2009) A comparative study of thermal behavior of iron and copper nanofluids. *J. Appl. Phys.*, **106**, 064307.

87 Mueller, F.H. and Houwink, R. (1942) Chemie und technologie der kunststoffe, Leipzig.

88 Angayarkanni, S.A. and Philip, J. (2013) Role of adsorbing moieties on thermal conductivity and associated properties of nanofluids. *J. Phys. Chem. C*, **117**, 9009–9019.

89 Das, S.K., Putra, N., and Roetzel, W. (2003) Pool boiling characteristics of nanofluids. *Int. J. Heat Mass Transf.*, **46**, 851–862.

90 Singh, D. *et al.* (2009) An investigation of silicon carbide–water nanofluid for heat transfer applications. *J. Appl. Phys.*, **105**, 064306.

91 Shima, P.D., Philip, J., and Raj, B. (2010) Synthesis of aqueous and nonaqueous iron oxide nanofluids and study of temperature dependence on thermal conductivity and viscosity. *J. Phys. Chem. C*, **114**, 18825–18833.

92 Gao, J.W., Zheng, R.T., Ohtani, H., Zhu, D.S., and Chen, G. (2009) Experimental investigation of heat conduction mechanisms in nanofluids: clue on clustering. *Nano Lett.*, **9**, 4128–4132.

93 Yu, W., Xie, H., Chen, L., and Li, Y. (2010) Enhancement of thermal conductivity of kerosene-based Fe_3O_4 nanofluids prepared via phase-transfer method. *Colloids Surf. A*, **355**, 109–113.

94 Duangthongsuk, W. and Wongwises, S. (2009) Measurement of temperature-dependent thermal conductivity and viscosity of TiO_2–water nanofluids. *Exp. Therm. Fluid Sci.*, **33**, 706–714.

95 Yang, B. and Han, Z.H. (2006) Temperature-dependent thermal conductivity of nanorod-based nanofluids. *Appl. Phys. Lett.*, **89**, 083111.

96 Zeng, Y.X., Zhong, X.W., Liu, Z.Q., Chen, S., and Li, N. (2013) Preparation and enhancement of thermal conductivity of heat transfer oil-based MoS_2 nanofluids. *J. Nanomater.*, **2013**, 1–6.

97 Yu, Q., Kim, Y.J., and Ma, H. (2008) Nanofluids with plasma treated diamond nanoparticles. *Appl. Phys. Lett.*, **92**, 103111.

98 Nasiri, A., Niasar, M.S., Rashidi, A., Amrollahi, A., and Khodafarin, R. (2011) Effect of dispersion method on thermal conductivity and stability of nanofluid. *Exp. Therm. Fluid Sci.*, **35**, 717–723.

99 Yu, W., Xie, H., Chen, L., and Li, Y. (2009) Investigation of thermal conductivity and viscosity of ethylene glycol based ZnO nanofluid. *Thermochim. Acta*, **491**, 92–96.

100 Bjorneklett, A., Halbo, L., and Kristiansen, H. (1992) Thermal conductivity of epoxy adhesives filled with silver particles. *Int. J. Adhes. Adhes.*, **12**, 99–104.

101 Branson, B.T., Beauchamp, P.S., Beam, J.C., Lukehart, C.M., and Davidson, J.L. (2013) Nanodiamond nanofluids for enhanced thermal conductivity. *ACS Nano*, **7**, 3183–3189.

102 Ponmani, S., William, J.K.M., Samuel, R., Nagarajan, R., and Sangwai, J.S. (2014) Formation and characterization of thermal and electrical properties of CuO and ZnO nanofluids in xanthan gum. *Colloids Surf. A*, **443**, 37–43.

103 Esfe, M.H., Saedodin, S., Wongwises, S., and Toghraie, D. (2015) An experimental study on the effect of diameter on thermal conductivity and dynamic viscosity of Fe/water nanofluids. *J. Therm. Anal. Calorim.*, **119**, 1817–1824.

104 Jiang, H., Xu, Q., Huang, C., and Shi, L. (2015) Effect of temperature on the effective thermal conductivity of *n*-tetradecane-based nanofluids containing copper nanoparticles. *Particuology*, **22**, 95–99.

105 Sun, C., Bai, B., Lu, W.-Q., and Liu, J. (2013) Shear-rate dependent effective thermal conductivity of H_2O+SiO_2 nanofluids. *Phys. Fluids*, **25**, 052002.

106 Xia, G., Jiang, H., and Ran Liu, Y.Z. (2014) Effects of surfactant on the stability and thermal conductivity of Al_2O_3/de-ionized water nanofluids. *Int. J. Therm. Sci.*, **84**, 118–124.

107 Lee, J.H., Lee, S.H., and Jang, S.P. (2014) Do temperature and nanoparticle size affect the thermal conductivity of alumina nanofluids? *Appl. Phys. Lett.*, **104**, 161908.

108 Choi, S.U.S., Zhang, Z.G., Yu, W., Lockwood, F.E., and Grulke, E.A. (2001) Anomalous thermal conductivity enhancement in nanotube suspensions. *Appl. Phys. Lett.*, **79**, 2252–2254.

109 Han, Z.H., Yang, B., Kim, S.H., and Zachariah, M.R. (2007) Application of hybrid sphere/carbon nanotube particles in nanofluids. *Nanotechnology*, **18**, 105701.

110 Shaikha, S., Lafdi, K., and Ponnappan, R. (2007) Thermal conductivity improvement in carbon nanoparticle doped PAO oil: an experimental study. *J. Appl. Phys.*, **101**, 064302.

111 Jha, N. and Ramaprabhua, S. (2009) Thermal conductivity studies of metal dispersed multiwalled carbon nanotubes in water and ethylene glycol based nanofluid. *J. Appl. Phys.*, **106**, 084317.

112 Jiang, W., Ding, G., and Peng, H. (2009) Measurement and model on thermal conductivities of carbon nanotube nanorefrigerants. *Int. J. Therm. Sci.*, **48**, 1108–1115.

113 Horton, M. *et al.* (2010) Magnetic alignment of Ni-coated single wall carbon nanotubes in heat transfer nanofluids. *J. Appl. Phys.*, **107**, 104320.

114 Wright, B. *et al.* (2007) Magnetic field enhanced thermal conductivity in heat transfer nanofluids containing Ni coated single wall carbon nanotubes. *Appl. Phys. Lett.*, **91**, 173116.

115 Ghozatloo, A., Rashidi, A.M., and Shariaty-Niasar, M. (2014) Effects of surface modification on the dispersion and thermal conductivity of CNT/water nanofluids. *Int. Commun. Heat Mass Transf.*, **54**, 1–7.

116 Li, D., Zhao, W., Liu, Z., and Zhu, B. (2011) Experimental investigation of heat transfer enhancement of the heat pipe using CuO–water nanofluid. *Adv. Mater. Res.*, **160–162**, 507–512.

117 Shima, P.D., Philip, J., and Raj, B. (2009) Magnetically controllable nanofluid with tunable thermal conductivity and viscosity. *Appl. Phys. Lett.*, **95**, 133112.

118 Laskar, J.M., Philip, J., and Raj, B. (2009) Experimental evidence for reversible zippering of chains in magnetic nanofluids under external magnetic fields. *Phys. Rev. E*, **80**, 041401.

119 Martin-Gallego, M. *et al.* (2011) Thermal conductivity of carbon nanotubes and graphene in epoxy nanofluids and nanocomposites. *Nanoscale Res. Lett.*, **6**, 1–7.

120 Nika, D.L., Askerov, A.S., and Balandin, A.A. (2012) Anomalous size dependence of the thermal conductivity of graphene ribbons. *Nano Lett.*, **12**, 3238–3244.

121 Kole, M. and Dey, T.K. (2013) Investigation of thermal conductivity, viscosity, and electrical conductivity of graphene based nanofluids. *J. Appl. Phys.*, **113**, 084307.

122 Hadadian, M., Goharshadi, E.K., and Youssef, A. (2014) Electrical conductivity, thermal conductivity, and rheological properties of graphene oxide-based nanofluids. *J. Nanopart. Res.*, **16**, 2788.

123 Zhang, H.-S., Guo, Z.-X., Gong, X.-G., and Cao, J.-X. (2012) Thermal conductivity of sawtooth-like graphene nanoribbons: a molecular dynamics study. *J. Appl. Phys.*, **112**, 123508.

124 Guo, Z., Zhang, D., and Gonga, X.-G. (2009) Thermal conductivity of graphene nanoribbons. *Appl. Phys. Lett.*, **95**, 163103.

125 Hu, J., Schiffli, S., Vallabhaneni, A., Ruan, X., and Chen, Y.P. (2010) Tuning the thermal conductivity of graphene nanoribbons by edge passivation and isotope engineering: a molecular dynamics study. *Appl. Phys. Lett.*, **97**, 133107.

126 Aravind, S.S.J. and Ramaprabhu, S. (2012) Graphene wrapped multiwalled carbon

nanotubes dispersed nanofluids for heat transfer applications. *J. Appl. Phys.*, **112**, 124304.

127 Baby, T.T. and Ramaprabhu, S. (2010) Investigation of thermal and electrical conductivity of graphene based nanofluids. *J. Appl. Phys.*, **108**, 124308.

128 Lee, G.-J. and Rhee, C.K. (2014) Enhanced thermal conductivity of nanofluids containing graphene nanoplatelets prepared by ultrasound irradiation. *J. Mater. Sci.*, **49**, 1506–1511.

129 Aravind, S.S.J. and Ramaprabhu, S. (2013) Graphene–multiwalled carbon nanotube-based nanofluids for improved heat dissipation. *RSC Adv.*, **3**, 4199–4206.

130 Gupta, S.S. *et al.* (2011) Thermal conductivity enhancement of nanofluids containing graphene nanosheets. *J. Appl. Phys.*, **110**, 084302.

131 Chu, K., Jia, C.-C., and Li, W.-S. (2012) Effective thermal conductivity of graphene-based composites. *Appl. Phys. Lett.*, **101**, 121916.

132 Chu, K., Li, W.S., Jia, C.C., and Tang, F.-l. (2012) Thermal conductivity of composites with hybrid carbon nanotubes and graphene nanoplatelets. *Appl. Phys. Lett.*, **101**, 211903.

133 Ma, L. *et al.* (2014) Viscosity and thermal conductivity of stable graphite suspensions near percolation. *Nano Lett.*, **15**, 127–133.

134 Abbasi, S.M., Rashidi, A., Nemati, A., and Arzani, K. (2013) The effect of functionalisation method on the stability and the thermal conductivity of nanofluid hybrids of carbon nanotubes/gamma alumina. *Ceram. Int.*, **39**, 3885–3891.

135 Chen, L., Yu, W., and Xie, H. (2012) Enhanced thermal conductivity of nanofluids containing Ag/MWNT composites. *Powder Technol.*, **231**, 18–20.

136 Botha, S.S., Ndungu, P., and Bladergroen, B.J. (2011) Physicochemical properties of oil-based nanofluids containing hybrid structures of silver nanoparticles supported on silica. *Ind. Eng. Chem. Res.*, **50**, 3071–3077.

137 Suresh, S., Venkitaraj, K.P., Selvakumar, P., and Chandrasekar, M. (2011) Synthesis of Al_2O_3–Cu/water hybrid nanofluids using two step method and its thermo physical properties. *Colloids Surf. A*, **388**, 41–48.

138 Paul, G., Philip, J., Raj, B., Das, P.K., and Manna, I. (2011) Synthesis, characterization, and thermal property measurement of nano-$Al_{95}Zn_{05}$ dispersed nanofluid prepared by a two-step process. *Int. J. Heat Mass Transf.*, **54**, 3783–3788.

139 Baghbanzadeh, M., Rashidi, A., Rashtchian, D., Lotfi, R., and Amrollahi, A. (2012) Synthesis of spherical silica/multiwall carbon nanotubes hybrid nanostructures and investigation of thermal conductivity of related nanofluids. *Thermochim. Acta*, **549**, 87–94.

140 Ho, C.J., Huang, J.B., Tsai, P.S., and Yang, Y.M. (2010) Preparation and properties of hybrid water-based suspension of Al_2O_3 nanoparticles and MEPCM particles as functional forced convection fluid. *Int. Commun. Heat Mass Transf.*, **37**, 490–494.

141 Suresh, S., Venkitaraj, K.P., Selvakumar, P., and Chandrasekar, M. (2012) Effect of Al_2O_3–Cu/water hybrid nanofluid in heat transfer. *Exp. Therm. Fluid. Sci*, **38**, 54–60.

142 Nine, M.J., Munkhbayar, B., Rahman, M.S., Chung, H., and Jeong, H. (2013) Highly productive synthesis process of well dispersed Cu_2O and Cu/Cu_2O nanoparticles and its thermal characterization. *Mater. Chem. Phys.*, **141**, 636–642.

143 Baby, T.T. and Sundara, R. (2011) Synthesis and transport properties of metal oxide decorated graphene dispersed nanofluids. *J. Phys. Chem. C.*, **115**, 8527–8533.

144 Baby, T.T. and Ramaprabhu, S. (2011) Experimental investigation of the thermal transport properties of a carbon nanohybrid dispersed nanofluid. *Nanoscale*, **3**, 2208–2214.

145 Nine, M.J., Batmunkh, M., Kim, J.H., Chung, H.S., and Jeong, H.-M. (2012) Investigation of Al_2O_3–MWCNTs hybrid dispersion in water and their thermal characterization. *J. Nanosci. Nanotechnol.*, **12**, 4553–4559.

146 Baby, T.T. and Ramaprabhu, S. (2011) Synthesis and nanofluid application of silver nanoparticles decorated graphene. *J. Mater. Chem.*, **21**, 9702–9709.

147 Baby, T.T. and Sundara, R. (2013) Synthesis of silver nanoparticle decorated multiwalled carbon nanotubes–graphene mixture and its heat transfer studies in nanofluid. *AIP Adv.*, **3**, 012111.

148 Chen, L., Cheng, M., Yang, D., and Yang, L. (2014) Enhanced thermal conductivity of nanofluid by synergistic effect of multi-walled carbon nanotubes and Fe_2O_3 nanoparticles. *Appl. Mech. Mater.*, **548**, 118–123.

149 Sari, A. and Karaipekli, A. (2007) Thermal conductivity and latent heat thermal energy storage characteristics of paraffin/expanded graphite composite as phase change material. *Appl. Therm. Eng.*, **27**, 1271–1277.

150 Ding, Y., Alias, H., Wen, D., and Williams, R.A. (2006) Heat transfer of aqueous suspensions of carbon nanotubes (CNT nanofluids). *Int. J. Heat Mass Transf.*, **49**, 240–250.

151 Lamas, B.C. *et al.* (2011) EG/CNTs nanofluids engineering and thermo-rheological characterization. *J. Nano Res.*, **13**, 69–74.

152 Patel, H.E., Anoop, K.B., Sundararajan, T., and Das, S.K. (2008) Model for thermal conductivity of CNT-nanofluids. *Bull. Mater. Sci.*, **31**, 387.

153 Gao, T. and Jelle, B.P. (2013) Thermal conductivity of TiO_2 nanotubes. *J. Phys. Chem. C*, **117**, 1401–1408.

154 He, Y., Men, Y., Zhao, Y., Lu, H., and Ding, Y. (2009) Numerical investigation into the convective heat transfer of TiO_2 nanofluids flowing through a straight tube under the laminar flow conditions. *Appl. Therm. Eng.*, **29**, 1965.

155 Duan, F., Kwek, D., and Crivoi, A. (2011) Viscosity affected by nanoparticle aggregation in Al_2O_3–water nanofluids. *Nanoscale Res. Lett.*, **6**, 248.

156 Faugeras, C. *et al.* (2010) Thermal conductivity of graphene in corbino membrane geometry. *ACS Nano*, **4**, 1889–1892.

157 Lee, S.D.P.S.W. *et al.* (2010) Effects of nanofluids containing graphene/ graphene-oxide nanosheets on critical heat flux. *Appl. Phys. Lett.*, **97**, 023103.

158 Barrau, S., Demont, P., Peigney, A., Laurent, C., and Lacabanne, C. (2003) DC and AC conductivity of carbon nanotubes–polyepoxy composites. *Macromolecules*, **36**, 5187–5194.

159 Bryning, M.B., Milkie, D.E., Islam, M.F., Kikkawa, J.M., and Yodh, A.G. (2005) Thermal conductivity and interfacial resistance in single-wall carbon nanotube epoxy composites. *Appl. Phys. Lett.*, **87**, 161909.

160 Cipriano, B.H. *et al.* (2008) Conductivity enhancement of carbon nanotube and nanofiber-based polymer nanocomposites by melt annealing. *Polymer*, **49**, 4846–4851.

161 Huang, X., Iizuka, T., Jiang, P., Ohki, Y., and Tanaka, T. (2012) Role of interface on the thermal conductivity of highly filled dielectric epoxy/AlN composites. *J. Phys. Chem. C*, **116**, 13629–13639.

162 Marconnet, A., Yamamoto, N., Panzer, M.A., and L.Wardle, B., and Goodson, K.E. (2011) Thermal conduction in aligned carbon nanotube polymer nanocomposites with high packing density. *ACS Nano*, **5**, 4818–4825.

163 Kuznik, F., David, D., Johannes, K., and Roux, J.J. (2011) A review on phase change materials integrated in building walls. *Renew. Sustain. Energy Rev.*, **15**, 379–391.

164 Shatikian, V., Ziskind, G., and Letan, R. (2008) Numerical investigation of a PCM-based heat sink with internal fins: constant heat flux. *Int. J. Heat Mass Transf.*, **51**, 1488–1493.

165 Karaipekli, A. and Sarı, A. (2009) Capric–myristic acid/vermiculite composite as form-stable phase change material for thermal energy storage. *Sol. Energy*, **83**, 323–332.

166 Ai, D., Su, L., Gao, Z., Denga, C., and Dai, X. (2010) Study of ZrO_2 nanopowders based stearic acid phase change materials. *Particuology*, **8**, 394–397.

167 Wu, S., Zhu, D., Li, X., Li, H., and Lei, J. (2009) Thermal energy storage behavior of Al_2O_3–H_2O nanofluids. *Thermochim. Acta*, **483**, 73.

168 Zeng, J.L., Cao, Z., Yang, D.W., Sun, L.X., and Zhang, L. (2010) Thermal conductivity enhancement of Ag nanowires on an organic phase change material. *J. Therm. Anal. Calorim.*, **101**, 385–389.

169 Babu, K. and Kumar, T.S.P. (2011) Effect of CNT concentration and agitation on surface heat flux during quenching in CNT nanofluids. *Int. J. Heat Mass Transf.*, **54**, 106–117.

170 Angayarkanni, S.A. and Philip, J. (2014) Tunable thermal transport in phase change materials using inverse micellar templating and nanofillers. *J. Phys. Chem. C*, **118**, 13972–13980.

171 Harish, S., Ishikawa, K., Chiashi, S., Shiomi, J., and Maruyama, S. (2013) Anomalous thermal conduction characteristics of phase change composites with single-walled carbon nanotube inclusions. *J. Phys. Chem. C*, **117**, 15409–15413.

172 Sun, P.C. *et al.* (2013) Room temperature electrical and thermal switching CNT/hexadecane composites. *Adv. Mater.*, **25**, 4938–4943.

173 Zheng, R., Gao, J., Wang, J., and Chen, G. (2011) Reversible temperature regulation of electrical and thermal conductivity using liquid–solid phase transitions. *Nat. Commun.*, **2**, 1–6.

174 Babaei, H., Keblinski, P., and Khodadadi, J.M. (2013) Thermal conductivity enhancement of paraffins by increasing the alignment of molecules through adding CNT/graphene. *Int. J. Heat Mass Transf.*, **58**, 209–216.

175 Sutherland, R., Davis, R., and Seyer, W. (1959) Heat transfer effects: molecular orientation of octadecane. *Ind. Eng. Chem.*, **51**, 585–588.

176 Powell, R.W., Challoner, A.R., and Seyer, W.F. (1961) Correspondence. Measurement of thermal conductivity of *n*-octadecane. *Ind. Eng. Chem.*, **53**, 581–582.

177 Lyeo, H.K. *et al.* (2006) Thermal conductivity of phase-change material $Ge_2Sb_2Te_5$. *Appl. Phys. Lett.*, **89**, 151904.

178 Schiffres, S., Harish, S., Maruyama, S., Shiomi, J., and Malen, J.A. (2013) Tunable electrical and thermal transport in ice-templated multilayer graphene nanocomposites through freezing rate control. *ACS Nano*, **7**, 11183–11189.

179 Patel, H.A., Garde, S., and Keblinski, P. (2005) Thermal resistance of nanoscopic liquid–liquid interfaces: dependence on chemistry and molecular architecture. *Nano Lett.*, **5**, 2225–2231.

Part Six
Technologies, Intellectual Property, and Markets

41

Nanomaterials for Li-Ion Batteries: Patents Landscape and Product Scenario

Md Shakeel Iqbal, Nisha C. Kalarickal, Vivek Patel, and Ratnesh Kumar Gaur

International Advanced Research Centre for Powder Metallurgy & New Materials (ARCI), Centre for Knowledge Management of Nanoscience & Technology, Balapur, Hyderabad 500005, Telangana, India

41.1
Introduction

Efficient energy storage systems have become the pivotal technology for the energy-driven modern society. Highly efficient, low cost, and environment-friendly energy storage media are extremely necessary to meet the diverse needs of modern society such as hybrid/all-electric automobiles, myriad of portable electronic devices, medical implants, and for the effective utilization of renewable energy sources in future smart grids, as well as power delivery systems. Among the various secondary battery technologies, lithium-ion batteries are on the forefront of the latest technology, and have quickly dominated the market for consumer electronics. Owing to the various advantages, such as high voltage, high energy-to-weight ratio (i.e., energy density), long cyclic life, no memory effect, and low self-discharge when not in service, lithium-ion batteries are also aggressively entering the more power-demanding applications such as power tool equipments, defense, aerospace, and even in the emerging electric vehicle market.

41.2
Lithium-Ion Battery: Basic Concepts

A lithium-ion battery cell consists of a positive electrode (cathode) as the source of lithium ions and a negative electrode (anode), which can accept lithium ions separated by a liquid or solid electrolyte. An ion permeable separator ensures no direct contact between the electrodes by blocking electron transport through it. During charging process, lithium ions deintercalate from the source material (cathode) and move through the electrolyte and intercalate in the anode

Nanotechnology for Energy Sustainability, First Edition. Edited by Baldev Raj, Marcel Van de Voorde, and Yashwant Mahajan.

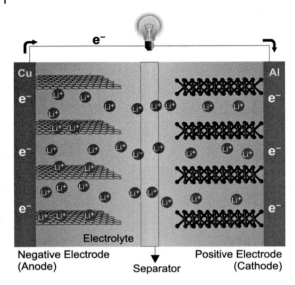

Figure 41.1 Schematic illustration of discharging process in a typical lithium-ion battery cell.

material. During discharging, this process is reversed and change in the charge distribution inside the host material skeleton and an overall change in the material causes electrons to flow in the external circuit, which will be available to perform electrical work. Figure 41.1 shows the schematic representation of discharging process occurring in a typical lithium-ion battery cell.

The main reactions occurring during the charging–discharging process in Li-ion batteries are (i) diffusion of Li ions within solid electrode materials, (ii) charge transfer at electrode–electrolyte interface, and (iii) movement of Li^+ ion in the electrolyte. Among these reactions, the rate-controlling step is considered to be the solid state diffusion of Li ions as it restricts the rate of charge/discharge, resulting in a low power output. A faster diffusion can be achieved in a system with higher diffusion coefficient and lower diffusion length. The basic property requirements for active materials used as electrodes are high reversible capacity, good structural flexibility and stability, fast Li-ion diffusion, long cycle life, improved safety, low cost, and environmental benignity. The key figures of merit that determine the performance of a rechargeable Li-ion battery are the energy density (amount of energy stored per unit mass or volume) and power density (the maximum practical sustained power output per unit mass or volume).

The first commercially available Li-ion battery was introduced by Sony in collaboration with Asahi Kasei in the year 1991 and since then a lot of research and development has happened in this area. Threefold increase in energy density has been achieved since the first commercialization of Li-ion battery and at the same time cost has come down by a factor of 10 during the same period [1]. Since lithium-ion batteries are used in diverse applications, they are customized

Table 41.1 Types of Li-ion batteries and their application areas.

S. No	Chemical name	Abbreviation	Active material	Applications	Advantages
1	Lithium cobalt oxide	LCO	$LiCoO_2$	Cell phones, laptops, cameras	High capacity
2	Lithium manganese oxide	LMO	$LiMnO_2$, $LiMn_2O_4$	Power tools, e-bikes, electric vehicles, medical devices	Safe, high specific power, and long life than LCO batteries
3	Lithium iron phosphate	LFP	$LiFePO_4$		
4	Lithium nickel manganese cobalt oxide	NMC	$LiNiMnCoO_2$		
5	Lithium molybdate	LiMo	Li_2MoO_4	Electronic vehicles and hybrid electronic vehicles	High charging rate capability and large capacity
6	Lithium titanate	LTO	$Li_4Ti_5O_{12}$	Gaining importance in electric vehicles, power train, and grid storage	High energy storage capacity, low-temperature performance
7	Lithium nickel cobalt aluminum oxide	NCA	$LiNiCoAlO_2$		
8	Lithium sulfur	Li-S	Li-S	Solar-powered airplane, electric vehicle, energy storage, defense	High energy density, lightweight

according to the demands of a particular application and different types of batteries with specific properties have been developed accordingly. Li-ion batteries are mainly classified into eight different types[1] according to the active material used in the electrodes, as described in Table 41.1.

The key performance parameters for lithium-ion batteries are mainly determined by the composition, morphology, and intrinsic properties of electrode materials. Hence, significant efforts have been made toward developing new and high-performance materials for battery components. However, the performance of currently available lithium-ion batteries only partially meets the requirements of different applications. For example, lithium-ion batteries for deployment in hybrid electric vehicles (HEV), plug-in hybrid electric vehicles (PHEV), and pure

1) http://batteryuniversity.com//learn/article/types_of_lithium_ion (January 4, 2016).

electric vehicles (PEV) need two to five times more energy density than the presently available lithium batteries technology. Some of the major challenges associated with developing next generation of lithium-ion batteries are improving power density or rate capability, increasing capacity, reducing capacity fading associated with cycling, increasing energy efficiency at higher cycling rates, and improving cycling life. However, irrespective of various approaches in designing electrode materials with high intercalation rates, micrometer-sized bulk materials are reaching their performance plateau owing to the limits imposed by the intrinsic diffusivity of lithium ions in solid-state (about 10^{-8} cm^2/s), lithium-ion intercalation capacities, and structural stability. Hence, design and development of nanostructured materials-based electrodes is considered to be a promising approach to overcome the current limitations and for improving the performance of the lithium-ion batteries.

41.3
Advantages of Nanostructured Materials

Scaling down the material dimension to nanosize reduces the diffusion length and thereby shortens the distance that Li ions and electrons need to travel during cycling in the solid state, which in turn significantly increases the rate of lithium insertion/removal at the electrodes. Thus, by nanostructuring, faster Li storage in electrode material and high rate capability (high power) can be achieved. The large surface area of nanostructured electrodes enhances the interface between electrode and electrolyte resulting in high lithium-ion flux across the interface. The increased electrode–electrolyte interface also increases the number of active sites for electrode reactions, which in turn reduces electrode polarization loss and improves power density (or rate capability), energy efficiency, and usable energy density [2–5].

Engineered nanostructures with tailored crystal structure and morphology can lead to new Li-storage and transport mechanisms that cannot take place in micrometer-sized bulk materials resulting in high capacities and rechargeability. Nanostructured materials also provide improved mechanical strength and structural integrity. Low-dimensional materials such as nanowires and nanotubes have higher mechanical strength and can be engineered to accommodate volume changes in specific directions. Similarly, in nanoporous electrodes, the internal pores can be utilized to cushion the large volume expansion and contraction occurring during the charge–discharge cycling. This prevents mechanical damage and ensures structural integrity, thereby minimizing the capacity fading otherwise caused by electrical isolation of active electrode materials due to the pulverization of electrodes.

Application of nanostructured materials in lithium-ion battery electrodes has invoked significant interest in recent years due to their various advantageous properties as discussed above. Literature reviews summarizing the state-of-the art and present day challenges of nanotechnology applications in lithium-ion battery have been published in recent years. The faster pace of research and development activities taking place in this area makes it critical for all stakeholders to be aware of the progress occurring across the globe. In this direction, patents are considered as a significant source of technological information that is not published elsewhere. Thus, detailed analysis of patents can be an effective tool to assess and quantify the research activities and technological progress happening in a particular technology domain. In this chapter, we aim to provide a comprehensive overview of the technological developments in the area of nanostructured materials for application in lithium-ion batteries using an in-depth patent landscaping analysis.

41.4
Patent Analysis

41.4.1
Methodology

The current patent analysis has been conducted to get an overview of patent landscape for the incorporation of nanomaterials in Li-ion batteries, with a focus on identifying the key nanomaterials used for property enhancement, emerging technology trends, major assignees, and application areas of nanotechnology-enabled lithium-ion batteries. All keywords relevant to nanomaterials for Li-ion batteries were collected and a search string was framed using all the possible keywords, synonyms, and abbreviations for the concept of nanotechnology-enabled lithium-ion battery with appropriate logical operators and truncations. Search was performed over a time span of January 1, 2004 to September 31, 2015 using Thomson Innovation database[2] resulting in 7746 patents, which were pertinent to the concept of incorporation of nanomaterials in Li-ion batteries.

41.4.2
Year-Wise Patent Publication Trend

Figure 41.2 shows the year-wise trends (2004 to July 2015) of patents worldwide with respect to publication year, which reflects the continuing technological progress in this area. As is evident from the figure, during 2004–2006 the patenting activity was low, but picked up momentum in the following years and a sharp increase in the number of published/granted patents was observed from the year 2011 onward. This is due to the increasing commercial success in using nanomaterials for batteries and entry of many corporate assignees in this field.

The number of published patents increased by more than sixfold, from 331 patents in 2010 to 2120 patents in 2014, which indicates an average annual growth rate of 59.3% during the period. The healthy trend in the number of published patents shows the increasing importance of nanotechnology for

2) www.thomsoninnovation.com/login (August, 2014).

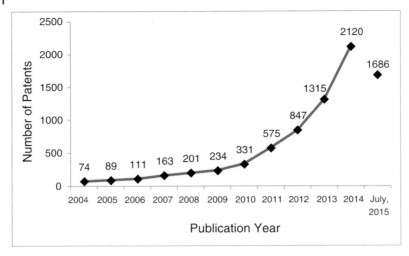

Figure 41.2 Annual growth of patents with respect to publication year.

improving the performance of lithium-ion batteries and nanotechnology-enabled lithium-ion batteries are expected to have a larger impact in this technology domain in the coming years.

41.4.3
Geographical Region-Wise Distribution of Patenting Activity

The worldwide distribution of the patents for Li-ion batteries in which nanomaterials are incorporated is shown in Figure 41.3. It can be seen that China is the world leader with the highest number of patent filings (3976 patents) followed by Japan (1185 patents), the United States (1097 patents), South Korea (941 patents), Germany (196 patents), and other countries, respectively. The top five most prolific countries in terms of published/granted patents in this area account for 95% of the global patent publications on nanomaterials-incorporated lithium-ion batteries. It is also interesting to note that India appears in the list of top 10 countries having highest number of published/granted patents.

41.4.4
Patenting Activity by Assignees

The patents were also analyzed to identify active players in the area of nanomaterials-incorporated Li-ion batteries. As depicted in the pie chart presented in Figure 41.4, the major players are categorized under three groups, namely, corporate, university/research institutes, and independent inventors. Out of the 7746 published patents/applications, major share of patents, that is, 4325 patents (56%), is owned by the corporate sector, followed by university/research institutes (2869 patents, 37%) and independent inventors (549 patents, 7%).

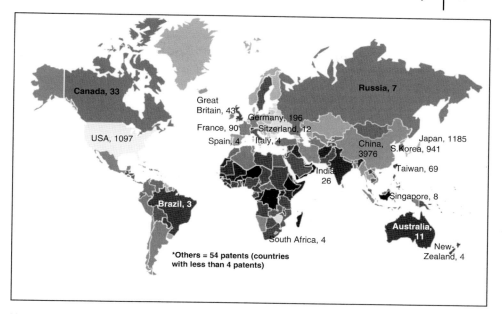

Figure 41.3 Priority country-wise distribution of nanomaterials-based Li-ion battery patents.

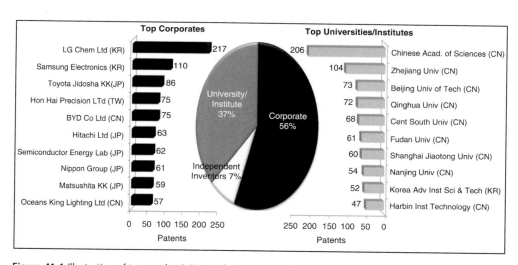

Figure 41.4 Illustration of top academic/research institutes and top corporate along with pie chart showing distribution of patents related to nanomaterials-incorporated Li-ion batteries.

Figure 41.4 also highlights the top 10 key players in the field of nanomaterials-incorporated Li-ion batteries as shown in the form of the bar chart.

In the corporate sector, LG Chem Ltd has emerged as the top most assignee having 217 patents, followed by Samsung Electronics Ltd (110 patents), Toyota Jidosha KK (86 patents), Hon Hai Precision Ind Co. Ltd (75 patents), BYD Co. Ltd (75 patents), and other leading corporate, respectively. Among the corporate players, Japanese companies dominate the list with five companies appearing in the top 10 list and remaining companies belong to South Korea, China, or Taiwan. In the universities/research institutes sector, organizations based in China overshadow the list by occupying 9 out of the top 10 positions. Chinese Academy of Sciences leads the list with 206 patents, followed by Zhejiang University (104 patents), Beijing University of Technology (73 patents), Qinghua University (72 patents), Central South University (68 patents), and other university/research institutes, respectively, as presented in Figure 41.4.

41.5
Technology Analysis

41.5.1
Classification of Li-Ion Batteries

Figure 41.5 shows 7746 relevant patents related to nanomaterials-incorporated Li-ion batteries segregated according to different types of Li-ion batteries. As we can see from the figure, majority of patents (1932 patents) belong to lithium iron phosphate battery (LFP) type with applications in electric vehicles (automotive) due its high power storage capability and relatively low cost of iron phosphate. Lithium manganese oxide (LMO) battery type is the second widely used one accounting for 1530 patents for mostly applications in power tools, e-bikes, electric vehicles, and medical devices with advantages of high specific power, safe usage, and longer life than LCO batteries. The number of patents belong to the lithium cobalt oxide (LCO) battery type is 1424, which is commonly used in the consumer electronics (portable electronics) as LCO offers high power densities along with longer life cycle.

Lithium titanate (LTO) is widely used for anode material, which corresponds to 996 patents. It has the advantages of operating even at low temperature as compared to other battery types and have high power densities. Lithium nickel manganese cobalt oxide (NMC) battery having high nominal voltage of about 3.6–3.8 V per cell and highest energy densities with wide range of applications in consumer electronics accounts for 832 patents. The other types of batteries such as NCA, Li-molybdate, and Li-sulfur cover the rest of the patent publications with 591, 375, and 46 patents, respectively.

Figure 41.6 displays the activities of top assignees having patenting activity across the wide spectrum of Li-ion battery technologies. Most of the top

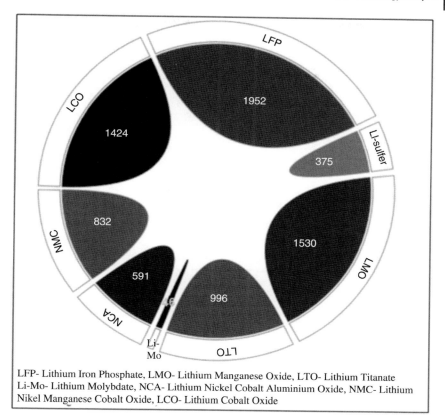

LFP- Lithium Iron Phosphate, LMO- Lithium Manganese Oxide, LTO- Lithium Titanate
Li-Mo- Lithium Molybdate, NCA- Lithium Nickel Cobalt Aluminium Oxide, NMC- Lithium
Nikel Manganese Cobalt Oxide, LCO- Lithium Cobalt Oxide

Figure 41.5 Segregation of patents with respect to different types of Li-ion batteries.

assignees have assimilated a broad portfolio incorporating patents related to important lithium-ion battery technologies.

In the figure, LG Chem Ltd is in the top position with most number of patents and about 44 patents are filed for LMO (lithium manganese oxide) type battery followed by 40 patents filed for NCA (lithium nickel cobalt aluminum oxide) type, 36 patents filed for LFP (lithium iron phosphate) type, 32 patents filed for NMC (lithium nickel manganese cobalt oxide) type, 29 patents filed for LCO (lithium cobalt oxide) type, 24 patents filed for LTO (lithium titanate) type, and 4 patents filed for Li–sulfur type battery. Similarly, other top assignees along with their respective patents published in various battery types is shown in the figure. It is also interesting to note that Samsung Electronics has a slightly higher focus on LMO battery type as compared other Li-ion battery technologies. Another noteworthy finding is the dominance of Chinese Academy of Sciences in Li–sulfur battery technology.

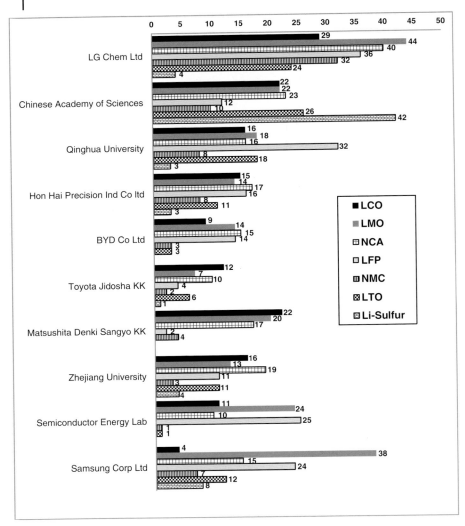

Figure 41.6 Distribution of patents for top assignees with respect to Li-ion battery type in which nanomaterials are used.

41.5.2
Application of Nanomaterials in Li-Ion Batteries

An in-depth analysis was performed to understand the role of nanomaterials in various components of Li-ion battery. Figure 41.7 shows the segregation of patents depending on the use of nanomaterials in various components of Li-ion battery. It has been observed that the major focus is on incorporating

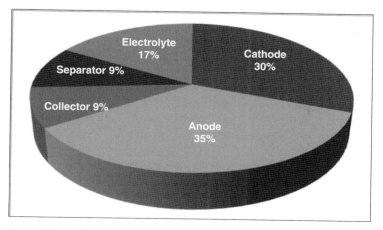

Figure 41.7 Application of nanomaterials in various components of Li-ion battery.

nanomaterials as conductive additive in electrodes and developing nanostructured electrodes to improve the performance of Li-ion batteries.

Various nanomaterials used in Li-ion batteries are segregated as shown in Figure 41.8. Carbon nanotube (CNT) is the widely used nanomaterial in Li-ion

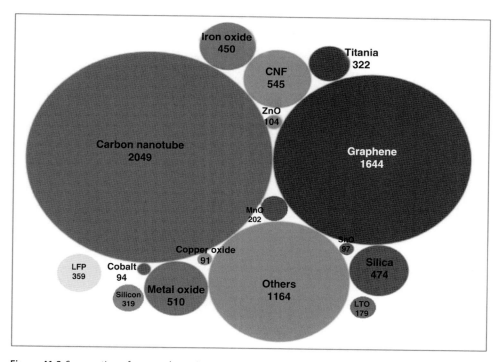

Figure 41.8 Segregation of patents by various nanomaterials used in Li-ion batteries.

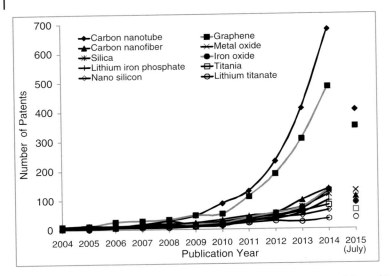

Figure 41.9 Historical trend of patenting activity for important nanomaterials used in lithium-ion batteries.

batteries accounting for 2049 patents. Graphene is the second most widely used nanomaterial in Li-ion batteries, resulting in 1644 patents. Carbon nanofibers are observed in 545 patents published in this area followed by silica represented in 474 patents, iron oxide with 450 patents, titania with 322 patents, manganese oxide with 202 patents, zinc oxide with 104 patents, cobalt oxide with 94 patents, and copper oxide with 91 patents.

The historical trend in patenting activity for most important nanomaterials used in lithium-ion batteries is shown in Figure 41.9. The importance of carbon nanotubes and graphene for lithium-ion batteries and their exceptional role in improving the performance of lithium-ion batteries is clearly evident from the figure. Looking at the annual number of patents filed on these materials, patents related to CNT have grown by close to eightfold from 2010 to 2014 and graphene witnessed ninefold increase in published/granted patents during the same period. The other nanomaterials that have shown similar increments in patenting activity are lithium iron phosphate, iron oxide, and silica. A brief overview of technological innovations related to these nanomaterials as observed in patents are as follows:

Carbon nanotubes: CNTs exhibit significant advantages for application as additive for both anode and cathode materials, or as the replacement material for anodes in LIBs. The unique feature of CNT having one-dimensional structure with saturated carbon bonds on the surface raises extra potentials beyond a mere lithium storage material. Large aspect ratio and high electrical conductivity enables the formation of conductive networks at a rather low loading of CNTs that is particularly beneficial for improving gravimetric capacity and power density of oxide-based composite electrodes [6–9]. Addition of 0.5–1 wt% CNTs can theoretically completely replace the commonly used conductive agents. They are used as conductive additive in most commonly used inorganic oxide cathode materials such as $LiCoO_2$, $LiMn_2O_4$, and $LiFePO_4$. Addition of CNTs into inorganic oxides effectively improves the electron transfer between the active material and current collector, which in turn improves rate capability and cyclic efficiency. The large specific surface area and a more accessible structure for Li interaction as well as excellent mechanical properties can be utilized to buffer the volumetric change in the electrodes during charging/discharging processes due to their superior electrical and mechanical properties. Significant patenting activity is observed on Si–CNT composites to mitigate the effects of pulverization of silicon anodes resulting from severe volume change during charging and discharging. Addition of CNTs into lithium titanate anode improves the conductivity and results in better rate capability and capacity retention. Lithium titanate–CNT core-sheath anode composite demonstrated enhanced lithium storage capacities and kinetics. CNT arrays are also increasingly being explored due to their large surface area and ordered electrode configuration [10]. They are used for the preparation of freestanding CNT paper anodes that are light, flexible, and highly conductive for application in flexible energy storage devices.

Graphene: The unique two-dimensional structure along with large specific surface area ($\approx$2630 m^2/g), excellent electrical conductivity ($\approx$15 000 cm^2/(V s) at 300 K and $\approx$60 000 cm^2/(V s) at 4 K), and high thermal conductivity ($\approx$3000 W/(m K) near room temperature) render graphene as suitable candidate for anode material in Li-ion batteries [11–13]. The 2D structure is advantageous for storing lithium ions on both surfaces of the hexagonal C-ring-based sheet and enhanced lithium storage capability is observed for double or multilayered graphene sheets. Graphene is used as a matrix to support metal anode materials, where the active material can be encapsulated with graphene, wrapped around, or anchored on graphene to form the composite electrode material. Incorporation of graphene in anodes of lithium alloying metals such as silicon and tin helps to prevent electrode disintegration due to electrochemical pulverization and improves Li-cycling performance. Moreover, direct deposition of such metal particles or films on graphene enables the fabrication of self-supporting electrodes without the need for binders or additional current collector, thereby minimizing the nonelectroactive material in the electrode [14]. Graphene–metal oxide composite materials such as graphene–TiO_2 and graphene–SnO_2 have been developed to improve the electrical conductivity of metal oxide anodes. In another invention, graphene–Fe_3O_4 nanocomposite anodes are devised to improve the rate performance and cycling stability of Fe_3O_4 anode [15].

Incorporation of graphene can significantly improve various properties of conventional cathode materials, including electrical conductivity and kinetics of electrons and lithium ion transportation as well as reduce particle agglomeration during cycling. Graphene is used as a conductive additive in cathode materials such as $LiFePO_4$, $LiMn_2O_4$, $LiMnO_2$, and V_2O_5. The effective three-dimensional conducting network formed by graphene enhances the rate capability and cyclability of composite cathodes.

Lithium iron phosphate (LiFePO₄): It is one of the most promising cathode materials due to various advantageous properties such as high theoretical capacity, thermal stability, low cost, and environment friendliness. However, its practical application is hindered due to the poor rate performance resulting from sluggish mass- and charge transport kinetics. Nanostructuring, along with coating/mixing with carbon and heteroatom doping are evolved as effective strategies toward overcoming the inherent limitations of pristine $LiFePO_4$. Most of the patents dealing with nano-$LiFePO_4$ discuss the preparation of nanopowders and achieving carbon coating on $LiFePO_4$. Carbon-coated $LiFePO_4$ nanopowder was prepared by supercritical solvothermal method using triethanolamine as reported by LG Chem Ltd [16]. Hydrothermal sovothermal or carbothermal methods using polyols under inert conditions also produce nano-$LiFePO_4$ with controlled particle size. Lithium iron phosphate hierarchical structure containing micron-sized spheres constructed by layered lithium iron phosphate nanosheets have been reported by Tsinghua University and Hon Hai Precision Industry Co., Ltd. The lithium iron phosphate having a sheet structure with a surface parallel to the ac crystal plane of the crystal lattice is the most advantageous to the Li ions traversing, whereas the hierarchical structure has a decreased specific surface area, and thus needs fewer conductive agents and binders. Additionally, the hierarchical structure of lithium iron phosphate with orderly arrangement has an increased tap density as compared to those having disorderly aggregated nanograins [17].

Iron oxides: Among various transition metal oxides, iron oxides (Fe_3O_4 and Fe_2O_3) are gaining increasing attention as anode materials due to their high theoretical capacity (≈ 1000 mA h/g), nontoxicity, fire resistance, high abundance, high corrosion resistance, and low processing cost [18]. However, thermodynamically unfavorable reversible lithium ion extraction, poor cyclability due to volume changes occurring during charging/discharging, and low conductivity act as deterrents in their application as anode materials. Nanostructuring of iron oxides has emerged as an effective approach to improve the structural integrity and enhance their electrochemical performance. Iron oxide–carbon hybrid nanostructures have also been explored for achieving high capacity and high rate capability. The carbon component of the nanocomposite promotes electron transport and acts as buffer layer or support in enhancing the structural stability of the material. A wide variety of nanocomposites, including carbon-coated Fe_3O_4 nanospindles [19], iron oxide/carbon core–shell particles [20], iron oxide/carbon porous nanocomposite [21], iron oxide/carbon yolk–shell nanocomposite [22], iron oxide–carbon xerogels [23], CNT/iron oxide composites, and graphene-coated iron oxide, have been reported as anode materials.

Nanosilicon: Among various alloying metals, silicon (Si) has the highest gravimetric capacity (4200 m Ah/g, lithiated to $Li_{4.4}Si$) and volumetric capacity (9786 mA h/cm³) in addition to having a relatively low discharge voltage. Silicon is also the second most abundant element on earth and is environment friendly [24–26]. The fundamental limitations of silicon as anode material in lithium-ion batteries are the poor capacity retention and large initial irreversible capacity resulting from the large volume changes occurring during the lithium insertion and extraction processes. Rational designing using various nanostructures has been envisaged to address the material challenges associated with silicon for electrode applications. High aspect ratio nanostructures of Si, such as fibers, sheets, flakes, tubes or, ribbons, having the smallest dimension approximately 10-fold smaller than the other two dimensions is expected to result in a 10-fold increase in the irreversible capacity loss [27]. Silicon anode formed from submicron diameter pillars of silicon with a fractional surface coverage (F) of ~0.5 fabricated on an end-type silicon wafer using island lithography showed structural integrity throughout the cycling [28]. Si nanoclusters and Si–graphite nanocomposites also showed improvements in the cycle life and lithium capacity as compared to the silicon powder with binder. Another type of a silicon nanomaterial, the porous silicon ("pSi"), has been shown to be a promising anode for lithium-ion rechargeable batteries. Electrodes made using porous silicon fabricated by electrochemical etching and subsequent coating with a passivating agent showed very high capacity (3000 mAh/g) for at least 60 cycles, which is 80% of theoretical value of silicon [29]. Anodes fabricated using CNT–silicon composites having a "plush ball" architecture formed by growing CNT on Si particles showed a reversible capacity of 800 mA h/g, and good dynamic behavior [30]. The better electrochemical properties are attributed to the unique structure that provides large specific surface area, large porosity, and structural integrity. Various other nanostructure morphologies such as hollow Si nanostructures, doped and undoped Si nanowires, Si–CNT composites, and hierarchical structures containing nanoparticles and pillars have been developed for improving the performance of anodes in lithium-ion batteries.

41.5.3
Distribution of Patents by Application Sector

The patents are further segregated based on application sector as shown in Figure 41.10. The major application area for patents publications for nanomaterials-incorporated Li-ion batteries is consumer electronics, which constitutes 55% (1681 patents) of total patents. Electric vehicles is in the second position with 31% (943 patents), followed by medical devices with 7% (221 patents), aerospace sector with 4% (114 patents), and military sector with 3% (89 patents).

Figure 41.11 displays the year-wise distribution of published patents with respect to application sectors. It can be observed from the bubble chart that significant patenting activity is observed for consumer electronics and electric vehicles sector in recent years and about twofold increase in the patent publications has been observed since 2011 in these two sectors. The other three sectors, aerospace, medical devices, and military, having comparatively lower number of patents published, showed a sequential year-wise increase in patent publications.

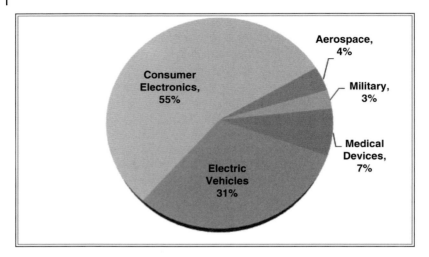

Figure 41.10 Distribution of patents by application sectors.

41.5.4
Top-20 Assignees versus Publication Year

The year wise patent publication trends of top-20 assignees in the field of nano-materials-incorporated Li-ion battery is shown in Figure 41.12.

LG Chem Ltd is having the most number of patents published (103 patents) in the year 2015 and has come out with unique fabrication technology along with

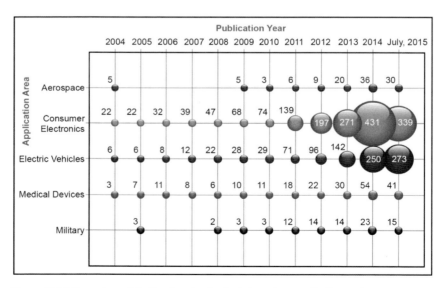

Figure 41.11 Year-wise publication trends of patents in various application sectors.

S. No.	Top 20-Assignees	Publication Year											
		July 2015	2014	2013	2012	2011	2010	2009	2008	2007	2006	2005	2004
1	CHINESE ACADEMY OF SCIENCES	36	65	51	28	15	8	1		2	1		
2	LG CHEM LTD	103	54	15	12	10	9	4	2	4		3	1
3	SAMSUNG	16	23	18	10	9	3	9	7	7	3	3	2
4	ZHEJIANG UNV	16	49	9	15	7	2	1	1	2	2		
5	TOYOTA JIDOSHA KK	17	27	15	9	4	7	2	4		1		
6	BYD CO LTD	10	11	8	10	11	9	5	9	1	1		
7	HON HAI PRECISION LTD	16	23	4	4	5	5	6		2	7	1	2
8	BEIJING UNIV	12	14	14	6	7	6	3	4	2	4		1
9	QINGHUA UNIV	12	12	14	11	5	3	2	2	4	3	2	2
10	UNIV CENT SOUTH	12	19	16	6	5	4	3	3				
11	HITACHI LTD	10	16	3	8	13	4		1	2	3	1	2
12	SEMICONDUCTOR ENERGY LAB	10	20	25	5	2							
13	FUDAN UNIV	3	6	7	8	4	6	4	15	5		3	
14	NIPPON GROUP	10	14	13	4	10	2	3	1	1		1	2
15	SHANGHAI JIAOTONG UNIV	7	16	12	6	3	6	4	3				3
16	MATSUSHITA DENKI SANGYO KK	2	4	5	3	8	5	4	9	15	2	1	1
17	HAIYANGWANG LIGHTING TECHNOLO	6	12	20	9	3							
18	BTR NEW ENERGY MATERIALS INC	13	14	15	2	5							
19	OCEANS KING LIGHTING SCI & TECHN	7	40	4	3								
20	BOSCH GMBH ROBERT	14	18	5	2	1	1						

Figure 41.12 Year-wise patent publication trends of top 20 assignees.

customized cathode and anode chemistries that optimize performance and cost without reducing battery life with expanding its production capacity from consumer electronics to electric vehicles[3]. Chinese Academy of Sciences has published highest number of patents (65 patents) during 2014 as compared to other top assignees in the same year. Zhejaing University has also published more number of patents (49 patents) during the year 2014. Similarly, the other assignees and their year-wise patent publication trends are shown in the Figure 41.12. As is evident from the figure, most of the assignees are showing increased patenting activity from year 2013 onward.

41.5.5
Correlation of Top Assignees versus Nanomaterial Usage

An in-depth analysis was performed to understand the focus of patenting activity of top assignees in terms of different nanomaterials, which are being used in Li-ion batteries. For this purpose, the patents for top 20 assignees in the area

3) http://www.lgchem.com/global/vehicle-battery/car-batteries (January, 2016)

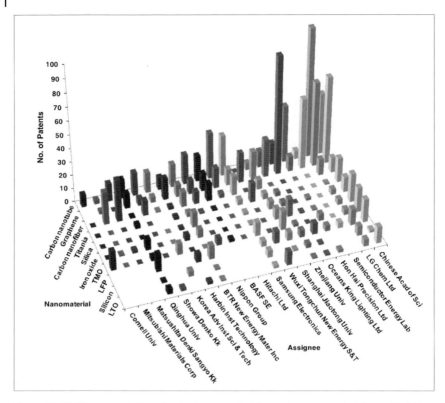

Figure 41.13 3D representation of patents segregated for various nanomaterials used in Li-ion batteries versus top 20 assignees.

of Li-ion batteries were selected and based on the claims study of each patent the various nanomaterials used in Li-ion batteries were segregated, as shown in Figure 41.13. For example, Hon Hai Precision Ind Co. Ltd, has specific focus on using carbon nanotubes in electrode materials because of their unique set of electrochemical and mechanical properties to create composite electrodes [31,32], binder-free electrodes, array electrodes [33], and flexible electrodes [34], as well as for modifying current collector [35,36]. LG Chem Ltd is developing cable-type battery, wherein CNT is used to coat on the mesh-type current collector and as conductive additive in electrodes to improve their conductivity [37,38]. They are also exploring CNT composite anodes, CNT composite binders, and CNT–sulfur composites for lithium–sulfur battery. Similarly, Matushita Electric Ind Co. Ltd and Mitsubshi Materials Corp are working on carbon nanofiber-based electrodes for Li-ion batteries as carbon nanofibers provide good insulation and controls large volume change during charging–discharging process.

Oceans King Lighting Sci & Tech Co. Ltd, Semiconductor Energy Labs, and Chinese Academy of Sciences are mainly working on graphene-based Li-ion batteries as graphene can provide high electrical conductivity and also high specific surface area. Oceans King Lighting Sci & Tech Co. Ltd has a large patent portfolio around graphene electrodes and various composites such as graphene–lithium titanate, graphene–silicon, graphene–cobalt oxide, graphene–LiFePO$_4$, graphene–sulfur, and V$_2$O$_5$–graphene for electrode applications. Patents published by Semiconductor Energy Labs describe graphene and graphene oxide as additive in Li-ion battery electrodes. Similarly, Wuxi Tongchun New Energy Sci & Tech Ltd is working on titania-based Li-ion batteries as it has high specific capacity, strong process adaptability, and is suitable for industrial production.

Figure 41.14 shows the distribution of patents for cathode and anode of Li-ion battery with respect to the use of nanomaterial. Carbon nanofiber, graphene, and silica are majorly used nanomaterials for anode of Li-ion batteries, whereas other nanomaterials are used for both cathode and anode of Li-ion battery. Extensive research on electrode materials suggests that size reduction of materials can dramatically enhance the battery performance. It is clearly evident from the above graphs that nanomaterials can play an important role in enhancing the battery performance when used in electrodes.

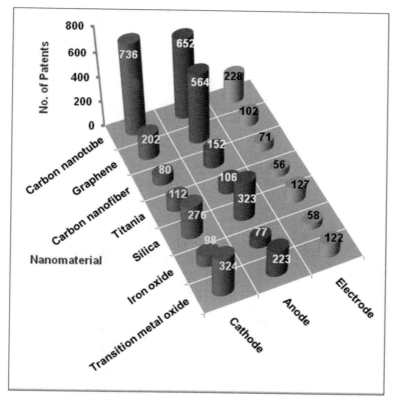

Figure 41.14 Segmentation of patents containing nanomaterials for cathode and anode of Li-ion battery for top 20 assignees.

41.6
Commercial Status of Nano-Enabled Li-Ion Batteries

Nano-enabled Li-ion batteries are growing in commercial use after more than 20 years of research and development. Japanese electronics giant Sony Corporation had brought out the first commercial nanotechnology-based "Nexelion" Li-ion battery in 2005, wherein a nanostructured anode made from a tin–cobalt–carbon (Sn–Co–C) composite was used to enhance the battery performance. As of now, Nexelion is not available on the market. Table 41.2 provides an overview of the commercial status of nano-enabled Li-ion batteries that are currently available on the market. As is evident from the table, nanosized $LiFePO_4$, lithium titanate, lithium manganese oxide, and lithium cobalt oxide have been incorporated in successfully commercialized battery products. In most of the cases, however, these materials are used as additives in 1–5% composition in either anode or cathode of respective bulk materials. There are several other raw

Table 41.2 An overview of commercial status of nano-enabled Li-ion batteries in market.

Sl. No.	Manufacturers	Products	Nanomaterials/ battery component	Country (headquarters)
1.	A123 Systems, LLC	Nanophosphate® AMP20M1HD-A, Nanophosphate AHP14-M1Ultra-A, Nanophosphate ANR26 650M1-B, Nanophosphate APR18 650M1-A, Nanophosphate AMP20 Energy Modules	$LiFePO_4$/ cathode	USA
2.	Advanced Battery Technologies, Inc.	Large rechargeable polymer lithium-ion (PLI) battery	Lithium titanate spinel/ anode	USA
3.	Altair Nanotechnologies Inc.	PowerRack, 24 V 60 Ah battery module, 60 Amp hour cell	Lithium titanate/anode and cathode	USA
4.	Electrovaya, Inc.	SuperPolymer®2.0	Lithiated manganese oxide	Canada
5.	Front Edge Technology, Inc.	NanoEnergy®	$LiCoO_2$ film/ cathode	USA
6.	NEC Energy Solutions, Inc.	ALM™ 12V35, ALM™12V7	$LiFePO_4$/ cathode	USA
7.	Planar Energy, Inc.	Solid-state lithium batteries	$LiMn_2O_4$	USA
8.	Toshiba Corp.	Super Charge Ion Battery (SCiB™)	Lithium titanate/anode	Japan

material manufactures such as CNano Technology Limited, USA, Nanocyl S.A., Belgium, US Research Nanomaterials Inc, USA, Novarials, USA, and XP Nano Material Co. Ltd, China, that are supplying CNT-based conductive additive products for lithium-ion battery application. These additives provide improved battery performance with less loading especially for high-power applications such as electrical vehicles and power tools. Various other technologies related to Li-ion batteries incorporating CNT-based electrode designs are currently under development. Atlanta-based Sila Nanotechnologies is actively pursuing silicon-coated vertically aligned carbon nanotube anodes for improving energy storage capacity and energy density of lithium-ion batteries for electric vehicle applications. Contour Energy Systems, Inc. based at Azura, California, has also exclusively licensed a carbon nanotube technology from Massachusetts Institute of Technology (MIT).

41.7
Market

Incorporation of nanomaterials into Li-ion batteries is already being translated from laboratory level into commercial products and this translation is ongoing with rapid growth. Introduction of nanotechnology in Li-ion batteries plays an important role in producing cleaner forms of energy with already large patent filings observed from many corporate companies such as LG Chem Ltd, Samsung Electronics, Matsushita Industrial Co., Ltd, BYD Co. Ltd, Altair Nanotechnologies, Sony, Hon Hai Precision Ind Co. Ltd, and IRICO Group Co. Ltd. The nanomaterials are incorporated according to size, grade, and applications. Within the applications segment, consumer electronics, that is, laptops/notebooks, cellular/mobile phones are the current major growth area for battery materials market. It is expected to maintain a dominating position over the next 5 years. The automotive applications are projected to show highest growth due to huge demand for electric vehicles in next 5 years[4]. The global market for Li-ion batteries in light duty vehicles is estimated to grow from $1.6 billion in 2012 to almost $22 billion in 2020. Power tools applications segment is poised for growth in Asia-Pacific and the rest of the world. In 2013, the estimated global battery material demand was worth nearly $5.1 billion and is expected to reach $11.3 billion by 2018[5], at a CAGR of 13% from 2013 to 2018. Asia-Pacific, with top position in battery materials market value, is expected to grow at nearly 14% CAGR in next 5 years.

41.8
Conclusions and Future Perspectives

The demand for lithium-ion batteries capable of providing high energy densities as well as high power densities to meet the future challenges of essential energy

4) http://www.navigantresearch.com/research/electric-vehicle-batteries (January, 2016).
5) http://www.marketsandmarkets.com/Market-Reports/battery-raw-materials-market-866.html?gclid=CPqtpI-Xir0CFewE4godVFkAKA (January, 2016).

storage is rapidly growing. In this regard, nanostructured materials have demonstrated significant potential to improve the performance of rechargeable lithium-ion batteries. We have conducted an in-depth patent landscaping analysis to discern the technological developments occurring in the rapidly evolving area of nanostructured materials for application in lithium-ion batteries. The detailed analysis of patents provided valuable insights regarding the growth in patenting activity, key players, major contribution by countries, and geographical regions, as well as technology innovation and trends.

The patenting activity in lithium-ion batteries showed sixfold increase in the filing of patents during 2010–2014 period, which corresponds to an average annual growth rate of 59.3% during the period. This is an important indication for the rapid growth in research and development activity and increased importance of nanomaterials for developing future generation of high-performance lithium-ion batteries. Most of the global patenting activity in this area is observed in only five countries, which account for 95% of the global patent publications on nanomaterials-incorporated lithium-ion batteries. China has emerged as the major contributor in this area with 3976 published/granted patents, followed by Japan (1185 patents), United States (1097 patents), South Korea (941 patents), and Germany (196 patents). Among the top five countries, except for China, all other countries have wider presence in various patent offices across the globe, whereas China's patenting activity is more domestic oriented.

Analysis of key assignees showed the active involvement of both corporate and academic/research institutions in developing nanomaterial-based battery components. The significant presence of corporate assignees, which contributed 56% of global published/granted patents shows the substantial commercial potential of nanomaterials-related technologies for lithium-ion batteries. Among the corporate players, Japanese companies dominate the list with five companies appearing in the top 10 list and remaining companies belong to South Korea, China, or Taiwan. However, South Korea-based LG Chem Ltd has emerged as the top most assignee having 217 patents, followed by Samsung Electronics Ltd, South Korea (110 patents), and Toyota Jidosha KK, Japan (86 patents). In the universities/research institutes sector, organizations based in China overshadow the list by occupying 9 out of the top 10 positions.

The incorporation of nanomaterials in lithium batteries has resulted in enhancement of energy storage capacity, withstanding extreme temperature conditions, extended life cycle of the batteries, and so on. Carbon nanotubes were observed as the widely used nanomaterial in the lithium-ion batteries, followed by graphene, lithium iron phosphate, nanosilicon, iron oxide, lithium titanate, and others. Diverse applications of Li-ion batteries for clean energy storage applications ensure that all major stakeholders, such as consumer electronics manufactures, electric and hybrid vehicle manufactures, and other high-end users are keenly interested in developing new manufacturing techniques, which will bring down the cost of production for widespread deployment. This is also evident from the commercial success in bringing nano-enabled lithium-ion batteries on the market, which is getting populated with new products having diverse anode and cathode chemistries. Some of the promising future electrode technologies that assure significant improvements in energy and power density are high-voltage (5 V) and/or high-capacity (>300 m Ah/g) cathode materials such as oxide spinels and poly-anion materials with Ni and Co cations; high-capacity metal alloy and composite anodes such as silicon, tin, their alloys, and silicon–carbon composites, and so on; and high-voltage and solid polymer nano-composite electrolytes.

However, there are significant technical challenges that need to be overcome before commercialization for various nanostructured anode and cathode materials that are under development for high-power lithium-ion battery technology. For example, even though nanostructured Si-based anodes have shown very promising performance, the complex synthesis processes and significant deterioration of the structural integrity during mass scaling up may hinder the application to commercial batteries. For nanostructured cathodes, the large surface area can increase the solubility of electrode materials in the electrolyte solution due to the increased chemical activity. Hence, nanostructured $LiCoO_2$, $LiNiO_2$, and $LiMn_2O_4$ need further nanoscale coating in order to suppress metal dissolution and increase the electronic conductivity. Overall, the development of low-cost scalable fabrication processes for nanomaterials with desired properties and performance parameters is important for commercial applications.

While Li-ion batteries still offer the best performance among all existing battery systems, various competing technologies such as lithium–sulfur and lithium–air batteries with high theoretical specific energy have gathered momentum in recent times with intense research and development activities being pursued in these areas. These alternative cathode chemistries are particularly important to meet the requirements of ~500 km range between charges for all electric vehicles and grid energy storage. However, there are several challenges concerning safety, extended life cycle, reactivity of the lithium anode, and the poor reversibility and efficiency of the oxygen electrode that need to be addressed for evolving these technologies to the current level of lithium-ion battery technology. Development of novel nanocomposite materials and unique nanoarchitectures to engineer novel electrode structures are expected to play important roles in addressing these challenges and advancing the next generation lithium batteries for powering portable electronics devices and in supporting the growing number of hybrid electric vehicles.

References

1 Crabtree, G., Kócs, E., and Trahey, L. (2015) The energy-storage frontier: lithium-ion batteries and beyond. *MRS Bull.*, **40**, 1067–1076.

2 Bruce, P.G., Scrosati, B., and Tarascon, J.-M. (2008) Nanomaterials for rechargeable lithium batteries. *Angew. Chem., Int. Ed.*, **47**, 2930–2946.

3 Guo, Y.-G., Hu, J.-S., and Wan, L.-J. (2008) Nanostructured materials for electrochemical energy conversion and

storage devices. *Adv. Mater.*, **20**, 2878–2887.

4 Song, M.-K., Park, S., Alamgir, F.M., Cho, J., and Liu, M. (2011) Nanostructured electrodes for lithium-ion and lithium-air batteries: the latest developments, challenges, and perspectives. *Mater. Sci. Eng. R*, **72**, 203–252.

5 Armstrong, M.J., O'Dwyer, C., Macklin, W.J., and Holmes, J.D. (2014) Evaluating the performance of nanostructured materials as lithium-ion battery electrodes. *Nano Res.*, **7** (1), 1–62.

6 de las Casas, C. and Li, W. (2012) A review of application of carbon nanotubes for lithium-ion battery anode material. *J. Power Sources*, **208**, 74–85.

7 Lahiri, I. and Choi, W. (2013) Carbon nanostructures in lithium-ion batteries: past, present, and future. *Crit. Rev. Solid State*, **38**, 128–166.

8 Liu, X.-M., Huang, Z.D., Oha, S.W., Zhang, B., Maa, P.-C., Yuen, M.M.F., and Kim, J.-K. (2012) Carbon nanotube (CNT)-based composites as electrode material for rechargeable Li-ion batteries: a review. *Compos. Sci. Technol.*, **72**, 121–144.

9 Landi, B.J., Ganter, M.J., Cress, C.D., DiLeo, R.A., and Raffaelle, R.P. (2009) Carbon nanotubes for lithium-ion batteries. *Energy Environ. Sci.*, **2** (6), 638–654.

10 Zhou, G., Li, F., and Cheng, H.-M. (2014) Progress in flexible lithium batteries and future prospects. *Energy Environ. Sci.*, **7**, 1307–1338.

11 Wu, S., Xu, R., Lu, M., Ge, R., Iocozzia, J., Han, C., Jiang, B., and Lin, Z. (2015) Graphene-containing nanomaterials for lithium-ion batteries. *Adv. Energy Mater.*, **5**, 1500400.

12 Xu, C., Xu, B., Gu, Y., Xionga, Z., Sunb, J., and Zhao, X.S. (2013) Graphene-based electrodes for electrochemical energy storage. *Energy Environ. Sci.*, **6**, 1388–1414.

13 Wua, Z.-S., Zhoua, G., Yina, L.-C., Rena, W., Lia, F., and Chenga, H.-M. (2012) Graphene/metal oxide composite electrode materials for energy storage. *Nano Energy*, **1** (1), 107–131.

14 Kung, H.H. and Lee, J.K. (2009) Electrode material comprising graphene composite materials in a graphite network formed from reconstituted graphene sheets. U.S. Patent Application US20110111303 A1, filed Nov. 6, 2009 and published May 12, 2011.

15 Fichtner, M., Hahn, H., and Prakash, R. (2012) Carbon encapsulated transition metal oxide nanocomposite, a method for its preparation and its use in Li-ion batteries. U.S. Patent Application US20140294981 A1, filed Sep. 26, 2012 and published Oct. 2, 2014.

16 Jun, I.K., Cho, S.B., Oh, M.H., and Jang, W. (2014) Method for producing carbon-coated lithium iron phosphate nanopowder. PCT Patent Application PCT/KR2014/000265, filed Jan. 9, 2014 and published Sep. 30, 2015.

17 Wang, L., He, X.-M., Sun, W.-T., Li, J.-J., Huang, X.-K., and Gao, J. (2012) Lithium-ion phosphate hierarchical structure, method for making the same, and lithium-ion battery using the same. U.S. Patent 8,962,186, filed Apr. 27, 2012 and issued Feb. 24, 2015.

18 Zhang, L., Wu, H.B., and Lou, X.W. (2014) Iron-oxide-based advanced anode materials for lithium-ion batteries. *Adv. Energy Mater.*, **4**, 1300958.

19 Kang, W., Shen, Q., Zhao, C., and Xu, F. (2012) High specific capacity spindle-shaped ferroferric oxide/carbon nano composite material for negative electrode of lithium-ion battery. China Patent CN102623693B, filed Apr. 13, 2012 and issued July 2, 2014.

20 He, C., Liu, E.-Z., Shi, C.-S., Zhao, N.-Q., and Wu, S. (2012) Carbon-coated iron oxide lithium-ion battery anode material and method. China Patent CN102790217B, filed July 26, 2012 and issued July 9, 2014.

21 Lvpeng, P., Wang, J., Zeng, Z., Zhao, H., and Zhang, T. (2013) Method for preparing iron oxide based high-performance lithium-ion battery anode materials. China Patent CN103208625B, filed April 24, 2013 and issued February 25, 2015.

22 Du, N., Liu, J., Yang, D., and Zhang, H. (2013) Preparation method of iron

trioxide/carbon yolk-eggshell nano-composite structure. China Patent CN103204490B, filed Mar. 5, 2013 and issued Dec. 17, 2014.

23 Liu, X., Lvpeng, P., Li, X., Wang, J., Zeng, Z., and Zhao, H. (2013) Preparation method of iron oxide cathode material for lithium-ion battery. China Patent CN103227324B, filed Apr. 24, 2013 and issued Apr. 1, 2015.

24 Wu, H. and Cui, Y. (2012) Designing nanostructured Si anodes for high energy lithium-ion batteries. *Nano Today*, **7**, 414–429.

25 Szczech, J.R. and Jin, S. (2011) Nanostructured silicon for high capacity lithium battery anodes. *Energy Environ. Sci.*, **4**, 56–72.

26 Zamfir, M.R., Nguyen, H.T., Moyen, E., Lee, Y.H., and Pribat, D. (2013) Silicon nanowires for Li-based battery anodes: a review. *J. Mater. Chem. A*, **1**, 9566–9586.

27 Green, M. (2008) Silicon anode for a rechargeable battery. U.S. Patent Application US20100190061A1, filed May 9, 2008 and published July 29, 2010.

28 Green, M. (2003) Structured silicon anode. U.S. Patent Application US20060097691A1, filed November 5, 2003 and published May 11, 2006.

29 Biswal, S.L., Wong, M.S., Thakur, M., Sinsabaugh, S.L., and Isaacson, M.J. (2010) Structured silicon battery anodes. PCT Patent Application PCT/US2010/054577, filed Oct. 28, 2010 and published May 5, 2011.

30 Shu, J., Li, H., and Huang, X.-J. (2005) Carbon silicon composite material, its preparation method and use. China Patent Application CN1903793A, filed July 26, 2005 and published Jan. 31, 2007.

31 Huang, X.-K., He, X.-M., Jiang, C.-Y., Wang, D., Gao, J., and Li, J.-J. (2011) Electrode composite material, method for making the same, and lithium-ion battery using the same. U.S. Patent 9219276B2, filed May 13, 2011 and issued Dec. 22, 2015.

32 Feng, C., Zhang, H.-X., Jiang, K.-L., and Fan, S.-S. (2008) Anode of a lithium battery and method for fabricating the same. U.S. Patent 8017272B2, filed Apr. 4, 2008 and issued Sep. 13, 2011.

33 Chen, G.-L. and Leu, C. (2003) Lithium-ion battery comprising nanomaterials. U.S. Patent 7060390B2, filed on Mar. 31, 2003 and issued Jun. 13, 2006.

34 Wu, Y., Wu, H.-C., Luo, S., Wang, J.-P., Jiang, K.-L., and Fan, S.-S. (2015) Lithium-ion batteries. U.S. Patent Application 20150207143A1, filed Jan. 15, 2015 and published July 23, 2015.

35 Wang, J.-P., Wang, K., Jiang, K.-L., and Fan, S.-S. (2014) Lithium-ion battery. U.S. Patent 9105932 B2, filed June 18, 2014 and issued Aug. 11, 2015.

36 Wang, J.-P., Jiang, K.-L., and Fan, S.-S. (2011) Current collector and lithium-ion battery. U.S. Patent 8785053B2, filed Dec. 23, 2011 and issued July 22, 2014.

37 Kwon, Y.-H., Chang, S.-K., Oh, B.-H., Kim, J.-Y., Jung, D.-S., and Woo, S.-W. (2012) Cable-type rechargeable battery. EP Patent 2793306B1, filed Dec. 14, 2012 and issued Oct. 21, 2015.

38 Kwon, Y.-H., Oh, B.-H., Kim, J.-Y., and Woo, S.-W. (2014) Anode for cable-type secondary battery and cable-type secondary battery comprising the same. U.S. Patent 9099747B2, filed Apr. 24, 2014 and issued Aug. 4, 2015.

42

Nanotechnology in Fuel Cells: A Bibliometric Analysis

Manish Sinha,[1] Ratnesh Kumar Gaur,[2] and Harshad Karmarkar[1]

[1]*Gridlogics Technologies Pvt. Ltd, Sunflower Commercial, 77/1 Baner Road, Baner, Pune 411045, Maharashtra, India*
[2]*ARCI, Centre for Knowledge Management of Nanoscience & Technology, Balapur, Hyderabad 500005, Telangana, India*

42.1
Introduction

A fuel cell generates electricity by converting chemical energy from a fuel, when positively charged hydrogen ions react with oxygen or another oxidizing agent. Fuel cells are different from batteries because to sustain the chemical reaction in fuel cells, it requires continuous source of fuel and oxygen or air, whereas in a battery electromotive force (emf) is generated by the reaction of chemicals present in the battery. As long as fuel oxygen and air is supplied, fuel cells can produce electricity continuously.

Nanotechnology can be used to provide significant advantages to various types of fuel cells. For example, in solid oxide fuel cells (SOFCs), nanomaterials can provide high surface-area-to-volume ratio, which results in the increase of the active electrode area. Researchers are working on various nanomaterials to bring down the operating temperature (900–1100 °C) of SOFCs to intermediate range (500–700 °C). Similarly, in case of direct methanol fuel cells (DMFCs), nanomaterials can be used as catalyst to increase the reaction kinetics that can significantly increase the performance and power output of DMFCs. Recently, research has focused on increasing the performance and activity of catalysts. Nano-enabled polymer electrolyte membrane fuel cell have several advantages: operating temperature is low, potential for low cost and volume, low weight, compactness, long stack life, sustained operation at a high current density, suitability for discontinuous operation, and fast start-ups.

In spite of these various advantages, there are certain limitations that prevent them from reaching widespread commercial use. Nanotechnology has the potential to alleviate many of these problems and can improve the efficiency of nano-enabled fuel cells in several ways. Nanotechnology and nanomaterials have

Nanotechnology for Energy Sustainability, First Edition. Edited by Baldev Raj, Marcel Van de Voorde, and Yashwant Mahajan.
© 2017 Wiley-VCH Verlag GmbH & Co. KGaA. Published 2017 by Wiley-VCH Verlag GmbH & Co. KGaA.

significant impact on improving various properties of electrocatalysts and membranes.

42.2
Literature Analysis

Research efforts on improving the performance of fuel cell using nanotechnology are reviewed in this section. The detailed literature analysis presented here covers the period from 2005 to August 2015. It gives an insight into the yearly publication trend, region-wise publication trend, top institutions and organizations engaged in this area of research, and the major funding agencies. The literature analysis was done using the Web of Science database. Various search strings with keyword sets have been used to extract the published journal literature on the role of nanotechnology in developing fuel cells. The obtained result of 15 572 records were further refined according to different types of fuel cells[1,2]: (i) proton exchange membrane fuel cells (PEMFCs); (ii) solid oxide fuel cells; (iii) direct methanol fuel cells (DMFCs); (iv) phosphoric acid fuel cells (PAFCs); (v) alkaline fuel cells (AFCs); (vi) molten carbonate fuel cells (MCFCs), and so on as well as components of fuel cells such as (i) membrane; (ii) catalyst; (iii) electrodes, and so on. Fuel cell support a range of applications, from automotive propulsion to industrial power supplies, to stationary residential, to portable power supplies for electronic equipment.

Present analysis provides the growth trend of publications in the past 10 years, the major countries, and research/academic institutions involved in the research activities for fuel cells. Majority of publications are for PEMFC and direct methanol fuel cells (DMFC) and evidently solid oxide fuel cells shown in Figure 42.1.

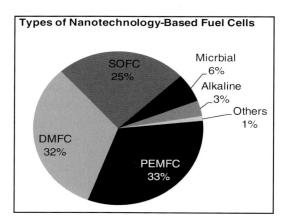

Figure 42.1 Focus of research activities related to MWCNT application in various fuel cells.

1) https://en.wikipedia.org/wiki/Fuel_cell.
2) http://energy.gov/eere/fuelcells/types-fuel-cells.

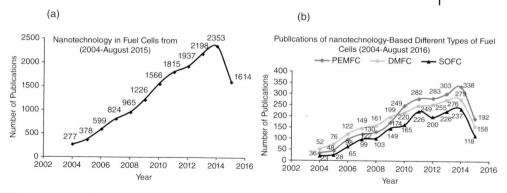

Figure 42.2 (a) Publication trend of research publications for nanotechnology in fuel cells. (b) Publication trend of research publications for nanotechnology in various fuel cells.

42.2.1
Literature Publication Trend

Literature publications for nanotechnology in fuel cells show steady growth (Figure 42.2a). However, focused research activities are still being pursued in various academic and research organizations to develop nanotechnology-based catalyst supports, composite membranes and coatings of bipolar plates, and so on to improve the overall performance of fuel cells. Also, we can see in Figure 42.2b that PEMFC has showed promising growth since 2009 and solid oxide fuel cell is the only type with a dip in number of publications in the last 3 years.

42.2.2
Literature Publications by Geographical Region

Figure 42.3a and b gives an overview of the scientific and technological activities of the main regions and countries in the world toward the use of nanotechnology

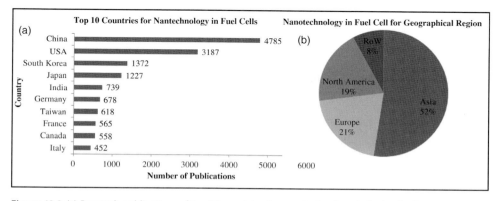

Figure 42.3 (a) Research publications of top 10 countries for nanotechnology in fuel cells. (b) Geographical distribution of scientific publications for nanotechnology in fuel cells.

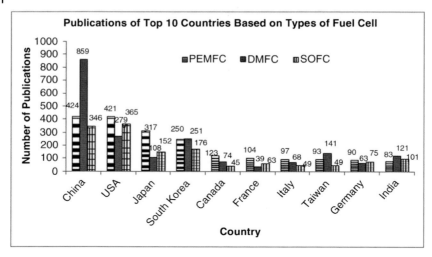

Figure 42.4 Top 10 countries having maximum number of scientific publications in different types of fuel cells.

in fuel cells. It is evident from Figure 42.3b that the Asian region is leading in terms of research activities with 52% of total global publications for nanotechnology in fuel cells followed by Europe and North America with 21 and 19% each of global publications, respectively.

Overall numbers of publications were used as the measure to assess the research activity of top 10 countries. It is quite clear from Figure 42.4 that China is the leader with 4785 publications followed by the United States (3187), South Korea (1372), Japan (1227), and India (739). Countries with largest shares of fuel cell publications tend to exhibit a degree of similarity in their publication profiles can be seen in Figure 42.4.

China, the United States, Japan, and South Korea accounted for 72–85% of their publications for different types of fuel cells such as PEMFCs, DMFCs, and SOFCs, but also have publications in almost all the subfields of fuel cell technology. We can see from Figure 42.4 that publications for these three types of fuel cells varied from country to country; although China has highest number of publications in DMFCs, the United States and Japan have higher shares of publications in PEMFCs.

42.2.3
Publication Trend by Institutes and Organizations

The contribution of various research institutes and organizations in the area of nanotechnology for fuel cells was studied to identify the key players actively pursuing research in this area, as shown in Figure 42.5.

Present analysis shows that Chinese Academy of Sciences and the US Department of Energy (DOE) have the highest number of publications. It is also evident

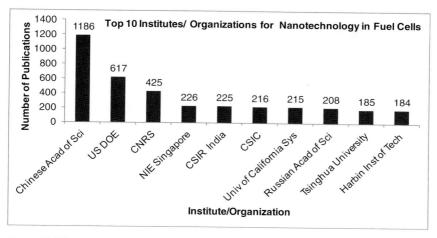

Figure 42.5 Top 10 research institutions/organizations for nanotechnology in fuel cells.

from the figure that academic/research organizations from Asian region popu-late the top 10 list with most of the top prolific research institutions located in China.

42.3
Patent Landscaping

The objective of this section is to explore patenting activity of nanotechnology in fuel cells through bibliometric analysis. Based on online research, keywords related to nanotechnology for fuel cells were identified. The keywords were combined with International Patent Classification (IPC) codes and Co-Operative Patent Classifica-tion (CPC) codes. The search was performed on the title, abstract, and claims section. Further refinement of the primary data set was performed through a visual/manual analysis of records and irrelevant terms were excluded to ensure that the data set was relevant. This refinement and data accumulation was done using PatSeer.

42.3.1
Publication Trend

Figure 42.6 represents publication trend for nanotechnology in fuel cells. It can be seen as innovations for nanotechnology in fuel cells and its resulting publica-tions started to show up from the late 1990s with continuous growth since 2007. It can also be seen that there has been a substantial rise in total number of publi-cations since 2007, which amounts to 72% of complete portfolio. Also, the trend shows there would be approximately a rise of 150 publications in the next 5 years. Innovation in the area of fuel cell technology is going to continue in the near future as shown by the publication trend.

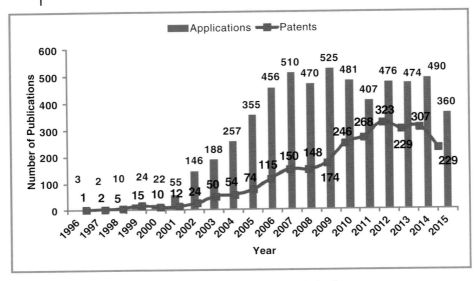

Figure 42.6 Publication trend across nanotechnology in fuel cells.

42.3.2
Top Assignees

Figure 42.7 illustrates distribution of patents among top 20 companies. The figure represents top 20 companies from 1323 assignees for a data set of 3330 records. Samsung SDI Co. Ltd leads the publications for nanotechnology in fuel

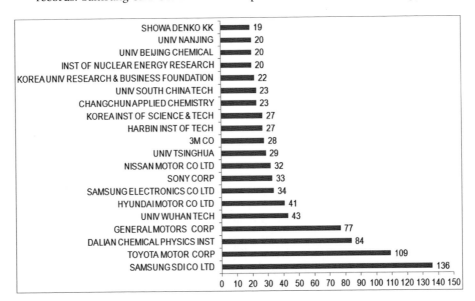

Figure 42.7 Top patent holders for use of nanotechnology in fuel cells.

cell with 136 records, followed by Toyota Motor, Dalian Chemical Physics Institute, General Motors, and University Wuhan Tech with 109, 84, 77, and 43 records, respectively.

It can be seen that maximum companies with Asian origin are among the top companies having research activity in nanotechnology in fuel cells, some of them being Samsung, Toyota, Dalian Chemical Physics Institute, Hyundai, and Sony. Among the top 20 companies, it is interesting to see 55% of companies are Research Institutes and Universities. This indicates despite considerable activity from the commercial organizations, much of the work in this area originates from universities. Samsung continues to be the dominant nonautomotive company in the list, whereas automotive companies are the major assignees. The continued presence of so many automotive manufacturers in the list each year serves to underline their efforts to commercialize fuel cell electric vehicles in the near future.

Note: Records for Samsung SDI and Samsung Electronics are analyzed separately as they hold significant patent portfolios.

42.3.3
Patents by Geographical Region

This section shows the geographical spread of patent applications filed globally for nanotechnology in fuel cells. It helps in identifying the geographies where major companies have chosen to protect their technologies, thus giving an indication as to where they foresee the real market for their products. It shows perfect indication of where innovation is taking place.

Figure 42.8 illustrates that filing in China is most active with 920 filings, followed by the United States and Japan with 754 and 615 filings, respectively. Among the European nations, Germany has 123 filings.

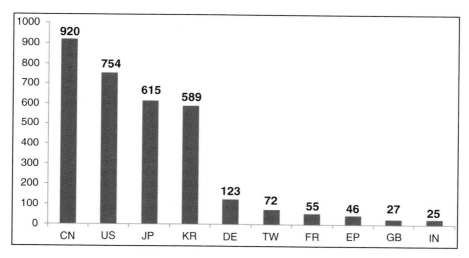

Figure 42.8 Top 10 patent offices based on total number of patent filings.

42.3.4
Classification of Fuel Cells

Classification of fuel cells is primarily based on the kind of electrolyte they employ. Type of chemical reaction, catalyst required, operating temperature ranges, fuel required, and various other factors depend on the electrolyte employed in fuel cells.

Various types of fuel cells are currently under development, each having its own advantages, limitations, and applications.

42.3.4.1 Classification Based on the Type of Electrolyte

1) Alkaline fuel cells (AFCs)
2) Phosphoric acid fuel cells (PAFCs)
3) Polymer electrolytic membrane fuel cells and proton exchange membrane fuel cells
4) Molten carbonate fuel cells (MCFCs)
5) Solid oxide fuel cells
6) Microbial fuel cells

42.3.4.2 Classification Based on Types of Fuel and Oxidant

1) Hydrogen (pure)–oxygen (pure) fuel cells
2) Hydrogen-rich gas–air fuel cells
3) Ammonia–air fuel cells
4) Synthesis gas–air fuel cells
5) Hydrocarbon (gas) –air fuel cells

Furthermore, primary components of nanotechnology-based fuel cells include the following[3]:

1) Electrodes
2) Gas diffusion layer
3) Membrane electrode assembly

Fuel cells can operate in a wide temperature range such as PEMFCs at room temperature, molten carbonate fuel cells at intermediate temperature (600 °C) solid oxide fuel cells at high temperature (SOFCs) (600–1000 °C). Figure 42.9 shows the classification of patents according to types of fuel cells, nanomaterial used as catalyst/supports.

As the demand for fuel cell technology is growing rapidly, fuel cells find commercial applications in hospitals, schools, portable electronic devices, and vehicles. Proton exchange membrane fuel cells are most researched of all types considering their characteristics such as light weight, small size, and more efficiency than other types of fuel cells. Figure 42.10 represents distribution of number of patents in

3) http://energy.gov/eere/fuelcells/parts-fuel-cell.

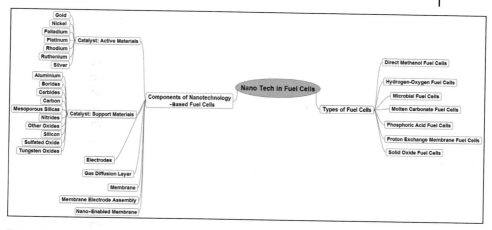

Figure 42.9 Segmentation of patents across nanotechnology-based fuel cells.

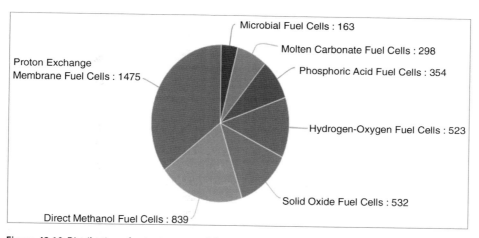

Figure 42.10 Distribution of patents among different fuel cell types.

different types of fuel cells. It can be seen that proton exchange membrane fuel cells with 1475 records (amounting for 65% of total number of publications for fuel cells) is the most favored type among researchers.

42.3.5
Publication Trend for Fuel Cell Types

Figure 42.11 compares trends across different fuel cell types. During 1970s, there were large amount of higher oil prices and energy shortages compelling research organizations to undertake research activities for developing more efficient ways for energy generation. It was not until the late 1980s and early 1990s that saw

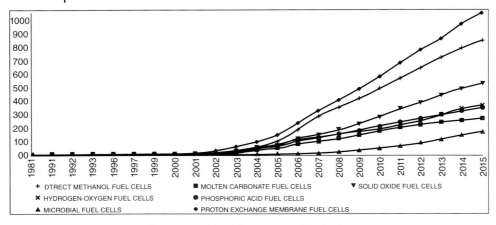

Figure 42.11 Publication trend for different types of fuel cells.

active research happening around proton exchange membrane fuel cells. This is because of their applications in transport, stationary fuel cell applications, and portable fuel cell applications. Also, solid oxide and phosphoric acid fuel cells show steady rise in publications. Microbial fuel cells have emerged since 2000 due to their use in power generation, wastewater treatment, and sensing devices.

Now we will see the focus of assignees working on the type of fuel cells. Among different types of fuel cells, a few have the potential of commercialization. From Figure 42.12 it can be seen that 60% of top 15 assignees are Research

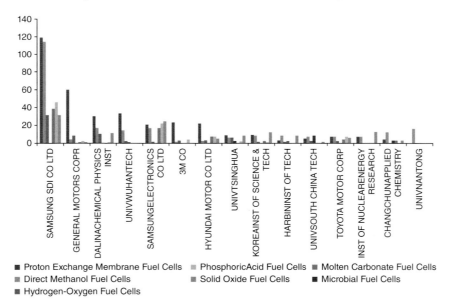

Figure 42.12 Top companies in different types of fuel cells.

Institutes or Universities. Also, Univ Nantong focuses only on direct methanol fuel cell. The corporate assignees have patents in PEMFCs, DMFCs, and SOFCs.

42.4
Proton Exchange Membrane Fuel Cells Patent Analysis

Nanotechnology provides better means to use hydrogen as a fuel. Using fuel cells in combination with fuel cells provides best solution to store hydrogen in automobiles resulting in lead to low and safe emission levels. Nano fuel cells are also used in medical devices, robotics, and microelectromechanical systems.

Among various types of fuel cells, we have seen that there is substantial number of patents belonging to PEMFCs, as shown in Figure 42.10. PEMFCs can be used for a wide variety of power applications: transportation, large-scale stationary power systems for buildings, portable/micropower and distributed generation due to low operating temperature, reduced size and weight, low cost and volume, sustained operation at a high current density and long stack life, suitability for discontinuous operation, and fast start-ups. We will observe some of the patent analysis in the area of PEMFCs. In Figure 42.13 it can be seen that total number of publications since 2007 has been constant.

42.4.1
Geographical Distribution of Patents for Nanotechnology in PEMFCs

Table 42.1 shows the number of new inventions filed in each priority jurisdiction during 2000–2015. China, Japan, and Korea exhibited a consistent growth for

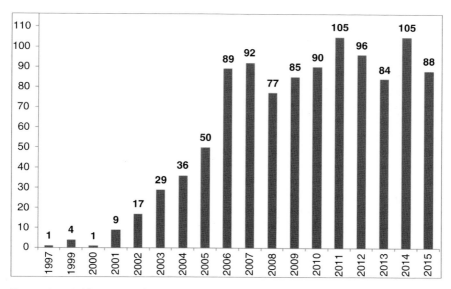

Figure 42.13 Publication trend across proton exchange membrane fuel cells.

Table 42.1 Filings for top 10 priority countries (2000–2015).

	Total	Priority country							
		2000–2001	2002–2003	2004–2005	2006–2007	2008–2009	2010–2011	2012–2013	2014–2015
CN	419	4	16	45	54	65	76	104	55
US	417	23	59	79	75	67	48	49	1
KR	315	3	15	67	74	47	53	50	6
JP	156	8	21	47	30	23	14	11	2
DE	47	7	7	5	11	2	2	5	0
FR	23	1	2	0	5	6	6	3	0
TW	23	0	0	5	4	6	5	3	0
EP	19	1	1	3	2	2	5	5	0
GB	11	0	0	1	1	3	4	2	0
IN	8	0	0	0	0	1	2	5	0

the last 5 years. Also, rate of applications in China have outpaced that of the United States in the last 5 years than in the 10 year period. This shows transition of research activity from the United States and European countries toward China, Japan, and Korea.

But when we consider published patents (Figure 42.14), we can see that the United States has maximum number of patents published in PEMFC with 485 of 1058 published records closely followed by China and Korea. It can be seen that many applicants have chosen PCT route (WO patents) and European Patent

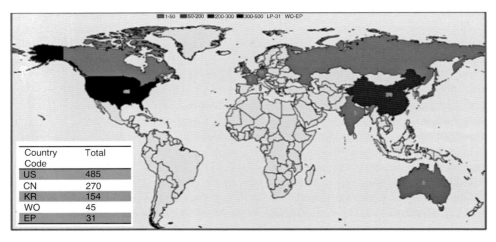

Country Code	Total
US	485
CN	270
KR	154
WO	45
EP	31

Figure 42.14 Geographical distribution of publications across proton exchange membrane fuel cells.

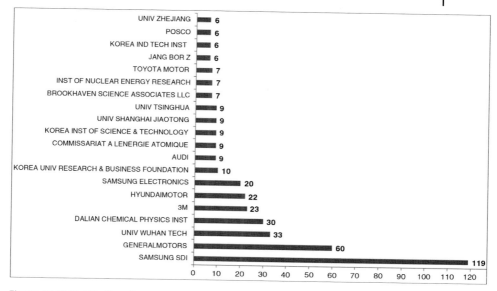

Figure 42.15 Distribution of patents among top 20 companies.

Office to protect their innovations. These authorities account for 76 of 1058 publications.

42.4.2

Top Assignees for Nanotechnology in PEMFCs

Figure 42.15 illustrates distribution of patents for research around PEMFCs among top 20 companies. Samsung SDI Co. Ltd. leads the publications for use of PEMFCs with 119 records, followed by General Motors, Univ Wuhan Tech, Dalian Chemical Physics Institute, and 3 M with 60, 33, 30, and 23 records. It can be seen that maximum companies with Asian origin are among the top companies having research activity in PEMFCs, some of them being Samsung, Hyundai, Dalian Chemical Physics Institute, and Toyota. As PEMFCs are widely used in transportation applications, many automotive companies such as General Motors, Hyundai Motors, Audi, and Toyota Motors appear in top 20 list.

42.4.3

Organization-Wise Distribution of Applications

Figure 42.16 shows distribution of applications for different types of organizations. In organization-wise distribution, different applicants were identified as firms, government organizations, and institutes (institute includes universities, academies, and hospitals) by manual research and classifying the

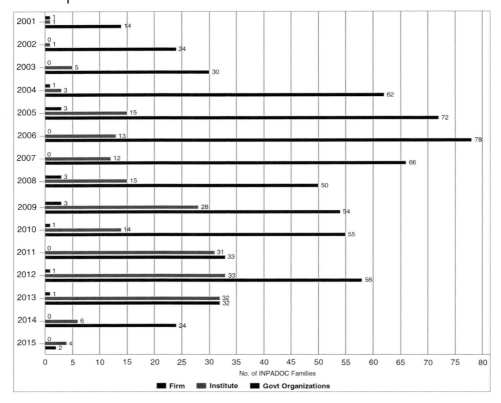

Figure 42.16 Distribution of applications across different types of organizational types.

extensions under different organizations. Firms refer to a business organization, such as a corporation, limited liability company, or partnership. Of the total 1468 applications, nearly 45% relate to firms, 14% have research institutes as their assignees. The timelines reveal that Institutes have increasing research activity since 2005.

42.5
Technology Analysis

Based on the detailed patent analysis, it was found that nanotechnology is predominantly used for developing components of PEMFCs. Most prominently, nanomaterials are used as catalyst support or active materials. The second most prominent application is in gas diffusion layer. Nanotechnology has also been explored for developing composite membranes, electrodes, and bipolar plates.

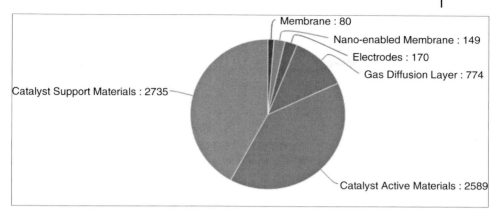

Figure 42.17 Distribution of patents in different components.

Figure 42.17 shows distribution of patents in different components used in fuel cells. It shows catalysts are the most favored components of nano-technology-based fuel cells with 5324 records. Furthermore, nano-enabled membrane, an upcoming term, also has presence in the chart with 149 records.

As we have already seen various developments in the area of PEMFCs, now we will look at how different components correlate with different types of fuel cells. Figure 42.18 represents matrix for different components used across fuel cell types.

42.5.1
Assignee Activity across Components of Fuel Cells

Figure 42.19 shows research activity of companies across different components. As membrane electrode assemblies have a wide range of applications ranging from automotive and other transport, electronics, and material handling, it can be seen that companies such as Toyota Motors, Samsung, and 3M are active in membrane electrode assembly.

42.5.2
Nanomaterials for Fuel Cell Catalyst Support/Active Materials

Despite the significant advances made by fuel cell technology, it still suffers from high-cost (mainly due to the catalyst) and durability issues. The high overpotential of oxygen reduction reaction at cathode also limit the perform-ance of fuel cells. Several avenues have been explored with the aim to reduce platinum loading and increase its utilization. From Figure 42.20a, platinum is the most effective catalysts for oxidation and reduction reactions. Nickel and

Fuel Cell Types (Rows) / Components (Column)	Proton Exchange Membrane Fuel Cells	Direct Methanol Fuel Cells	Microbial Fuel Cells	Hydrogen-Oxygen Fuel Cells	Molten Carbonate Fuel Cells	Phosphoric Acid Fuel Cells	Solid Oxide Fuel Cells
Membrane Electrode Assembly	1052	846	171	363	272	345	512
Catalyst Support Materials	944	784	150	335	234	307	361
Carbon	919	771	145	329	234	303	342
Silicon	322	242	54	114	69	96	134
Aluminium	180	117	24	72	47	60	86
Carbides	133	89	11	44	27	52	42
Nitrides	115	80	16	43	36	41	43
Other Oxides	105	73	11	29	22	32	37
Tungsten Oxides (WOx)	32	28	3	9	7	10	9
Borides	19	8		6	1	2	5
Mesoporous Silicas	17	14	4	3	3	7	4
Sulfated Oxide	4	4			3	1	3
Catalyst Active Materials	912	777	128	329	227	302	398
Platinum	837	733	99	293	198	274	281
Nickel	541	405	63	204	140	183	277
Ruthenium	527	503	35	178	129	187	174
Palladium	446	381	35	155	99	147	163
Gold	367	290	58	131	81	115	120
Silver	318	285	60	158	90	112	150
Rhodium	328	253	18	87	63	98	84
Gas Diffusion Layer	374	218	12	107	108	135	125
Nano-Enabled Membrane	64	42	10	10	8	15	20
Electrodes	5	7	5	6	15	8	51
Membrane	25	14		1	2	2	7

Figure 42.18 Distribution of patents across components and different fuel cell types.

palladium are also used as active materials for catalysts and carbon-based materials such as CNTs are used as support for catalyst-active materials as can be seen from Figure 42.20b. Various combination of the materials mentioned in both figures have been used as catalyst-support/active materials in fuel cells.

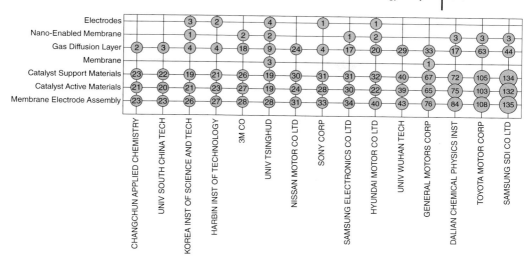

Figure 42.19 Assignee activity across components.

42.5.2.1 Fuel Cell: Types versus Catalysts-Active Materials

Fuel cells are of different types; however, they all work in the same general manner. They are made up of three components: electrodes (anode and cathode), the electrolyte, and the cathode. Also, the choice of catalyst for oxidizing the fuel in a fuel cell depends on the operating temperature. Fuel cell types therefore all have different catalysts. Platinum is the most active catalyst for proton exchange membrane fuel cell with 837 records followed by direct methanol fuel cell and hydrogen–oxygen fuel cell with 733 and 293 records, respectively. Figure 42.21 illustrates how different active catalyst materials relate to different fuel cell types.

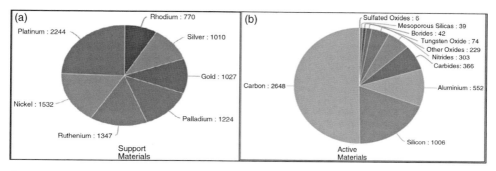

Figure 42.20 (a) Distribution of patents across different support materials for catalysts. (b) Distribution of patents across different active materials for catalysts.

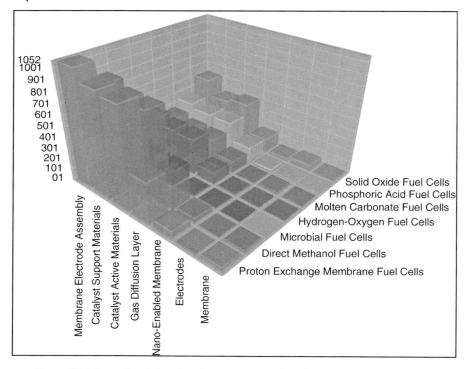

Figure 42.21 Research activity of catalyst active materials with respect to different fuel cell types.

42.5.3
Nano-Enabled Membranes

From Figure 42.17, we have seen nano-enabled membrane, an upcoming term that is also present in the chart with 149 records. We will see the patents filing trend in the area of nano-enabled membranes for fuel cells. Nanotechnology can improve the mechanical strength of polymer membrane matrices without adversely affecting the ionic conductivity. We can see from Figure 42.22 that there is a constant growth in filings since 2005 with maximum records in 2008 with 14 applications.

42.5.4
Nanomaterials in PEMFCs Gas Diffusion Layer

Various activities such as distribution of reactants to active site of electrode, enhancement of electrical contact between the electrodes and bipolar plates, and the management of water supplied and/or generated are done by gas diffusion layer. Microporous layer (MPL) in GDL is conventionally made from composites

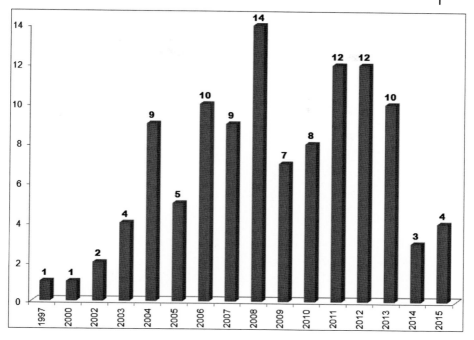

Figure 42.22 Filing trend for nano-enabled membrane.

of fluorinated polymer such as PTFE, FEP, or PVDF containing nanomaterials. But these nanomaterials should provide the required conductivity and hydrophobicity and antiflooding properties are imparted by the fluoropolymer. However, these polymers face structural stability issues during long-term operation under warm and humid conditions. Nanotechnology has been utilized to solve these issues related to gas diffusion layer. Also, similar to GDLs, various nanomaterials have been used to develop nanomembranes that can be used for longer durations without getting damage and can sustain high temperatures up to 200–250 °C.

42.6
Scenario of Commercial Products Can Be Moved after Future Perspectives

As there are many drawbacks of using fuel cells in commercial applications, nanotechnology overcomes many of these problems. Recent research activity around nanotechnology has resulted in promising nanomaterials that could make fuel cells cheaper, lighter, and more efficient. Researchers are trying to improve hydrogen membrane and create more efficient membranes that will allow them to build more lightweight and longer lasting fuel cells.

Table 42.2 lists the products associated with different fuel cell types and components involved.

Table 42.2 Nanotechnology-based products associated with different fuel cell types and components.

S. No.	Product	Types of fuel cell	Component	Function	Manufacturer
1.	BEI cathode	PEMFC	Electrode	Patented construction of BEI cathodes reduces the amount of platinum required also improves the overall power output and durability	Bing Energy
2.	BEI standard MEAs	PEMFC	Membrane electrode assembly	Carbon nanotube "buckypaper" cathode/anode technology in five or seven-layer configurations improves gas flows and effectiveness of catalyst utilization	Bing Energy
3.	CAT-110 fuel cell catalyst	Various	Catalyst	Replaces the primary catalyst and reduces the loading by 95%. Primary catalyst is carbon-supported platinum nanoparticles (Pt/C) inside PEMFC. Also, increases power density by up to 77×	Blue Nano
4.	Dynalene FC	PEMFC	Coolant	Excellent heat transfer properties and maintains a low level of electrical conductivity for at least 2 years	Dynalene Inc.
5.	HiSPEC®	Various	Catalyst	Single component, alloy, supported and unsupported catalysts	Johnson Matthey Fuel Cells
6.	NSTF™	PEMFC	Catalyst	Extreme thinness of the electrode helps achieve very high power density.	3M Company
7.	CNT-GDL	PEMFC	Gas diffusion layer	The growth of CNTs on the surface of carbon fiber or cloth (GDL) serves unique advantage as it supports the dispersion of noble metal nanoparticles (e.g., Pt catalyst for fuel cell application) in a three-dimensional electrode structure, thus providing higher catalyst utilization and more efficient mass transport	ElectroChem Inc
8.	EC-CNT Paper	PEMFC	GDLs	CNT paper is manufactured using carbon nanotube. It is also referred to as Bucky paper. It is highly hydrophobic (contact angle of 113°), highly porous (90%), and has a thermal conductivity of 2.7 kW/(m²h).	ElectroChem Inc
9.	NT-FM30K-H-P2/NT-NITMOSPA-H/NT-NIOCT-H	PEMFC	Gas diffusion layer	Highly conductive gas diffusion layer with excellent mass transport characteristics with superb hydrophobic performance	Nano-Tek (UK) Technology Ltd

42.7
Future Perspectives

The overall literature survey reveals that intense research is being pursued for developing improved PEMFCs using nanotechnology. The immense potential of nanotechnology in improving the performance and durability of fuel cells while decreasing their cost is the drive behind the increasing trend in the research efforts.

Catalyst support made of one-dimensional (1D) nanostructured inorganic oxide materials (i.e., nanowire, nanorod, nanotube, and nanofibers) are also being considered as one of the emerging area due to their physicochemical properties such as open mesoporous morphology and high specific surface area. The strong electronic interaction between the support and catalyst because of the semiconducting properties of these nanostructured inorganic oxide materials can improve the catalytic activity for electrochemical reactions. Nonplatinum metal catalysts have inferior durability at higher potentials that limits their application for automotive purposes, where performance above 0.6 V is highly desired. Platinum as a cathode catalyst has low selectivity for reduction of the oxidant and hence efforts are directed toward developing nonplatinum cathode catalysts that can address this particular problem. Extensive research with an objective to improve the durability of the catalyst and the electrode by delaying the corrosion rate of carbon has led to the development of crystalline carbon supports consisting of carbon nanotubes, carbon nanofibers, and carbon nanohorns displaying excellent corrosion resistance. Modification of conventional membranes using nanomaterials is an ideal approach to overcome their limitations and to improve the overall performance. Various types of nanomaterials such as inorganic oxides, clays, heteropolyacids, acid–base complexes, carbon-based nanostructures, and nanofibers have been used to improve mechanical stability, thermal stability, water retention, and proton conductivity. Thus, it is important that the emerging catalyst systems not only comply with the required catalytic performance but also should be attractive for large-scale automotive MEA production. Nanostructuring of the membrane has created new opportunities that may lead to designing of membranes that are highly efficient and can operate also in anhydrous condition at high temperature. Fuel cell technology is evolving rapidly with the advent of new nanostructured materials that can be considered as potential replacements for conventional materials used in various components. At present, the major thrust of nanomaterials is directed toward developing nanostructured electrodes and nanocomposite membranes, with the emerging prospect of expanding the use of nanomaterials to improve the performance of gas diffusion layer and bipolar plates.

42.8
Conclusion

Although fuel cells enjoy many advantages such as eliminating pollution caused by burning fossil fuels, low heat transmission, high efficiency levels than diesel or gas engines, high-quality DC power, operating silently compared to internal combustion engines, it has been found that these are expensive to produce and difficult to produce, transport, distribute, and store hydrogen.

A lot of research around this area is being made and a large number of promising nanomaterials are being developed that could make fuel cells cheaper, lighter, and more efficient.

This chapter provides an overview of the patenting activities related to fuel cells. It explains the trends in fuel cells and also describes and analyses various types of fuel cells such as proton exchange membrane fuel cells, direct methanol fuel cells, and components such as catalysts, electrodes, and gas diffusion layers. The chapter also covers key companies involved and their trends. From the analysis, it can be inferred that patenting activity (both filing and publications) in the area of nanotechnology in fuel cells has increased over the last decade. There is steady increase of number of applications published as against granted records. Asia-Pacific is the most favored geographical location for filing patents in the area of nanotechnology-based fuel cells among assignees with 1906 patents out of the total 3330 patents. It is due to rising integration of fuel cell technology in various applications such as automobiles, telecom towers, material handling equipment, and power generation and distribution. Proton exchange membrane fuel cells are widely used cell types as they enhance the electrochemical kinetics; increase carbon monoxide tolerance, and the generated steam can be utilized as a heat source when operated at high temperatures. Similarly, nano-enabled membranes have shown constant filings since 2005. Platinum is the most actively used component in fuel cells.

Also, microbial fuel cells are one of the emerging fuel cell types with applications across power generation, wastewater treatment, and sensing devices industries. Using nanotechnology in fuel cells could make a dramatic impact by improving the performance and fulfilling materials requirements.

Various methods for production and purification of hydrogen from hydrocarbon fuels have also been developed using nanotechnology. Recently, researchers at the University of Copenhagen have come up with the solution to reduce the amount of platinum needed as a catalyst in fuel cells. Some other researchers have come up with catalysts that do not use platinum. Recently, a proton exchange membrane having pores of about 5 nm in diameter capped by a layer of porous silica has been developed by researchers at the University of Illinois.

Acknowledgments

The authors are grateful to Director-International Advanced Research Center for Powder Metallurgy and New Materials (ARCI), Hyderabad, Department of Science and Technology (DST), Government of India, for granting permission to publish this chapter.

The authors express sincere gratitude to Dr. Y.R. Mahajan, Technical Adviser and Project Director (CKMNT), for their encouragement and support while preparing this chapter.

43

Techno-Commercial Opportunities of Nanotechnology in Wind Energy

Vivek Patel and Y.R. Mahajan

International Advanced Research for Powder Metallurygy and New Materials (ARCI), Centre for Knowledge Management of Nanoscience and Technology (CKMNT), Balapur P.O., Hyderabad 500005, Telangana, India

43.1
Introduction

Wind energy has been around us for several thousand years and it is the oldest large-scale source of power that has been used by mankind. Wind energy has become a mainstream renewable source of energy and delivered the crucial inputs for socioeconomic development throughout the world. It has been globally recognized that harvesting of wind energy is of considerable importance for the achievement of economic growth, energy security, sustainable development, and protection of the environment. Among various sources of alternate energy, wind is the world's fastest growing clean, emissions-free, and low-cost energy source. In 2014, around 51.4 GW of new wind power capacity was installed all over the world, bringing worldwide installed capacity to 369.6 GW.[1] This was a sharp rise of 44.2% in comparison to the 2013 (35.6 GW) market, and represents an overall increase in the global installed capacity of about 16% from 318.6 GW (2013) to 369.6 GW.[1] According to the market research report from BCC Research, the global wind energy market was worth US$165.5 billion in 2014 and the market is expected to grow at a compound annual growth rate (CAGR) of 7.2% between 2015 and 2020, resulting in US$176.2 billion in 2015 and US$250 billion in 2020.[2] With the increasing reliability on wind power, reducing cost of electricity generated from wind, as well as actionable government policy and decision making, the wind energy sector will continue to grow by leaps and bounds in the coming years.

1) http://www.gwec.net/wpcontent/uploads/2015/03/GWEC_Global_Wind_2014_Report_LR.pdf.
2) http://www.bccresearch.com/market-research/energy-and-resources/wind-energy-global-markets-report-egy058b.html.

Nanotechnology for Energy Sustainability, First Edition. Edited by Baldev Raj, Marcel Van de Voorde, and Yashwant Mahajan.
© 2017 Wiley-VCH Verlag GmbH & Co. KGaA. Published 2017 by Wiley-VCH Verlag GmbH & Co. KGaA.

The installing cost of a utility scale wind turbine ranges from about US$1.4 to 2.5 million. Most of the commercial-scale turbines installed today are of 2 to 5 MW capacities. The lifespan of a wind turbine is pegged at around 120 000 h in their estimated lifespan of 20–25 years. The maintenance, repair, and overhaul (MRO) cost constitutes a sizable share (20–25%) of the total levelized cost per kWh produced over the lifespan of a wind turbine. Nanotechnology has the potential to extend turbine longevity, reduce operation and maintenance requirements, and reduce lifecycle cost of wind energy. Hence, it is obvious that the exploitation of nano-enabled composites (prepregs), nanolube additives, grease, nanocoatings, nano-based adhesive/sealants, nano-enabled ultracapacitor/battery, nanosensors, and nano-based metal matrix composites (submarine power cable, windings, fasteners) in wind energy have attracted considerable attention among stakeholders over the last decade. Nanotechnology can play a vital role in offering nano-enabled components to both offshore and onshore wind energy operations. Potentiality and markets of nanotechnology is enormous in wind energy, particularly for offshore wind sources.

In this chapter, we review the impact of nanotechnology on wind energy sector such as the current status and perspectives of nanotechnology, advantages, limitations, growth drivers, materials, and applications, as well as future directions. This is not meant to be an overview of the field, but rather a comprehensive analysis with particular focus on nanotechnology strengths and potential in the wind energy.

43.2
Wind Energy Industry Requirements

Similar to other sustainable energy technologies, such as solar energy, wave power, geothermal energy, tidal power, and so on, wind energy is also capital intensive but it is one of the most cost-effective renewable technologies in terms of the cost per kWh of electricity generated. The wind energy industry requirements may be framed in the following broad categories as follows:

i) *Optimization of wind farm:* The top priorities may include reducing levelized cost of energy (LCOE), minimizing cost of energy per kWh, and enhancing wind plant reliability and durability.

ii) The key driver for cost reductions is the availability and use of affordable materials with superior strength-to-weight ratios, which would enable larger size rotors to be more cost-effective. Nano-enhanced materials have the potential to achieve this goal.

iii) *Demand for high-performance nanostructured materials with multifunctionality:* Wind energy industry has experienced significant demand for nanostructured materials to improve performance, reliability, and durability of turbine components and provide access to better wind resources with enhanced energy throughput.

The largest cost components of the turbine are rotor blades, gearbox, and tower, which together account for around 50–60% of the turbine cost. Among all the components of wind turbines (blades, nacelle, bearing, shaft, hub, gearbox, generator, battery, cables, tower, etc.), nanostructured materials are being used in blades, supercapacitors, and battery, while nano-enabled coatings, fasteners, lubricants, and greases are gaining momentum in the gearbox parts. Blades in a wind turbine are the largest consumer of nanomaterials, the properties of which mainly determine the performance, maintenance, repair, and overhaul (MRO), and life cycle of the turbine. Nanotechnology opportunities exist to improve the overall performance, lower the MRO costs, and accelerate the deployment of offshore wind power technology.

43.3
Growth Drivers

A changing wind energy industrial landscape from onshore to offshore installation is creating demand for lighter and durable blades, higher quality and better performing lubricants for drivetrain, and anti-icing and super hydrophobic coatings. Nano-enabled lubes, sensors, coolants, coatings, cables, fasteners, battery, and supercapacitor could be a game changer in the wind energy industry, if they could achieve better cost-performance ratio as compared to the conventional components. Growth drivers for nanotechnology-enabled wind energy components are as follows:

i) Increased traction for offshore wind farms (increase in installation and acceptance in the European Union)
ii) Concrete government policy
iii) Acceptance and demand of nano-enabled products by wind energy industry stakeholders
iv) Reduction in greenhouse gases and carbon footprint

43.4
Challenges

Unfortunately, the use of nanostructured materials in the wind energy industry can lead to a number of challenges too, which are depicted in Figure 43.1 and highlighted as follows:

i) *Dispersion:* It is still a challenging issue for nanostructured materials, as extra cost is associated with functionalization.
ii) *High aspect ratio:* The loading of nanomaterials varies from one application to another. For example, 0.1–1 wt% loading of carbon nanomaterials, such as carbon nanotubes and grapheme in resin formulations for composites, is sufficient to achieve desired mechanical properties, whereas loadings of

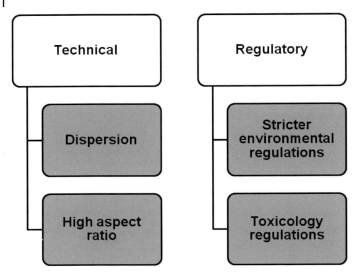

Figure 43.1 Commercialization challenges of nanomaterials in wind energy.

1–2 wt% in coatings, batteries, and supercapacitor are typically required to achieve desired properties. The high aspect ratio of carbon nanotubes (CNTs) and graphene means that a very low loading is needed to form a connecting network in a matrix resin that hinders high-volume consumption of CNTs and graphene in the wind energy industry.

iii) *Stricter environmental regulations and toxicity issues:* The stringent environmental regulations on toxicity of nanomaterials, such as silica and carbon nanotubes, especially from the developed nations, such as European Union and the United States, are limiting uses of these nanomaterials in a wide variety of application. National Institute for Occupational Safety and Health (NIOSH), USA has found the evidence on the carcinogenic potential of CNTs and silica, based on laboratory studies. In 2013, NIOSH released new guidelines for CNTs that there could be health risks at an exposure somewhere between 0.5 and just above $1 \, \mu g/m^3$ of air. In the United States, CNTs are mandated as chemical substances under the Section 5 of the Toxic Substances Control Act (TSCA). It requires manufacturers and importers of CNTs to conduct specific toxicity tests and to notify the Environmental Protection Agency (EPA) at least 90 days prior to their manufacture or import for commercial use in the United States.

In 2008, the European Commission made an amendment to its Registration, Evaluation, Authorization and Restriction of Chemicals (REACH) regulation to remove the exemption from REACH's registration and evaluation requirements for CNTs. As per the REACH regulation, it is mandatory for those CNTs manufacturers or importers who are manufacturing or importing CNTs in quantities greater than 1 metric ton per year to register CNTs under chemical substances before they can be manufactured, imported, or placed in the European Union. As part of these requirements, manufacturers and importers need to identify properties, hazards, and uses of CNTs and assess how risks to human health and the environment can be mitigated by applying suitable risk management measures.

43.5
Applications

Nanostructured materials are having superior physicomechanical and chemical properties and therefore possess the potential for very broad range of applications, namely, structural and functional in many areas of the wind energy industry. Table 43.1 shows the potential applications of nanostructured materials in the wind energy sector.

The wind turbine components, potential of nanotechnology to improve the performance of these components, and status of commercialization are shown in Figure 43.2.

Table 43.2 describes various nanomaterials and their specific contributions to component performance in wind turbines.

43.5.1
Structural Applications

43.5.1.1 Blades
In the wind industry, nanoscale materials-based prepregs are applied in the long rotating blades, which transform kinetic energy from wind into 3 phase AC electrical energy, which is then transformed into electricity. Due to the high wind speeds (15–25 m/s) and the size of the blades, the blade material has to bear high external loading, high gravitational forces and torque load, and high stress, which can be reduced by using high-performance nanomaterials.

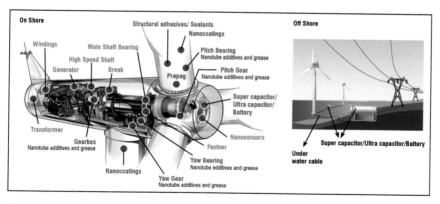

Figure 43.2 Wind turbine components, commercialization, and potential of nanotechnology.

Table 43.1 Potential applications of nanostructured materials in the wind energy components.

S. No.	Wind energy components	Potential applications	Nanostructured materials
1	Blades	Prepreg, coatings, structural adhesive, sealants, and fasteners	Graphene, CNTs, hybrid of graphene and CNTs, silica, metal matrix composite of Al and CNTs and graphene
2	Nacelle	Coatings, prepreg	Graphene, CNTs, hybrid of graphene and CNTs, and silica
3	Bearing	Lubricants, coatings, and seals	Silica, copper, tungsten disulfide, molybdenum disulfide, boron nitride, graphene, CNTs, and hybrid of graphene and CNTs
4	Main shaft	Coatings	Silica, CNTs, and hybrid of graphene and CNTs
5	Gearbox	Lubricants and coatings	Silica, copper, tungsten disulfide, molybdenum disulfide, boron nitride, graphene, CNTs, and hybrid of graphene and CNTs
6	Generator	Sensors, windings, coolants	Graphene, silicon, CNTs, and hybrid of graphene and CNTs, metal matrix composite of Cu–CNTs/graphene, alumina, copper oxide, silica
7	Power takeoff	Cables	Metal matrix composite of Cu–CNTs/graphene
8	Control system	Sensors	Graphene, CNTs, silicon, hybrid of graphene and CNTs
9	Yaw system	Sensors	Graphene, CNTs, hybrid of graphene and CNTs
10	Yaw bearing	Grease	Tungsten disulfide, molybdenum disulfide, boron nitride
11	Small engineering components	Antivibration mounts	Graphene, CNTs, hybrid of graphene and CNTs
12	Fasteners	Coatings and metal matrix composite	Graphene, CNTs, hybrid of graphene and CNTs, metal matrix composite of Al and CNTs and graphene
13	Condition monitoring system	Sensors	Graphene, silicon, CNTs, hybrid of graphene and CNTs
14	Lightning protection systems	Coatings	Graphene, CNTs, hybrid of graphene and CNTs
15	Tower	Cement, concrete, and fasteners	Graphene, CNTs, hybrid of graphene and CNTs
16	Energy storage and transmission	Battery, ultracapacitors, underwater cable	Graphene, CNTs, hybrid of graphene and CNTs

Table 43.2 Nanomaterials and their specific contributions to component performance in wind turbines.

Component	Nanomaterial used	Specific benefits or improvement for component	Reference
Turbine blade (structural)	CNT composite fibers	This technology offers reduced weight and increased density over first-generation carbon nanotube-strengthened glass fibers. CNT-infused fiber composites allow blade start-up in lower wind speeds and extended life of turbine blades and components[a)]	[1]
	MWCNT/carbon fiber-reinforced hybrid composites	Addition of MWCNTs to the matrix of a carbon fiber-reinforced composite (CFRC) can significantly decrease the interlaminar fatigue crack propagation rate while resulting in more than threefold increase in fatigue life	[2]
	MWCNT/glass fiber hybrid composite	The addition of CNTs to glass fiber epoxy hybrid composites can increase fatigue life under tension–tension and compression–compression loading conditions. These nanocomposites have the potential to act as self-sensing material for damage monitoring in fiber-reinforced composite structures	[3]
	Graphene platelets infiltrated into the epoxy resin or directly spray-coated onto the glass microfibers	Fiberglass/epoxy composites incorporated with only 0.02 wt% graphene platelets can increase the fatigue life of the composite in the flexural bending mode by up to 1200-fold while improving it by more than three to fivefold in uniaxial tensile fatigue conditions	[3]
	Epoxy incorporated with nanostructured amphiphilic block copolymer	The use of Dow's toughening technology was found to give a better balance of improved toughness and fatigue resistance for wind turbines without sacrificing other key performance properties and processability	[4]
Gearbox bearings (functional)	Graphene as a lubricant substitute	The graphene is able to remarkably reduce the wear rate and the coefficient of friction (COF) of steel. It also resists the oxidation (tribocorrosion) of the steel surfaces when present at sliding contact interfaces[b)]	[5]
	A novel combination of graphene wrapped around nanodiamond particles	It can reduce friction to almost zero resulting in significant energy savings	[5]

(continued)

Table 43.2 (Continued)

Component	Nanomaterial used	Specific benefits or improvement for component	Reference
Wind turbine cooling system (functional)	Tungsten sulfide nanoparticles (WS_2 NPs) as lubricant additive	Improvement of the rolling contact fatigue (RCF) life of rolling element bearings due to the ability of WS_2 NPs to reduce hydrogen permeation, thereby resulting in longer life	[6]
	Nanofluid based on water-containing alumina nanoparticles	Nanofluids are able to dissipate a large amount of heat and hence reduce the temperature rise of the electrical and mechanical components, thereby preventing the premature failure of components and improving overall efficiency	[7]
Electrical power transmission cables for wind farms (functional)	Carbon nanotube-based cables	Carbon nanotube-based cables are able to carry as much electric current as copper, and are lighter. These have potential sustainability, durability, and efficiency advantages[c]	
Turbine blades (sand erosion-resistant coatings)	Nanopapers based on carbon nanofibers (CNFs)	Carbon nanofiber-based nanopaper coatings are able to enhance the resistance of solid particle erosion of glass fiber (GF)/epoxy composites turbine blades due to the high strength of CNFS and their nanoscale structure	[8]
Turbine blades (anti-ice coatings)	Plasma treated nanotextured self-adhesive PU films	These nanotextured films impart hydrophobic and anti-icing characteristics to wind turbine blades. Prevention of Ice formation on surfaces excludes the possibility of higher energy consumption, thereby enhancing the energy efficiency[d]	
Turbine blades (UV-resistant coatings)	CNT and graphene-based nanocomposite coatings	These coatings are able to resist UV degradation and corrosion while retaining their hydrophobicity. It is observed that a small weight percentage of carbon nanotubes and graphenes in the polymeric coating increase the film resistance against UV light[e]	
Turbine blade (multifunctional coatings)	Carbon nanofiber (CNF) paper-based nanocomposite coating	Multifunctional nanocomposite coating material shows great promise for wind turbine blade applications due to its outstanding damping properties, excellent friction resistance, and superhydrophobicity	[9]

a) https://nice.asu.edu/nano/carbon-nanotube-cnt-infused-composite-fibers.
b) http://www.anl.gov/.../graphene-layers-dramatically-reduce-wear-and-friction-sliding-steels.
c) http://cleantechnica.com/2011/10/13/carbon-based-cable-achieves-milestone.
d) http://www.igb.fraunhofer.de/content/dam/igb/en/documents/sheets/gtm/1404_PB-gtm_Anti-Eis-Beschichtungen_en.pdf.
e) http://webs.wichita.edu/depttools/depttoolsmemberfiles/windenergy/FINAL%20TECHNICAL%20REPORT%20TASK42_3_110.pdf.

The aerodynamic design and length of blades have a significant effect on total energy output. The latest developments for blades reach 83.5 m and weigh about 35 tons. Larger wind turbines have greater energy output per unit rotor area; mainly driven by longer wind blade lengths capable of harnessing more wind energy. As blade length grows, the wind area swept increases exponentially and so does the energy output. In fact, with every doubling in blade length, the power output increases by a factor of four. On an average, the blade length has scaled up by 10% annually and doubled every 5 years.

The increase in blade length translates greater demand for nanocomposites based on epoxy and carbon nanomaterials prepregs to impart excellent mechanical properties, and better environmental stress-cracking resistance properties as well as reduced blade weight. However, as the blade length increases, the blade tends to deflect more and, consequently, the flexural stress generated is higher for the same force. Prepregs (ready to use materials) made of carbon nanomaterials such as multiwalled carbon nanotubes (MWCNTs), graphene, and hybrid of MWCNTs and graphene are being used by blade manufacturers. Nanocomposites not only impart enhanced mechanical properties but they also provide excellent electrical conductivity. A few of wind turbine manufacturers currently use nano-enabled prepregs as structural materials for their blades.

Wind turbines are vulnerable to lightning strike because of the nonconductive nature of composite material. Nanocomposites made of graphene and MWCNTs offer protection against damages occurring from lightning strikes, which are responsible for 10–15% of the failures of wind generators. They also provide electromagnetic shielding to the wind generators. In the longer term, design innovation coupled with hybrid nanomaterial is expected to improve efficiency and reliability of the wind turbine blades. Figure 43.3 shows growth drivers for nanocomposites demand in the wind energy sector.

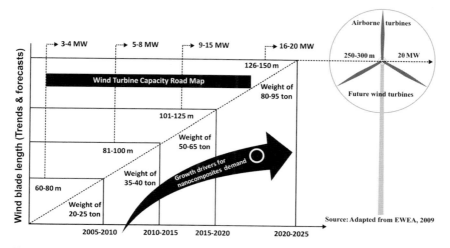

Figure 43.3 Growth drivers for nanocomposites demand in wind energy.

43.5.1.2 Other Applications

The research and development on materials with high strength, stiffness, electrical conductivity, thermal conductivity, tensile strength, lightweight, and high temperature stability has got a major boost with the advent of wind turbine blades. Usage of these materials can offer many advantages such as increase in payload for turbine blades, and lightweight submarine transmission cable for carrying electric power below the surface of the water (offshore wind farm). The inadequacy of metals and alloys in providing high strength and stiffness to a structure has led to the development of metal matrix nanocomposites. It is designed to possess features such as high thermal conductivity, high specific strength, stiffness, high strength at elevated temperatures, and low coefficient of thermal expansion.

Cu–carbon nanotubes metal matrix nanocomposites-enabled submarine power cable is another potential application for offshore wind energy. Submarine power cables are used primarily to transmit electricity from offshore to onshore grid connection. For fast and effective transmission of electricity from offshore to onshore, power cable should possess excellent electrical conductivity, dimensional stability, excellent tensile strength, and high abrasion resistance.

43.5.2
Functional Applications

43.5.2.1 Nanocoatings

The wind turbines (towers, blades, and nacelle) lifetime can be increased by using nano-based anti-icing coatings (offshore) and self-cleaning coatings (onshore). Low winter temperatures and cold climate in offshore locations routinely create ice on the rotor blades. The accumulation of ice produces additional weight on the rotor blades with increased aerodynamic drag. Increase in weight transmits more stress resulting in greater strain on the blades and impaired function of equipment (damages on bearings and gear boxes), thus leading to higher energy consumption, lower energy output, and higher maintenance costs. Heating system is an established method to reduce ice formation on the blade, but during the deicing period the wind turbine actually consumes power rather than generating power. Nanocoatings on the rotor blades could be an optimal solution to prevent ice formation at the outset.

The currently available commercial anti-icing nanocoatings (nanosilica) effectively prevent ice formation on the rotor blades. In 2014, Gamesa Corporación Tecnológica (formerly Grupo Auxiliar Metalúrgico) launched nanosilica-based anti-icing nanocoating (Bladeshield®). It is a hydrophobic and rain/sand erosion resistant coating that not only prevents the formation of ice but also improves antierosion performance. It is estimated that the use of anti-icing nanocoatings can increase rotor blade life by at least 2–5 years more than a conventional blade and maintenance cost can be reduced by 15–20%.

The consortium comprising SP, MW Innovation, Vattenfall R&D, Pegil Innovations, and Re-Turn AS have been developing technology for deicing of wind turbine blades using microwaves in combination with suitable hydrophobic top-coats. The real-life solution is enabled by Re-Turn's technology, which involves using efficient microwave absorbers such as CNTs and now the technology has been transferred to a new start-up company, Icesolution AS.

Vestas Wind Systems A/S in US Patent Application 20110020134 has presented a solution for improved lightning strike protection system for wind turbines by incorporating rotor blade surface with CNTs-based metal matrix composite.

The rotor blades are subjected to withstand extreme environmental conditions, both operational and environmental. Self-cleaning nanocoatings can prevent the accumulation of airborne particles, microbes, and debris on wind blades, thereby increasing the rotor blades' life, reducing the drag on the blades and improving their energy output. It can also reduce wear and tear that enables turbines to run longer without maintenance, and thus reducing operating costs.

Coatings with ceramic nanoparticles (ZnO, TiO_2, ZrO_2, SiO_2, Al_2O_3, and CeO_2) and carbon nanomaterials (CNTs and graphene) are designed to prevent airborne particles accumulation, UV degradation, and corrosion on the rotor blades, which can last 10–15 years. However, self-cleaning nanocoatings have not yet been commercialized.

The technology requirements for wind farm are very diverse and it is affected by differences in geographical and meteorological conditions. For example, offshore wind farm in particular have to deal with corrosion due to salt and dampness as compared to onshore wind farm. Thin-film nanocoatings over power train components (bearings, gears and hydraulic components) can be deposited by physical vapor deposition or plasma-assisted chemical vapor deposition.

43.5.2.2 Nanolubricants

The gearbox is a critical component in the wind turbine drivetrain, which converts rotor torque generated at a speed of 5–15 rpm, which could go up to a speed of around 1500 rpm for efficient conversion to electrical energy by the generator. The gearbox is subject to the external and internal loading, which includes the bending loads, gravitational loads, torque loads, as well as torsional loading.

The average lifespan of a wind turbine is pegged at around 20 years, but before attaining their full lifespan repair or overhaul is required. Gearbox is a key power-generating component in a wind turbine system, wherein generator is attached. The average lifespan of wind turbine gearboxes and bearings range from 7 to 13 years and vary by turbine type, gearbox subcomponent manufacturer, and gearbox manufacturer.[3] The cost of annual maintenance and gearbox replacement accounts for 2–3% of the original wind turbine cost. Gearboxes primarily fail due to micropitting (surface fatigue), bearing failures, and foaming. The conventional lubricants form sludge and deposits on the drivetrain

---

3) http://www.nawindpower.com/issues/NAW1505/FEAT_01_Meet-The-Achilles-Heel-Behind-Most-Gearbox-Failures.html.

components and adversely impact the performance of gearboxes. It is important to note that bearings are critical components for wind turbine gearbox reliability. The drivetrain components (transmission, generator, rotor shaft with bearings, coupling, and brake) have different requirements, and the right choice of lubricant additives is crucial to the smooth running of gearboxes, and therefore, maximizing overall performance of wind turbines.

Nanostructured materials have recently been considered as potential lubricant additives mainly due to specific advantages such as enhanced thermal stability, variety of particle chemistries, desired sizes to enter into contact asperities, and appropriate reaction rate with the surface without induction period (required for conventional lubricant additives) [10–14]. The nanolube additives can help to enhance performance, lifetime of wind turbine and productivity (energy output) as well as reduce downtime. It is estimated that the potential annual savings of switching to nanolube additives could be up to US$5000–7000 per wind turbine operated.

Various types of nanomaterials, predominantly of inorganic type, have been used as additives in lubricating oils to impart the tribological properties of lubricants. Nanolubes are based on the principles of solid lubrication and they are mostly used for antiwear, antifriction, and extreme pressure applications. The reduction of friction and wear are dependent on the characteristics of nanoparticles such as size, shape, and concentration. These advantages translate into longer equipment operation, increased efficiency, and extended maintenance intervals. In particular, the heavy metals, sulfur, and phosphorus-containing additives can be eliminated from the lubricant formulations by the addition of nanostructured materials.

The oil drain intervals for conventional lubricants range from 8 and 16 months and expectations for nanolube additives for offshore applications could be a drain interval of 3–5 years. Broadly, nanoparticles are being used as additives in lubricating oils and it can be classified into metal chalcogenides, soft metals, boron-based materials, carbon-based materials, single/mixed oxides, and organic/polymer materials. Nanostructured self-lubricating tribological coatings and coatings based on metal matrix composites with nanoparticles for the reduction of coefficient of friction and high resistance to wear is one strategy to enhance the lifespan of bearings. Ceramic-based nanocomposites can also be used for the production of more resistant bearings to extend the gearbox lifespan. Recently, Zhengzhou Dongshen Petrochemical Technology Co., Ltd, China has launched nano-Cu alloy-based lubricant additives for wind turbine' gearbox applications. Nano-copper being a soft metal has the advantage of the low shear intensity and coefficient of friction, easy slipping to move within the asperities, and relatively easy to mend the defects on the friction surface. Nano-copper has good ductility as well as low shear strength, which provide excellent friction-reducing and antiwear behavior as well as good self-repairing ability for steel–steel contacts.

Generator bearings are subjected to electric discharge damage failures due to passage of electrical current, and therefore special emphasis needs to be given on lubrication aspects. Rewitec GmbH has introduced nanostructured (silica and graphite) lubricating oil-based coatings (DuraGear®) to improve tribological properties, which reduces abrasion and wear of metal surfaces, as well as increases the load bearing capacity.

Silica in lubricating oil reduces friction; improves the wear resistance and load-carrying capacity of lubricating oils. It has been proposed that the ball-bearing effects of SiO_2 nanoparticles between the rubbing surfaces, and surface polishing by SiO_2 nanoparticles are the main causes of the reductions in wear and friction.

43.5.2.3 Nanofluids

The cooling system is being used in the wind turbines operation to keep the temperature of the generator, gear oil, and electronic components constant because during operation, wind turbines are bound to dissipate a large amount of heat, which may lead to a significant temperature rise of the mechanical and electrical components, resulting in a further reduction of the overall efficiency and unexpected MRO cost.

High temperatures, particularly during the summer season, could damage the electric generator and mechanical parts of the turbine; reliable efficiency of cooling system is critical for wind turbines.

The cooling system of most wind turbines uses either air cooling system or liquid cooling system and air cooling system can be further divided into natural ventilation cooling and forced air cooling. According to industry statistics, about 95% of the wind turbine manufacturers are using liquid cooling (water) for their larger turbines that operate in harsh environment. The cooling system of larger wind turbines usually requires high electric consumption to establish and sustain the airflow and, therefore it increases the amount of dissipated heat. In order to ensure that wind turbines can operate efficiently on a long-term basis, the research and development of nanotechnology-based cooling system with high efficiency and low energy consumption becomes particularly important.

The use of heat transfer fluids or nanofluids represents a possible solution to enhance the performance of water-cooled systems for onshore and offshore wind turbines.

Nanofluid, a term coined by Choi and his coworkers at Argonne National Laboratory of the United States in 1995 [15], is a new class of heat transfer fluids, which is developed by suspending nanoparticles with average sizes below 100 nm in the host fluids such as water, oil, diesel, ethylene glycol, and so on. The highest possible thermal properties can be achieved at the smallest possible concentrations (preferably <1% by volume) by uniform dispersion.

Nanofluids have attracted great interest recently because of the high demand of advanced coolants from the wind energy industry [16]. Cu nanoparticles are being used to develop nanofluids because the thermal conductivity of copper at room temperature is 700 times greater than that of water and 3000 times greater than of engine oil; for example, a small amount (<1% volume fraction) of Cu nanoparticles dispersed in ethylene glycol or oil could increase the inherent poor thermal conductivity of the liquid by 30–40%, while conventional particle–liquid suspensions require high concentrations (>10%) of particles to achieve such enhancement [17]. In addition, copper nanoparticles provide extremely high surface areas for heat transfer and therefore have great potential for use in heat transfer. The much larger relative surface areas of nanophase powders, when compared with those of conventional micrometer-sized powders, should markedly improve the heat transfer capabilities and stability of the suspensions. The other potential nanostructured materials in this field are alumina, carbon nanotubes, and graphene due to their inherent large surface area and higher thermal conductivities. As of now, commercial products related to nanofluids for wind energy are yet to enter the markets.

43.5.2.4 Ultracapacitors

Ultracapacitors are bridging the gap between batteries and capacitors as they are able to hold hundreds of times the amount of electrical energy as standard capacitors and batteries do. Ultracapacitors are used in electric vehicles, electronics, and wind energy applications. Batteries have a higher energy density but they take more time to recharge, whereas ultracapacitors are characterized by the lower energy density, but take shorter time to recharge.

In the past 5 years, wind turbine manufacturers have started to incorporate the pitch control system in a wind turbine, which is located in the rotor hub and controls the turbine blades. The main purpose for the pitch control systems is to change the angle of rotor blades according to wind speed and direction in order to capture the maximum wind energy as well as to protect the rotor blades against high speed winds.

A nano-enabled ultracapacitor is light in weight (one-fifth the weight of a battery); it goes in the hub of the blade and provides backup power and uninterrupted power systems for pitch control systems for reliable and safe operation of the wind farms. Wind turbine manufacturers, particularly for offshore systems, are quickly moving toward ultracapacitors adaptation because they have a longer lifecycle than batteries as well as low maintenance cost.

Graphene and hybrid carbon nanomaterials (graphene and MWCNTs) are promising candidates to replace activated carbons as the electrode materials in high-performance ultracapacitors due to their large surface area, high mesoporosity and electrolyte accessibility, and good electrical properties. Recently, Skeleton Technologies, Germany has launched graphene-enabled SkelCap line of ultracapacitors for wind turbines.

43.5.2.5 Nanosensors

Wind turbine sensors designed with the nanostructured materials possessing excellent electrical properties can be used for the detection of blade's internal structure defects, which can provide a new way to forecast wind turbine blade's service life. A nano-enabled sensor has tremendous potential to regulate the energy output and rotor speed to prevent overloading of structural components, when it is used in drivetrains.

Nanostructured materials can be used in the fabrication of sensor, where it utilizes carbon nanomaterials and ferromagnetic/nonmagnetic layered structures made up of atomically thin films, in the range of 2–5 nm thickness. The nano-enabled sensors have the ability to sense physical parameters, such as temperature, gas, pressure, humidity, wind speeds, and magnetic fields, by changing their electrical or magnetic properties.

Nanosensors, based on polymer nanocomposites (CNTs, graphene, silicon), have the ability to detect changes in environmental conditions using their electrical properties, which allow wind turbine operators to remotely but precisely monitor the condition and structural integrity of the rotor blades.[4]

43.5.2.6 Structural Adhesives

A wind blade consists of the structural top and bottom shell halves, which are bonded together with a structural adhesive around a shear web. Therefore, the integrity of the wind turbine blade is highly dependent on bonding quality of the structural adhesive because a significant amount of shear load needs to be transferred from the web to spar cap in the shell halves, and between the shells at the leading and trailing edges.

Nanomaterials are of great interest as fillers for structural adhesives; having several advantages such as enhanced fatigue resistance or toughness, outstanding shear and peel strength with excellent impact resistance, and durability compared to conventional micrometer-sized fillers-based structural adhesives at the same loading level. A limited quantity of nanosilica and carbon nanomaterials-based structural adhesives and sealants are used in the manufacturing of rotor blades. Even at higher loading (40–50 wt%) levels of nanoscale SiO_2 particles, the viscosity of resin is not increased much because there is absence of agglomeration tendency.[5] Concentrates of surface-modified SiO_2 nanoparticles in epoxies and acrylates have become available in industrial quantities. NANOPOX®A (Evonik Industries, Germany) series are high-performance concentrates of nanosilica (40 wt%) in epoxy resins and have been especially designed for the structural adhesive applications. The silica phase consists of surface-modified SiO_2 nanospheres with a particle size of 20 nm and a very narrow particle-size distribution, which prevents agglomeration and enables cross-linking into the resin matrix during curing process. It permeates and impregnates not only epoxy resin but also glass and carbon fiber fabric. The use of nanosilica in epoxy resin improves several properties of the structural adhesive such as increased toughness, modulus and hardness, significantly enhanced fatigue performance, dramatically improved compressive property, and reduced coefficient of thermal expansion.

NANOCRYL®A product series from Evonik Industries is a versatile dispersion of colloidal silica (20 nm) in a monofunctional, difunctional, and trifunctional

4) www.nanomagazine.co.uk/index.php?option=com_content&view=article&id=1265:university-of-houston-gets-12-million-grant-for-alternative-energy-research&catid=38:nano-news&Itemid=159.
5) http://composites.evonik.com/product/composites/en/markets/windenergy/pages/default.aspx.

acrylate monomer or oligomers and are used in the structural adhesive. Despite the high loading of SiO_2 (50 wt%), NANOCRYLA products possess low viscosity, high transparency, and do not show any sedimentation due to the agglomerate-free dispersion of the silica nanoparticles in the acrylate. It can be used in any common UV-curable acrylate systems. The performance of structural epoxy adhesives toughened with reactive liquid rubbers or core–shell materials can be significantly improved by modifying them with silica nanoparticles.[6]

43.5.2.7 Other Applications

Nanosilica is being used as reactive resin modifiers in epoxy resins (principally used in rotor blades) to enhance the mechanical properties, reduce the tendency for microcrack formation, and improves thermal expansion, shrinkage, and electrical properties.

Tower and foundation of wind turbines require a grouting or concrete, which should be able to absorb the enormous vibrations, wave actions, and torque during the whole life cycle of the wind turbines. Carbon nanomaterials-enabled concrete imparts high strength, fatigue resistance, and high modulus to the tower and foundation, and it can also withstand under very harsh conditions such as temperature ranges as low as 2–30 °C.

43.6
Intellectual Property Scenario

Patents play a pivotal role in innovation and economic performance, as well as in the development of technology transactions from research and development laboratories to market place. The increasing commercial exploitation of nanotech-enabled components in the wind energy sector has resulted in the surge in patenting activity to protect inventions by public and private research organizations and corporates, which is closely connected to recent evolution in innovation processes and product development. In this section, some of the most important patents have been highlighted as depicted in Table 43.3.

The US-published Patent Application 2013/0216390 [18] discloses a process for the production of carbon nanotube-reinforced wind turbine blade. In this invention, CNTs containing dispersion with resin (polyurethane/epoxy/vinyl ester) of 0.05–0.7 wt% is incorporated into fibrous material by vacuum infusion process or pultrusion process. It provides advantages such as uniform dispersion of CNTs, short demold times, reduced shrinkage, and improved fracture toughness.

Chinese-published Patent Application 2014/104087059 [19] discloses microwave-absorbing wind power blade coating, which utilizes graphene and carbon nanotubes and other carbon-based microwave-absorbing materials to accommodate the traditional blade coating. It comprises of uniformly mixing

---

6) http://hanse.evonik.com/sites/hanse/Documents/Hanse-Structural-Adhesives-en-web.pdf.

Table 43.3 Intellectual property scenario: nanotechnology in wind energy.

S. No.	Title	Patent number and publication date and year	Assignee	Key features/benefits	Nanomaterial used
1	Hydrophobic coating compositions for drag reduction	US 8258206 B2, Sep. 4, 2012	Ashland Licensing and Intellectual Property, LLC, USA	The invention involves hydrophobic coating for the application of wind turbine blade surfaces to reduce drag in fluid flow, thus providing energy savings	A blend of organic and/or inorganic polymers with hydrophobic nanoparticles of fumed silica and/or titania in a solvent
2	Lubrication of fluid turbine gearbox during idling or loss of electric grid	US20110168495 A1, March 8, 2011	General Electric Company, USA	An efficient lubrication system for lubricating a fluid turbine gearbox during loss of electric grid, idling, or any other emergency conditions	The lubricating oil contains nanoparticle additives such as aluminum oxide, titanium oxide, silver oxide, or their combinations
3	Reinforced composites produced by a vacuum infusion or pultrusion process	US20130216390 A1, Aug. 22, 2013	Bayer Material Science LLC, USA	The present invention enables production of large wind turbine blades characterized by short demold times, reduced shrinkage, and enhanced fracture toughness	CNTs containing dispersion with PU/epoxy/vinyl ester resin is incorporated into fibrous composite
4	Microwave-absorbing wind power blade coating as well as preparation method and application thereof	CN201410347844, Oct. 8, 2014	Paint Co., Ltd, China	The microwave absorbing wind turbine blade coating has anti-icing and deicing effects and can effectively reduce radar reflection. It also achieves a certain stealth effect	Graphene and carbon nanotubes and other carbon-based microwave-absorbing material to accommodate the traditional blade coating
5	Erosion-resistant impregnating resin systems and composites	WO2009038971 A1, March 26, 2009	3M Innovative Properties Co., USA	Improved rain erosion resistance of composite articles such as wind turbine blade or wind tower	The use of surface-modified silica nanoparticles in impregnating resin systems

(continued)

Table 43.3 (Continued)

S. No.	Title	Patent number and publication date and year	Assignee	Key features/benefits	Nanomaterial used
6	Deicing of a surface of structures in general such as wind turbine blades, aircraft wings using induction or radiation	US20150083863 A1, March 26, 2015	JKA Kemi AB, Sweden	The invention discloses a method that enables the facile deicing of a surface of a structure by the application of EM induction or IR/microwave radiation to heat up a layer or a coating on said surface of the structure in general	The layer contains conductive particles such as carbon, nanoparticles such as graphite, carbon nanotubes, carbon nanocones, and so on
7	Lightning receptors comprising carbon nanotubes	US20110020134 A1, Jan. 27, 2011	Vestas Wind Systems A/S, Denmark	The invention relates to a wind turbine rotor blade comprising a lightning protection system	The lightning receptors of rotor blades comprise CNTs. In embodiments, the nanotubes are located at an exterior surface, which is provided with a layer of CNTs and/or at least part of the lightning receptor is of a CNT–metal matrix composite
8	Sensing system for monitoring the structural health of composite structures	US7921727B2, April 12, 2011	University of Dayton, USA	A sensing system for use in monitoring the structural health of a structure such as a polymeric matrix composite structure (e.g., wind turbine) is provided	The system includes a sensor formed from a conductive ink-containing carbon nanofibers and a polymeric resin
	Nanocomposite microcapsules for self-healing of composite articles	WO2015171429 A1, Nov. 12, 2015	Wichita State University, USA	To prevent catastrophic failure and increase the service life of composites, such as wind turbines by incorporating nanocomposite microcapsules for self-healing of composite	The self-healing microcapsules comprise nanoparticulates such as graphene nanoflakes, single- and multiwalled-carbon nanotubes, carbon fibers/nanofibers, carbon black, nanoclay, nanotalc, boron nitride

					nanotubes, and boron nitride nanoflakes
10	Tethered Wind Turbine	US20080048453 A1, Feb. 28, 2008	Douglas J. Amick, Troy, USA	High-altitude wind turbine harnesses power of winds high in the sky by use of tether and cable technology	The tethered wind turbine of this invention utilizes carbon nanotube materials in its tether for both structural and conductive purposes
11	Cured composite composition	US20100249277 A1	General Electric Company, USA	The uncured polymer matrix shows low-enough viscosity for efficient use under resin transfer molding conditions, but which upon curing provide fiber-reinforced composite materials displaying excellent toughness and other properties	Fiber-reinforced composite having organic matrix, which comprises a cured epoxy continuous phase and a nanoparticulate thermoplastic block copolymeric discontinuous phase

70–99.9 parts of the traditional blade coating and 0.1–30 parts of a carbon nanomaterials microwave-absorbing material; and coating the blade with the obtained mixture by the traditional coating method. The patent claims about anti-icing and deicing effects and reduced radar reflection compared to the conventional wind blade coating systems.

WO-published Patent Application 2009/038971 [20] relates to the use of surface-modified nanoparticles of silica having a particle size between 5 and 500 nm in impregnating resin systems to improve the erosion resistance of composite components for wind energy. The patent claims that nanosilica-modified resin systems can be used in infusion processes with fibrous matrices, which offers great design flexibility without drastically affecting cure kinetics or resin viscosity.

The research and development on CNTs/graphene reinforced with copper metal matrix composites (MMCs) with high electrical conductivity, thermal conductivity, strength, stiffness, tensile strength, low coefficient of thermal expansion, lightweight, and high temperature stability has got a major boost with the advent of underwater cables, conductors, and windings applications in the wind energy sector. Usage of these materials can offer many advantages such increase in payload for turbines and lightweight underwater cables and wires. The inadequacy of metals and alloys in providing high strength and stiffness, and electrical and thermal conductivity to a structure, has led to the development of CNTs/graphene-reinforced copper MMCs. It can be prepared by a variety of processing techniques, which include powder metallurgy, electrochemical deposition, melting and solidification, and thermal spraying techniques.

43.7
Products Outlook

There have been significant research and development activities in nanotechnology for the wind energy industry, with nanocomposites (prepreg), nanolubes, nanocoatings, and ultracapacitor finding commercial applications since 2012 in blades, gearbox, and drivetrain components. The commercialized products are listed in Table 43.4.

It is interesting to note that silica, copper, and graphene nanomaterials-based products have more commercial success in the wind energy sector, as is evident in Table 43.2. Due to the absence of any larger aggregates, silica, copper, and graphene nanoparticles can be easily dispersed in composite structures without compromising the impregnation in the absence of excessive viscosity. These specific nanoparticles are not prone to agglomeration and are compatible with the state-of-the-art blade manufacturing technologies such vacuum-assisted resin transfer molding (VARTM), resin transfer molding, resin infusion, and prepreg.

Nanotechnology elicits great interest not only among researchers and government funding agencies but also among business houses and venture capitalists in the wind energy sector around the world. Graphene is touted as a miracle

Table 43.4 Nano-enabled commercialized products in the wind energy sector.

S. No.	Products	Nanomaterials	Applications	Company/weblink	Country
1.	DuraGear® [21]	Silica	Lubricants and grease	Rewitec GmbH www.rewitec.com	Germany
2.	DBDS061 [22]	Copper	Lubricants	Zhengzhou Dongshen Petrochemical Technology Co., Ltd	China
3.	SkelCap [23]	Graphene	Ultracapacitor	Skeleton Technologies www.skeletontech.com	Germany
4.	Bladeshield® [24]	Silica	Coatings	Gamesa Corporación Tecnológica www.gamesacorp.com	Spain
5.	NANOPOX® A [25]	Silica	Structural adhesive	Evonik Industries corporate.evonik.com	Germany
6.	NANOCRYL® A[a]	Silica	Structural adhesive	Evonik Industries corporate.evonik.com	Germany
7.	Nanostrength®[b]	ABA-type triblock copolymer with soft center block of poly butyl acrylate (PBA) and hard side block of poly methyl methacrylate (PMMA)	Structural adhesive	Arkema Inc. www.arkema.com	France
8.	EP-290[c]	Nano-toughened epoxy	Prepreg	GMS Industrial Pty Ltd www.gmscomposites.com	Australia

a) http://hanse.evonik.com/_layouts/websites/internet/SecuredLinksHandler.ashx?TargetUrl=%2fsites%2fhanse%2fen%2fproducts%2fnanosilicaconcentrates%2fnanocryl%2fpages%2f&PropertyId=231&cpvid=15917.

b) http://www.arkema.co.jp/en/products/plastic-japan/brochure-library.

c) http://www.gmscomposites.com/wp-content/uploads/2014/02/GMS-Composites-Prepreg-data-sheet_EP-290-v1.pdf.

material because of its extraordinary properties with the potential to revolutionize nanocomposites, nanolubes, ultracapacitors, nanoadhesives, nanocoatings, and metal matrix nanocomposite products in the wind energy application. It is 200 times stronger than steel, more durable than diamond, highly transparent, lighter than paper, highly flexible, and has 5 times more thermal and electrical conductivities than copper.

Matrix modification using micro-or nanostructured fillers (e.g., fumed silica, exfoliated silicate, and carbon nanotubes) has been utilized to enhance the mechanical properties of fiber-reinforced composite [26,27]. Up to ~16% improvement in interlaminar shear strength of composites has been reported by adding ~0.3% weight of carbon nanotubes to the matrix [26]. The effect of incorporation of graphene platelets on the fatigue life of traditional glass fiber–epoxy composites is reported by various researchers [28–34].

Carbon nanotubes (CNTs) are considered to be one of the most important materials of the twenty first century owing to their unique combination of mechanical, electrical, and thermal properties. An addition of a small amount (<1 wt%) of CNTs in the polymer matrix of composites can result in a significant improvement of polymer–matrix composites (PMC) component properties. For example, the addition of 0.3 wt% CNTs in a typical turbine blade epoxy and carbon fiber prepreg will increase the material thermal conductivity by more than 50–70%.[7]

Many studies have reported improvements of composites properties by adding small amounts (at the level of 0.1–0.5 wt%) of nanostructured materials, such as CNTs, graphene, and hybrid of CNTs and graphene, in the polymer matrix of composites; fiber sizing or interlaminar layers can allow the increase of interfacial shear strength, improved fatigue strength, compressive strength, and flexural strength, as well as fracture toughness of the components by 30–60%. Graphene nanoplatelets coatings on wind blades can increase the fatigue life up to 100 times.[8]

43.8
Future Development and Directions

Currently, wind energy is a major contributor to the world's electricity supply and is a driving force in the renewable energy sector. Nanotechnology is poised to play a key role in the spheres of onshore and offshore wind installations worldwide.

The wind turbine blade accounts for more than 20% of total installed costs onshore, while cost of offshore rotor blade can account for between 25 and 30%. In recent years, with the exponential growth in wind energy development

7) http://www.huntsman.com/advanced_materials/Applications/itemrenderer?p_rendertitle=no&p_renderdate=no&p_renderteaser=no&p_item_id=997718783&p_item_caid=1223.
8) thebulletin.org/myth-renewable-energy.

worldwide, particularly offshore, large size wind turbine blades (with a length more than 80 m) has become the mainstream direction for the development of wind energy sector to produce more electric power. Wind turbine blades are subject to multiple-mode external loading in service, which includes bending loads, compression loads, gravitational loads, cyclic loads, and torque loads; so only materials with very high strength, stiffness, fatigue resistance, and dimensional stability, that is, nanocomposites, can be used to fabricate wind turbine blades.

There is increasing interest from manufacturers to produce advanced turbine blades with enhanced physicomechanical properties. It is extremely important to design and fabricate blades with superior fatigue resistance and enhanced stiffness properties to ensure operational longevity and reduce MRO cost. The fiber-reinforced composites dominate the wind turbine blade market because of their high strength-to-weight ratio, superior fatigue characteristics, and ability to make complex geometries. The lifetime performances of rotor blades can be improved through the use of nanostructured materials and aerodynamic performance can be maximized.

The future perspectives of nanotechnology utilization depend on the development, reliability, and durability of offshore wind farms and their components. For example, wind turbine blades are subject to long-term mechanical, cyclic, and environmental loading, involving flapwise and edgewise bending loads, gravitational loads, and compressive and shear loading. The development of advanced wind turbines requires nanostructured materials that can sustain mechanical, thermal, environmental, and cyclic loadings over the years while retaining their high stiffness and integrity. Nanostructured materials selection should not only consider the mechanical performance but also the socio-economic factors such as reliability, workability, durability, and environmental stability. Carbon nanotubes and graphene have been explored extensively in the area of wind energy in a variety of functional components.

In the short term, nanosensor will become more commonplace in the offshore wind turbines to monitor blade conditions and rotor hub. Nanosensors can be installed on the blade to remotely detect any unseen ice formation, damage, and faults (microcracks) in the rotor blades. It can also be mounted on the top of the nacelle for early detection of high winds, rain, and lightning. Nano-enabled sensors should be rugged and capable of operating over a wide range of temperatures with good electromagnetic shielding protection.

Offshore wind farms are more susceptible to the formation of ice layer on the rotor blades as temperatures drop below 0 °C, thus reducing energy production and there could be a chance of breakdown. Therefore, nano-enabled coatings would be able to prevent the ice formation over wind turbine blades surface.

The performance of structural epoxy-enabled adhesives can be significantly improved by modifying them with nanostructured particles. The upper and lower shells of the rotor blades are manufactured separately and are glued together by the structural adhesives. Nanostructured materials provide excellent thixotropic and specific slump characteristics as well as excellent fatigue and load bearing properties to the structural adhesives. In March 2016, Huntsman Corporation collaborated with Haydale Composite Solutions Ltd to commercialize graphene-enhanced Araldite® adhesives for a range of applications in the wind energy market [35].

Permanent magnets, predominantly manufactured from neodymium, iron, and boron (Nd–Fe–B), seem to be the most important rare earth elements that are employed in highly efficient wind generators to maximize the energy production. Regarding the use of permanent magnets in wind turbine generators, neodymium, terbium, and dysprosium availability, cost and environmental impacts will be highly critical both in the short and medium term. According to the Bulletin of the Atomic Scientist, the gearbox of a 2 MW wind turbine contains about 363 kg of neodymium and 59 kg of dysprosium [36].

Nanocomposite magnet materials would be the next futuristic materials in this space. It is made up of nanoparticles of the metals, for example, neodymium-based nanoparticles mixed with iron-based nanoparticles. Nano-enabled permanent magnets offer greater magnetic properties as compared to the conventional magnetic alloys and use lesser rare-earth metals.

Scientific communities are gearing up to develop high-altitude wind turbines, called airborne wind energy systems (AWES), that are capable of harnessing stronger and more consistent winds higher in the atmosphere without a tower and gearbox. In contrast to conventional wind turbines, AWES might be able to reach higher altitudes, tapping into a large and so far unused wind power resource [37]. In contrast to conventional wind turbines, AWES are either flying freely in the air, or are connected by a tether to the ground, such as kites or tethered balloons. For example, Laddermills, which are composed of a series of kiteplanes on a long string, that use wind energy at an altitude of 9000 m, where wind speed can be 20 times higher than at sea level.

Ideally, nanostructured materials, such as graphene, CNTs, and hybrid of graphene and CNTs, can be used to produce tethered AWES for both structural and conductive purposes. Carbon nanomaterials would make the tether itself many times lighter, which allows AWES to fly much higher using less lifting gas. CNTs wires and metal matrix composite of Cu/CNTs could be possible application areas of the electric generator in this system, due to the higher electrical conductivity of CNTs wires than copper and it would enhance overall efficiency of the electric generator greatly.

43.9
Conclusion

Renewable energy sources, such as wind energy, are an essential component of socioeconomic development and economic growth. It can help in reducing the global warming and dependency on traditional fossil fuels. Nanotechnology has tremendous potential to transform the wind energy sector, although application of nanotechnology in the wind energy industry is at an early stage. The future prospects of incorporating nanostructured materials into wind energy technology look promising.

Nanoreinforced composite has attracted large interest of industry and research community as a promising material for wind energy applications. The use of nanomaterials can improve the performance and durability of wind turbine blades and gearbox components. CNTs, graphene nanoplatelets, silica, and hybrid of CNTs and graphene are primarily used for the performance enhancement of the composites system called prepregs.

Composite materials such as nano-enabled prepregs (glass/carbon fiber/ epoxy/polyester resin) represent a material of choice in the making of lighter, stronger, conductive, and more durable rotor blades and nacelle, whereas towers and hubs are predominantly made from steel and cast iron.

Nanosilicia and nanocarbon materials (CNTs and graphene) are being developed and commercialized to protect blades, nacelles, and towers to extend their service life. The impact of nanolube additives, mainly metal oxide, and nanoceramic (metal, metal oxide, and nonmetal oxide) coatings on gearboxes and bearings are also being investigated and used to reduce friction and to extend their service life. Researchers and wind turbine stakeholders have begun investigating Cu–MWCNTs metal matrix composites for use in submarine power cables and windings as well as nanostructured materials for use in sensor technologies and fasteners.

At present, the market of nano-enabled components in commercial wind energy applications does not appear voluminous but they can take a big leap onto onshore and offshore wind energy market in the next 5–10 years if wind turbine developers, blades manufacturers, and OEMs will be able to make a breakthrough in volume production of low-cost, high-quality materials with superior performance.

Acknowledgment

The authors are grateful to the Director, ARC International, for his kind permission to publish this chapter.

References

1 Romhány, G. and Szebényi, G. (2012) Interlaminar fatigue crack growth behavior of MWCNT/carbon fiber reinforced hybrid composites monitored via newly developed acoustic emission method. *Express Polym. Lett.*, 7, 572–580.

2 Böger, L., Wichmann, M.H.G., Hedemann, H., and Schulte, K. (2011) Influence of CNT modification on the fatigue life of glass fibre reinforced epoxy composites. Available at http://www.iccmcentral.org/ Proceedings/ICCM17proceedings/ Themes/Behaviour/Behaviour/FATIGUE% 20OF%20COMPOSITES/F13%2011% 20Boeger.pdf.

3 Yavari, F., Rafiee, M.A., Rafiee, J., Yu, Z.-Z., and Koratkar, N. (2010) Dramatic increase in fatigue life in hierarchical graphene

composites. *ACS Appl. Mater. Interfaces*, **2**, 2738–2743.

4 Jacob, G.C., Hoevel, B., Pham, H.Q., Dettloff, M.L., Verghese, N.E., Turakhia, R.H., Hunter, G., Mandell, J.F., and Samborsky, D. (2009) Technical advances in epoxy technology for wind turbine blade composite fabrication. Available at http://www.montana.edu/composites/documents/SAMPE%202007%20paper%20from%20DOW.pdf (accessed October 19–22).

5 Sealy, C. (2015) Diamond puts a new shine on friction-free graphene. *Nano Today*, **10**, 411–532.

6 Niste, Vlad (2015) WS₂ nanoparticles as lubricant additives. Doctoral thesis, Faculty of Engineering and the Environment, University of Southampton.

7 De Risi, A. *et al.* (2014) High efficiency nanofluid cooling system for wind turbines. *Therm. Sci.*, **18**, 543–554.

8 Zhang, N., Yang, F., Guerra, D., Shen, C., Castro, J., and Lee, J.L. (2013) Enhancing particle erosion resistance of glass-reinforced polymeric composites using carbon nanofiber-based nanopaper coatings. *J. Appl. Polym. Sci.*, **129**, 1875–1881.

9 Lianga, F., Goua, J., Kapatb, J., Guc, H., and Song, G. (2011) Multifunctional nanocomposite coating for wind turbine blades. *Int. J. Smart Nano Mater.*, **3**, 120–133.

10 Martin, J.M., Mogne, T.L., Chassagnette, C., and Gardos, M.N. (1992) Friction of hexagonal boron nitride in various environments. *Tribol. Trans.*, **35**, 462.

11 Rapoport, L., Leshchinsky, V., Lvovsky, M., Lapsker, I., Volovik, Y., and Tenne, R. (2002) Load bearing capacity of bronze, iron and iron–nickel powder composites containing fullerene-like WS₂ nanoparticles. *Tribol. Int.*, **35**, 47–53.

12 Mosuang, T.E. and Lowther, J.E. (2002) Relative stability of cubic and different hexagonal forms of boron nitride. *J. Phys. Chem. Solids*, **63**, 363–368.

13 Erdemir, A. (2005) Review of engineered tribological interfaces for improved boundary lubrication. *Tribol. Int.*, **38**, 249–256.

14 Joly-Pottuz, L., Dassenoy, F., Vacher, B., Martin, J.M., and Mieno, T. (2004) Ultralow friction and wear behaviour of Ni/Y-based single wall carbon nanotubes (SWNTs). *Tribol. Int.*, **37**, 1013–1018.

15 Das, S.K., Choi, S.U.S., Yu, W., and Pradeep, T. (2007) *Nanofluids: Science and Technology*, 1st edn, John Wiley & Sons, Inc., Hoboken, NJ.

16 Yu, W., France, D.M., Routbort, J.L., and Choi, S.U.S. (2008) Review and comparison of nanofluid thermal conductivity and heat transfer enhancements. *Heat Transf. Eng.*, **5**, 432–460.

17 Eastman, J.A., Phillpot, S.R., Choi, S.U.S., and Keblinski, P. (2004) Thermal transport in nanofluids. *Annu. Rev. Mater. Res.*, **34**, 219–246.

18 Usama, Y. and Serkan, U. (2013) Reinforced composites produced by a vacuum infusion or pultrusion process. U.S. Patent application US20130216390A1, filed Feb. 20, 2012 and published Aug. 22, 2013.

19 Qiu, D.-B. and Liu, Z.-W. (2014) Microwave-absorbing wind power blade coating as well as preparation method and application thereof. CN published patent application CN104087059A, filed July 21, 2014 and published Oct. 8, 2014.

20 Nelson, J.M. and Marx, R.E. (2009) Erosion resistant impregnating resin systems and composites. WO published patent application WO2009038971 A1, filed Sep. 3, 2008 and published Mar. 26, 2009.

21 http://www.rewitec.com/index.php/en/duragear.html.

22 http://dssh.en.alibaba.com/collection_product/wind_generator_maintenance/1.html.

23 www.skeletontech.com/prismatic-cells.

24 http://www.gamesacorp.com/en/communication/news.

25 http://hanse.evonik.com/_layouts/websites/internet/SecuredLinksHandler.ashx?TargetUrl=%2fsites%2fhanse%2fen%2fproducts%2fnanosilicaconcentrates%2fnanopox%2fpages%2f&PropertyId=231&cpvid=16604.

26 Wichmann, M.H.G., Sumfleth, J., Gojny, F.H., Quaresimin, M., Fiedler, B., and

Schulte, K. (2006) Glass-fibre-reinforced composites with enhanced mechanical and electrical properties: benefits and limitations of a nanoparticle modified matrix. *Eng. Fract. Mech.*, **73**, 2346–2359.

27 Vlasveld, D.P.N., Parlevliet, P.P., Bersee, H.E.N., and Picken, S.J. (2005) Fibre-matrix adhesion in glass fibre reinforced polyamide-6 silicate nanocomposites. *Compos. Part A*, **36**, 1–11.

28 Stankovich, S., Dikin, D.A., Dommett, G.H.B., Kohlhaas, K.M., Zimney, E.J., Stach, E.A., Piner, R.D., Nguyen, S.T., and Ruoff, R.S. (2006) Graphene-based composite materials. *Nature*, **442**, 282–286.

29 Ramanathan, T., Abdala, A.A., Stankovich, S., Dikin, D.A., Herrera-Alonso, M., Piner, R.D., Adamson, D.H., Schniepp, H.C., Chen, X., Ruoff, R.S., Nguyen, S.T., Aksay, I.A., Prud'Homme, R.K., and Brinson, L.C. (2008) Functionalized graphene sheets for polymer nanocomposites. *Nat. Nanotechnol.*, **3**, 327–331.

30 Williams, G., Seger, B., and Kamat, P.V. (2008) TiO_2–graphene nanocomposites: UV-assisted photocatalytic reduction of graphene oxide. *ACS Nano*, **2**, 1487–1491.

31 Rafiee, M.A., Rafiee, J., Wang, Z., Song, H., Yu, Z.-Z., and Koratkar, N. (2009) Enhanced mechanical properties of nanocomposites at low graphene content. *ACS Nano*, **3**, 3884–3890.

32 Rafiee, M.A., Rafiee, J., Wang, Z., Song, H., Yu, Z.-Z., and Koratkar, N. (2010) Fracture and fatigue in graphene nanocomposites. *Small*, **6**, 179–183.

33 Wang, D.-W., Li, F., Zhao, J., Ren, W., Chen, Z.-G., Tan, J., Wu, Z.-S., Gentle, I., Lu, G.Q., and Cheng, H.M. (2009) Fabrication of graphene/polyaniline composite paper via *in situ* anodic electropolymerization for high-performance flexible electrode. *ACS Nano*, **3**, 1745–1752.

34 Gong, L., Kinloch, I.A., Young, R.J., Riaz, I., Jalil, R., and Novoselov, K.S. (2010) Interfacial stress transfer in a graphene monolayer nanocomposite. *Adv. Mater.*, **22**, 2694–2697.

35 Shin, Y.C., Novin, E., and Kim, H. (2015) Electrical and thermal conductivities of carbon fiber composites with high concentrations of carbon nanotubes. *Int. J. Precis. Eng. Manuf.*, **3**, 465–470.

36 Yavari, F., Rafiee, M.A., Rafiee, J., Yu, Z.-Z., and Koratkar, N. (2010) Dramatic increase in fatigue life in hierarchical graphene composites. *ACS Appl. Mater. Interfaces*, **10**, 2738–2743.

37 Archer, C. and Caldeira, K. (2009) Global assessment of high-altitude wind power. *Energies*, **2**, 307–319.

**Part Seven
Environmental Remediation**

44

Nanomaterials for the Conversion of Carbon Dioxide into Renewable Fuels and Value-Added Products

Ibram Ganesh

International Advanced Research Centre for Powder Metallurgy and New Materials (ARCI), Department of Artificial Photosynthesis, Hyderabad 500 005, Andhra Pradesh, India

44.1
Introduction: Dealing with the Waste Stream Greenhouse CO_2 Gas

Recently, the Intergovernmental Panel on Climate Change (IPCC) has concluded that the fossil fuel burning and deforestation are responsible for increased CO_2 concentrations in the atmosphere. The civilization and industrialization have not only brought technology, modern life, and convenience to human life but also the pollution and emissions from factories, vehicles, and chemical plants. Several national governments have signed and ratified the Kyoto Protocol of the United Nations Framework Convention on Climate Change aiming at reducing CO_2 gas emissions [1]. Recently, the International Energy Agency (IEA) in its latest *World Energy Outlook* (*WEO*) journal revealed that, based on policies being practiced at the moment, by 2030 CO_2 emissions will attain 63% from today's level, which is almost 90% higher than those of 1990 [1]. The technology being employed (in certain places) at present to mitigate the CO_2-associated global warming problem is the *CO_2 sequestration* process, which is also called carbon capturing and storage (CCS) process. In this latter process, CO_2 is reversibly adsorbed from a flue gas mixture into a 20 vol% aqueous monoethanol amine solution, which releases back the captured CO_2 gas when heated at 80 °C. This process can be repeated for about 1000 cycles at almost 80% efficiency. At present, the CCS process is employed at almost more than 20 sites all over the world [1]. As CCS process is expensive and laborious, and as it does not contribute to any of the beneficial activities of human life, there has been a quest for other alternative processes for scalable synthesis of carbon-containing fuels using this CO_2 with the help of renewable energy and H_2O as source of electrons and protons. The reusing of the captured CO_2 as a renewable C_1 carbon resource can in fact contribute to a sustainable chemical industry, and as a consequence it reduces CO_2 emissions into the atmosphere [2–7]. CO_2 also has several direct applications in the industry such as a supercritical fluid. In fact, the

Nanotechnology for Energy Sustainability, First Edition. Edited by Baldev Raj, Marcel Van de Voorde, and Yashwant Mahajan.

expenditure incurred for CCS process has been found to be sufficient to convert the captured CO_2 into methanol following the thermochemical hydrogenation process being practiced at present in the petrochemical industry. This latter process is termed carbon capturing and utilization (CCU) process [2–5]. The value of the product formed in CCU process could be considered as a bonus. Furthermore, if any of the renewable energy resources are integrated with the CCU process effectively, it can indeed address the three major problems as mentioned earlier in the abstract. Besides economic benefits, the sociopolitical benefits come in terms of a positive image for companies adopting policies of utilizing CO_2 formed from fossil fuels [2–5].

Normally, most of the CO_2-producing plants are considerably far from the sites used for CCS process [4,5]. In Europe, for example, transporting CO_2 in pipelines to storage sites is quite difficult, and it will increase the cost of CCS process by at least 15–20%, which is considered to be unacceptable. The Department of Energy (DOE), USA, estimated that transportation of CO_2 in tankers on road is not acceptable if distance is more than 100 km [4,5]. In such cases, CCU process is the better option to CCS. In the total cost of CCS process, about 35–40% goes for transportation and storage. It is estimated that it would be between €35 and 50/ton for early commercial phase (after year 2020) and between €60 and 90/ton during demonstration phase, when transportation of CO_2 is made by pipelines for distances not over 200–300 km. If the transportation of CO_2 is by road, this cost would further increase. In such cases, CCU is further beneficial to the CCS option. The cost of methanol in Europe is about €225–240/ton, which would further increase by about 15% (about €20–30/ton) if methanol is produced from a mixture of CO_2 and H_2 instead of from syngas $(CO + H_2)$, following the existing industrial thermochemical routes. For CCS process, an average cost is >€20–30/ton. Surprisingly, today, the social cost of carbon has been estimated to be about US$2000/ton CO_2 [8]. This cost data unequivocally suggests that conversion of the captured CO_2 into methanol or to any other value-added product could be a beneficial option to the CCS process.

The challenges of converting CO_2 into value-added chemicals are great, but the potential rewards are also enormous. Methanol, one of the products of CO_2 conversion processes, can be directly employed in place of gasoline and diesel in the present existing energy distribution infrastructure without any major changes; hence, there would not be any sever economical consequences while transforming from fossil fuel energy dependency to nonfossil fuel, renewable or solar energy dependency [9]. Nature converts CO_2 into bioenergy via *natural photosynthesis* using exclusively solar energy. In the NP process, somewhat less than 1% of the sunlight is converted into bioenergy in the form of plant materials, which when accumulated and transformed over geologic ages yields fossil fuels [10]. Thus, AP has a tremendous potential, and it is a scientific challenge, and upon successful development of it, the market would be gigantic. Owing to CO_2's extremely stable nature, converting it back to a useful value-added chemical in an endothermic reaction, on the same scale and with the same rate currently it is being produced, is out of today's scientific and technological ability.

However, a close study of the existing literature on this subject hints that the successful development of AP is no longer an unrealistic dream [6,7]. Furthermore, this process could be developed quite efficiently in comparison to the NP. For example, there are certain endothermic reactions that are being practiced at industry, such as production of syngas, H_2, methanol, and so on, over certain metal oxide-based catalysts in thermochemical routes in the petrochemical industry. A considerable effort has been made to convert CO_2 into several industrially important chemicals through various methods in which different forms of energy were utilized to drive this endothermic CO_2 reduction reaction to yield different kinds of value-added chemicals and chemical intermediates and fuels [11–25]. On the other hand, the large-scale implementation of solar and other renewable sources of electricity requires improved means for energy storage.

Although at present the utilization of CO_2 volume is very meager in the industrial sector, it is important to realize that several such small volumes can make a real considerable impact on the total CO_2-related global warming mitigation strategy [4,5]. CO_2 can be a feedstock to produce several useful chemicals. This in fact saves the money being spent for CCS process [4,5]. Furthermore, there are also several other considerable advantages if CO_2 is utilized as a chemical feedstock in some of the existing chemical processes. For example, (i) CO_2 is a renewable feedstock unlike oil or coal, (ii) CO_2 is an inexpensive and nontoxic gas, hence, it could be a substitute to replace several toxic chemicals such as, phosgene of isocyanates, (iii) production of chemicals from CO_2 can lead to a all new industrial productivity, and (iv) new routes to existing chemical intermediates and products could be economical than current methods [4,5]. Thus, the research on CO_2 conversion and utilization could be a proactive approach to the sustainable industrial and energy development.

If the chemical activation of the inert CO_2 is achieved with only the required thermodynamic energy inputs without much of overpotentials requirement, there could be a boost for large-scale industrial applications of CO_2 as a chemical feedstock [11–21]. In the laboratory, a great variety of catalysts and methods have been employed so far to activate CO_2 to react with and form several value-added chemicals [11–21]. At present, the major chemicals being produced from CO_2 are *urea, salicylic acid, inorganic carbonates, ethylene/propylene carbonates*, and *polycarbonates* [26]. Furthermore, it is also used as an additive to CO in the production of methanol from syngas [6,7]. Some of the major products produced from CO_2 industrially are shown in Figure 44.1 [26].

The production of *sodium carbonate* (calcined soda) by the *Solvay* method also consumes considerable volumes of CO_2. As of today, the annual production of sodium carbonate is about 30 million tons in entire industry. In the *salicylic acid* synthesis also, reasonably high amount of CO_2 is consumed. CO_2 is also employed in the carboxylation of phenol under pressure (*Kolbe–Schmitt* reaction). The cooligomerization of unsaturated hydrocarbons and CO_2 results in the formation of various synthetic intermediates including *acids, esters, lactones*, and *pyrones* [26–30]. The reaction of alkynes with CO_2 (to form 2-pyrones

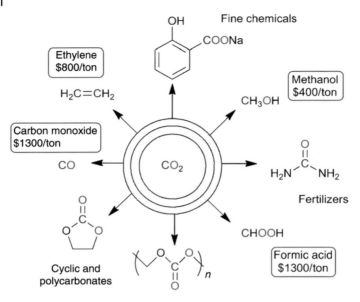

Figure 44.1 Carbon dioxide reduction to value-added products [26].

catalyzed by 3D metal complexes) is one of the few examples of a homogeneous catalytic reaction commercialized, which leads to the formation of C—C bond on CO_2 insertion. The variation of alkyne substituent could result into a wide variety of 2-pyrones [6,7]. However, the widespread use of CO_2, as a C_1 feedstock for the synthesis of methanol in processes such as electrochemical reduction of CO_2, needs a higher thermodynamic energy input (>1.5 V); hence, these processes will not be economical if fossil fuels are utilized as energy aid. This indicates the need of solar energy and judicious usage of catalysts for activating CO_2. Microalgae use solar energy to activate CO_2 [10].

As already mentioned, although there are six different types of reactions available for converting CO_2 into value-added chemicals, at present only the electrochemical routes appear to have reached a stage where they can be converted into commercial plants [6,7]. In view of this, the latest state of the art on electrochemical reduction of CO_2 to CO and to other value-added renewable chemical fuels over different kinds of nanostructured materials or nanoparticles-based catalytic and cathodic systems have been presented and discussed in this chapter.

44.2
Theoretical Potentials for Electrochemical Reduction of CO_2

Practically, the main criteria for an electrochemical system of CO_2 fixation are (i) the capability of using electrons derived from water as an abundant, inexpensive source of reductant; and (ii) the availability of an inexpensive and durable catalyst [6,7]. Therefore, adequate electrocatalysts are needed to forward the electrode-driven chemical reactions in electrochemical reduction of CO_2 (henceforth this process is abbreviated as ERC). Among many different approaches developed thus far for CO_2 activation, ERC is considered as a potentially "clean" method since the reduction proceeds at the expense of a sustainable supply of electric energy. Theoretically, CO_2 can be reduced in an electrochemical cell in an aqueous solution (pH = 7, 1 M electrolyte at 25 °C and 1 atm. CO_2) to form CO, HCOOH, CH_4, or other hydrocarbons at potentials around +0.2 to −0.2 V (versus NHE). Experimentally, however, very negative potentials must be applied to have ERC. These large overpotentials not only consume more electrical energy but also promote the uncontrolled formation of competitive reduction products, such as, H_2, causing low energetic efficiency (EE) and poor selectivity of the desired product.

The ERC comprises of an anodic half-cell reaction of water oxidation, and a cathodic half-cell reaction of CO_2 reduction, as illustrated by Eqs. (44.1) and (44.2), respectively, for example [7]:

$$\text{Cathodic reaction}: \quad CO_2 + 2\,H^+ + 2\,e^- \rightleftharpoons CO + H_2O \qquad (44.1)$$

$$\text{Anodic reaction}: \quad 2\,H_2O \rightleftharpoons O_2 + 4\,H^+ + 4e^- \qquad (44.2)$$

Briefly, this process can be described as simultaneous oxidation of water and reduction of CO_2. The electrical potential supplied to the electrochemical cell causes the water oxidation reaction to occur on the surface of anode into molecular oxygen, protons (H^+) and electrons (e^-). From these products, oxygen is evolved at the surface of anode into the headspace of the compartment, whereas the electrons flow through the outer circuit to reach cathode, where they combine either with protons (H^+) that have traveled through the membrane/junction/salt bridge to reach the cathode to form molecular hydrogen (H_2) by consuming electrons at cathode reached through outer circuit or with CO_2 to form any value-added product including CO, HCOOH, HCHO, CH_3OH, C_2H_5OH, C_3H_7OH, C_2H_6, C_2H_4, and CH_4. The product to be formed out of CO_2 would be decided by the involved reaction conditions and the catalyst in the cathodic system. The CO_2 reduction reaction usually competes with the hydrogen evolution reaction (HER) and if suitable reaction conditions and catalysts are not involved in the reaction, the major reduced product would be only H_2 gas. A great number of electrochemical conditions have been established by the scientific community to reduce CO_2 selectively to form several products with high selectivity and yield as those catalysts inhibit or suppress HER [7]. As the CO_2 reduction reaction is also influenced by its concentration in the electrolyte, several methods have been developed to increase its concentration in the electrolyte [7]. In order to develop an economical ERC process, it is first and foremost important to understand the theoretical reduction potentials of the involved CO_2 reduction reaction. The standard cell potential can be estimated following Eq. (44.3):

$$E^{\circ}_{cell} = E_{cathode} - E_{anode}. \qquad (44.3)$$

The standard cell potential can be determined from the Gibbs free energy of the reaction (Eq. (44.4)), which in turn can be determined from the changes that occur in the values of enthalpy and entropy of a particular reaction (Eq. (44.5)): for example, the standard reduction potential of CO_2 reduction to CO can be calculated following Eq. (44.6), which is the net reaction of Eqs. (44.1) and (44.2). The data of the enthalpy of formation and the entropy of formation of reactants and products associated to CO_2 reduction reactions have been taken from standard data available at [31,32]. The ΔG of reaction (Eq. (44.6)) can be estimated to be 257.2171 kJ/mol, which in turn facilitates the estimation of $E°$ as -1.33 V (versus NHE):

$$\Delta G = -nFE°, \tag{44.4}$$

where n is the number of electrons involved in the reaction and F is the Faraday constant (96,485 C/mol):

$$\Delta G = \Delta H - T\Delta S. \tag{44.5}$$

Overall reaction: $CO_2 + H_2O \rightarrow CO + H_2O + 1/2\, O_2$ (44.6)

Thus, by knowing the values of the overall reaction cell potential ($E°$) and the anodic potential for water oxidation reaction (in this case, 1.23 V versus NHE), the reduction potential of CO_2 to CO can be calculated, which is -0.1 V (versus NHE) (Eqs. (44.7) and (44.8)).

$$-1.33 = E_c - 1.23 \text{ V(versus NHE)}, \tag{44.7}$$

$$E_c = -0.1 \text{ V(versus NHE)}. \tag{44.8}$$

The effect of pH on a reaction potential can be determined by using Eq. (44.9), which leads to the value of -0.53 V (versus NHE) at pH $= 7.0$ (Eq. (44.10)).

$$E_c = E_c - 0.059 \,(\text{pH}) \text{ V versus NHE}, \tag{44.9}$$

$$E_c = -0.53 \text{ V versus NHE}. \tag{44.10}$$

Tables 44.1–44.4 depict the standard reduction potentials of various reactions of ERC at pH $= 0$ and at pH $= 7.0$.

Although the ERC to fuel chemicals and chemical feedstock powered by intermittent renewable electricity is an attractive route for the conversion of CO_2 and for the production of renewable energy resource, there are several fundamental challenges yet to be resolved, including high overpotential and low FE due to the associated competitive HER [11–25]. In thermodynamics, the Gibbs free energy of ERC is always positive at medium and high pH range; hence, the theoretical potentials are negative. In kinetics, the ERC normally requires overpotentials >1.0 V to get reasonable amounts of reduced products. In an aqueous electrolyte, the H_2O also undergoes reduction and releases H_2 as a major by-product. Thus, water reduction is always in competition with CO_2 reduction in electrochemical cells when aqueous-based electrolytes are employed. Certain metals, such as Hg that possesses high H_2 overvoltage (overpotential), can suppress

Table 44.1 Various products to be formed in electrochemical CO_2 reduction reaction and their standard Gibbs energy of formation.

Net reaction	Anodic half-cell reaction (oxidation process)	Cathodic half-cell reaction (reduction process)	$\Delta G_{rxn}^\circ = [\Delta H_{rxn}^\circ - T\Delta S_{rxn}^\circ]$ (kJ/mol)	ΔG_{rxn}° (kJ/mol)
$2H_2O \leftrightarrows 2H_2\uparrow + O_2\uparrow$	$2H_2O \rightarrow 4H^+ + 4e^- + O_2\uparrow$	$4H^+ + 4e^- \rightarrow 2H_2\uparrow$	$[571.66 - 298 \times 0.3266]$	474.33
$CO_2 + H_2O \leftrightarrows CO + H_2O + 1/2O_2\uparrow$	$H_2O \rightarrow 2H^+ + 2e^- + 1/2O_2\uparrow$	$CO_2 + 2H^+ + 2e^- \rightarrow CO\uparrow + H_2O$	$[283.01 - 298 \times 0.08655]$	257.38
$CO_2 + H_2O \leftrightarrows HCOOH + 1/2O_2\uparrow$	$H_2O \rightarrow 2H^+ + 2e^- + 1/2O_2\uparrow$	$CO_2 + 2H^+ + 2e^- \rightarrow HCOOH$	$[254.34 + 298 \times 0.05215]$	269.86
$CO_2 + H_2O \leftrightarrows HCHO + O_2\uparrow$	$2H_2O \rightarrow 4H^+ + 4e^- + O_2\uparrow$	$CO_2 + 4H^+ + 4e^- \rightarrow HCHO + 1/2O_2$	$[570.74 - 298 \times 0.14025]$	528.94
$CO_2 + 2H_2O \leftrightarrows CH_3OH + 3/2O_2\uparrow$	$3H_2O \rightarrow 6H^+ + 6e^- + 3/2O_2\uparrow$	$CO_2 + 6H^+ + 6e^- \rightarrow CH_3OH + H_2O$	$[725.97 - 298 \times 0.08085]$	701.87
$2CO_2 + 3H_2O \leftrightarrows C_2H_5OH + 3O_2\uparrow$	$6H_2O \rightarrow 12H^+ + 12e^- + 3O_2\uparrow$	$2CO_2 + 12H^+ + 12e^- \rightarrow C_2H_5OH + 3H_2O$	$[1366.90 - 298 \times 0.13875]$	1325.56
$3CO_2 + 4H_2O \leftrightarrows C_3H_7OH + 9/2O_2\uparrow$	$9H_2O \rightarrow 18H^+ + 18e^- + 9/2O_2\uparrow$	$3CO_2 + 18H^+ + 18e^- \rightarrow C_3H_7OH + 5H_2O$	$[2021.24 - 298 \times 0.19565]$	1962.94
$CO_2 + 2H_2O \leftrightarrows CH_4\uparrow + 2O_2\uparrow$	$4H_2O \rightarrow 8H^+ + 8e^- + 2O_2\uparrow$	$CO_2 + 8H^+ + 8e^- \rightarrow CH_4 + 2H_2O$	$[890.57 - 298 \times 0.2429]$	818.18
$2CO_2 + 3H_2O \leftrightarrows C_2H_6\uparrow + 7/2O_2\uparrow$	$7H_2O \rightarrow 14H^+ + 14e^- + 7/2O_2\uparrow$	$2CO_2 + 14H^+ + 14e^- \rightarrow C_2H_6 + 4H_2O$	$[1560.51 - 298 \times 0.3098]$	1468.18
$2CO_2 + 2H_2O \leftrightarrows C_2H_4\uparrow + 3O_2\uparrow$	$6H_2O \rightarrow 12H^+ + 12e^- + 3O_2\uparrow$	$2CO_2 + 12H^+ + 12e^- \rightarrow C_2H_4 + 4H_2O$	$[1411.08 - 298 \times 0.2673]$	1331.42

Table 44.2 The calculations involved in the arrival of enthalpy and entropy of various electrochemical CO_2 reduction reactions.

Product	products – reactants	ΔH_{rxn}° = [enthalpy of products – reactants] (kJ/mol)	ΔH_{rxn}° (kJ/mol)	ΔS_{rxn}° = [entropy of products – reactants] (J/(K mol))	ΔS_{rxn}° (J/(K mol))
Hydrogen	$[(2\,H_2 + O_2) - (2\,H_2O)]$	$[((2\times0) + (0)) - (2\times-285.83)]$	571.66	$[((2\times130.7) + (205.1)) - (2\times69.95)]$	326.60
Carbon monoxide	$[(CO + 1/2\,O_2) - (CO_2)]$	$[(-110.5) - (-393.51)]$	283.01	$[(197.7 + 1/2\times205.1) - (213.7)]$	86.55
Formic acid	$[(HCOOH + 1/2\,O_2) - (CO_2 + H_2O)]$	$[(-425) - (-393.51 - 285.83)]$	254.34	$[(129 + 102.22) - (69.95 + 213.7)]$	−52.15
Formaldehyde	$[(HCHO + O_2) - (CO_2 + H_2O)]$	$[(-108.6) - (-393.51 - 285.83)]$	570.74	$[(129 + 102.22) - (69.95 + 213.7)]$	140.25
Methanol	$[(CH_3OH + 3/2\,O_2) - (CO_2 + 2\,H_2O)]$	$[(-239.2) - ((-393.52) - (3\times285.83))]$	725.97	$[(126.8 + 307.65] - (2\times69.95 + 213.7)]$	80.85
Ethanol	$[(C_2H_5OH + 3\,O_2) - (2CO_2 + 3\,H_2O)]$	$[(277.6) - ((-2\times393.52) - (3\times385.83))]$	1366.9	$[(160.7 + 615.3) - (3\times69.95 + 2\times213.7)]$	138.75
Propanol	$[(C_3H_7OH + 9/2\,O_2) - (3CO_2 + 4\,H_2O)]$	$[(-302.6) - ((-3\times393.52) - (4\times285.83))]$	2021.24	$[(1116.55] - (3\times213.7 + 4\times69.95)]$	195.65

Table 44.3 The half-cell reactions of cathode during electrochemical CO_2 reduction.

Cathodic half-cell reaction (reduction process)	$\Delta G^\circ_{rxn} = -nFE^\circ_{cell} \Rightarrow E^\circ_{cell} = -\dfrac{\Delta G^\circ}{nF}$ (V versus NHE at pH = 0)	E_c (at pH = 0) = E_{cell} + E_a (V versus NHE)	E_c (at pH = 7) = $E^\circ c$ − 0.059 × 7 (V versus NHE)
$4H^+ + 4e^- \rightarrow 2H_2\uparrow$	−1.23	−1.23 + 1.23 = 0	−0.41
$CO_2 + 2H^+ + 2e^- \rightarrow CO\uparrow + H_2O$	−1.33	−1.33 + 1.23 = −0.10	−0.51
$CO_2 + 2H^+ + 2e^- \rightarrow HCOOH$	−1.39	−1.39 + 1.23 = −0.16	−0.57
$CO_2 + 4H^+ + 4e^- \rightarrow HCHO + 1/2O_2$	−1.37	−1.37 + 1.23 = −0.14	−0.55
$CO_2 + 6H^+ + 6e^- \rightarrow CH_3OH + H_2O$	−1.21	−1.21 + 1.23 = 0.02	−0.39
$2CO_2 + 12H^+ + 12e^- \rightarrow C_2H_5OH + 3H_2O$	−1.14	−1.14 + 1.23 = 0.09	−0.32
$3CO_2 + 18H^+ + 18e^- \rightarrow C_3H_7OH + 5H_2O$	−1.13	−1.13 + 1.23 = 0.10	−0.31
$CO_2 + 8H^+ + 8e^- \rightarrow CH_4 + 2H_2O$	−1.06	−1.06 + 1.23 = 0.17	−0.24
$2CO_2 + 14H^+ + 14e^- \rightarrow C_2H_6 + 4H_2O$	−1.08	−1.08 + 1.23 = 0.15	−0.26
$2CO_2 + 12H^+ + 12e^- \rightarrow C_2H_4 + 4H_2O$	−1.15	−1.15 + 1.23 = 0.08	−0.33

Table 44.4 Thermodynamic properties of reactants and products of various CO_2 reduction reactions.

Chemical	Enthalpy of formation (ΔH°_f) (kJ/mol)	Entropy of formation (S°) (J/K mol)
O_2	0	205.1
H_2	0	130.7
CO_2	−393.509	213.7
H_2O	−285.83	69.95
CO	−110.5	197.7
$HCOOH$	−425.0	129.0
$HCHO$	−108.6	218.8
CH_3OH	−239.2	126.8
C_2H_5OH	−277.6	160.7
C_3H_7OH	−302.6	193.6
CH_4	−74.6	186.3
C_2H_6	−84.0	229.2
C_2H_4	52.4	219.3

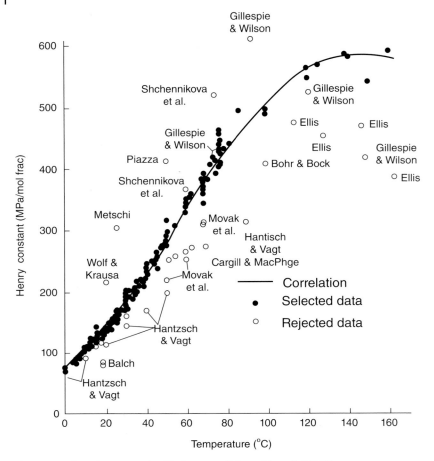

Figure 44.2 Henry's constant for CO_2 in water. (Adapted from Ref. [33].)

HER, but lead to the formation of only formate ions ($HCOO^-$) at very high over-potentials. Involvement of appropriate catalytic systems can promote the reaction rates, and directs the selective pathways with minimum excess energy requirements and facilitate the formation of desired CO_2 reduction products. Furthermore, the free energy change for CO_2 reduction is a function of electrode nature and the electrolyte pH (Figure 44.2) [33].

As can be seen from Figure 44.2, the reactivity of ERC is very low, but the equilibrium potentials are not very negative when compared with HER in aqueous electrolyte solutions under standard reaction conditions (Tables 44.1–44.4). The reaction potentials of water oxidation that occurs at anode surface in electrochemical cells are also presented in these tables. The primary reactions that occur on the surface of electrode in an aqueous solution at 25 °C can be explained as follows.

The single-electron reduction of CO_2 to $CO_2^{\bullet-}$ (Eq. (44.11)) occurs at -1.90 V (versus NHE). This step has also been considered to be a *rate-determining step* (RDS) in the CO_2 reduction process. According to *Nernst* equation, the theoretical equilibrium potentials decrease with increasing solution pH [33].

$$CO_2 + e^- \rightarrow CO_2^{\bullet-} \quad (-1.90 \text{ V versus SHE}) \tag{44.11}$$

Considering their low equilibrium potentials, thermodynamically, the CO_2 reduction products of methane and ethylene should form at less cathodic potentials than HER; however, due to kinetics of these reasons this does not happen. This suggests that the former reactions need additional high overpotential energies when compared to HER. Similarly, the electrochemical reduction of CO_2 to methanol and methane also requires high electrode overpotentials, which make these processes uneconomical as the amount of energy consumed during these products formation is much larger than the energy stored in them.

Two-electron reduction reactions that occur at pH 7 and 25 °C versus SHE are shown in Eqs. (44.12) and (44.13). Both H_2O and CO_2 reductions are very difficult to occur via the one-electron process, since the one-electron reduction products, $H^{\bullet}$ atom and $CO_2^{\bullet-}$ (formate radical), are extremely energetic species [33]. By using pulse-radiolytic methods, Schwarz and Dodson have determined E° for $CO_2/CO_2^{\bullet-}$ couple to be -1.9 V versus NHE in water, and the intrinsic barrier for this one-electron reduction was estimated to be about 0.6 V. The $CO_2^{\bullet-}/CO_2^{2-}$ potential has been estimated to be about -1.2 V versus NHE [33]. One-electron reduction potentials at pH 7 and 25 °C versus NHE are also shown in Eqs. (44.14) and (44.15). In the H_2O/H_2 reaction, certain metal hydride complexes provide catalytic routes [33]. In the CO_2/CO or CO_2/HCO_2^- reaction, catalysis by metal complexes involve coordination of CO_2 or its insertion into a metal hydride bond to yield formate species by consuming a minimum overpotential (or kinetic) energy.

$$2\,H_2O_{(l)} + 2\,e^- \rightarrow H_{2(g)} + 2\,OH^-_{(aq)} \quad (E^{\circ} = -0.41 \text{ V versus SHE}) \tag{44.12}$$

$$CO_{2(l)} + H_2O_{(l)} + 2\,e^- \rightarrow HCO^-_{2(g)} + 2OH^-_{(aq)}$$
$$(E^{\circ} = -0.31 \text{ V versus SHE}) \tag{44.13}$$

$$H + -2.7\,V\,H \cdot +0.03\,V\,H^- \quad \text{(versus SHE)} \tag{44.14}$$

$$CO_2 - 1.9\,V\,CO_2^{\bullet-} - 1.2\,V\,CO_2^{2-} \quad \text{(versus SHE)} \tag{44.15}$$

Despite studying electrochemical reduction of CO_2 for more than a century, neither an electrode material nor a catalytic system is identified (or developed) for exclusive reduction of CO_2 into any of the liquid fuel chemicals without allowing HER [7]. The large thermodynamic and kinetic energy requirements associated with electrochemical CO_2 reduction reactions demand use of certain efficient catalytic systems to suppress HER and to speed up utilization of all generated protons and electrons at anode only in the CO_2 reduction purposes that occur on the cathode surface.

Recently, the Department of Energy, USA, has concluded that the major obstacle that is preventing efficient conversion of CO_2 into several energy-bearing products is the lack of efficient catalytic systems [34]. Hence, DOE has called for research that investigates systems to reduce the overpotentials associated with CO_2 conversion while maintaining high current efficiencies. The catalysts reported for this CO_2 reduction reaction generally suffer from low energetic efficiency, poor product selectivity, and rapid deactivation. Although a few homogeneous catalytic systems showed initial activities at overpotentials below 600 mV [22–25], most quickly they lose their activity under reaction turnover conditions. Although pyridine-catalyzed reaction has been found to be an exception, it is yet to be tested for the required extended time periods [22–25]. A promising catalyst for efficient CO_2 conversion would need to exhibit both high energy efficiency (i.e., high Faradaic efficiency (FE) for CO production at low overpotential) and high current density (i.e., high rate, turnover frequency (TOF), or turnover number (TON)). Furthermore, the reaction efficiency (i.e., the overpotential associated with a particular reduction process), rate of the reaction (i.e., current density), product selectivity, stability of the process, and so on have been found to be a strong function of electrocatalytic and cathodic system (nature as well as physical state, nano-, or bulk-type) and reaction media/electrolyte (aqueous (protic) or nonaqueous (aprotic)) involved. The speciation of CO_2 versus electrolyte pH is presented in the following section as this information enables one to design the reaction parameters for efficient electrochemical conversion of CO_2 into value-added chemicals.

44.3
CO_2 Speciation versus Electrolyte pH

The dissolved CO_2 in aqueous media exists in three to four different forms (or species), including CO_2 gas ($CO_{2(g)}$), liquid (solvated) CO_2 ($CO_{2(liq)}$), carbonic acid (H_2CO_3), bicarbonate ion (HCO_3^-), and carbonate ion (CO_3^{2-}), whose concentrations vary with aqueous electrolyte pH [35,36]. As it is very difficult to distinguish $CO_{2(liq)}$ between H_2CO_3, they are usually considered as one component. At 25 °C, about $90 \, cm^3$ of CO_2 dissolves in 100 ml water ($C_l/C_g = 0.8$) (Eq. (44.16) and Scheme 44.1).

$$CO_{2(gas)} \rightleftharpoons CO_{2(aq)}, [CO_{2(aq)}] = K_H P_{CO_2}; \quad pK_H = 1.464 \, (P \text{ in atm.})$$

$$(44.16)$$

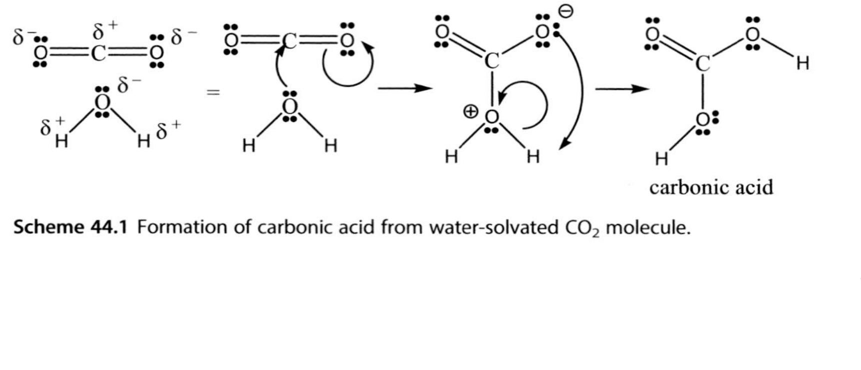

Scheme 44.1 Formation of carbonic acid from water-solvated CO_2 molecule.

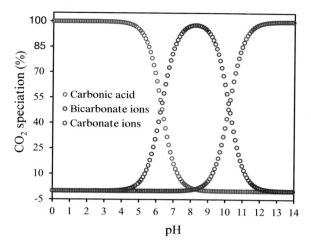

Figure 44.3 CO_2 speciation versus solution pH. (Adapted from Refs [35–37].)

The formation of H_2CO_3 takes place by the nucleophilic addition of H_2O molecule to the CO_2 molecule. As carbon atom in the CO_2 molecule possesses slightly positive charge, and as oxygen atoms possess slightly negative charge due to the prevalent differences in their electronegativity values, a lone pair of electrons of oxygen atom of water molecule nucleophilically attacks the carbon atom in the CO_2 molecule, leading to the formation of carbonic acid after a few rearrangements of electrons and protons (Scheme 44.1) [35,36].

The reaction of converting CO_2 into carbonic acid is a simple dissolution process, which is governed by Henry's law, which states that the CO_2 concentration in the solution is proportional to the partial pressure of CO_2 in the gas phase that is in contact with the solution phase (Eq. (44.17)), where ρ_{CO_2} is the partial pressure of the gas in the bulk atmosphere (Pa), K is the constant, and X_{CO_2} is the equilibrium mole fraction of CO_2 solute in liquid phase. According to Carrol and Mather, a form of Henry's law can be used for modeling the solubility of CO_2 in water for pressures up to about 100 MPa (Figure 44.3) [35–37].

$$\rho_{CO_2} = K \times X_{CO_2}.$$ (44.17)

The solubility of CO_2 is a temperature-dependent parameter (Table 44.5) [38]. Normally, the dissolution of CO_2 decreases with increasing temperature of the solution. Equilibrium is established between the dissolved CO_2 and H_2CO_3

Table 44.5 Solubility of CO_2 at a partial pressure for CO_2 of 1 bar [38].

Temperature (°C)	0	10	20	30	40	50	80	100
Solubility (cm³ CO^2/g of water)	1.8	1.3	0.88	0.65	0.52	0.43	0.29	0.26

Table 44.6 Equilibrium constants for CO_2 [35–37].

Log of equilibrium constant	25 °C, $I = 0$
$[CO_2] = K_H P_{CO_2}$	−1.464
$[HCO_3^-][H^+] = K_{a1} [CO_2]$	−6.363
$[CO_3^{2-}][H^+] = K_{a2}[HCO_3^-]$	−10.329
$[OH\text{-}][H^+] = K_w$	−13.997

Table 44.7 Selected values of equilibrium constants extrapolated to $I = 0$ [38].

Temperature (°C)	pK_H° (mol/(kg atm))	pK_{a1}° (mol/kg)	pK_{a2}° (mol/kg)	pK_w (mol/kg)	pK_{so}° (mol/kg)
30	1.521	6.336	10.290	13.83	8.53
35	1.572	6.317	10.250	13.68	8.58
37	1.591	6.312	10.238	13.62	8.60
40	1.620	6.304	10.220	13.53	8.63
45	1.659	6.295	10.195	13.39	8.69

(Eq. (44.18)). This reaction is kinetically slow as its equilibrium constant (K_r) is only about 1.7×10^{-3}. At equilibrium, only a small fraction (about 0.2–1.0%) of the dissolved CO_2 is actually converted to H_2CO_3. Most of the CO_2 remains as solvated molecular CO_2 [35–37].

Table 44.6 lists the equilibrium constants for CO_2 dissolution and acid dissociation constants (pK_a values) of the resultant molecules at 25 °C and zero ionic strength ($I = 0$) [35–37]. The variation of pK_a values with temperature is shown in Table 44.7 [38].

H_2CO_3 is a weak acid that dissociates in two steps (Eqs. (44.19) and (44.20)). As the rate of hydration of dissolved CO_2 (Eq. (44.18)) is very slow (about 0.1 s), this step can be easily separated from the much faster (10^{-6} s) H_2CO_3 dissociation reaction into H^+ and HCO_3^- (Eq. (44.19)). If any cations are present in the electrolyte, these carbonate anions will react with them to form insoluble carbonates. For example, seawater contains Ca^{2+} and Mg^{2+}, which are responsible for the formation of limestone ($CaCO_3$) and $MgCO_3$ upon reaction with CO_2. The formation of these deposits usually drives reaction equilibrium positions more toward right side, resulting in acidification of the seawater (Eqs. (44.21) and (44.22)) (Scheme 44.2) [35–37].

$$CO_{2(g)} \overset{+H_2O}{\rightleftharpoons} CO_{2(liq)} \overset{+H_2O}{\rightleftharpoons} H_2CO_3 \rightleftharpoons HCO_3^- + H_3O^+ \overset{+H_2O}{\rightleftharpoons} CO_3^{2-} + H_3O^+ \overset{+Ca^{2+}}{\rightleftharpoons} CaCO_3 \downarrow$$

Scheme 44.2 Reactions of different forms of CO_2.

$$CO_{2(liq)} + H_2O_{(l)} \rightleftharpoons H_2CO_{3(liq)} \quad (K_r = 1.7 \times 10^{-3}) \tag{44.18}$$

$$[CO_{2(aq)}] \rightleftharpoons H^+ + HCO_3^- ([H^+][HCO_3^-] = K_{a1}[CO_{2(aq)}])$$
$$(pK_{a1} = 6.363 \quad \text{at} \quad 25\,°C \quad \text{and} \quad I = 0) \tag{44.19}$$

$$HCO_3^- \rightleftharpoons H^+ + CO_3^{2-} ([H^+][CO_3^{2-}] = K_{a2}[HCO_3^-])$$
$$(pK_{a2} = 10.329 \quad \text{at} \quad 25\,°C \quad \text{and} \quad I = 0) \tag{44.20}$$

$$Ca^{2+} + CO_3^{2-} \rightleftharpoons CaCO_3 \downarrow \quad (S = 4.96 \times 10^{-9};$$
$$S = \text{solubility constant}) \tag{44.21}$$

$$Mg^{2+} + CO_3^{2-} \rightleftharpoons MgCO_3 \downarrow \quad (S = 6.82 \times 10^{-6}) \tag{44.22}$$

If one considers the room temperature to be below 40 °C (for example, in Hyderabad, India), at this temperature the Henry coefficient for CO_2 in water is about 220 MPa/mol fraction [35–37]. At 1 bar pressure of CO_2 over the aqueous electrolyte, the mole fraction of CO_2 in water would be about 0.00 045 (= 0.1 MPa/ 220 MPa/mol fraction). Since at 40 °C the molar density of water is about 55.18 (= 992.2/18.02), the concentration of dissolved CO_2 in the electrolyte can be 25.077 mM/l (= 0.00045 × 55.18), which is equal to 1.510×10^{22} (25.077 × Avogadro number, 6.023×10^{23}) CO_2 species in total ($H_2CO_3 + HCO_3^- + CO_3^{2-}$) [35–37].

The distribution of different CO_2 species in aqueous electrolyte as a function of its pH under 1 bar CO_2 pressure at 40 °C and an ionic strength of zero ($I = 0$) can be derived using two acid dissociation constants of carbonic acid ($K_{a1} = 4.97 \times 10^{-7}$ and $K_{a2} = 6.03 \times 10^{-11}$), which can be generated from their corresponding pKa values ($pK_{a1} = 6.304$ and $pK_{a2} = 10.220$) [35–37]. The basic equations required to calculate the distribution of CO_2 species can be derived from the equilibrium reactions of carbonic acid dissociation (Eqs. (44.23) and (44.24)), which can be reorganized as shown in Eqs. (44.25) and (44.26) [35–37]. If the total initial concentration of carbonic acid is assumed as C_T (= $[H_2CO_3] + [HCO_3^-] + [CO_3^{2-}]$), then the concentrations of $[H_2CO_3]$, $[H_2CO_3]$ and $[CO_3^{2-}]$ can be derived using Eqs. (44.27)–(44.29).

$$K_{a1} = \frac{[H_3O^+][HCO_3^-]}{[H_2CO_3]} \left(\approx \frac{[H_3O^+][HCO_3^-]}{[CO_2 + H_2CO_3]} \approx \frac{[H_3O^+][HCO_3^-]}{[CO_2]_{liq}} \right) \tag{44.23}$$

$$K_{a2} = \frac{[H_3O^+][CO_3^{2-}]}{[HCO_3^-]} \tag{44.24}$$

$$C_T = \frac{[H_3O^+]^2[CO_3^{2-}]}{K_{a1}K_{a2}} + \frac{[H_3O^+][CO_3^{2-}]}{K_{a2}} + [CO_3^{2-}] \tag{44.25}$$

$$\Rightarrow [CO_3^{2-}]\left(1 + \frac{[H_3O^+]}{K_{a2}} + \frac{[H_3O^+]^2}{K_{a1}K_{a2}}\right) \tag{44.26}$$

$$\Rightarrow \left[CO_3^{2-}\right] = \frac{K_{a1}K_{a2}C_T}{[H_3O^+]^2 + K_{a1}[H_3O^+] + K_{a1}K_{a2}} \qquad (44.27)$$

$$[HCO_3^-] = \frac{K_{a1}[H_3O^+]C_T}{[H_3O^+]^2 + K_{a1}[H_3O^+] + K_{a1}K_{a2}} \qquad (44.28)$$

$$[H_2CO_3] = \frac{C_T[H_3O^+]^2}{[H_3O^+]^2 + K_{a1}[H_3O^+] + K_{a1}K_{a2}} \qquad (44.29)$$

The distribution curves obtained upon plotting the values of $[H_2CO_3]$, $[HCO_3^-]$, and $[CO_3^{2-}]$ versus solution pH values from 0 to 14 are shown in Figure 44.4 [35–37]. Above pH 10 and below pH 4, the solutions contain mainly CO_3^{2-} and H_2CO_3/CO_2, respectively. Near pH 7, all three species HCO_3^-, CO_2^{2-}, and H_2CO_3 are present in the solution.

Depending upon the experimental procedures (and time scales), the above equilibrium positions can lead to substantial changes in the CO_2 partial pressure above a solution and in the concentration(s) of dissolved species. For example, in stoichiometric (neutralization) reactions, either solvated or molecular CO_2 gas reacts directly with monoethanolamine during separation of CO_2 from the exhaust flue gas of thermal power plants when it is passed through an aqueous solution of 20 vol% monoethanolamine (Eq. (44.30)) or with epoxides in copolymerization reactions [39,40]. However, in ERC any one of the three CO_2 species (H_2CO_3, HCO_3^-, or CO_3^{2-}) can undergo reduction depending on the solution pH.

$$CO_2 + HOCH_2CH_2NH_2 \rightarrow HOCH_2CH_2NH_2^+CO_2^- \quad \text{(zwitterion)} \qquad (44.30)$$

As reduction of already negatively charged HCO_3^- and CO_3^{2-} ions is normally difficult, the ERC experiments are preferentially conducted in electrolytes with pH ≤ 6 so that the neutral carbonic acid $[H_2CO_3]$ species undergo reduction

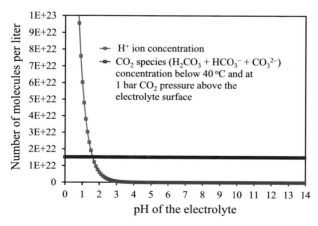

Figure 44.4 Variation of H^+ and CO_2 species concentration as a function of solution pH. (Adapted from Ref. [23].)

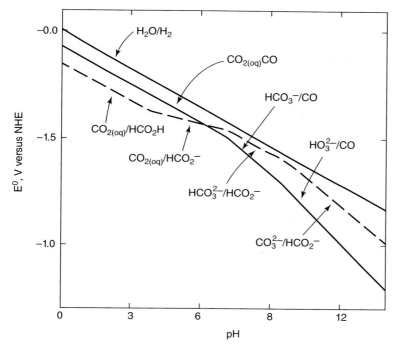

Figure 44.5 $E°$ values (versus normal hydrogen electrode (NHE)) for two-electron reduction of H_2O and CO_2 as a function of pH and dominant forms of reactants and products. The following standard state conventions are used: H_2O, liquid; H_2 and CO, gas (i.e., 1 atm. pressure); CO_2, HCO_2H, HCO_2^-, HCO_3^-, and CO_3^{2-}, aqueous (i.e., 1 M). (Adapted from Refs [35–37].)

relatively preferentially. However, at this pH, CO_2 reduction reaction competes with HER. Since above pH 2 the concentration of CO_2 species is always much higher than that of H^+ concentration (Figure 44.5), a suitable CO_2 reduction catalyst should enhance the CO_2 reduction reaction over HER [35–37].

44.4
Effect of Particle Size on Electrode Performance in Electrochemical CO_2 Reduction Reaction

To succeed in ERC, highly efficient catalysts must be required to lower the reduction overpotentials and to control the energy pathways of reaction intermediates. The key points in developing such catalysts include the successive CO_2 adsorption, intermediates formation, and product removal from active sites during the ERC. Various metal electrocatalysts have been screened experimentally and found that the trends on product composition and distribution are strongly dependent on the binding energy of reaction intermediates on metal electrodes. For instance, CO_2 is converted into CO on Au and Ag, and into hydrocarbons on Cu due to the adsorption strength of CO^* intermediates on metals [41]. Apart from that the particle size metal electrode also plays a great role in the selectivity of the product formation and energy efficiency of ERC [13–18]. Varying the particle size shows a size effect on the ERC, where the dependence of FE and current density on particle size has been observed on Ag, SnO_2, Cu, and Au nanoparticles. Recent advances in the synthesis of NPs have allowed for testing of potentially increased reaction kinetics due to the controlled surface area and surface morphology achieved. This is demonstrated by ERC into hydrocarbons on Cu NPs, or into CO on gold-based NPs/clusters [41]. Recently, a new form of Au nanostructured catalyst made by anodization and electroreduction of an Au electrode has been employed to show high selectivity for catalyzing CO_2 reduction to CO with a FE of ~96% at −0.35 V (versus NHE) and current densities between 2 and 4 mA/cm^2 [41].

There are also few studies looking into the size effect among NPs smaller than 10 nm, which are most commonly used in catalysis. Studies on the CO_2 reduction reactivity over Au NPs within a very narrow range of ~1–3 or 4–10 nm have been conducted [41]. Current density for both CO and H_2 production increased with decreasing the size of Au NPs along with decreasing FE toward CO production. Density functional theory (DFT) calculations revealed that the edge site on Au NPs is much more active than terrace and corner sites for CO_2 reduction to CO, while corner sites are highly active for competitive HER [41].

44.5
Effect of Particle Size on the Efficiency of Aqueous-Based CO_2 Reduction Reactions

The ERC to CO in an aqueous solution depends on the energetic stabilization of reduction intermediates by catalytically active surfaces. The reaction is suggested to include the following steps (Eqs. (44.31)–(44.35)) [41]:

$$CO_2 + H^+(aq) + e^- + {}^* \rightarrow COOH^* \tag{44.31}$$

$$COOH^* + H^+(aq) + e^- \rightarrow CO^* + H_2O \tag{44.32}$$

$$CO^* \rightarrow CO + {}^* \tag{44.33}$$

$$H^+(aq) + e^- + {}^* \rightarrow H^* \tag{44.34}$$

$$H^* + H^+(aq) + e^- \rightarrow H_2 + {}^* \text{ or } H^* + H^* \rightarrow H_2 + 2^* \tag{44.35}$$

where the asterisk (*) denotes either a surface-bound species or a vacant catalytically active site. The formation of CO not only depends on the stabilization and reduction of a $COOH^*$ but also on the ability of the catalyst to liberate the CO product. For high CO selectivity, the catalyst needs to balance steps 1–3 while inhibiting the HER via H^+ reduction, a major side reaction that is often observed in studying ERC.

The size-dependent ERC activity over 2.4–10.3 nm size palladium (Pd) nanoparticles has been demonstrated by performing a controlled potential electrolysis (CPE) experiment in an H-type electrochemical cell separated by a Nafion 115 membrane in CO_2-saturated 0.1 M $KHCO_3$ catholyte (pH 6.8) at room temperature and under atmospheric pressure, and the formation of CO was found with a FE of 5.8% at −0.89 V (versus NHE) over 10.3 nm Pd NPs and with a FE of 91.2% over 3.7 nm Pd NPs, along with 18.4-fold increase in the current density [42]. The Gibbs free energy diagrams built from DFT calculations revealed that the adsorption of CO_2 and the formation of key reaction intermediate $COOH^*$ are much easier on edge and corner sites than on terrace sites of Pd NPs. In contrast, the formation of H^* for competitive HER has been found to be similar on all the sites. A volcano-like curve of the TOF for CO production within the size range observed suggests that CO adsorption, $COOH^*$ formation, and CO removal during CO_2 reduction can be tuned by varying the size of Pd NPs due to the changing ratio of corner, edge, and terrace sites [42].

The large-surface-area tin oxide (SnO_2) nanocrystals prepared by a facile hydrothermal method exhibited excellent activity for ERC to formate [43]. These novel nanostructured tin (Sn) catalysts could reduce CO_2 selectively to formate at overpotentials as low as ~340 mV. In aqueous $NaHCO_3$ solutions, maximum FE for formate production of >93% has been observed with high stability and current densities of >10 mA/cm^2 on graphene supports. The notable reactivity toward CO_2 reduction achieved here can arise from a compromise between $CO_2^{\bullet-}$ and nanoscale Sn surface and subsequent kinetic activation toward protonation and further reduction. The importance of tin oxide (SnO_x) nanoparticles in improving the efficiency of ERC on Sn has been demonstrated by comparing the activity of Sn electrodes subjected to different pre-electrolysis treatments [21]. In aqueous $NaHCO_3$ solution saturated with CO_2, a Sn electrode with a native SnO_x layer exhibited potential-dependent CO_2 reduction activity consistent with literature findings. In contrast, an electrode etched to expose fresh Sn^0 surface exhibited higher overall current densities but almost HER over the entire 0.5 V range of potentials examined. However, when a thin-film catalyst obtained by simultaneous electrodeposition of Sn^0 and SnO_x on titanium (Ti) electrode was employed for ERC under identical reaction conditions, an eightfold higher partial current density (CD) and fourfold higher FE in comparison to the one noted on a Sn electrode with a native SnO_x layer has been noted. These results implicate the participation of SnO_x in ERC pathway on Sn electrodes and suggest that metal–metal oxide composite materials are promising catalysts for sustainable fuel synthesis.

The gold (Au) NPs have been found to selectively reduce CO_2 to CO in electrochemical cell (EC) in 0.5 M $KHCO_3$ at 25 °C [41]. Among monodisperse 4, 6, 8, and 10 nm NPs tested, the 8 nm Au NPs showed the maximum FE (up to 90% at −0.67 V versus NHE). DFT calculations suggested that more edge sites (active for CO evolution) are present than corner sites (active for the competitive H_2 evolution reaction) on the Au NP surface, which facilitates the stabilization of the reaction intermediates, such as COOH*, and the formation of CO. This mechanism has also been established by Au NPs embedded in a matrix of butyl-3-methylimidazolium hexafluoro-phosphate (BMIM-PF$_6$) ionic liquid for more efficient COOH* stabilization that exhibited higher reaction activity (3 A/g mass activity) and selectivity (97% FE) at −0.52 V (versus NHE). This work has demonstrated the importance of using monodisperse Au NPs to optimize the available reaction intermediate binding sites for efficient and selective ERC to CO. The Au NPs derived from a thick Au oxide film have exhibited high selectivity to CO in ERC reaction when performed in 0.5 M NaHCO$_3$ solution (pH = 7.2) at overpotentials as low as 140 mV and retained their activity for at least 8 h [20]. Under identical reaction conditions, polycrystalline Au electrodes and several other nanostructured Au electrodes obtained under alternative methods required at least an additional 200 mV overpotential to attain comparable CO_2 reduction activity and rapidly lost their activity. Electrokinetic studies indicated that the improved catalysis is linked to dramatically increased stabilization of the $CO_2^{\bullet-}$ intermediate on the surfaces of the oxide-derived Au electrodes [41].

The Cu NPs formed by the reduction of Cu_2O layers developed during annealing of Cu foil in air at 500 °C exhibited a strong dependence on the initial thickness of the Cu_2O layer for ERC [19]. Thin Cu_2O layers formed upon annealing of Cu foil at 130 °C resulted in electrodes whose activities were indistinguishable from those of polycrystalline Cu. In contrast, Cu_2O layers formed at 500 °C that were ≥3 μm thick resulted in electrodes that exhibited large roughness factors and required 0.5 V less overpotential than polycrystalline Cu to reduce CO_2 at a higher rate than H_2O. The combination of these features resulted in CO_2 reduction geometric current densities >1 mA/cm^2 at overpotentials <0.4 V, a higher level of activity than all previously reported metal electrodes evaluated under identical reaction conditions. Moreover, the activity of the modified electrodes was found to be stable over the course of several hours, whereas a polycrystalline Cu electrode exhibited deactivation within 1 h under identical reaction conditions. These electrodes may be particularly useful for elucidating the structural properties of Cu that determine the distribution between CO_2 and H_2O reduction and provide a promising lead for the development of practical catalysts for electrolytic fuel synthesis [19].

As noted earlier, although several improvements have been made in ERC to produce CO and HCOOH in terms of lowering the associated overpotentials, improving the product selectivity, and so on over Au, Pd, and Sn NPs containing cathodic systems in 0.5 M NaHCO$_3$-based aqueous electrolytes, the desired FE, rate of reaction (i.e., current density up to several hundreds of mA/cm^2), and stability for commercial practices are yet to be achieved. One of the key challenges for water-based electrolytes to be overcome is low FE due to the involvement of competitive HER, as this latter reaction is much easier than CO_2

electrolysis on most transition metal catalysts. Certain investigations have eliminated water conversion by eliminating water from the system, but this has not proven to be a practical choice. However, when the room-temperature ionic liquids (RTILs) are employed in conjunction with water as an electrolyte and metal nanoparticles as catalysts or cathodic systems, further improvements in ERC process are noted. Section 44.6 deals with the results obtained in ERC processes performed by involving RTILs.

44.6
Effect of Particle Size on the Efficiency of Nonaqueous-Based CO_2 Reduction Reactions

Of late, the ECR involving RTILs have been found to show more than an order of magnitude higher current density in comparison to those systems not involving any RTIL [13,14,44]. When RTILs are employed together with the NP-based metal cathodic systems in acetonitrile (MeCN) electrolyte, improved results are noted in comparison to those involving the bulk metals as cathodes. Like strong acids (H_2SO_4, HNO_3, HCl, etc.), RTILs also completely dissociate into ions and are not diluted by any bulk solvent. Moreover, because of their poor vapor pressures at room temperature unlike acids, such as HCl, the RTILs are thermally and electrically quite stable. RTILs consist of a bulky organic cation (e.g., imidazole) and an organic/inorganic anion (e.g., tratrafluoroborate, hexafluorophosphate, etc.). Furthermore, owing to their poor coordinating nature, they are highly polar solvents without coordinating strongly to solutes. Due to their low volatile nature, RTILs are environment-friendly unlike certain organic solvents, which possess significant vapor pressures. Furthermore, RTILs do not need any supporting electrolyte in electrochemical reactions as they themselves can act like supporting electrolytes and bear up to 3–6 V charge [13,14,44].

The large overpotentials associated with ERC reactions have been identified to be due to the formation of a high energy one electron reduced intermediate $CO_2^{\bullet-}$ with a standard redox potential of −1.9 V versus NHE. In a study, 1-ethyl-3-methylimidazolium tetrafluoroborate (EMIM-BF₄), (Eq. (44.36)) was employed to reduce the overpotential associated with the formation of $CO_2^{\bullet-}$ intermediate during ERC [13,14,44]. In this latter study, the involved electrocatalytic system has reduced CO_2 to CO at overpotentials below 0.2 V [14]. This system relies on an ionic liquid electrolyte to lower the energy of the $(CO_2)^-$ intermediate, most likely by complexation, and thereby lowering the initial reduction barrier. The silver (Ag) NPs as cathodic system could catalyze the formation of CO at 1.5 V (versus NHE), just slightly above the minimum (i.e., thermodynamically predicted equilibrium potential) voltage of 1.33 V. This system continued producing CO for at least 7 h at FE greater than 96% [14]. In this case, first EMIM-BF₄ RTIL converts CO_2 into $CO_2^{\bullet-}$ intermediate, which is then catalyzed over a metal cathode to form CO. When Ag was employed as cathode, CO was formed at overpotential of less than 200 mV, suggesting that CO_2 conversion to CO can occur without the large energy loss.

$$(44.36)$$

Interestingly, owing to their ionic characteristics, RTILs can stabilize the intermediate $CO_2^{\bullet-}$ *in situ* during CO_2 reduction by columbic complexation. This stabilization considerably lowers the electrode potential required for CO_2 reduction. The activation energy (overpotential) needed to form the $EMIM^+\text{-}CO_2^-$ intermediate has been found to be lower than the one required to form $CO_2^{\bullet-}$ in the absence of stabilization by RTIL. In the presence of $EMIM^+\text{-}BF_4^-$, CO_2 reduction starts at potentials as low as -250 mV versus SHE, and this RTIL has reduced about 80% of overpotential associated with CO_2 activation by one electron reduction [14]. As CO_2 cannot survive in basic solution, neutral to slightly acidic media is normally employed for ERC reactions. The HER is a strong function of solution pH, and the equilibrium potential decreases with increase of pH. In acidic conditions, HER is a thermodynamically more preferred reaction over reduction of CO_2 [14]. In aqueous medium, CO_2 undergoes reduction to various products and releases hydroxide ions. Due to the release of hydroxide ions during CO_2 reduction reactions, the pH near to the surface of the electrode is different from the equilibrium value as the rate of neutralization between the hydroxide anion and CO_2 is considerably slow in aqueous solution under ambient conditions [14]. For this reason, supporting electrolytes, such as, $KHCO_3$, K_2HPO_4, and so on, are normally employed to facilitate CO_2 reduction reactions smoothly as these electrolytes supply anions with buffering action, thereby nullify the effect of pH change at the electrode surface due to the release of hydroxide ions (Eqs. (44.37) and (44.38)) [14]. The supporting electrolytes, such as, KCl, $NaClO_4$, K_2SO_4, and so on, do not possess the ability to release protons, hence to exhibit buffer action to nullify the effect of pH change by the formed hydroxide ions during CO_2 reduction reaction. It can be inferred from these results that $EMIM\text{-}BF_4$ has several useful characteristics for reducing CO_2 at relatively less negative potentials and low temperatures in nonaqueous medium such as MeCN [14].

$$OH^- + HCO_3^{3-} \rightleftharpoons H_2O + CO_3^{2-} \qquad (44.37)$$

$$OH^- + HPO_4^{2-} \rightleftharpoons H_2O + PO_4^{3-} \qquad (44.38)$$

Interestingly, unlike earlier studies in which water-based electrolyte has suppressed the efficiency of ERC reaction due to competitive HER, the addition of water to $EMIM\text{-}BF_4$ electrolyte has actually been found to increase the efficiency of CO_2 conversion to CO [13]. In fact, little HER was noted over Ag electrode in EMIM–water mixture electrolyte, until water concentration did not exceed 90% by mole. An improvement in the rate of CO production was noted on Ag and platinum (Pt) when water was added to the electrolyte. The actual reason behind the increase in rate of CO formation upon addition of water to the EMIM-BF$_4$ electrolyte is yet to be determined [13]. The FE to CO is increased when water was added to the dry EMIM-BF$_4$ electrolyte reaching nearly 100% at 89.5 mol% water. The 89.5% water mixture contains 49.7 mol/l water [13]. At higher water concentrations, HER began; consequently, the FE of CO was dropped. Nevertheless, the use of Ag and Au (noble metals) as cathodes for production of CO has been hampered by the exorbitant cost of these materials, which eliminates their practical use on the scale required for alternative fuel synthesis. Indeed, the dearth of cost-effective systems that can efficiently and selectively drive ERC reaction highlights the need for new electrode–catholyte pairings that can selectively promote the ERC to CO at appreciable rate (high current density) and low overpotentials [13].

The development of affordable electrocatalysts that can drive ERC to CO with high selectivity, efficiency, and large current densities is a critical step on the path to production of liquid carbon-based fuels. In a recent study, inexpensive bismuth (Bi)-based carbon monoxide (CO) evolving catalyst (Bi-CMEC) has been developed for reducing CO_2 to CO in conjunction with EMIM-BF$_4$ or BMIM-BF$_4$ ionic liquid with appreciable current density at overpotentials below 0.2 V [15]. This catalyst is selective for CO production operating with a FE of approximately 95%. This activity has been on far with those historically observed over expensive Ag and Au cathodes [15]. A CPE experiment, when conducted at −1.95 V (versus SCE) in MeCN saturated with CO_2 and 20 mM EMIM-BF$_4$ RTIL using the bimodified glassy carbon electrode (GCE) as a cathode for 60 min, the measured CO in the headspace was corresponding to a FE of nearly 95% for the $2e^-/2H^+$ conversion of CO_2 to CO, with an average partial current density of about 3.77 ± 0.7 mA/cm^2. This inexpensive system has catalyzed the evolution of CO with current density as high as 25–30 mA/cm^2 and attendant energy efficiencies of ∼80% for this cathodic half-cell reaction. These metrics highlight the efficiency of Bi-CMEC, since only noble metals have been previously shown to promote this fuel-forming half-cell reaction with such high energy efficiency. Moreover, the rate of CO production by Bi-CMEC ranges from approximately 0.1 to 0.5 mmol/(cm^2 h) at an applied overpotential of 250 mV for a cathode with surface area equal to 1.0 cm^2. This CO evolution activity is much higher than that afforded by other nonnoble metal cathode materials and distinguishes Bi-CMEC as a superior and inexpensive platform for electrochemical conversion of CO_2 to fuel [15].

Inexpensive triflate salts of Sn^{2+}, Pb^{2+}, Bi^{3+}, and Sb^{2+} have been employed as precursors for the electrodeposition of CO_2 reduction cathode NPs from MeCN solutions, providing a general and facile electrodeposition strategy, which streamlines catalyst synthesis [16]. The ability of these four platforms to drive the formation of CO in ERC reaction in the presence of [BMIM]OTf has also been established. The electrochemically prepared Sn and Bi catalysts proved to be highly active, selective, and robust platforms for CO evolution, with partial current densities of 5–8 mA/cm^2 at applied overpotential of less than 250 mV.

By contrast, the electrodeposited Pb and Sb NP catalysts could not promote rapid CO generation with the same level of selectivity. The Pd NP material was only ~10% as active as the Sn and Bi NP systems at an applied potential of $E = -1.95$ V (versus SCE) and is rapidly passivated during catalysis. The Sb-comprised NP cathode material showed no activity for conversion of CO_2 to CO under analogous conditions. It can be concluded that the 1,3-dialkylimidazo-liums promote CO production, but only when they are used in combination with an appropriately chosen electrocatalyst material. More broadly, these results suggest that the interactions between CO_2, the imidazolium promoter, and the cathode surface are all critical to the observed catalysis. Nevertheless, the development of inexpensive systems with an efficient CO_2 reduction capability remains a challenge as the observed current densities in these systems are also not enough high.

A layer-stacked bulk molybdenum disulfide (MoS_2) with molybdenum (Mo)-terminated edges exhibited highest ERC performance (i.e., a current density of about 120 mA/cm^2) reported to date, which could be a cost-effective substitute for noble metal catalysts [17]. This performance was demonstrated in a 4 mol% EMIM-BF$_4$ ionic liquid containing aqueous solution. *In the same diluted electro-lyte, commonly used Ag NPs exhibited moderate performance, whereas bulk silver (Ag) catalyst was unable to reduce CO_2.* This significantly higher CO_2 reduction current density (relative to noble metal catalysts) has been mainly attributed to a high density of d-electrons in Mo-terminated edges and also to its low work function.

One of the weaknesses of all the investigated RTIL-based ERC systems to produce CO using NPs-based cathodic systems has been the lower reaction rates than those required for a commercial process. Typically, the commercial processes run at a turnover rate of about $1-10$ s^{-1} in contrast to the rate of 1 s^{-1} or less noted in these studies [14]. Indeed, a rate of about 60 turnovers/s with a rotating disk electrode at a cathode potential of 2 V (versus SCE) has been noted, although there is a scope for upscaling the process. At present, the employed cathodes have a surface area of about 6 cm^2, whereas the commercial electrochemical cells for the chloralkali process possesses an electrochemical surface area on the order of 10^9 cm^2. At 2 V (versus SCE), the reported cells produced CO at a rate of only about <10 μmol/min, whereas commercial processes require thousands of moles per minute per cell. Nevertheless, there is a scope for further development of the reactor configuration and exact operating conditions, for example, to increase the turnover number and product yields.

A CPE experiment for electrocatalytic synthesis of low-density polyethylene (LDPE) from CO_2 over the nanostructured (ns) TiO$_2$ film in a solvent mixture of water and ionic liquid, EMIM-BF$_4$ at room temperature under ambient pressure has been developed recently [45]. Under the employed reaction conditions, the ns TiO$_2$ film has been found to be remarkably efficient and selective for the electroreduction of CO_2. The current efficiency for the formation of the electrolytic product is about $8-14$% at -1.50 V (versus SCE).

44.7
Reverse Microbial Fuel Cells: The Practical Artificial Leaves

One of the most exciting applications of microbial interactions with electrochemistry is microbial electrosynthesis (MES) [46–50]. In fact, the process can be defined as an artificial form of photosynthesis in which microorganisms utilize electrons derived from an electrode to reduce CO_2 and H_2O to multicarbon extracellular products and O_2. The production of liquid transportation fuels with MES is particularly attractive because electricity generation by renewable technologies is neither continuous nor always compatible with demand and its storage is quite difficult. Recently, the application of biocatalysts has been gradually increased in *electrosynthetic* processes because of their *higher specificity* and *versatility* relative to existing chemical catalysts. Generally, *bioelectrosynthesis* relies on the interaction between biocatalysts and electrodes and mainly employs immobilized enzymes or organelles on the electrode surfaces. Recently, MES has been introduced to describe the electricity-driven reduction of CO_2 using the whole microorganisms as electrocatalysts. Several microorganisms (e.g., *Sporomusa ovata*, *Clostridium ljungdahlii*, *Clostridium aceticum*, and *Moorella thermoacetica*) have been identified, which can grow on CO_2 as the electron acceptors and reduce it into organic chemicals through an *anaerobic respiration process* [46,47]. Among them, acetogenic microorganisms are strict anaerobic bacteria that can couple H_2O oxidation with CO_2 reduction to produce acetate. The capability of acetogenic bacteria, such as *S. ovate,* in acquiring electrons from graphite electrodes to reduce CO_2 to acetate has proved the possibility of the bioelectrosynthesis concept. It was reported that the *S. ovata biofilm* on the electrode surface of a reverse-microbial fuel cell (R-MFC) system can produce acetate as well as small amounts of 2-oxobutyrate concomitantly with current consumption so that the electron recovery for these products may be over 85%.

Substantial acetate production has been reported in recent MES studies [46]. However, this approach is intrinsically limited by several factors including the low energy capture and transfer efficiency of photosynthesis [46]. Therefore, there has been a great deal of interest in the development of alternative technologies that simultaneously improve energy capture and transfer to biosynthetic pathways optimized for production of useful compounds. One way to address this challenge is through the development of R-MFCs. In a standard *microbial fuel cell*, organisms oxidize organic fuels and transfer electrons into an electrochemical system so that fuels are converted to electrical energy. In an R-MFC, this process is reversed so that electrical energy is used by cells to drive CO_2 *fixation* to high-energy organics. The critical challenge for this approach is that energy must be efficiently transferred from an electrode into a biological host that is capable of using this energy for biosynthesis. R-MFC platforms could have a significant impact in the biofuel and fuel chemical arena as they would be capable of using electricity generated from all renewable sources including *wind, geothermal, hydroelectric, nuclear,* and *solar.* They could be used in a variety of global locations, and they could be used for long-term storage of excess electrochemical energy.

The most of the successful applications of microbe electrode interactions for bioelectrosynthesis reported in the literature have been the R-MFCs, which involve the direct supply of electrons to microorganisms at the cathode surface to activate the biocatalysis process [46–50]. Microbes in the cathodic chamber consume electrons to reduce the substrate molecules, and generate the final products of the process. In addition, the energetics of the living system is provided by the electron transfer process. One approach to creating R-MFCs is to use direct electron transfer from an electrode to the cells. This has been termed microbial electrosynthesis. While this energy transfer can be accomplished directly, where electroactive cells in a biofilm utilize electrons from an electrode for anabolism, diffusion issues and the requirement for 2D biofilms make direct microbial electrosynthesis a challenging proposition.

An alternative approach to creating R-MFCs is to use soluble electron mediators that can shuttle electrons from the electrode to the cells. The use of a mediator enables the utilization of planktonic cells in the bioreactor and facilitates easy 3D scale-up of the individual components. Furthermore, the use of mediators can also enable separate stage designs that afford spatial and temporal decoupling of energy capture and bioproduction. This can allow both processes to be operated and optimized separately. In the mediated approach, electrons are first transferred from an electrode to a soluble mediator and then the mediator would be oxidized by the cell. Inorganic compounds that are linked with chemoautotrophy and can be electrolytically regenerated, such as hydrogen (H_2) as well as those involved with the nitrogen, iron, and sulfur *biogeochemical cycles* (i.e., ammonia (NH_3), nitrite (NO_2^-), iron (Fe^{2+}), and hydrogen sulfide (H_2S)), are attractive options for use in this platform since they can facilitate the construction of multicarbon organics from CO_2 using naturally occurring carbon fixation pathways. These inorganic compounds naturally yield sufficient energy to support biomass growth and can be reduced via electrolysis.

Successful R-MFC operation has recently been demonstrated using electrochemically produced formate coupled with genetically modified *Ralstonia eutropha* cells that were able to produce isobutanol. In this version of the R-MFC, CO_2 was electrochemically fixed into formate at the electrode, which was subsequently used by cells to produce a biofuel. In a recent study, the feasibility of using an alternative electron transfer mediator and chemolithoautotrophy for primary production has also been investigated. They demonstrated a sustained biomass production in an R-MFC that had two components: (i) an electrochemical reactor that produces ammonia from nitrite using electrical energy, and (ii) a biological reactor containing a naturally chemolithoautotrophic, ammonia-oxidizing bacterium, *Nitrosomonas europaea* (Figure 44.6) [46]. This organism was chosen as it is a well-studied autotrophic ammonia-oxidizing bacterium whose genome has been sequenced. Furthermore, a stable long-term operation (>15 days) of the separate stage R-MFC, which facilitated fixation of CO_2 via the Calvin–Benson–Bassham (CBB) cycle using energy derived solely from ammonia, has been reported. These results suggest that this approach can be expanded

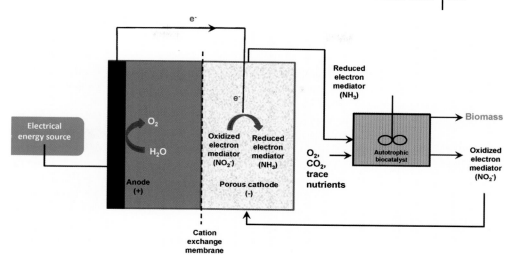

Figure 44.6 Overview of a reverse microbial fuel cell that uses the ammonia/nitrite redox couple as a mediator and ammonia-oxidizing *Nitrosomonas europaea* cell as biocatalysts [46].

to produce biofuels and other chemicals via genetic modification of the *N. europaea* cells in the bioreactor without the need for photosynthesis. For this purpose, an electrochemical reactor was designed for the regeneration of ammonia from nitrite, and current efficiencies of 100% were achieved [46]. Calculations indicated that overall bioproduction efficiency could approach $2.7 \pm 0.2\%$ under optimal electrolysis conditions. The application of chemolithoautotrophy for industrial bioproduction has been largely unexplored, and results suggest that this and related R-MFC platforms may enable fuel chemical and related biochemical production.

Recently, the protons and electrons released by exoelectrogenic bacteria in specially designed reactors have catalyzed to form H_2 through the addition of a small voltage to the circuit [50]. By improving the materials and reactor architecture, hydrogen gas was produced at yields of 2.01–3.95 mol/mol (50–99% of the theoretical maximum) at applied voltages of 0.2–0.8 V (versus NHE) using acetic acid, a typical dead-end product of glucose or cellulose fermentation. At an applied voltage of 0.6 V, the overall energy efficiency of the process was 288% based solely on electricity applied, and 82% when the heat of combustion of acetic acid was included in the energy balance, at a gas production rate of $1.1 \, m^3$ of H_2 per cubic meter of reactor per day. Direct high-yield hydrogen gas production was further demonstrated by using glucose, several volatile acids (acetic, butyric, lactic, propionic, and valeric), and cellulose at maximum stoichiometric yields of 54–91% and overall energy efficiencies of 64–82%. This electrohydrogenic process thus provides a highly efficient route for producing hydrogen gas from renewable and carbon-neutral biomass resources. Nevertheless, the R-MFC-based ERC processes have to be developed further before practicing them at commercial plants.

44.8
Concluding Remarks and Future Perspectives

The conversion of CO_2 to value-added chemicals has been found to be very important from the point of view of addressing the problems related to (i) the CO_2-associated global warming, (ii) renewable energy production, and (iii) storing of energy in high-energy-density chemical fuels. Among the various methods developed for converting CO_2 into value-added chemicals, the electrochemical routes have been found to be more promising as it has been developed to the required extent and it uses stable and reliable systems required for commercial practices. Although these electrochemical routes have been developed to a greater extent in comparison to other routes, they suffer from very poor process efficiencies and product selectivity. Of late, the nanoparticle-based catalytic and cathodic systems in conjunction with RTILs have been found to be effective in solving some of the problems associated with electrochemical CO_2 reduction processes. Although the results reported in the open literature are not sufficient to practice these processes at industry level, there are already several industries all over the world producing value-added chemicals from waste stream greenhouse CO_2 to value-added chemicals using renewable energy resources. It suggests that the some of the reported electrochemical CO_2 conversion routes can be closely observed for further development using nanostructured materials and nanoparticle-based catalytic/cathodic systems using RTILs so that they can be practiced for commercial purposes.

Acknowledgments

Author wishes to specially acknowledge all the researchers whose work has been referred to in this chapter. Thanks are also due to Dr. G. Sundararajan, Director, ARCI, for his kind encouragement and permission to publish this chapter.

References

1 Robinson, A.B., Robinson, N.E., and Soon, A. (2007) Environmental effects of increased atmospheric carbon dioxide. *J. Am. Physicians Surg.*, **12**, 79–90.

2 Olah, G.A., Goeppert, A., and Prakash, G.K.S. (2009) Chemical recycling of carbon dioxide to methanol and dimethyl ether: from greenhouse gas to renewable, environmentally carbon neutral fuels and synthetic hydrocarbons. *J. Org. Chem.*, **74**, 487–498.

3 Olah, G.A. (2005) Beyond oil and gas: the methanol economy. *Angew. Chem., Int. Ed.*, **44**, 2636–2639.

4 Centi, G. and Perathoner, S. (2011) CO_2-based energy vectors for the storage of solar energy. *Greenhouse Gas Sci. Technol.*, **1**, 21.

5 Centi, G. and Perathoner, S. (2009) Opportunities and prospects in the chemical recycling of carbon dioxide to fuels. *Catal. Today*, **148** (3–4), 191.

6 Ganesh, I. (2015) Solar fuels vis-á-vis electricity generation from sunlight: the current state-of-the-art (a review). *Renew. Sustain. Energy Rev.*, **44**, 904–932.

7 Ganesh, I. (2014) Conversion of carbon dioxide into methanol – a potential liquid fuel: fundamental challenges and

opportunities (a review). *Renew. Sustain. Energy Rev.*, **31**, 221–257.

8 Moore, F.C. and Diaz, D.B. (2015) Temperature impacts on economic growth warrant stringent mitigation policy. *Nat. Clim. Change*, **5** (2), 127–131.

9 Arakawa, H., Aresta, M., Armor, J.N., Barteau, M.A., Beckman, E.J., Bell, A.T., Bercaw, J.E., Creutz, C., Dinjus, E., Dixon, D.A. *et al.* (2001) Catalysis research of relevance to carbon management: progress, challenges, and opportunities. *Chem. Rev.*, **101** (4), 953–996.

10 Barber, J. (2007) Biological solar energy. *Philos. Trans. A Math. Phys. Eng. Sci.*, **365** (1853), 1007–1023.

11 Halmann, M. (1978) Photoelectrochemical reduction of aqueous carbon dioxide on *p*-type gallium phosphide in liquid junction solar cells. *Nature*, **275** (5676), 115–116.

12 Inoue, T., Fujishima, A., Konishi, S., and Honda, K. (1979) Photoelectrocatalytic reduction of carbon dioxide in aqueous suspensions of semiconductor powders. *Nature*, **277** (5698), 637–638.

13 Rosen, B.A., Zhu, W., Kaul, G., Salehi-Khojin, A., and Masel, R.I. (2013) Water enhancement of CO_2 conversion on silver in 1-ethyl-3- methylimidazolium tetrafluoroborate. *J. Electrochem. Soc.*, **160** (2), H138–H141.

14 Rosen, B.A., Salehi-Khojin, A., Thorson, M.R., Zhu, W., Whipple, D.T., Kenis, P.J.A., and Masel, R.I. (2011) Ionic liquid-mediated selective conversion of CO_2 to CO at low overpotentials. *Science*, **334** (6056), 643–644.

15 Medina-Ramos, J., Pupillo, R.C., Keane, T.P., DiMeglio, J.L., and Rosenthal, J. (2015) Efficient conversion of CO_2 to CO using tin and other inexpensive and easily prepared post-transition metal catalysts. *J. Am. Chem. Soc.*, **137** (15), 5021–5027.

16 Medina-Ramos, J., DiMeglio, J.L., and Rosenthal, J. (2014) Efficient reduction of CO_2 to CO with high current density using *in situ* or *ex situ* prepared Bi-based materials. *J. Am. Chem. Soc.*, **136** (23), 8361–8367.

17 Asadi, M., Kumar, B., Behranginia, A., Rosen, B.A., Baskin, A., Repnin, N., Pisasale, D., Phillips, P., Zhu, W., Haasch, R. *et al.* (2014) Robust carbon dioxide reduction on molybdenum disulphide edges. *Nat. Commun.*, **5**, doi 10.1038/ncomms5470.

18 DiMeglio, J.L. and Rosenthal, J. (2013) Selective conversion of CO_2 to CO with high efficiency using an inexpensive bismuth-based electrocatalyst. *J. Am. Chem. Soc.*, **135** (24), 8798–8801.

19 Li, C.W. and Kanan, M.W. (2012) CO_2 reduction at low overpotential on Cu electrodes resulting from the reduction of thick Cu_2O films. *J. Am. Chem. Soc.*, **134** (17), 7231–7234.

20 Chen, Y., Li, C.W., and Kanan, M.W. (2012) Aqueous CO_2 reduction at very low overpotential on oxide-derived Au nanoparticles. *J. Am. Chem. Soc.*, **134** (49), 19969–19972.

21 Chen, Y. and Kanan, M.W. (2012) Tin oxide dependence of the CO_2 reduction efficiency on tin electrodes and enhanced activity for tin/tin oxide thin-film catalysts. *J. Am. Chem. Soc.*, **134** (4), 1986–1989.

22 Morris, A.J., McGibbon, R.T., and Bocarsly, A.B. (2011) Electrocatalytic carbon dioxide activation: the rate-determining step of pyridinium-catalyzed CO_2 reduction. *ChemSusChem*, **4** (2), 191–196.

23 Barton Cole, E., Lakkaraju, P.S., Rampulla, D.M., Morris, A.J., Abelev, E., and Bocarsly, A.B. (2010) Using a one-electron shuttle for the multielectron reduction of CO_2 to methanol: kinetic, mechanistic, and structural insights. *J. Am. Chem. Soc.*, **132** (33), 11539–11551.

24 Barton, E.E., Rampulla, D.M., and Bocarsly, A.B. (2008) Selective solar-driven reduction of CO_2 to methanol using a catalyzed *p*-GaP based photoelectrochemical cell. *J. Am. Chem. Soc.*, **130**, 6342–6344.

25 Seshadri, G., Lin, C., and Bocarsly, A.B. (1994) A new homogeneous electrocatalyst for the reduction of carbon dioxide to methanol at low overpotential. *J. Electroanal. Chem.*, **372** (1–2), 145–150.

26 Finn, C., Schnittger, S., Yellowlees, L.J., and Love, J.B. (2012) Molecular approaches to the electrochemical reduction of carbon dioxide. *Chem. Commun.*, **48** (10), 1392–1399.

27 Darensbourg, D.J. (2010) Chemistry of carbon dioxide relevant to its utilization: a personal perspective. *Inorg. Chem.*, **49** (23), 10765.

28 Abbott, D. (2010) Keeping the energy debate clean: how do we supply the world's energy needs? *Proc. IEEE*, **98** (1), 42–66.

29 Aruchamy, A., Aravamudan, G., and Rao, G.V.S. (1982) Semiconductor based photoelectrochemical cells for solar energy conversion: an overview. *Bull Mater. Sci.*, **4**, 483–426.

30 Bak, T., Nowotny, J., Rekas, M., and Sorrell, C.C. (2002) Photo-electrochemical hydrogen generation from water using solar energy: materials-related aspects. *Int. J. Hydrogen Energy*, **27**, 991–1022.

31 Dean, J.A. (1979) *Lange's Handbook of Chemistry*, vol. **9**, 11th edn, McGraw-Hill, New York, pp. 4–9: 128.

32 Lide, D.R. (2003) *CRC Handbook*, vol. **5**, 84th edn, CRC Press, Boca Raton, FL, pp. 5–5, 60, 65: 85–65: 86.

33 Keene, F.R., Creutz, C., and Sutin, N. (1985) Reduction of carbon dioxide by tris (2,2′-bipyridine)cobalt(I). *Coord. Chem. Rev.*, **64**, 247–260.

34 Lewis, N., Crabtree, G., Nozik, A.J., Wasielewski, M.R., and Alivisatos, P. (2005) Basic Research Needs for Solar Energy Utilization – Report on the Basic Energy Sciences Workshop on Solar Energy Utilization, pp. 1–276. Available at http://wwwscdoegov/bes/reports/files/SEU_rptpdf (accessed April 18–21).

35 Stelmachowski, P., Sirotin, S., Bazin, P., Mauge, F., and Travert, A. (2013) Speciation of adsorbed CO_2 on metal oxides by a new 2-dimensional approach: 2D infrared inversion spectroscopy (2D IRIS). *Phys. Chem. Chem. Phys.*, **15** (23), 9335–9342.

36 Carroll, J. and Mather, A. (1992) The system carbon dioxide–water and the Krichevsky–Kasarnovsky equation. *J. Solution Chem.*, **21** (7), 607–621.

37 Kohn, S.C., Brooker, R.A., and Dupree, R. (1991) ^{13}C MAS NMR, a method for studying CO_2 speciation in glasses. *Geochim. Cosmochim. Acta*, **55** (12), 3879–3884.

38 Butler, J.N. and Cogley, D.R. (1998) *Ionic Equilibrium: Solubility and pH Calculations*, John Wiley & Sons, Inc., New York, 559 pp.

39 Gupta, M., Coyle, I., and Thambimuthu, K. (2003) CO_2 capture technologies and opportunities in Canada. 1st Canadian CC&S Technology Roadmap Workshop 2003, September 18.

40 Metz, B. and Thambimuthu, K. (2002) Workshop on carbon dioxide capture and storage. Proceedings 2002, Regina, Canada, November 18–21, p. 1.

41 Zhu, W., Michalsky, R., Metin, Ö., Lv, H., Guo, S., Wright, C.J., Sun, X., Peterson, A.A., and Sun, S. (2013) Monodisperse Au nanoparticles for selective electrocatalytic reduction of CO_2 to CO. *J. Am. Chem. Soc.*, **135** (45), 16833–16836.

42 Gao, D., Zhou, H., Wang, J., Miao, S., Yang, F., Wang, G., Wang, J., and Bao, X. (2015) Size-dependent electrocatalytic reduction of CO_2 over Pd nanoparticles. *J. Am. Chem. Soc.*, **137** (13), 4288–4291.

43 Zhang, S., Kang, P., and Meyer, T.J. (2014) Nanostructured Tin catalysts for selective electrochemical reduction of carbon dioxide to formate. *J. Am. Chem. Soc.*, **136** (5), 1734–1737.

44 Rosen, B.A., Haan, J.L., Mukherjee, P., Braunschweig, B., Zhu, W., Salehi-Khojin, A., Dlott, D.D., and Masel, R.I. (2012) *In situ* spectroscopic examination of a low overpotential pathway for carbon dioxide conversion to carbon monoxide. *J. Phys. Chem. C*, **116** (29), 15307–15312.

45 Chu, D., Qin, G., Yuan, X., Xu, M., Zheng, P., and Lu, J. (2008) Fixation of CO_2 by electrocatalytic reduction and electropolymerization in ionic liquid–H_2O solution. *ChemSusChem*, **1** (3), 205–209.

46 Khunjar, W.O., Sahin, A., West, A.C., Chandran, K., and Banta, S. (2012) Biomass production from electricity using ammonia as an electron carrier in a reverse microbial fuel cell. *PLoS One*, **7** (9), e44846.

47 Kazemi, M., Biria, D., and Rismani-Yazdi, H. (2015) Modelling bio-electrosynthesis in a reverse microbial fuel cell to produce acetate from CO_2 and H_2O. *Phys. Chem. Chem. Phys.*, **17** (19), 12561–12574.

48 Luo, X., Zhang, F., Liu, J., Zhang, X., Huang, X., and Logan, B.E. (2014) Methane production in microbial reverse-electrodialysis methanogenesis cells (MRMCs) using thermolytic solutions. *Environ. Sci. Technol.*, **48** (15), 8911–8918.

49 Nam, J.-Y., Cusick, R.D., Kim, Y., and Logan, B.E. (2012) Hydrogen generation in microbial reverse-electrodialysis electrolysis cells using a heat-regenerated salt solution. *Environ. Sci. Technol.*, **46** (9), 5240–5246.

50 Cheng, S. and Logan, B.E. (2007) Sustainable and efficient biohydrogen production via electrohydrogenesis. *Proc. Natl. Acad. Sci. USA*, **104** (47), 18871–18873.

45

Nanomaterial-Based Methods for Cleaning Contaminated Water in Oil Spill Sites

Boris I. Kharisov,[1] H.V. Rasika Dias,[2] Oxana V. Kharissova,[1] and Yolanda Peña Méndez[1]

[1]Universidad Autónoma de Nuevo León, Facultad de Ciencias Quimicas, Ave. Universidad s/n, Ciudad Universitaria, San Nicolás de los Garza, N.L. 66455, Mexico
[2]Department of Chemistry and Biochemistry, The University of Texas at Arlington, Arlington, TX 76019, USA

45.1
Introduction

Nanotechnology as an emerging field provides ways to remediate the environmental issues in a more sustainable way with reduced consumption of energy and materials, in comparison to the conventional routes. The term "nanoremediation" consists of the application of reactive nanomaterials for removal, transformation, and detoxification of contaminants. Nanotechnology is applied for environmental purposes to resolve the existing problems and to prevent possible future problems, caused by the "materials–energy" interactions [1]. Different methods can be applied for water treatment [2], depending on robustness, mobility, flexibility, reliability, modularity, cost, chemical and energy demand, and brine level or residual disposal requirements. Overall, a final choice for the water purification method should be based on several specific factors. Various aspects of nanotechnology-assisted water treatment techniques were discussed in recent reviews [3–6].

Crude oil composition consists of saturated hydrocarbons, aromatics, resins, asphaltenes, naphthenic acids and bases, among others. Contamination of groundwater and soil with petroleum and/or its refinery products has existed from the beginning of the era of oil extraction. In particular, pollutions are caused by fuel leakage from old underground storage tanks, processes of oil extraction, refineries, terminals for fuel distribution, and incorrect disposal, as well as oil spills in all steps of oil and gasoline production, storage, and transportation. Polycyclic aromatic hydrocarbons (PAHs), a part of oil, are persistent organic pollutants with certain toxicity, low biodegradability, and carcinogenic activity, tending to bioaccumulate, affecting human health and ecosystems [7]. Large quantities of water containing PAHs are being produced as a result of

Nanotechnology for Energy Sustainability, First Edition. Edited by Baldev Raj, Marcel Van de Voorde, and Yashwant Mahajan.
© 2017 Wiley-VCH Verlag GmbH & Co. KGaA. Published 2017 by Wiley-VCH Verlag GmbH & Co. KGaA.

petroleum extraction, leading to a considerable damage of the surrounding environment. In this respect, this contaminated water must be purified before being disposed back to the environment under control of environmental agencies. Oil spills and industrial oil wastes need to be removed from seas, lakes, or rivers, and adequately treated.

When accidents occur during oil extraction and processing in real conditions, oil/water/solid particle mixtures could appear and, being hardly separated, affect considerably the environment. The hydrocarbon and other products, contained in crude oil, after mixing with water, can change their initial state very rapidly. In saline water (seas and oceans), petroleum can be present in distinct migration phases, for instance as surface films (slicks), oil-in-water and water-in-oil emulsions, lumps and aggregates, or in the form of dissolved compounds, adsorbed by bottom sediments and suspensions, or consumed by aquatic organisms. In the conditions of cold weather, oil biodegradation, evaporation, and dissolution are very slow. In the presence of snow and ice in winter, the oil viscosity is considerably affected by cold water, converting it to very sticky and thick material. In addition, oil lumps can also be formed. However, in these icy conditions, the petroleum can stay as in a "natural container," which prevents its expansion and spreading, so necessary operations to save an environment can have more time.

Conventional routes for water treatment include manual cleanup, filtration (using filters such as ceramic, biosand, charcoal or activated carbon, granular media, fiber, and fabric), heat and UV radiation, *in situ* burning, chemical treatment and dispersion (coagulation–flocculation, chemical disinfection, flocculant disinfection), and desalination (by reverse osmosis, distillation, or adsorption) [8]. The methods mentioned above serve well for separating oil–water mixtures as well, in particular the use of microfibrous sorbents (i.e., nylon fibers), froth flotation technique, and greener biodispersants (i.e., emulsan, sopholipids, and rhamnolipids). As an example, in case of chemical dispersants, their surfactant molecules (surface-active agents) migrate to the oil–water interface and reduce the oil–water interfacial tension. In permanently moving water, oil droplets break away from the oil slick, disperse, and remain in suspension, becoming biodegradable targets. As an example of the use of modified natural products, hydrophobic organoclays can be produced from bentonites (natural hydrophilic clays) by their hydrophobization with quaternary amines. Resulting organoclays are very efficient in selectively adsorbing the oil from water.

Despite more than two decades of intensive growth of the nanotechnology field, the nanomaterials were started to be applied for water purification/cleanup only some years ago (see comprehensive reviews [9–19] and a recent book [20] in this area). *Nanotechnology-based* water treatment technologies consist of CNT (carbon nanotube) based methods (including CNT membranes, nanomesh), nanofiltration techniques (membranes and devices, nanofibrous alumina filters, nanofiber gravity-flow devices), nanoporous ceramics, clays, and other adsorbents and composites, such as polypyrrole–CNT nanocomposite and cyclodextrin nanoporous polymer, zeolites, use of nanocatalysts (nZVI (zerovalent iron), nano-iron oxides, nano-TiO_2 as photocatalysts or adsorbents), and magnetic nanoparticles (for instance, magnetoFerritin). Magnetic nanoparticle-based methods have been utilized recently, although in a low scale, for remediation of the sites, contaminated with oil. These methods caused a considerable interest to several petroleum industries, because of an insufficient number of good results in water cleanup and complete removal of petroleum applying conventional technologies of water–oil separation.

Several conventional materials for oil capture have various disadvantages, for example, low degradability (synthetic polymers (PPE, butyl rubber, etc.)) and low sorption capacity (inorganic hydrophilic minerals such as vermiculite (oil absorbance 3 g/g), exfoliated graphite, zeolites, and so on; natural fibers, for instance corn stalk, nonwoven wool (oil absorbance about 11 g/g), cotton fibers, etc.). On the contrary, nanotechnology-based methods can be much more effective. Thus, CNTs possess large surface area, oleophilic property, low density, and high strength (oil absorbance 112 g/g). Some nanoremediation projects use nanoscale elemental iron to render contaminants benign by either reducing or absorbing them. Iron nanoparticles oxidize in water and turn to rust, releasing free radicals, which are able quickly to break down harmful contaminants into safer substances [21].

In this chapter, we discuss selected aspects of the most recent achievements in the field of separation of water and oil (petroleum or products of its reelaboration) from their mixtures (oil spills) by nanotechnology-based technologies. These processes, with some exceptions, are, overall, basically developed in laboratory scale and are currently rarely applied for large-scale treatment of soil or water surfaces.

45.2
Inorganic Nanomaterials and Composites

Zero-valent iron is being currently utilized for remediation purposes alone and/or in combination with oxidants, mainly peroxides (Fenton-like reactions). nZVI [22,23] and iron nanometric composites (for instance, with MWCNTs or activated carbon) are currently used for water purification from petroleum impurities too. As an example, the synthesized Fe nanoparticles (with BET surface area $35.8 \, m^2/g$ and 40–80 nm in particle size) were proven to be a good Fenton-type catalyst for oil degradation (removal of hydrocarbon pollutions in soil), under ultrasound-assisted technique [24]. Influence of such factors as nZVI and H_2O_2 concentrations, temperature effects, and ultrasound power were studied. Taking into account the obtained results, it was observed that petroleum degradation does not increase when the power of ultrasound increases, but surprisingly the degradation reaches an optimum value and further decreases with the increased ultrasound power. The petroleum degradation increased with temperature, H_2O_2, and nZVI concentration. It was observed that the efficiency of hydrocarbon elimination by this Fenton-like oxidation ranges from 80 to 98%. The optimum degradation conditions are as follows: temperature 40 °C, H_2O_2

concentration 30 mM, pH 3.5, nZVI concentration 0.4 g/L, ultrasonic power 500 W. In a related report [25] on Fenton-like reactions applying reductant and oxidant system for oil capture and further chemical treatment, for a similar process of crude oil oxidation in soil by H_2O_2 using nZVI as catalyst, an optimal molar ratio H_2O_2:Fe^0 is ~33 : 1 and reaction time of 4 h, leading to 91% removal of organic phase. An excess of H_2O_2 rapidly decomposes in soil due to its inorganic and/or organic components, acting as decomposition catalysts. A similar technique (nZVI/UV/H_2O_2), a photo-Fenton-like method, was also applied to treat diesel fuel (which has created many problems for water resources) for TPH reduction from diesel fuel in the aqueous phase [26]. The experimental removal rate was 95–100% (predicted 95.8). This 60 min process can be used as a pretreatment step for the biological removal of TPH from diesel fuel in the aqueous phase.

In addition to the iron nanoparticles acting without supporting phase, 3D Fe/C nanocomposites with macroporous structure (their composition depends on the pore size (0.35–3.0) and diameter of polystyrene (PS) microsphere (0.67–4.2)) proved to be very selective absorption materials for the elimination of oils from water phase. These macroporous nanomaterials were fabricated by sinteration of a mixture of ferric nitrate precursor and closely packed polystyrene microspheres. According to the X-ray diffraction data, *iron carbide* (Fe_3C) phase and the α-Fe metallic phase were discovered, testifying that the Fe–C nanocomposites were synthesized. These nanomaterials showed superoleophilic and superhydrophobic properties without alteration of low-surface-energy chemicals. These macroporous nanomaterials absorbed a wide range of oils and nonpolar organic solvents on water surface rapidly and selectively. Due to magnetic properties of these nanocomposites, the elimination of the absorbed organic phase from the water surface was easily reached applying a magnetic field. In comparison to Fe_2O_3@C nanoparticles, the capacity to adsorb oil for thus prepared nanocomposites was found to be much higher. In addition, even after repeated cycles these nanomaterials retained highly oleophilic and hydrophobic properties after removing oils from water surface [27].

As seen earlier, *peroxides* are frequently used for the creation of Fenton-like system on nZVI basis. Hydrogen peroxide is frequently used for this purpose. In our opinion, its high solubility in water affects the oil removal process, since H_2O_2 may not stay in contact with nZVI for prolonged periods under real conditions. CaO_2, which has low solubility in water and could serve as a source of slow formation of H_2O_2, is also attractive (see below). Researchers also pay attention to the persulfate anion (PS, $S_2O_8^{2-}$), which, combined with the nZVI, could serve for PS activation (alone or together with Pd(0) [28]) or for water disinfection [29] or efficient decomposition of a series of organic molecules, for instance, trichloroethylene or naphthalene. As an example, Fe^0 was utilized for activated PS oxidation of trichloroethylene (TCE) as a source of ferrous ions [30]. It was established that in the TCE/Fe^0/PS system, fast TCE degradation was accompanied by chloride ion formation and the rapid $S_2O_8^{2-}$ decomposition as a result of TCE mineralization. When naphthalene was used, it was similarly shown [31] that its decomposition using nZVI-activated persulfate is considerably more effective than nZVI alone. The naphthalene degradation by nZVI-activated persulfate was shown to be >99% in comparison with <10% using nZVI alone. However, due to the presence of such strong oxidant anion as $S_2O_8{}^{2-}$, the nanoparticles of nZVI exposed to PS were observed to be passivated very fast. So, the reaction rate was reduced to a magnitude representative of an unactivated persulfate system. Additionally, nZVI particle was covered with an iron sulfate layer following exposure to persulfate (Figure 45.1). Similar layer of FeOOH appears in the surface of fresh recently obtained nZVI particles. Fe^{2+} ions, appearing in the system as a result of the direct reaction pathway (Eq. (45.1)), activate persulfate anion resulting in generation of sulfate free radicals (Eq. (45.2)). Then, Fe^{2+} can be regenerated by reaction with Fe^{3+} ions, located on the ZVI surface (Eq. (45.3)). Despite the nZVI particle surfaces are passivated, nZVI is considered as a promising persulfate activator compared to the traditional persulfate activators such as ion Fe^{2+} and granular zero-valent iron. Other investigations utilizing PS and Fe^{2+} ion (used instead of nZVI) showed that $SO_4{}^{\bullet-}$ was potentially effective in selectively degrading toluene in surfactant flushing effluents. We note that the future development of this technology could

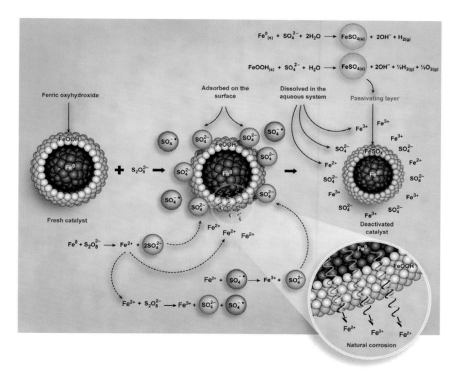

Figure 45.1 Conceptual model that illustrates the formation of an iron sulfate complex on the surface of a nZVI particle. (Reproduced with permission from American Chemical Society.)

lead to elaboration of persulfate-based approach for oil removal:

$$Fe^0 + S_2O_8^{2-} \rightarrow Fe^{2+} + 2\,SO_4^{2-} \tag{45.1}$$

$$Fe^{2+} + S_2O_8^{2-} \rightarrow Fe^{3+} + SO_4^{2-} + SO_4^{\bullet-} \tag{45.2}$$

$$2\,Fe^{3+} + Fe^0 \rightarrow 3\,Fe^{2+} \tag{45.3}$$

In the last decade, a material based on nanometric-size oxides and peroxides (generally those of calcium [32,33]) was used *in situ* to eliminate oil spills from underground oil tanks. The obtained results from this technology based on redox processes confirmed cheaper and faster methods, producing lower overall contaminant levels in comparison to the previous remediation techniques. Most of these contaminated sites were in New Jersey, where the cleanup was carried out in consultation with the New Jersey Department of Environmental Protection[1]. For the calcium peroxide application (50–200 nm particles of CaO_2 with a metal catalyst), peculiarities and advantages of this nano-peroxide for treatment of soil and groundwater polluted petroleum hydrocarbons are as follows: (a) injection was performed through 1–1¼ in. holes in soil or paving. Soil and paving resealed at the end of day with almost no disruption of lawns. (b) The number of points installed is 5–20 injection points per day depending on soil conditions, access, and depth of treatment needed. There is no limit to the depth that is treatable as long as a well or injection point can be installed to the depth. (c) Process will continue bioremediation for up to 6 months if necessary. (d) Petroleum odors eliminated within minutes of injection or application. The contaminants, confirmed to be treatable, are heating oil, #6 oil, gasoline, MTBE (methyl tertiary butyl ether), ethylene glicol, solvents, and coal tar, in geological conditions as sand, silt, fractured rock, fill materials, landfills, and sediments. The oxygen, formed as a result of the reaction of calcium peroxide with water, leads to a creation of an aerobic environment that supports natural bioremediation by aerobic organisms present in the soil.

In case of *iron oxides*, the Fenton system Fe_2O_3–H_2O_2 was used for decomposition of pyrene, which is very low biodegradable, carcinogenic, and highly persistent PAH pollutant [34]. Its 99% removal was reached applying sodium pyrophosphate as chelating agent for 6 h at 300 mM H_2O_2 concentration and 30 mM of nano-iron concentration and at pH 3. Applying magnetic fields, magnetite (Fe_3O_4) nanoparticles capped with natural product rosin amidoxime (bioactive nanosystems having antimicrobial effects) derivatives R-AN and RK-AN (Figure 45.2), fabricated at 45°C by a coprecipitation reaction (Figure 45.3), were examined as a crude oil collector [35]. Applying an external magnetic field, the magnetic amidoxime can be easily separated from aqueous phase. The data of oil recovery efficiency using coated and uncoated magnetite are listed in Table 45.1, showing the good oil collection of oil without magnetite precipitation in water at MOR (magnetite–oil ratio) 1 : 5. Low oil spill collector properties for uncoated magnetite are due to its low hydrophobicity.

---

1) www.continentalremediation.com/HomeHtgOIl.htm.

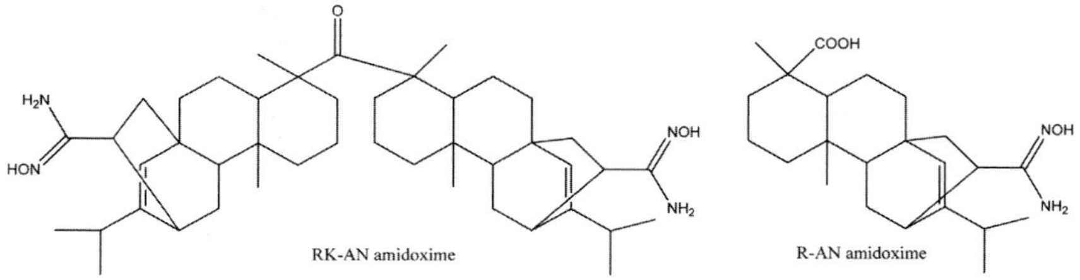

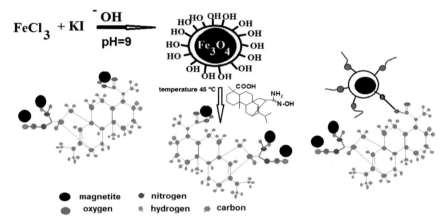

Figure 45.2 R-AN and RK-AN amidoximes.

Figure 45.3 Synthesis of magnetite-coated rosin amidoxime. (Reproduced with permission from MDPI.)

Table 45.1 Efficiency of magnetite nanoparticles at different MOR.

Magnetite	MOR	Efficiency indexes (% EI)
Uncoated magnetite	1:1	45
	1:5	20
	1:10	10
Magnetite/R-AN amidoxime	1:1	80
	1:5	75
	1:10	70
Magnetite/RK-AN amidoxime	1:1	95
	1:5	90
	1:10	80

Reproduced with permission from MDPI.

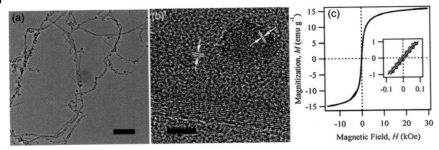

Figure 45.4 Physical properties of magnetic carbon nanotubes. (a) Transmission electron micrograph (TEM) of MCNTs. (b) TEM showing lattice fringes for carbon nanotubes and superparamagnetic iron oxide nanoparticles (SPIONs). (c) Superparamagnetic loop for MCNTs at 300 K with the inset showing a negligible coercivity. Scale bars: (a) 200 nm; (b) 5 nm. (Reproduced with permission from Elsevier Science.)

As an example of a *nano-iron oxide* composite, multiwalled carbon nanotubes decorated with superparamagnetic SPIONs (Figure 45.4) were evaluated on a laboratory scale for oil–water separation (Figure 45.5) [36]. MWCNTs were fabricated by TCVD (thermal CVD) method with $CoMg_{10}Mo_{10}O$ as a catalyst and methane as carbon source. The decoration process was carried out by the coprecipitation of ferrous and ferric iron ions on MWCNTs. The studied composite was found to eliminate oil droplets in two steps: (i) dispersion of MWCNTs at the water–oil interface and (ii) drag the droplets by a magnet along with them out of water phase. Spent MCNTs can be easily regenerated using EtOH after separation process.

For CNTs, without artificially introduced magnetic impurities, CNT-based filters were applied for the separation of heavy hydrocarbons from petroleum during crude oil postdistillation and, in addition, for the removal of microbes from drinking water [37]. Also, two types of CNTs, agglomerated CNTs and vertically aligned CNTs (VACNTs, prepared using ferrocene as the catalyst precursor), were used for oil adsorption [38]. It was revealed that oil sorption process by CNTs does not depend on the surface area of these types of carbon nanotubes. The high sorption effect of VACNTs is based on the intertube space containing large-sized macropores. In addition, CVD-fabricated carbon nanofibers and CNTs, located on the surface of expanded vermiculite ($(Mg^{+2}, Fe^{+2}, Fe^{+3})_3$ $[(Al, Si)_4O_{10}](OH)_2 \cdot 4H_2O$, this is a hydrous, silicate mineral, classified as a phyllosilicate and that expands greatly when heated) were applied for removal of oil spills in water [39]. Good results were also shown for sponges on the basis of carbon nanomaterials (CNTs and graphene). Thus, superhydrophobic CNTs sponges (Figure 45.6) were applied for cleaning oil slicks on sea waters [40]. Being compared with woolen felt and polypropylene fiber fabric, classic sorbents, the CNT sponges were shown to have a larger oil sorption capacity (92.30 g/g versus $< \sim 8$ g/g). High-surface area nanoporous spongy graphene (SG), being applied for oil elimination (Figure 45.7), showed highly efficient absorption of

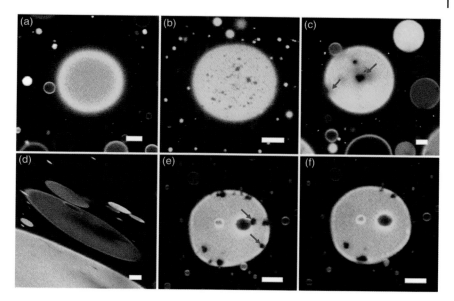

Figure 45.5 Oil–water separation in action captured by confocal laser-scanning fluorescence microscopy, presented in snapshots. (a) Oil-in-water emulsion made of diesel. (b) Pristine CNTs dispersed at the oil–water interface. (c) MCNTs dispersed at the oil–water interface and immersed inside the oil droplet. Examples are marked by red arrows. (d) MCNT-wrapped oil droplets under the influence of an external magnetic field placed at the lower right corner. The snapshot reveals the change of droplet shape in the direction of their movement toward the magnetic source. (e) SPIONs dispersed at the oil–water interface. (f) SPIONs removed from the oil–water interface by an external magnetic field. Examples of removed SPIONs are marked by red arrows in (e). Shapes and positions of oil droplets are marginally influenced by the external magnetic field. Scale bars: 5 mm. (Reproduced with permission from Elsevier Science.)

petroleum products, fats, and toxic solvents such as toluene and chloroform (up to 86 times of its own weight) without any further pretreatment [41]. SG was shown to be regenerated (Figure 45.8) up to 10 times by heating (99% of absorbed phase can be released). In general, carbon-based nanomaterials, in that number functionalized, are most known and investigated in nanotechnology, can eliminate a series of various types of other pollutants from water, having both hydrophilic and hydrophobic nature. MWCNTs and SWCNTs, in additional to crude oil above, can eliminate also 1,2-DCB, phenanthrene, pyrene, benzene, toluene, ethylbenzene, p-xylene, and naphthalene, some of which are components of crude oil. Other carbon nanomaterials can eliminate these and other products, for instance, β-cyclodextrin-grafted MWCNTs (they eliminate PCBs), fullerene C_{60}, OH-functionalized MWCNTs (eliminate phenanthrene, lindane, chlorobenzene), and so on.

Sponges for oil recovery purposes are known not only for carbon allotropes but also for other compounds, for instance, titanium dioxide UV-responsive

Figure 45.6 Morphology of the CNT sponges. (a) Photograph of a macroscopic CNT sponge with hollow cylindrical structure. (b and c) SEM images of different magnifications. (d) TEM image. (Reproduced from Springer with permission.)

nanosponge. Thus, a reusable nanosponge, composed of hydrophobic hydrocarbon and hydrophilic TiO_2 nanoparticles, was prepared and applied for oil absorption or desorption, being responsive to UV irradiation [42]. The main effect is as follows: the hydrocarbon in the nanosponge selectively absorbs oil from water, meanwhile this absorbed oil can be released back (up to 98%) into the water by TiO_2 in response to UV irradiation. We note that this compound, whose use in nanotechnology is overall indeed very high, has been studied in a series of experimental works dedicated to oil recovery and water purification [43]. Thus, a superoleophobic mesh with self-cleaning underwater properties was fabricated by the layer-by-layer assembly of sodium silicate and TiO_2 nanoparticles on the stainless steel mesh [44]. Applying UV illumination, oil can be separated, allowing facile recovery of the separation ability of the contaminated mesh. In a related research, a wastewater from petroleum refinery, containing a series of aromatic and aliphatic organic compounds, was treated using conic-shape, circulating, and upward mixing reactor with nano-TiO_2 particles, as the photocatalyst in UV/TiO_2 process, at 45°C [45]. Maximum reduction in chemical oxygen demand of >78% was observed after ~120 min and 72% after 90 min. TiO_2-based membranes are also known; for an effective oil–water separation membrane, a selective wettability and excellent chemical and thermal stability is required. Thus, as an example of such a membrane, solvothermally grown TiO_2

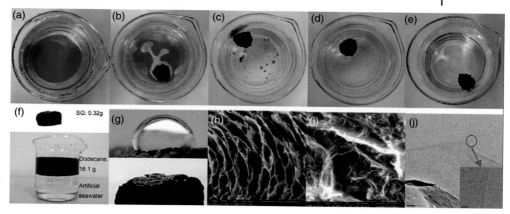

Figure 45.7 Oil absorption and characterization of SG. (a–e) Absorption of dodecane in spongy graphene (SG) at intervals of 20 s. Dodecane (stained with Sudan red 5B) floating on artificial seawater was completely absorbed within 80 s. (f) Efficiency of oil absorption. SG can be molded into any shape. An SG bulk with the shape of a triangular prism was obtained in this case with a mass of 0.32 g. This block absorbed 16.1 g dodecane floating on water, corresponding to a weight gain of 50.3. (g) Contact angle of SG surface (upper panel) and fast absorption of dodecane (lower panel). The contact angle to water was 114°, but dodecane was quickly absorbed without any residue remaining on the surface. (h) SEM image of the microporous (fusiform) structure of SG. The scale bar represents 1 mm. SG has a typical specific surface area of 432 m^2/g. (i) SEM image of the graphene skeleton. Scale bar: 1 μm. (j) Transmission electron microscopy image of the graphene skeleton. The scale bar is 50 nm; the scale bar of the inset is 5 nm. (Reproduced with permission from Wiley.)

nanosheet-anchored carbon nanofibers (hierarchical TiO$_2$ with nano/microstructure was grown on CNF surface) can separate very efficiently (>99%) water and oil by gravity [46]. The membrane is stable in thermal, extreme pH conditions and ultrasound; additionally, 400–700 L/m^2 flux can be reached. Overall, ceramic membranes have been confirmed to be a good media for separating oil–water mixtures, acting as a barrier to oil in the aqueous stream [47]. However, the separation process could be limited due to the decay of permeate flux during operation because of concentration polarization and fouling. In case of hierarchical nanoporous membranes, fabricated by the phase inversion process, they were capable of separating water-in-oil emulsions well below 1 mm in size [48]. It was demonstrated that the thickness of the nanoporous layer determines the flow resistance.

Indeed, the data on TiO$_2$ uses for environmental purposes are contradictory. It is known that crystalline TiO$_2$ (the anatase phase) has been extensively reported as an excellent material for oil spill remediation, possessing such attractive characteristics as the activity under ambient temperatures, absorption in the near ultraviolet range, resistance to deactivation, high photocatalytic efficiency, stability in acidic and basic aqueous media, and showing, in its bulk form, an absence of biological toxicity [49–56]. However, despite these impressive material and photocatalytic properties, at this moment the titania has *not* been selected as

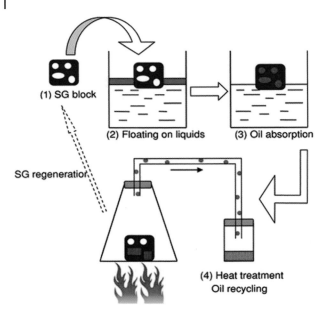

(1) SG block

(2) Floating on liquids

(3) Oil absorption

SG regeneration

(4) Heat treatment
Oil recycling

Figure 45.8 Four-step schematic diagram of spongy graphene (SG) recycling process. SG can be regenerated and reused without affecting its performance when heated up to the temperature around the boiling point of absorbate. The liquid could be evaporated, condensed, and recollected elsewhere. Simply after heat treatment, the SG material was ready to be used in the next cycle of absorption without further process. (Reproduced with permission from Wiley.

one of commercial products for oil spill remediation because of a row of complications [57–59]. The main disadvantages are as follows. The powder titania has a strong tendency to cluster; the formed agglomerations possess a reduced catalytic activity. Another problem has a relation with final recovery and separation of TiO_2 suspensions containing nanoparticles and/or microparticles after being applied for oil spill remediation. Unrecovered nanoparticles of titania can enter animal or human cells, exhibiting toxic effects. For instance, TiO_2 micro- and nanoparticles were found to inhibit 3H-thymidine incorporation by human monocyte–macrophage cells. Despite this problem, for future applications, it is predicted that titania-containing nanomaterials, particularly those containing TiO_2 incorporated within environmentally nonhazardous bulk supports, can be commercially used for oil spill remediation purposes.

Other inorganic nanomaterials are considerably less represented in the techniques for oil recovery. Among them, ZnO nanorod arrays (Figure 45.9), which are dense, vertically aligned, with a large area, superoleophilic and superhydrophobic properties (even without surface modification), were CVD-prepared on the stainless steel [60]. Separation efficiencies were shown to be >97% in the water–oil filtration process. Also, the best superhydrophobic coated mesh was found when the stainless steel mesh pore size was ∼75 μm. Nonphotosensitive

Figure 45.9 SEM images of the as-grown ZnO nanorod arrays on the stainless steel mesh. (a) Large-area view of the coated mesh. (b) top images of the ZnO nanorod arrays on a stainless steel wire. (c) high-magnification ZnO nanorod arrays on a stainless steel wire. (d) SEM side views of the ZnO nanorod arrays with height about 4 μm. (Reproduced with permission from Springer.

nano-magnesium oxide MgO was found to purify water from oil pollutants [61], showing that its degradation effect on the oil pollutants is related with pH, dosage of nano-MgO, and water temperature. Maximum registered removal rate of oil was determined as 93.92%. Also, porous boron nitride nanosheets with very high specific surface area showed good sorption for a variety of oils (pump oil, engine oil), solvents (ethanol, toluene, ethylene glycol), and dyes [62]. This BN material is resistant to oxidation, absorbs up to 33 times its own weight in oils and organic solvents, and can be easily regenerated by heating or burning in air.

45.3
Nanosized Natural and Synthetic Polymers

Coating nanoparticles, sponges, and other nanostructured materials with corresponding polymers could lead to an increase in their hydrophobicity, and, as a consequence, a possibility to improve oil recovery from water. Applications of such polymer-coated nanoparticles (PCNPs) for enhanced oil recovery were recently reviewed [63]. It was noted that, in particular, among other challenges for future research in this area, it will be necessary to study the thermodynamics of polymer rearrangement on the PCPNPs surface and effect of different types of polymer coatings on wettability alteration. Similar to the inorganic sponges above, superhydrophobic and superoleophilic sponges were fabricated by the method based on polymerized octadecylsiloxane (PODS) coatings by immersion in an ethanol solution of octadecyltrichlorosilane [64]. Resulting material is able to have absorption capacities of 42–68 times for toluene, light petroleum, and methylsilicone oil for 50 cycles (Figure 45.10). Hydrophobic foams, relatively close to sponges, also were obtained on the polymer basis. Thus, a water-repellent, water-floating, and oil-absorbing magnetic nanocomposite on the basis of polyurethane foams (commercially available) functionalized with submicrometer polytetrafluoroethylene particles and colloidal SPIONs was found to separate effectively oil from water [65]. Such combined functionalization of the foam surfaces considerably increases the speed of oil absorption. It was found that the nanoparticle capping molecules play a major role in this mechanism, that is, the samples, obtained by deposition of nanoparticle dispersion first and then with polymer particles, are the most effective in "oil from water" separation process. These systems can be used in the selective removal of large oil spills from water by means of a magnetic field.

Natural polymers, even being hydrophilic, such as cellulose, can be applied for petroleum removal after corresponding hydrophobization of their internal surface. Thus, vapor deposition of hydrophobic silanes on ultraporous cellulose (UPC) led to the fabrication of nanomaterials sponge-like aerogels [66,67]. The precursor matrix, UPC, was first prepared from dispersion of cellulose nanomaterial in water by freeze-drying. The absorption capacity of this material was found to be up to 45 times its own weight in oil; in addition, the aerogel can be

(a) PODS-modified sponge

(a) unmodified sponge

Figure 45.10 Toluene/water collection process with the PODS-modified sponge (a) and unmodified sponge (b). A few drops of ink were added into the mixture for clear observation. Reproduced with permission from American Chemical Society.

reused after draining oil off. Similarly, the potential of chemically modified starch nanocrystals for the adsorption of 2-naphthol, trichlorobenzene, xylene, quinoline, nitrobenzene, dichlorobenzene, and chlorobenzene was reported [68]. In addition, sugarcane bagasse (waste material from sugar mill; a combination of cellulose, lignin, and other minor components) was shown to be favorable for elimination of petroleum by-products from water at different pH, adsorbent dose, and initial concentration [69].

45.4
Nanomaterials-Based Membranes

The selectivity for a wide variety of oils and/or organic solvents is a very important requisite for nanomaterials, applied for fabrication of filters, membranes, or adsorbents. Polypropylene, raw cotton, fibers on silicon-coated glass basis, and other related conventional materials are used in classic adsorbents and absorb both organic solvents and water. Meanwhile, the nanomaterials, such as nanowire membranes, hydrophobic core–shell magnetic $Fe_2O_3@C$ nanoparticles, Recam@CNTs sponges, and so on, adsorb oil selectively from its mixtures with water due to a unique combination of superoleophilic and superhydrophobic properties. In case of Recam@CNTs aerogels (see below) and sponges, Gigasorb, and so on, they are highly flexible, robust, and capable of taking part in a high number of compressive cycles (oil absorption and desorption), maintaining their properties. Such peculiarities are of a high value and possess a series of benefits for regeneration processes in oil remediation steps.

45.5
Aerogels

Aerogels have since long become an important research area for water purification technologies [70], showing an importance of this material for environmental cleanup. In particular, the aerogels containing the group $CF_3(CH)_2$- (CF_3-aerogel) were prepared from $(CH_3O)_4Si$ and 30, 10, 1.5 mol% $CF_3(CH_2)_2Si(OCH_3)_3$ in methanol by a $NH_3 \cdot H_2O$-catalyzed reaction with further supercritical extraction of CH_3OH [71]. These aerogels are highly hydrophobic and, consequently, are capable of absorbing hydrocarbons very efficiently. When oil and salt–water mixtures are treated with such an aerogel at an oil:aerogel ratio of 3.5, it was established that all CF_3–aerogels cleanly separated the water from oil, regardless of $CF_3(CH)_2$- group concentration. These CF_3-aerogels can absorb oil as much as 237 times their weight, much more than previously determined for high absorptive materials. More recent achievements in this area are also exciting. For instance, silica aerogel granules, obtained acid–base-catalyzed sol–gel process for two steps, including drying of alcogels at normal pressure, were investigated for desorption and absorption studies of different organic liquids [72]: four aromatic compounds, four alkanes, three oils, and four alcohols. The granules of this hydrophobic aerogel revealed a very high uptake capacity and rate. The desorption processes using oils and solvents were carried out by maintaining aerogel granules with absorbed organic compounds at distinct temperatures and determining their weight in regular intervals of time until complete desorption of liquid organic phase. It was also established that the structure of this aerogel was not seriously affected during the process of solvent adsorption. At the same time, the absorption of oil led to the shrinkage; a dense structure was formed after desorption.

45.6
Toxicity, Cost, and Selection of Nanomaterials for Water Cleanup from Oil

The treatment of oil spills using nanomaterials has shown a host of advantages and successful attempts. At the same time, however, in case of any nanomaterial for any wide application, definite risks exist, which are related with their uses. Due to small particle size and high reactivity, the nanomaterials frequently affect human health through inhalation, ingestion, or absorption through skin. From smaller to larger organisms in aqueous systems, they can migrate through food chains, reaching human organisms and being able to damage microbial societies in the environment. Nanoparticles' and nanomaterials' toxicity is well known [73–76]. It was suggested that, in general, small-size nanoparticles possess more toxic properties in comparison to larger agglomerates, despite differences in particle types. Textbooks on the nanotechnology generally contain classic information that the toxicity data cannot be extrapolated from bulk materials to nanosized particles. The sand in the beach is nonhazard, while nanosized SiO_2 is toxic. Activated carbon is nontoxic and can be used for gas absorption in human digestive tract, while pristine carbon nanotubes are toxic (however, functionalizing them, their toxicity can change). So, definite precautions are needed prior to using nanomaterials for environmental purposes, in particular for water cleanup.

A large variety of absorbents, nanocomposites, nanomaterials, and oxidants for petroleum are known to date. However, it is still difficult to choose the most appropriate material and method for purification of water from oil impurities. Several factors, sometimes contradictory, can influence the final choice: the cost of production of a nanomaterial, availability or simplicity of its fabrication, possible decrease of nanomaterial activity with time or change of natural conditions, stability in water and air, toxicity of final products of its application, storage conditions, and type of application. As an example, the preparation of carbon nanotubes is still expensive, although a possibility to form magnetic materials with iron, its oxides, ferrites, and so on and high absorption capacity presents a great advantage. Strong magnets can be applied for final oil recovery and possible reuse. CNT sponges are superhydrophobic and cannot adsorb any quantity of water during oil spill remediation. It is known that nZVI is readily used for oil contaminated soils. In case of its application, the following details should be taken into account: nZVI reactivity (its nanoparticles oxidize fairly fast being contacted with air), cost, possible agglomeration, further loss of activity, and easy adherence to solid surfaces limiting its applications. On the contrary, low toxicity and final dissolution of Fe nanoparticles are in favor of nZVI uses. In case of aerogels, their hydrophobic granules revealed a very strong uptake capacity. In case of TiO_2, as it was discussed above, despite impressive photocatalytic and material properties, titania has not been selected as a commercial oil spill remediation product due to a series of complications.

All these real and possible problems and misgivings, although with advantages of the proposed methods, lead to the necessity to increase the number of available techniques and nanomaterials for water purification and oil remediation. Application conditions for nanomaterials (scale of problem, temperature, humidity, lake or sea environments (type and salinity of water), distance from cities or villages, and geological conditions of oil wells, among many others) influence this selection, so more opportunities should be available, especially in urgent situations. In addition, taking into account nanotoxicology problems, the nZVI or its composites seem attractive; in some cases, the treatment with magnetic nanoparticles or use of absorbents could be more attractive. In any case, concrete conditions and scale influence on possible application of nanomaterials. Their cost is also important. Thus, according to our own calculations for distinct chemicals for nanomaterials preparation, in order to separate successfully 1 L of crude oil from water, a wide range from US\$11 to 400 was observed, mainly based on the cost of precursors for nanomaterials, lesser on the synthesis difficulties. Aerogels and xerogels, showing high absorption capacity for oil, showed high cost at the same time; however, making their specific production for the only purpose – oil remediation – it will be possible to decrease considerably their price. A similar opinion can be expressed for lesser cost CNTs: a considerable decrease of their cost is annually observed. Therefore, one can observe a commercial availability for certain nanotechnological methods for water cleanup from oil pollutants. Other important factors are rate and capacity of sorption of crude oil, recyclability, regeneration factor, disposal costs, use of natural materials (jute, cotton, wood), and even those considered sometimes as wastes (scoria) as absorbents.

45.7
Conclusions and Further Outlook

The use of nanotechnology in water treatment in oil-containing wastewaters currently includes the use of the nanoparticles (noble metals (Au, Ag; their use is considered as something exotic), other zero-valent metals (Fe), ceramic and metal oxides (alumina, titania, ZnO), magnetic nanoparticles and nanocomposites (Fe_2O_3, Fe_3O_4, core-shell $Fe_xO_y@C$, and related carbon nanocomposites, among others), magnetic polymer nanocomposites, magnetic liquid foams

(EcoMag® and CleanMag®), carbon nanotubes and sponges, exfoliated graphite, graphene nanosheets and nanoworms, RECAM® technology on the carbon basis, dendrimers (poly(amidoamine) PAAM), nanocomposites (CeO_2–CNTs, graphene–polypyrrole, etc.), organoclays with magnetic Fe_xO_y nanoparticles, nanofilters, nanosensors, nanowire membranes, and nanodispersants [77]. Overall, the methods described above are applied in a laboratory and pilot plant scale, although in comparison with classic methods for oil recovery (see Section 45.1), they seem to be much more effective due to special processes, which are less likely for larger nonnanoscale particles. This special performance of nanomaterials and nanocomposites is caused by their high surface area and reactivity. In addition, an *in situ* treatment of oil phases is possible. Applying magnetic nanomaterials, absorbed oil can be magnetically eliminated from water–oil mixtures. Less known hazard properties of common nanomaterials can be an obstacle for scaling up their use for water treatment from petroleum, despite their high potential. In our opinion, as future research in this area, we consider the development of triple systems such as "nanosorbent/nZVI/strong organic oxidant" (for instance, inorganic or organic peroxides), "oxidant/aerogel/magnetic nanoparticles," or "natural low-cost porous hydrophobic materials/oxidants"; they are promising and are currently being developed in our laboratories.

References

1 Mansoori, G.A., Rohani Bastami, T., Ahmadpour, A., and Eshaghi (2008) Environmental application of nanotechnology. *Annu. Rev. Nano Res.*, **2**, 73.

2 Duraisamy, R.T., Heydari Beni, A., and Henni, R. (2013) State of the art treatment of produced water. *Water Treatment*, INTECH, pp. 199–222.

3 Pendergast, M.T. and Hoek, E. (2011) A review of water treatment membrane nanotechnologies. *Energy Environ. Sci.*, **4**, 1946–1971.

4 Pandey, B. and Fulekar, M.H. (2012) Nanotechnology: remediation technologies to clean up the environmental pollutants. *Res. J. Chem. Sci.*, **2** (2), 90–96.

5 Qu, X., Alvarez, P.J.J., and Li, Q. (2013) Applications of nanotechnology in water and wastewater treatment. *Water Res.*, **47**, 3931–3946.

6 Pradeep, T. and Anshup (2009) Noble metal nanoparticles for water purification: a critical review. *Thin Solid Films*, **517**, 6441–6478.

7 Ayanda, O.S. (2014) Occurrence, fate and treatment methods of polycyclic aromatic hydrocarbons, polychlorinated biphenyls, dioxins and furans: a mini review. *Res. Rev. J. Mater. Sci.*, **2** (4), 14–21.

8 Somasundaran, S., Patra, P., Farinato, R.S., and Papadopoulos, K. (eds) (2014) *Oil Spill Remediation: Colloid Chemistry-Based Principles and Solutions*, John Wiley & Sons, Inc., New York, 374 pp.

9 Rajan, C.S. (2011) Nanotechnology in groundwater remediation. *Int. J. Environ. Sci. Dev.*, **2** (3), pp. 182–187.

10 Matlochova, A., Placha, D., and Rapantova, N. (2013) The application of nanoscale materials in groudwater remediation. *Pol. J. Environ. Stud.*, **22** (5), 1401–1410.

11 Dallas, P., Kelarakis, A., and Giannelis, E.P. (2013) Nanostructured materials for environmentally conscious applications. *Sustainable Nanotechnology and the Environment: Advances and Achievements*, vol. 1124, ACS Symposium Series, American Chemical Society, pp. 59–72.

12 Bhattacharya, S., Saha, I., Mukhopadhyay, A., Chattopadhyay, D., Chand Ghosh, U.,

and Chatterjee, D. (2013) Role of nanotechnology in water treatment and purification: potential applications and implications. *Int. J. Chem. Sci. Technol.*, **3** (3), 59–64.

13 EPA (2008) Nanotechnology for Site Remediation: Fact Sheet, United States Environmental Protection Agency, 542-F-08-009, 17 pp.

14 Agarwal, A. and Joshi, H. (2010) Application of nanotechnology in the remediation of contaminated groundwater: a short review. *Recent Res. Sci. Technol.*, **2** (6), 51–57.

15 Savage, N. and Diallo, M.S. (2005) Nanomaterials and water purification: opportunities and challenges. *J. Nanopart. Res.*, 7, 331–342.

16 Jain, K.K. (2012) Nanotechnology and water. *Contemp. Mater.*, **III** (1), 26–30.

17 Prabhakar, V. and Bibi, T. (2013) Nanotechnology, future tools for water remediation. *Int. J. Emerg. Technol. Adv. Eng.*, **3** (7), 54–59.

18 Tratnyek, P.G. and Johnson, R.L. (2006) Nanotechnologies for environmental cleanup. *Nano Today*, **1** (2), 44–48.

19 Yunus, I.S., Harwin, Kurniawan, A., Adityawarman, D., and Indarto, A. (2012) Nanotechnologies in water and air pollution treatment. *Environ. Technol. Rev.*, **1** (1), 136–148.

20 Street, A., Sustich, R., Duncan, J., and Savage, N. (eds) (2008) Nanotechnology applications for clean water. *Solutions for Improving Water Quality*, Elsevier Science, 700 pp.

21 Dadrasnia, A., Salmah, I., Emenike, C.U., and Shahsavari, N. (2015) Remediation of oil contaminated media using organic material supplementation. *Pet. Sci. Technol.*, **33**, 1030–1037.

22 Taghizadeh, M., Yousefi Kebria, D., Darvishi, G., and Golbabaei Kootenaei, F. (2013) The use of nano zero valent iron in remediation of contaminated soil and groundwater. *Int. J. Sci. Res. Environ. Sci.*, **1** (7), 152–157.

23 Watlington, K. (2005) Emerging Nanotechnologies for Site Remediation and Wastewater Treatment. Report for the U.S. Environmental Protection Agency Office of Solid Waste and Emergency

Response Office of Superfund Remediation and Technology Innovation Technology Innovation and Field Services Division Washington, DC. Available at www.epa.gov, www.clu-in.org.

24 Roozbeh Jameia, M., Reza Khosravib, M., and Anvaripour, B. (2013) Degradation of oil from soil using nano zero valent iron. *Sci. Int. (Lahore)*, **25** (4), 863–867.

25 Ershadi, L., Ebadi, T., Ershadi, V., and Rabbani, A.R. (2011) Chemical oxidation of crude oil in oil contaminated soil by Fenton process using nano zero valent iron, in *2nd International Conference on Environmental Science and Technology IPCBEE*, vol. 6, IACSIT Press, Singapore.

26 Dehghani, M., Shahsavani, E., Farzadkia, M., and Reza Samaei, M. (2014) Optimizing photo-Fenton like process for the removal of diesel fuel from the aqueous phase. *J. Environ. Health Sci. Eng.*, **12** (87), 7.

27 Chu, Y. and Pan, Q. (2012) Three-dimensionally macroporous Fe/C nanocomposites as highly selective oil-absorption materials. *ACS Appl. Mater. Interfaces*, **4** (5), 2420–2425.

28 Al-Shamsi, M.A. and Thomson, N.R. (2013) Treatment of a trichloroethylene source zone using persulfate activated by an emplaced Nano-Pd–Fe^0 zone. *Water Air Soil Pollut.*, **244** (1780), 12.

29 hn, S., Peterson, T.D., Righter, J., Miles, D.M., and Tratnyek, P.G. (2013) Disinfection of ballast water with iron activated persulfate. *Environ. Sci. Technol.*, **47**, 11717–11725.

30 Liang, C.. and Lai, M.-C. (2008) Trichloroethylene degradation by zero valent iron activated persulfate oxidation. *Environ. Eng. Sci.*, **25** (7), 1071–1078.

31 Al-Shamsi, M.A.. and Thomson, N.R. (2013) Treatment of organic compounds by activated persulfate using nanoscale zerovalent iron. *Ind. Eng. Chem. Res.*, **52**, 13564–13571.

32 Karn, B., Kuiken, T., and Otto, M. (2009) Nanotechnology and *in situ* remediation: a review of the benefits and potential risks. *Environ. Health Persp.*, **117** (12), 1823–1831.

33 Mueller, N.C.. and Nowack, B. (2010) Nanoparticles for remediation: solving big

problems with little particles. *Elements*, **6**, 395–400.

34 Jorfi, S., Rezaee, A., Moheb-Ali, G.-A., and Jaafarzadeh, N.A. (2013) Pyrene removal from contaminated soils by modified Fenton oxidation using iron nano particles. *J. Environ. Health Sci. Eng.*, **11** (17), 8.

35 Atta, A.M., Al-Lohedan, H.A., and Al-Hussain, S.A. (2015) Functionalization of magnetite nanoparticles as oil spill collector. *Int. J. Mol. Sci.*, **16**, 6911–6931.

36 Wang, H., Lin, K.-Y., Jing, B., Krylova, G., Sigmon, G.E., McGinn, P., Zhu, Y., and Na, G. (2013) Removal of oil droplets from contaminated water using magnetic carbon nanotubes. *Water Res.*, **47**, 4198–4205.

37 Srivastava, A., Srivastava, O.N., Talapatra, S., Vajtai, R., and Ajayan, P.M. (2004) Carbon nanotube filters. *Nat. Mater.*, **3**, 610–614.

38 Fan, Z.J., Yan, J., Ning, G.O., Wei, T., Qian, W.Z., Zhang, S.J., Zheng, C., Zhang, Q., and Wei, F. (2010) Oil sorption and recovery by using vertically aligned carbon nanotubes. *Carbon*, **48**, 4197–4200.

39 Moura, F.C.C. and Lago, R.M. (2009) Catalytic growth of carbon nanotubes and nanofibers on vermiculite to produce floatable hydrophobic "nanosponges" for oil spill remediation. *Appl. Catal. B Environ.*, **90**, 436–440.

40 Zhu, K., Shang, Y.-Y., Sun, P.-Z., Li, Z., Li, X.-M., Wei, J.-Q., Wang, K.-L., Wu, D.-H., Cao, A.-Y., and Zhu, H.-W. (2013) Oil spill cleanup from sea water by carbon nanotube sponges. *Front. Mater. Sci.*, **7** (2), 170–176.

41 Bi, H., Xie, X., Yin, K., Zhou, Y., Wan, S., He, L., Xu, F., Banhart, F., Sun, L., and Ruoff, R.S. (2012) Spongy graphene as a highly efficient and recyclable sorbent for oils and organic solvents. *Adv. Funct. Mater.*, **22** (21), 4421–4425.

42 Kim, D.H., Jung, M.C., Cho, S.-H., Kim, S.H., Kim, H.-Y., Lee, H.J., Oh, K.H., and Moon, M.-W. (2015) UV-responsive nanosponge for oil absorption and desorption. *Sci. Rep.*, **5** (12908), 12.

43 Cheraghian, G., Hemmati, M., Masihi, M., and Bazgir, S. (2013) An experimental investigation of the enhanced oil recovery and improved performance of drilling fluids using titanium dioxide and fumed silica nanoparticles. *J. Nanostruct. Chem.*, **3** (78), 9.

44 Zhang, L., Zhong, Y., Cha, D., and Wang, P. (2013) A self-cleaning underwater superoleophobic mesh for oil–water separation. *Sci. Rep.*, **3** (2326), 5.

45 Saien, J. and Shahrezaei, F. (2012) Organic pollutants removal from petroleum refinery wastewater with nanotitania photocatalyst and UV light emission. *Int. J. Photoenergy*, **2012**, 703074.

46 Tai, M.H., Gao, P., Tan, B.Y.L., Sun, D.D., and Leckie, J.O. (2015) A hierarchically-nano structured TiO_2–carbon nanofibrous membrane for concurrent gravity-driven oil–water separation. *Int. J. Environ. Sci. Dev.*, **6** (8), 590–595.

47 Vieira, T.M., de Souza, J.S., Barbosa, E.S., de Lima Cunha, A., de Farias Neto, S.R., and Barbosa de Lima, A.G. (2012) Numerical study of oil/water separation by ceramic membranes in the presence of turbulent flow. *Adv. Chem. Eng. Sci.*, **2**, 257–265.

48 Brian, R., Solomon, Md., Nasim, H., and Kripa, K.V. (2014) Separating oil–water nanoemulsions using flux-enhanced hierarchical membranes. *Sci. Rep.*, **4** (5504), 6.

49 Narayan, R. (2010) Titania: a material-based approach to oil spill remediation? *Mater. Today*, **13** (9), 58–59.

50 Liu, S., Sun, X., Li, J.G., Li, X., Xiu, Z., Yang, H., and Xue, X. (2010) Fluorine- and iron-modified hierarchical anatase microsphere photocatalyst for water cleaning: facile wet chemical synthesis and wavelength-sensitive photocatalytic reactivity. *Langmuir*, **26**, 4546–4653.

51 Beydoun, D., Amal, R., Low, G., and McEvoy, S. (1999) Role of nanoparticles in photocatalysis. *J. Nanopart. Res.*, **1**, 439–458.

52 Salem, I. (2003) Recent studies on the catalytic activity of titanium, zirconium, and hafnium oxides. *Catal. Rev.*, **45**, 205–296.

53 Hoffmann, M.R., Martin, S.T., Choi, W.Y., and Bahnemann, D.W. (1995) Environmental applications of

semiconductor photocatalysis. *Chem. Rev.,* **95**, 69–96.

54 Anpo, M. and Takeuchi, M. (2003) The design and development of highly reactive titanium oxide photocatalysts operating under visible light irradiation. *J. Catal.,* **216**, 505–516.

55 Na, P., Zhao, B., Gu, L., Liu, J., and Na, J. (2009) Deep desulfurization of model gasoline over photoirradiated titanium-pillared montmorillonite. *J. Phys. Chem. Solids,* **70**, 1465–1470.

56 Neatu, S., Sacaliuc-Parvulescu, E., Levy, F., and Parvulescu, V.I. (2009) Photocatalytic decomposition of acetone over dc-magnetron sputtering supported vanadia/TiO₂ catalysts. *Catal. Today,* **142**, 165–169.

57 O'Regan, B., Moser, J., Anderson, M., and Graetzel, M. (1990) Vectorial electron injection into transparent semiconductor membranes and electric field effects on the dynamics of light-induced charge separation. *J. Phys. Chem.,* **94**, 8720–8726.

58 Xuzhuang, Y., Yang, D., Huaiyong, Z., Jiangwen, L., Martins, W.N., Frost, R., Daniel, L., and Yuenian, S. (2009) Mesoporous structure with size controllable anatase attached on silicate layers for efficient photocatalysis. *J. Phys. Chem. C,* **113**, 8243–8248.

59 Shanbhag, A.S., Jacobs, J.J., Black, J., Galante, J.O., and Glant, T.T. (1994) Macrophage/particle interactions: effect of size, composition and surface area. *J. Biomed. Mater. Res.,* **28**, 81–90.

60 Li, H., Li, Y., and Liu., Q. (2013) ZnO nanorod array-coated mesh film for the separation of water and oil. *Nanoscale Res. Lett.,* **8** (183), 6.

61 Zhu, M.-F., Deng, C., Su, H.-B., You, X.-D., Zhu, L., Chen, P., and Yuan, Y.-H. (2013) Oil pollutants degradation of nano-MgO in micro-polluted water. *Comput. Water, Energy Environ. Eng.,* **2013** (2), 12–15.

62 Lei, W., Portehault, D., Liu, D., Qin, S., and Chen, Y. (2013) Porous boron nitride nanosheets for effective water cleaning. *Nat. Commun.,* **4**, 1777.

63 ShamsiJazeyi, H., Miller, C.A., Wong, M.S., Tour, J.M., and Verduzco, R. (2014) Polymer-coated nanoparticles for enhanced oil recovery. *J. Appl. Polym. Sci.,* **131**, 40576.

64 Ke, Q., Jin, Y., Jiang, P., and Yu., J. (2014) Oil/water separation performances of superhydrophobic and superoleophilic sponges. *Langmuir,* **30**, 13137–13142.

65 Calcagnile, P., Fragouli, D., Bayer, I.S., Anyfantis, G.C., Martiradonna, L., Cozzoli, P.D., Cingolani, R., and Athanassiou, A. (2012) Magnetically driven floating foams for the removal of oil contaminants from water. *ACS Nano,* **6** (6), 5413–5419.

66 Cervin, N., Aulin, C., Larsson, P., and Wagberg, L. (2012) Ultra porous nanocellulose aerogels as separation medium for mixtures of oil/water liquids. *Cellulose,* **19** (2), 401–410.

67 Korhonen, J.T., Kettunen, M., Ras, R.H., and Ikkala, O. (2011) Hydrophobic nanocellulose aerogels as floating, sustainable, reusable, and recyclable oil absorbents. *ACS Appl. Mater. Interfaces,* **3** (6), 1813.

68 Alil, S., Aloulou, F., Thielemans, W., and Boufi, S. (2011) Sorption potential of modified nanocrystals for the removal of aromatic organic pollutant from aqueous solution. *Ind. Crops Prod.,* **33**, 350–357.

69 Sarkheil, H. and Tavakoli., J. (2015) Oil-polluted water treatment using nano size bagasse optimized-isotherm study. *Eur. Online J. Nat. Soc. Sci.,* **4** (2), 392–400.

70 Reynolds, J.G., Coronado, P.R., and Hrubesh, L.W. (2001) Hydrophobic aerogels for oil-spill cleanup? Intrinsic absorbing properties. *Energy Source,* **23** (9), 831–843.

71 Reynolds, J.G., Coronado, P.R., and Hrubesh, L.W. (2001) Hydrophobic aerogels for oil-spill clean up: synthesis and characterization. *J. Non Cryst. Solids,* **292** (1–3), 127–137.

72 Parale, V.G., Mahadik, D.B., Kavale, M.S., Venkateswara Rao, A., Wagh, P.B., and Gupta, S.C. (2011) Potential application of silica aerogel granules for cleanup of accidental spillage of various organic liquids. *Soft Nanosci. Lett.,* **1**, 97–104.

73 Sharifi, S., Behzadi, S., Laurent, S., Laird Forrest, M., Stroevee, P., and Mahmoudi, M. (2012) Toxicity of nanomaterials. *Chem. Soc. Rev.,* **41**, 2323–2343.

74 De Stefano, D., Carnuccio, R., and Chiara Maiuri, M. (2012) Nanomaterials toxicity and cell death modalities. *J. Drug Deliv.*, **2012**, 14.–167896.

75 Uo, M., Watari, F., Sato, Y., and Tohji, K. (2011) Toxicity evaluations of various carbon nanomaterials. *Dent. Mater. J.*, **30** (3), 245–263.

76 Khan, H.A. and Arif, I.A. (eds) (2012) *Toxic Effects of Nanomaterials*, Bentham e-Books.

77 Mahajan, Y.R. (2011) Nanotechnology-based solutions for oil spills. *Nanotechnol. Insights*, **2** (1), 1–19.

46

Nanomaterials and Direct Air Capture of CO_2

Dirk Fransaer

VITO, Flemish Institute for Technological Research, Boeretang 200, 2400 Mol, Belgium

46.1
Introduction

Global warming, air pollution, and ocean deterioration are three main climate challenges the earth is facing today. All of these key challenges are expected to have a major impact on both livability and biodiversity worldwide.

The global warming challenge is essentially linked to the CO_2 emissions of our future energy and production system. Emissions ranges for baseline scenarios and mitigation scenarios limiting CO_2 equivalent concentrations to levels of about 450 ppm CO_2-eq are likely to limit warming to 2 °C above preindustrial levels. These mitigation scenarios require substantial emissions reductions over the next few decades and near-zero emissions of greenhouse gases by 2100 (see Figure 46.1) [1].

"Theoretical" key measures to achieve such mitigation goals include decarbonizing electricity generation as well as efficiency enhancements and behavioral changes, to reduce energy demand without compromising development. The global energy demand is therefore expected to show only a slight increase due to an increased awareness on energy-efficient behavior largely counteracted by enhanced living standards in developing countries.

In scenarios reaching 450 ppm CO_2-eq concentrations by 2100, global CO_2 emissions from the energy supply sector are projected to decline over the next decade and are characterized by reductions of 90% or more below 2010 levels between 2040 and 2070.

IEA [2], however, predict a fossil fuel use that will continue to play a predominant role in the next decades. Especially for the transport sector, this is expected to be the case. As a result, the share of decarbonization of the energy demand will only happen gradually and probably not to a level of 80 or 90% in the next decades.

Nanotechnology for Energy Sustainability, First Edition. Edited by Baldev Raj, Marcel Van de Voorde, and Yashwant Mahajan.
© 2017 Wiley-VCH Verlag GmbH & Co. KGaA. Published 2017 by Wiley-VCH Verlag GmbH & Co. KGaA.

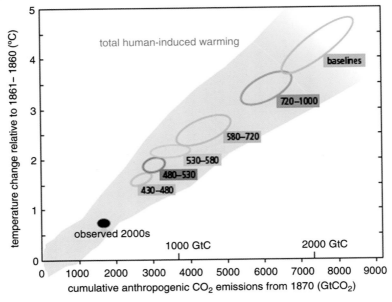

Figure 46.1 Warming versus cumulative CO_2 emissions under various scenarios. Colored plume shows the spread of past and future projections from a hierarchy of climate carbon cycle models driven by historical emissions and the four RCPs over all times out to 2100. Ellipses show total anthropogenic warming in 2100 versus cumulative CO_2 emissions from 1870 to 2100 from a climate model. The width of the ellipses in terms of temperature is caused by the impact of different scenarios for non-CO_2 climate drivers. The filled black ellipse shows observed emissions to 2005 and observed temperatures in the decade 2000–2009 with associated uncertainties [1].

The present CO_2 emissions amount to approximately 8 Gton CO_2, significantly larger than the present CO_2 market at 0.1–0.3 Gton CO_2. However, if the market can be increased 10–100-fold, a vastly different picture will ensue. Also, only half of the emitted emissions end up in the atmosphere. The other half is absorbed in vegetation, the soil, and the seas, leading to unwanted acidification. Fully closing the carbon cycle by capturing carbon dioxide from the atmosphere thus reduces theoretically the carbon dioxide concentrations.

As stated by the newly appointed chairman of IPCC, Hoesung Lee[1]: "We need more investment in carbon capture and storage technology. As Working Group III of the Fifth Assessment pointed out, it will be very difficult to reach zero carbon emissions without it." This chapter will discuss both elements: the carbon capture process and the carbon reuse aspects and the necessary interlinking between both from a "value" point of view.

---

1) http://www.energypost.eu/interview-hoesung-lee-new-chairman-ipcc-enormous-value-carbon-capture-storage/.

46.2

CO$_2$ as a Resource

Two essentially opposing mechanisms are at the moment considered to lower the carbon dioxide emissions or concentrations. One is more of "a penalty system" using a carbon price, a carbon tax, or an emission trading scheme targeted at reducing CO$_2$ emissions.

An emission trading system is used in the European Union and a number of other countries based on the carbon market that was written into the United Nations Kyoto Protocol and is now international law for the signatories to the Kyoto protocol. In September 2015, even China announced plans to launch a national carbon trading scheme in 2017.

A number of local and national governments are considering or have already implemented a carbon price. This carbon price varies from $1 in Mexico to $168/ton CO$_2$ in Sweden [3]. Even higher CO$_2$ prices are circulating and used, for example, in a forecast of the EIB [4]. In "The economic Appraisal of Investment Projects at the EIB," the principles underlying the incorporation of external cost and particularly the cost of carbon are set out. The estimates of carbon pricing over the period 2010–2030 are based on a study of the Stockholm Environment Institute (SEI), but are extended to 2050 as shown in Figure 46.2.

Here carbon dioxide costs are foreseen to even reach over €200/ton CO$_2$ by the year 2045.

The other mechanism by which carbon dioxide emissions and ultimately concentrations could be lowered is a positive model whereby carbon dioxide is used as a resource and thus susceptible to all market and industrial mechanisms.

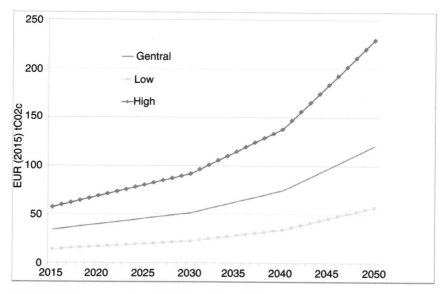

Figure 46.2 EIB cost of carbon till 2050 from EIB draft [4].

There already existed a CO$_2$ market in 2014 to the tune of 126 million tons CO$_2$ with a total market value in excess of \$10 billion. The used CO$_2$ stems mainly from natural sources or from concentrated CO$_2$ sources (with a purity in excess of 90%) such as industrial fluxes from sugar, natural gas by-product, ethylene, and ammonia production: 70 million tons. A new ammonia production plant could be scheduled for the port of Antwerp with an estimated CO$_2$ exhaust of 3 million tons CO$_2$. Here the cost of CO$_2$ lies approximately between 5 and \$30/ton CO$_2$.

This carbon dioxide is at present used in the following:

- *EOR (enhanced oil recovery):* 66 million tons CO$_2$ per year in the United States since 1973. This market is under pressure lately due to the low oil prices between \$25 and 35/barrel.
- *Fertilizers and urea production:* 36 million tons CO$_2$
- *Food and beverages:* 14–15 million ton CO$_2$
- Others such as metal welding, carbonate, and refrigerant.

Between these two opposing mechanisms lies CCS (carbon capture and storage/sequestration), whereby carbon dioxide is captured and stored or sequestered. EOR or carbonization of materials is a storage technique that is already in use because of its economic value. Otherwise, carbon dioxide is treated in a CCS project as a waste product only to be disposed of.

Geological storage can be provided in deep saline aquifer layers or in geological structures containing, for example, peridotite or basalt rock cliffs. Peridotite is a mixture of serpentine and olivine rock that effectively seals the carbon dioxide as a stable magnesium carbonate mineral. Vast geological quantities of peridotite exist as it is the dominant rock of the upper part of the Earth's mantle. Peridotite is named for the gemstone peridot, a glassy green gem mined in a.o. Arizona (Peridot Cove), while the Samail Ophiolite, in Oman, is assumed to contain some 30 000 km^3 of peridotite.

Basalt rock cliffs contain holes that are solidified gas bubbles from the basalt's formation from volcanic lava flows up to millions of years ago. Pumping carbon dioxide into these bubbles causes it to react to form stable limestone –calcium carbonate. To speed up these absorption processes, the carbon dioxide is dissolved in water and injected into the rocks under high pressures.

The dividing line between CCS and CCU (carbon capture and use–reuse) largely runs along this "positive" and "penalty" system, reflecting in part CO$_2$ as either a resource, raw material ("optimistic," CCU), or as waste material ("penalty"-system) to be dumped in geological storage sites or as part of an EOR solution. As recent studies point out, even modest targets for CCU allow amounts of CO$_2$ to be mitigated that exceed current CCS targets [5] – certainly as EOR is losing most of its importance due to falling oil prices, more political attention should be given to CCU as assisting, in an economic manner, the climate change challenge.

CCU does not only bring the environmental benefit of emission reductions or concentration reductions but also delivers CO$_2$ as a raw material in new products or energy carriers, thus creating a circular CO$_2$ economy.

46.3
Circular CO$_2$ Economy

This new circular CO$_2$ economy has to be based on renewable energy sources that directly reduce the emitted CO$_2$ but also provide, at least some of the time, electricity at near-zero or at low prices. In addition, it contributes to lowering both the already emitted CO$_2$ in the past and CO$_2$ emissions in the near future, which according to recent findings of the IPCC (2013) [6] will have a long-lasting (hundreds of years) effect on the atmosphere. The IPCC [1] acknowledges the necessity to remove these emissions in addition to mitigation measures avoiding CO$_2$ emissions.

Therefore, a global solution to the climate change problem must be found when both CCU and renewable energy sources act in unison and is fully complemented with all efforts on decarbonization of the energy supply with renewable energy, flexible energy systems, and energy efficiency.

This leads to the key solution of direct air capturing and integration of these emitted CO$_2$ emissions in the fuel and product process chains. In this context, carbon capture and use becomes an integrated business solution combining storage issues for the future energy system with innovative fuel and product processing. This idea is strongly supported in the new integrated Strategic Energy Plan (SET) of the European Commission as released in September 2015 [7].

For CCU to play this role, two main obstacles need to be overcome: the (high) cost of capturing the carbon dioxide and the additional energy often needed to convert CO$_2$ into useful materials, that is, energy. Novel technologies are being explored to tackle both issues [8], but by appropriately addressing direct air capture, the first part of the equation, that is, the high cost of capturing carbon dioxide, can be overcome.

46.4
CO$_2$ Capture or Separation Technologies

Direct air capture, apart from its physical occurrence in vegetation and trees (the REDD+ programme from the United Nations), was and is until now the least investigated, but might carry a huge potential, although the American Physical Society's most optimistic calculations estimate the cost of direct air capture at $600/ton of carbon dioxide removed.

Natural CO$_2$ capture and reuse is seen daily in the function of trees and vegetation. The efficiency of these natural systems are small, but forests also provide other advantages to nature and humankind such as biodiversity, food and fiber, novel drugs, and so on and they are already playing an important role in reducing the overall CO$_2$ concentrations as only half of the emitted carbon dioxide ends up in the air. The rest is captured in soil, water, and trees and vegetation,. This natural vegetation CO$_2$ capture process is being exploited in UN programs such as REDD+ and the "Coalition of Rainforest Nations." Tropical forests thus play a major role in current atmospheric concentration of carbon dioxide [9]. Through photosynthesis, trees use CO_2, water, and sunlight to grow, while storing carbon dioxide. Man-made materials such as MEA, sodium carbonate, and others under development have a far higher efficiency in capturing CO_2 but require economically expensive energy to absorb and release the captured CO_2.

CO_2 can be captured using different technologies and at different stages of the combustion/power generation process and also directly from air: postcombustion, precombustion, oxyfuel combustion, chemical looping combustion, cryogenic separation, and direct air capture technologies (Figure 46.3). Excellent overview articles exist on all of these CO_2 capture technologies [10].

Precombustion is mainly applied to coal gasification plants, while postcombustion and oxyfuel combustion can be applied to both coal- and gas-fired plants. Precombustion used in coal gasification plants is a fully developed technology where high CO_2 concentration enhances the sorption efficiency. Existing power plants can be retrofitting with this technology, but the temperature of the exhaust gases induces efficiency decay issues and associated high operational costs.

Postcombustion technology is currently the most mature process for CO_2 capture [11,12]. Postcombustion can be retrofitted into existing plants, but carries

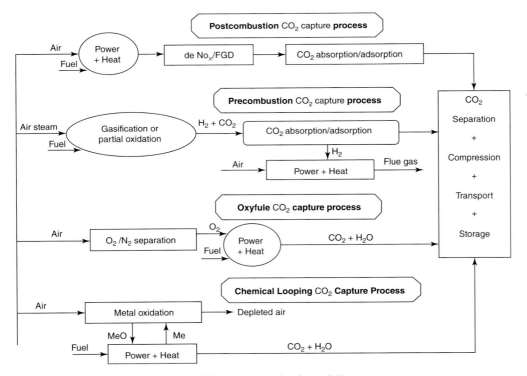

Figure 46.3 Overview of CO_2 capture technologies [10].

with it a penalty on the overall efficiency of the power plant (approximately 6% for the gas-fired power plants and 8% for coal-fired power plants [13,14]), while "low" CO_2 concentration affects the capture efficiency.

As for the two other processes, chemical looping is still under research investigation and often tied to specific patented situations and/or materials and is not yet implemented at an industrial scale but only at pilot scale [15] at the University of Darmstadt [15] and oxyfuel combustion carry, at present, a high cost price as, for example, cryogenic O_2 production is a costly process.

There exist different CO_2 separation technologies that can be applied to isolate the CO_2 from the flue/fuel gas stream prior to transportation. Advanced technologies, such as wet scrubber, dry regenerable sorbents, membranes, cryogenics, pressure and temperature swing adsorption, and other concepts have been developed. These technologies are compared in Ref. [16] and will be discussed subsequently.

Cryogenic separation captures carbon dioxide under extremely low temperatures so that it is immediately available under liquid form, thereby saving compression energy. It is viable only for very high CO_2 concentrations and needs special attention to reach its promise of energy savings. Consecutively, compressing, refrigeration, and separation should avoid large energy penalties from the refrigeration processes as often impurities are present in the gas mixture causing a lowering of the phase transition temperature of CO_2 even under $-80\,°C$ [17]. The separation process might be approximately 50% cheaper than MEA and other absorption methods [18,19].

While novel materials and technologies are being developed to capture CO_2 from flue gas and directly from air, most present-day and tested CO_2 capture technologies depend on the affinity of amines and their derivatives with CO_2 (monoethanolamine (MEA), diethanolamine (DEA), and potassium carbonate [20]) to obtain efficiency and selectivity for the implemented CO_2 capture technology. MEA and DEA are reported to have efficiencies over 90% with regard to CO_2 absorption [21]. Where it is clear that CO_2 is much more abundant in flue gases from, for example, coal- or gas-fired power plants (over 10% of the exhausted gas), the high temperatures of flue gases cause a deterioration of the used amines, resulting in increased operational costs and are potentially harmful to human health and the environment [22].

However, from an overview of different CO_2 capture technologies [23] including absorption using solvents or solid sorbents, pressure- and temperature-swing adsorption using various solid sorbents, cryogenic distillation, and membrane separation, it was concluded that the most promising method for CO_2 separation was liquid absorption using MEA. Ceramic and metallic membranes for membrane diffusion should produce technologies that are significantly more efficient at separation than liquid absorption, but as of now are not yet available. And the search is still largely ongoing.

A literature overview of the costs associated with these technologies exists, for example, Gibbins and Chalmers [24] compared the three technologies for both gas- and coal-fired plants. They reported that for coal-fired plants, the precombustion technology presented the lowest cost per ton of CO_2 avoided, while the postcombustion and oxyfuel technologies are of similar costs. However, for gas-fired plants, the cost per ton of CO_2 avoided for the postcombustion capture was almost 50% lower than the other two capture technologies.

46.5
New Roads into CO$_2$ Capture: Direct Air Capture and Nanomaterials

At present, only 20% of the worldwide CO_2 emissions from coal-fired power plants is captured, resulting in increased CO_2 emissions in the atmosphere. CO_2 capture and reuse can only contribute as a partly solution to the climate change challenge if CO_2 can effectively be captured in significant quantities and at an economically viable cost. Until now, an economic viable route was only available for EOR (enhanced oil recovery) and for smaller quantities, carbon dioxide was used in carbonization processes or as feedstock to greenhouses and ingredients in food and beverages.

In all of these cases, the cost of capturing CO_2 was not prohibitive for its later use, but most often the operational cost of capturing carbon dioxide still proves too high, mostly because the impact on the efficiency of the power plant and related cost makes it prohibitive or at least very expensive. The capturing methodology itself is also often prone to significant operational costs.

Mostly conventional methods such as alkyl amine solutions are used to scrub the CO_2 from exhaust gas streams, which is costly and inefficient. The major drawback of the existing technology include low CO_2 wt%, degradation of the solvent, and the high temperature required to regenerate the adsorbed gas [25].

Amines deteriorate faster or get contaminated sooner due to the high temperatures of the flue gas, thereby invoking repair or renewal costs and negatively affecting the efficiency of the capturing process.

Research into the capabilities of amines while looking at some other critical aspects of CO_2 capture and desorption such as material stability, kinetics of adsorption and desorption, improved sorbent adsorption efficiency, and the effect of water on sorbent adsorption behavior [26] showed that amine adsorbents with a methyl alkyl group between the amine and silicon atom have no thermal stability and display a severe loss of amine content upon heating. Adsorbents with ethyl and propyl alkyl chains demonstrate oxidative and thermal stability, but are prone to deactivation via urea formation in the presence of high-concentration and high-temperature, dry CO_2 for prolonged periods. Primary amines are found the best candidates for CO_2 capture from air as they possess both the highest amine efficiency for CO_2 adsorption and enhanced water affinity compared to other amine types. Also recently, it became clear that at lower temperatures (as low as 65 °C), CO_2 can be recuperated from the capturing amine solvent.

Therefore, two options seem worth pursuing: direct air capture and the development of new better capturing materials and technologies, for both flue gas capturing and for direct air capture.

In terms of improved processes for flue gas capturing, the best candidate for an ideal capturing process at present would probably be a membrane solution provided the efficiency of the power plant is not too much influenced or cryogenic separation provided the energy cost of the cooling is kept under control.

Where at first sight direct air capture seems to be far less effective, the CO_2 concentration in air (at 400 ppm) is only a fraction of the CO_2 concentration in flue gases (over 10%), but the operational and capital costs can under a number of conditions be largely reduced. As an example, the most frequent applied technology today for DAC uses an ion-exchange resin where adsorption and desorption are based on sorption in dry air and desorption in water vapor under vacuum [27]. The energy consumption in the installation is 50 kJ/mol and based on a natural wind speed of 1 m/s, and a basic installation based on 60 filters in a housing of 2.5 m $\times$ 1 m $\times$ 0.4 m captures 1 ton CO_2/day, initially at \$200/ton CO_2 and an ultimate goal of \$30/ton CO_2.

In terms of combined materials and processes, Ref. [28] reports the development of a combined process that uses a polyamine to capture carbon dioxide from the atmosphere in conjunction with a ruthenium-based catalyst to reduce CO_2 to methanol. A 79% conversion rate for the captured CO_2 was reported.

Overall, in terms of new materials, attention is focused on materials that are low cost both in production and in operation, have a high carbon dioxide capture efficiency, and are not prone to temperature degradation or contamination. The materials should be easy to apply and the carbon dioxide should be released at low cost from the formed substance. Often these materials show a carbon dioxide efficiency that is not only temperature related but also depend on relative humidity. For direct air capture, these materials should ultimately perform best in a range of 45–80% RH.

A number of research institutes are working on these materials, whereby nanomaterials often form an important ingredient.

46.6
Nanomaterials

At first silica-supported poly(ethyleneimine) (PEI) materials are demonstrated [29] to be promising adsorbents for CO_2 capture from ambient air. The materials have an enhanced thermal stability with extremely high CO_2 adsorption capacities under simulating ambient air conditions (400 ppm CO_2 in inert gas), exceeding 2 mol CO_2/kg$_{sorbent}$, as well as enhanced adsorption kinetics compared to conventional class 1 sorbents. The adsorbents show excellent stability in cyclic adsorption–desorption operations, even under dry conditions in which aminosilica adsorbents are known to lose capacity due to urea formation [30].

Subsequent research using, for example, fully polymeric and bio-based CO_2 sorbents [31] produced based on oxidized nanofibrillated cellulose (NFC) and a high molar mass polyethylenimine (PEI) shows similar CO_2 adsorption capacity, that is, 2,2 mol/CO_2/kg$_{sorbent}$) with a low adsorption half-time of 10.6 min. Some of these materials, however, show a loss of CO_2 adsorption capacity at lower relative humidity, for example, in the 20–50% RH range. This compromises their use for direct air capture.

The amines are often applied to honeycomb structures that attribute significantly to the investment cost. Foam-like materials [31] based on amines or incorporating amines from the start could therefore be very useful to lower the cost of air capture. In this sense, the above NFC/PEI material in foam form has a 97% porosity with a specific surface area of 2.7–8.3 m^2/g.

46.6.1
Metal–Organic Frameworks (MOF)

Zeolites, microporous aluminosilicate materials, are among the most commercially available adsorbents studied for CO_2 capture [32]. Their pores have the ability to selectively sort molecules based on a size exclusion process [33]. From numerous studies, it is apparent that zeolites are a more effective adsorbent at lower temperatures and higher pressures [34]. It is expected that the disadvantages associated with zeolites prevent them from becoming a major contributor.

Therefore, attention is shifted to MOFs that are constructed by self-assembly of metal ions and organic linkers with large surface areas, adjustable pore sizes, and controllable surface properties they provide among others such as high selectivity for CO_2 [35]. An overview of amine-supported metal–organic frameworks for carbon dioxide capture in direct air capture can be found in Ref. [36]. Here the reported carbon dioxide uptake is well above 2.2 mol CO_2/kg, and even approaching and exceeding 2.5 mol CO_2/kg, making it among the highest carbon dioxide adsorption capacities cited in literature.

Microporous metal–organic frameworks have also received considerable attention due to the high mass flux, thermal stability, adjustable chemical functionalities, extra high porosity, and availability of hundreds of well-characterized materials. A number of MOFs were developed for gas storage and separation applications, for example, MOF-177 exhibits a CO_2 sorption capacity of 1.4 g of CO_2/g of sorbent material [26], while chem-4Li has shown a high selectivity compared to other MOF materials for CO_2 in a range of gas mixtures (CO_2/H$_2$, CO_2/N$_2$, CO_2/CO, and CO_2/O$_2$). The interest in MOFs stems largely from the ability to design them as adsorbents that can be fine-tuned with regard to pore size, shape, and chemical functionality [37].

Limiting the pore size of MOFs to the ultramicroporous region increases significantly the selectivity for CO_2 capture at relatively moderate adsorption enthalpies compared to N$_2$ that show a larger kinetic diameter than CO_2. Reference [38] describes an MOF material with a dynamic structure that acts as a single molecule trap for CO_2.

A series of materials based on MOFs using zinc nitrate and organic building blocks have been constructed with a capturing capacity for carbon dioxide two to three times higher than conventional solvents. The crystals have a diameter of approximately 35 nm and a wall thickness of approximately 40 nm. The specific surface area for 1 g of material is approximately 1 ha [39].

Based on their framework properties, MOFs are classified as first, second, and third generation [40,41]. Second-generation materials have rigid permanently porous structures that are not altered upon the removal or introduction of guest molecules. In contrast, third-generation materials exhibit a degree of flexibility or dynamic behavior that allows them to tailor their pore structure in accordance with the type of guest molecule.

46.6.2
Gas Separation

The largest component of air at atmospheric conditions is N_2. Therefore, any solution that effectively separates CO_2 from N_2 might be of great value for the direct air capture problem.

CO_2 and N_2 can also be separated by virtue of their different kinetic diameters (3.30 and 3.64 A, respectively). Kinetic separations are not intrinsically associated with a large enthalpy difference and therefore offer a lower energy penalty for regeneration of the adsorbent. To effectively separate adsorbates based on size, precise control of the limiting pore diameter is necessary.

A specific approach to carefully tune the pore dimensions of MOFs is to utilize third-generation materials that have dynamic or flexible structures as this provides an opportunity to closely tailor the required pore diameters. Only certain gases can enter the pores of such materials and once inside, they induce a selective pore expansion or "gate opening." A potential downside of this "gate-opening" process in gas separations is that once "opened," these materials may act as adsorbents for all components of the gas mixture.

A significant drawback of a number of MOF usages is that the enthalpy of adsorption of CO_2 is directly proportional to the energy cost for regeneration of the material subsequent to saturation [42]. For example [43], MOFs with pores functionalized by alkylamines show enthalpies of adsorption as high as 96 kJ/mol and typically require temperatures in excess of 100 °C to fully regenerate [44].

46.7
Carbon Nanotubes

CO_2 separation through nanostructure membranes is believed to be energy efficient. Several types of nanomembranes are being developed. These include polymer nanostructure membrane, carbon molecular sieve membranes, and ceramic nanomembranes, whereby ceramic nanomembranes are more stable compared to polymer membranes, having greater tolerance to extreme conditions such as high temperatures.

Nanotubes are a logical extension of this search for suitable structures with proper chemical and physical properties. CO_2 adsorption capacity of graphene was demonstrated with a maximum adsorption capacity of 21.6 mmol/g at 11 bar pressure and room temperature [45].

Carbon nanotubes allow a wide range of functionalization for selective CO_2 adsorption, for example, with organic nitrogen-containing bases. An additional advantage of carbon nanotubes is the possibility of direct ohmic heating of the material to remove the captured CO_2.

However, the introduction of a functional group that specifically binds one species and improves on the adsorption selectivity will simultaneously decrease the diffusion of these molecules [46]. This inverse relationship between the adsorption and diffusion selectivity has been investigated by Krishna [47] in a broad range of meso- and microporous materials, including zeolites, carbon nanotubes, carbon molecular sieves, and metal–organic frameworks.

Next to amine adsorption processes, alternative concepts based on chemical adsorption of CO_2 onto metal oxides and physical adsorption on activated carbons, silicas, zeolites, and nonporous calixarenes [48,49] were also proposed.

46.7.1
Nanoporous Membranes

Nanoporous membranes are developed and selected to function at high temperatures ($T > 700$ K) typical for combustion or other high-temperature process gases. Very high CO_2 separation factors can be obtained with specific nanomembranes having well-defined thicknesses and porosities.

46.7.2
Nanocrystals for Carbon Capture

Basic hydroxides and oxides when converted into nanosized particles show a higher reactivity toward reaction with CO_2 and thus more efficient storage. Colloidal magnesium oxide nanocrystals have been synthesized for just this, that is, long-term storage of carbon dioxide. It is estimated that such magnesium oxide nanocrystals are promising candidates for a test system to model the kinetics of dissolution and mineralization in a simulated fluid-rock reservoir and thus probe a pathway in carbon dioxide sequestration.

46.7.3
Nanoparticle Ionic Materials for Carbon Capture

Nanoparticle ionic materials (NIMS) are lately considered for novel carbon dioxide capture technology. They consist of a hard nanoparticle core functionalized with organic and sometimes polymeric corona. NIMS as solvent-free particle fluids are highly tunable and functional materials with considerable miscibility and dispersion capabilities [50].

NIMS are therefore the nanoscale analog of ionic liquids.

46.7.4
CuO Nanoparticle-Loaded Porous Carbons

Porous carbons loaded with copper oxide nanoparticles yield high CO_2 capture efficiencies where the nanoparticles are responsible for the measured behavior. Copper oxide nanoparticles have electron donor properties, while CO_2 molecules are generally electron acceptors that results in a strong interaction and efficient capture [51].

46.7.5
Selectively Permeable Membranes

CO_2 can also be trapped effectively using selectively permeable membranes. To obtain an efficient separation process, the membrane should be made of ionic or polymeric nanoparticles that have greater affinity toward CO_2 while repelling undesired gases such as oxygen, methane, carbon monoxide, or nitride. These membranes are ideally suited to separate CO_2 from air. The driving force in the separation process is either electrochemical, temperature, or pressure difference to enhance the diffusion of CO_2 across the membrane.

46.7.6
Cellulose-Based Porous Nanomaterials

Porous carbons, in general, provide promising methods for long-term storage of carbon dioxide. These carbons bear a number of advantages in terms of cost, availability, large surface area, and design. Biomass such as cellulose could be used for the production of carbon sorbents for CO_2 capture [52]. The porosity of these materials determine CO_2 uptake. Porosity can quite easily be modified by modifying the activation temperature and the amount of the activating agent, typically potassium hydroxide.

46.7.7
Nanocomposites for Carbon Capture

Multiple component polymer-based nanocomposites might also provide a cost-effective and energy-efficient method of direct CO_2 capture and storage.

An extension is the inclusion of nanocomposite sorbents based on amines immobilized in porous solids. These should offer significant advantages such as potential elimination of corrosion problems and lower energy cost for sorbent regeneration. To support these nanocomposite sorbents/particles-specific structures with well-defined pore size, pore volumes based on a high surface area are needed. Mesoporous silica loaded with amine molecules with pores in the nanometer range increases significantly the amount of sorption sites for CO_2 [53].

46.8
Conclusion

One additional route to combat climate change is direct air capture of carbon dioxide. To use this technology on a significantly larger scale than today for either storing or sequestering the carbon dioxide, for example, underground or converting it to useful chemicals and biofuel, it should be captured at low cost.

Although carbon dioxide capture from flue gas is in essence an old technology, costs remain relatively high due to the high capital investment costs and significant operational cost, for example, the energy cost for capturing CO_2 and the maintenance of the capture material, often amine based.

Direct air capture thus comes in the picture as a possible better alternative provided the deterioration of the material can be prevented and the energy cost to remove the carbon dioxide from air can be circumvented.

Novel materials and analysis allow the selection of capturing material with a high dioxide uptake exceeding 2.5 mol CO_2/kg at medium to high levels of relative humidity.

Combined new processes and new materials are developed to directly convert the captured CO_2 into useful chemicals. Nanomaterials might play a significant role in fine-tuning the selectivity process and the capturing process.

References

1 IPCC (2014) Climate Change 2014: Synthesis Report. Contribution of Working Groups I, II, and III to the Fifth Assessment Report of the Intergovernmental Panel on Climate Change (core writing team, R.K. Pachauri and L.A. Meyer (eds)). IPCC, Geneva, Switzerland, 151 pp. Available at http://www.ipcc.ch/.

2 IEA (2014) World Energy Outlook 2014, International Energy Agency, London, November 12, 2014. Available at http://www.worldenergyoutlook.org/.

3 World Bank Report (2014) Carbon Pricing, May.

4 EIB (2013) EIB Climate Strategy.

5 Styring, P. and Armstrong, K. (2015) Assessing the potential of utilization and storage strategies for post-combustion CO_2 emissions reduction. *Front. Energy Res.*, **3**, 1–9.

6 IPCC, Ciais, P., Sabine, C., Bala, G., Bopp, L., Brovkin, V., Canadell, J., Chhabra, A., DeFries, R., Galloway, J., Heimann, M., Jones, C., Le Quéré, C., Myneni, R.B., Piao, S., and Thornton, P. (2013) Carbon and other biogeochemical cycles, in *Climate Change 2013: The Physical Science Basis. Contribution of Working Group I to the Fifth Assessment Report of the Intergovernmental Panel on Climate Change* (eds T.F. Stocker, D. Qin, G.-K. Plattner, M. Tignor, S.K. Allen, J. Boschung, A. Nauels, Y. Xia, V. Bex, and P.M. Midgley), Cambridge University Press, Cambridge, UK.

7 EC (2015) Communication from the Commission, Towards an Integrated Strategic Energy Technology (SET) Plan: Accelerating the European Energy System Transformation. Available at http://ec.europa.eu/energy/sites/ener/files/documents/1_EN_ACT_part1_v8_0.pdf.

8 Dowson, G.R.M., Dimitriou, I., Owen, R.E., Reed, D.G., Allen, R.W.K., and Styring, P. (2015) Kinetic and economic analysis of reactive capture of dilute carbon dioxide with Grignard reagents. *Faraday Discuss.*, **183**, 47–65.

9 Malhi, Y. and Grace, J. (2000) Tropical forests and atmospheric carbon dioxide. *TREE*, **15** (8), 332–337.

10 Leung, D.Y.C., Caramanna, G., and Maroto-Valer, M.M. (2014) An overview of current status of carbon dioxide capture and storage technologies. *Renew. Sustain. Energy Rev.*, **39**, 426–443.

11 Zero Emissions Resource Organisation (2013) Available at http://zeroCO2.no.

12 Bhown, A.S. and Freeman, B.C. (2011) Analysis and status of post-combustion carbon dioxide capture technologies. *Environ. Sci. Technol.*, **45**, 8624–8632.

13 IEA (2007) International Energy Agency Report: Capturing CO_2. IEA Greenhouse Gas R&D Programme. ISBN 978-1-898373-41-4.

14 IEA (2006) International Energy Agency Report, CO_2 Capture as a Factor in Power Station Investment Decisions. Report No. 2006/8, May.

15 Gunnarson, A. (2014) Process simulation of a 1 MWth chemical looping pilot plant with coal and biomass as fuel. M.Sc. thesis, Department of Chemical and Biological Engineering, Chalmers University Technology, Gothenburg, Sweden.

16 Cuéllar-Franca, R.M. and Azapagic, A. (2015) Carbon capture, storage and utilisation technologies: a critical analysis and comparison of their life cycle environmental impacts. *J. CO2 Utilization*, **9**, 82–102.

17 Xu, G., Liang, F., Yang, Y., Hu, Y., Zhang, K., and Liu, W. (2014) An improved CO_2 separation and purification system based on cryogenic separation and distillation theory. *Energies*, **7**, 3484–3502.

18 Gottlicher, G. and Pruschek, R. (1997) Comparison of CO_2 removal systems for fossil fuelled power plants. *Energy Convers. Manag.*, **38**, S173–S178.

19 Tuinier, M.J., Annaland, M.V.S., Kramer, G.J., and Kuipers, J.A.M. (2010) Cryogenic CO_2 capture using dynamically operated packed beds. *Chem. Eng. Sci.*, **65**, 114–119.

20 Hendriks, C. (1995) *Energy Conversion: CO2 Removal from Coal-Fired Power Plant*, Kluwer Academic Publishers, The Netherlands.

21 Veawab, A., Aroonwilas, A., and Tontiwachwuthiku, P. (2002) CO_2 absorption performance of aqueous alkanolamines in packed columns. *Fuel Chem.*, **47**, 49–50.

22 Rochelle, G.T. (2012) Thermal degradation of amines for CO_2 capture. *Curr. Opin. Chem. Eng.*, **1–2**, 183–190.

23 Aaron, D. and Tsouris, C. (2005) Separation of CO_2 from flue gas: a review. *Sep. Sci. Technol.*, **40**, 321–348.

24 Gibbins, J. and Chalmers, H. (2008) Carbon capture and storage. *Energy Policy*, **36**, 4317–4322.

25 Department of Energy (1993) A research Nees assessment for the capture, utilization and disposal of carbon dioxide from fossil fuel-fired power plants. DOE/ER-30194, Massachusetts Institute of Technology, Cambridge, MA, p. 61.

26 Millward, A.R. and Yaghi, O.M. (2005) Metal−organic frameworks with exceptionally high capacity for storage of carbon dioxide at room temperature. *J. Am. Chem. Soc.*, **127**, 17998–17999.

27 Lackner, K.S. (2010) Washing carbon out of the air. Scientific American, June.

28 Kothandaraman, J., Goeppert, A., Czaun, M., Surya Prakash, G.K., and Olah, G.A. (2015) Conversion of CO_2 from air into methanol using a polyamine and a homogeneous ruthenium catalyst. *J. Am. Chem. Soc.* doi: 10.1021/jacs.5b12354.

29 Choi, S., Gray, M.L., and Jones, C.W. (2011) Amine-tethered solid adsorbents coupling high adsorption capacity and regenerability for CO_2 capture from ambient air. *ChemSusChem*, **4** (5), 628–635.

30 Didas, S.A. (2014) Structural properties of aminosilica materials for CO_2 capture. Ph.D. dissertation, Georgia Institute of Technology.

31 Sehaqui, H., Elena Galvez, M., Becatinni, V., Ng, Y.C., Steinfeld, A., Zimmermann, T., and Tingaut, P. (2015) Fast and reversible direct CO_2 capture from air onto all-polymer nanofibrillated cellulose–polyethylenimine foams. *Environ. Sc. Technol.*, **49**, 3167–3174.

32 Ashley, M., Magiera, C., Ramidi, M., Blackburn, G., Scott, T.G., Gupta, R., Wilson, K., Ghosh, A., and Biswas, A. (2012) Nanomaterials and processes for carbon capture and conversion into useful by-products for a sustainable energy future. *Greenhouse Gas Sci. Technol.*, **2**, 419–444.

33 Delgado, M.R. and Arean, C.O. (2011) Carbon monoxide, dinitrogen, and carbon dioxide adsorption on zeolite H-beta:IR spectroscopic and thermodynamic studies. *Int. J. Hydrogen Energy*, **36** (8), 5286–5291.

34 Liu, Z., Grande, CA., Li, P., Yu, J., and Rodrigues, A.E. (2010) Adsorption and desorption of carbon dioxide and nitrogen on zeolite 5A. *Sep. Sci. Technol.*, **46** (3), 434–451.

35 Wu, D., Xu, Q., Liu, D., and Zhong, C. (2010) Exceptional CO_2 capture capability and molecular-level segregation in a Li-modified metal organic framework. *J. Phys. Chem. C*, **114** (39), 16611–16617.

36 Shekhah, O., Belmabkhout, Y., Chen, Z., Guillerm, V., Cairns, A., Adil, K., and Eddaoudi, M. (2014) Made-to-order metal-organic frameworks for trace carbon dioxide removal and air capture. *Nat. Commun.* doi: 10.1038/ncomms5228

37 Ortiz, G., Brandès, S., Rousselin, Y., and Guilard, R. (2011) Selective CO_2 adsorption by a triazacyclononane-bridged microporous metal–organic framework. *Chem. Eur. J.*, **17** (4), 6689–6695.

38 Wriedt, M., Sculley, J.P., Yakovenko, A.A., Ma, Y., Halder, G.J., Balbuena, P.B., and Zhou, H.-C. (2012) Low-energy selective capture of carbon dioxide by a pre-designed elastic single-molecule trap. *Angew. Chem., Int. Ed.*, **51**, 1–6.

39 Pacific Northwest National Laboratory (2010) Molecularly organized nanomaterials for carbon dioxide capture. December.

40 Kitagawa, S., Kitaura, R., and Noro, S. (2004) Functional porous coordination polymers. *Angew. Chem., Int. Ed.*, 2334–2375.

41 Uemura, K., Matsuda, R., and Kitagawa, S. (2005) Flexible microporous coordination polymers. *J. Solid State Chem.*, **178**, 2420–2429.

42 Mason, J.A., Sumida, K., Herm, Z.R., Krishna, R., and Long, J.R. (2011) Evaluating metal–organic frameworks for post-combustion carbon dioxide capture via temperature swing adsorption. *Energy Environ. Sci.*, **4**, 3030–3040.

43 Bloch, W.M., Babarao, R., Hill, M.R., Doonan, C.J., and Sunby, C.J. (2013) Post-synthetic structural processing in metal–organic framework material as a mechanism for exceptional CO_2/N_2 selectivity. *J. Am. Chem. Soc.* doi: 10.1021/ja4032049

44 McDonald, T.M., D'Alessandro, D.M., Krishna, R., and Kong, J.R. (2011) Enhanced carbon dioxide capture upon incorporation of *N,N'*-dimethylethylenediamine in the metal–organic framework CuBTTri. *Chem. Sci.*, 2022–2028.

45 Mishra, A.K. and Ramaprabhu, S. (2011) Carbon dioxide adsorption in graphene sheets. *AIP Adv.*, **1**, 032152.

46 D'Alessandro, D.M., Smit, B., and Long, J.R. (2010) Carbon dioxide capture: prospects for new materials. *Angew. Chem., Int. Ed.*, **49**, 6058–6082.

47 Krishna, R. (2009) Describing the diffusion of guest molecules inside porous structures. *J. Phys. Chem. C*, **113**, 19756.

48 Thallapally, P.K., McGrail, P.B., Dalgarno, S.J., Schaef, H.T., Tian, J., and Atwood, J.L. (2008) Gas-induced transformation and expansion of a non-porous organic solid. *Nat. Mater.*, **7**, 146–150.

49 Dalgarno, S.J., Thallapally, P.K., Barbour, L.J., and Atwood, J.L. (2007) Engineering void space in organic van der Waals crystals: calixarenes lead the way. *Chem. Soc. Rev.*, **36**, 236–245.

50 Bourlines, B.A., Herrera, R., Chalkias, N., Jiang, D.D., Zhang, Q., Archer, A.L., and Giannelis, P.E. (2005) Surface functionalized nanoparticles with liquid-like behavior. *Adv. Mater*, **17**, 234–237.

51 Kim, B.J., Cho, K.S., and Park, S.J. (2010) Copper-oxide decorated porous carbons for carbon dioxide adsorption behaviors.

J. Colloid Interface Sci., **243** (2), 575–578.

52 Sevilla, M. and Fuertes, A.B. (2011) Sustainable porous carbons with a superior performance for CO_2 capture. *Energy Environ. Sci.*, **4**, 1765–1771.

53 Qi, G., Wang, Y., Estevz, L., Duan, X., Anako, N., Park, A.-H.A. *et al.* (2011) High efficiency nanocomposite sorbents for CO_2 capture based on amine functionalized mesoporous capsules. *Energy Environ. Sci.*, **4**, 444–452.

Index

Nanotechnology for Energy Sustainability, First Edition. Edited by Baldev Raj, Marcel Van de Voorde,
and Yashwant Mahajan.
© 2017 Wiley-VCH Verlag GmbH & Co. KGaA. Published 2017 by Wiley-VCH Verlag GmbH & Co. KGaA.